STUDENT SOLUTIONS MANUAL

Sen-Ben Liao
Lawrence Livermore National Laboratory

to accompany

FUNDAMENTALS OF PHYSICS

Tenth Edition

STUDENT SOLUTIONS MANUAL

Sen-Ben Liao
Lawrence Livermore National Laboratory

to accompany

FUNDAMENTALS OF PHYSICS

Tenth Edition

David Halliday
University of Pittsburgh

Robert Resnick
Rensselaer Polytechnic Institute

Jearl Walker
Cleveland State University

WILEY

ISBN 978-1-11823066-4

Printed in the United States of America

10 9 8 7 6 5 4 3 2 1

Printed and bound by Bind-Rite Graphics.

PREFACE

This Student Solutions Manual is designed for use with the textbook *Fundamentals of Physics*, Tenth Edition, by David Halliday, Robert Resnick, and Jearl Walker. Its primary purpose is to show students by example how to solve various types of problems given at the ends of chapters in the text.

The solutions are prepared using a four-step problem-solving strategy called TEAL, which stands for Think, Express, Analyze, and Learn. This approach starts from basic definitions or fundamental relationships, highlights the key equations involved, and guide students through the mathematical steps required to obtain a solution, which is then assessed for its correctness and implications. By avoiding the mere plugging of numbers into equations derived in the text, we hope students will learn to examine any assumptions that are made in setting up and solving each problem. Please send comments, suggestions, and any errors that you may find to my email address: senben_liao@yahoo.com.

I am extremely grateful to Jearl Walker for providing answers to all the problems, and to Stuart Johnson and Geraldine Osnato for all their support over the years on the solutions manual projects. I would also like to thank Aly Rentrop for her superb job overseeing the entire SSM project, and the team of people involved in the production of this manual at Wiley. Finally, I want to thank my wife, Sioin, and my children for all their encouragement and loving support during the writing process.

Sen-Ben Liao
Tracy, CA 95377

TABLE OF CONTENTS

Chapter 1

1-1

THINK In this problem we're given the radius of the Earth, and asked to compute its circumference, surface area and volume.

EXPRESS Assuming Earth to be a sphere of radius

$$R_E = \left(6.37 \times 10^6 \text{ m}\right)\left(10^{-3} \text{ km/m}\right) = 6.37 \times 10^3 \text{ km},$$

the corresponding circumference, surface area and volume are:

$$C = 2\pi R_E, \quad A = 4\pi R_E^2, \quad V = \frac{4\pi}{3} R_E^3.$$

The geometric formulas are given in Appendix E.

ANALYZE (a) Using the formulas given above, we find the circumference to be

$$C = 2\pi R_E = 2\pi(6.37 \times 10^3 \text{ km}) = 4.00 \times 10^4 \text{ km}.$$

(b) Similarly, the surface area of Earth is

$$A = 4\pi R_E^2 = 4\pi\left(6.37 \times 10^3 \text{ km}\right)^2 = 5.10 \times 10^8 \text{ km}^2,$$

(c) and its volume is

$$V = \frac{4\pi}{3} R_E^3 = \frac{4\pi}{3}\left(6.37 \times 10^3 \text{ km}\right)^3 = 1.08 \times 10^{12} \text{ km}^3.$$

LEARN From the formulas given, we see that $C \sim R_E$, $A \sim R_E^2$, and $V \sim R_E^3$. The ratios of volume to surface area, and surface area to circumference are $V/A = R_E/3$ and $A/C = 2R_E$.

1-5

THINK This problem deals with conversion of furlongs to rods and chains, all of which are units for distance.

EXPRESS Given that $1 \text{ furlong} = 201.168 \text{ m}$, $1 \text{ rod} = 5.0292 \text{ m}$ and $1 \text{ chain} = 20.117 \text{ m}$, the relevant conversion factors are

$$1.0 \text{ furlong} = 201.168 \text{ m} = (201.168 \cancel{\text{m}})\frac{1 \text{ rod}}{5.0292 \cancel{\text{m}}} = 40 \text{ rods},$$

and

$$1.0 \text{ furlong} = 201.168 \text{ m} = (201.168 \cancel{\text{m}})\frac{1 \text{ chain}}{20.117 \cancel{\text{m}}} = 10 \text{ chains}.$$

Note the cancellation of m (meters), the unwanted unit.

ANALYZE Using the above conversion factors, we find

(a) the distance d in *rods* to be $d = 4.0 \text{ furlongs} = (4.0 \text{ furlongs})\frac{40 \text{ rods}}{1 \text{ furlong}} = 160 \text{ rods},$

(b) and in *chains* to be $d = 4.0 \text{ furlongs} = (4.0 \text{ furlongs})\frac{10 \text{ chains}}{1 \text{ furlong}} = 40 \text{ chains}.$

LEARN Since 4 furlongs is about 800 m, this distance is approximately equal to 160 rods ($1 \text{ rod} \approx 5 \text{ m}$) and 40 chains ($1 \text{ chain} \approx 20 \text{ m}$). So our results make sense.

1-17

THINK In this problem we are asked to rank 5 clocks, based on their performance as timekeepers.

EXPRESS We first note that none of the clocks advance by exactly 24 h in a 24-h period but this is not the most important criterion for judging their quality for measuring time intervals. What is important here is that the clock advance by the same (or nearly the same) amount in each 24-h period. The clock reading can then easily be adjusted to give the correct interval.

ANALYZE The chart below gives the corrections (in seconds) that must be applied to the reading on each clock for each 24-h period. The entries were determined by subtracting the clock reading at the end of the interval from the clock reading at the beginning.

CLOCK	Sun. -Mon.	Mon. -Tues.	Tues. -Wed.	Wed. -Thurs.	Thurs. -Fri.	Fri. -Sat.
A	−16	−16	−15	−17	−15	−15
B	−3	+5	−10	+5	+6	−7
C	−58	−58	−58	−58	−58	−58
D	+67	+67	+67	+67	+67	+67
E	+70	+55	+2	+20	+10	+10

Clocks C and D are both good timekeepers in the sense that each is consistent in its daily drift (relative to WWF time); thus, C and D are easily made "perfect" with simple and predictable corrections. The correction for clock C is less than the correction for clock D, so we judge clock C to be the best and clock D to be the next best. The correction that must be applied to clock A is in the range from 15 s to 17s. For clock B it is the range from −5 s to +10 s, for clock E it is in the range from −70 s to −2 s. After C and D, A has the smallest range of correction, B has the next smallest range, and E has the greatest range. From best to worst, the ranking of the clocks is C, D, A, B, E.

LEARN Of the five clocks, the readings in clocks A, B and E jump around from one 24-h period to another, making it difficult to correct them.

1-23

THINK This problem consists of two parts: in the first part, we are asked to find the mass of water, given its volume and density; the second part deals with the mass flow rate of water, which is expressed as kg/s in SI units.

EXPRESS From the definition of density: $\rho = m/V$, we see that mass can be calculated as $m = \rho V$, the product of the volume of water and its density. With 1 g = 1 × 10^{-3} kg and 1 cm^3 = (1 × 10^{-2}m)^3 = 1 × 10^{-6}m^3, the density of water in SI units (kg/m^3) is

$$\rho = 1\ \text{g/cm}^3 = \left(\frac{1\ \text{g}}{\text{cm}^3}\right)\left(\frac{10^{-3}\ \text{kg}}{\text{g}}\right)\left(\frac{\text{cm}^3}{10^{-6}\ \text{m}^3}\right) = 1 \times 10^3\ \text{kg/m}^3.$$

To obtain the flow rate, we simply divide the total mass of the water by the time taken to drain it.

ANALYZE (a) Using $m = \rho V$, the mass of a cubic meter of water is

$$m = \rho V = (1 \times 10^3\ \text{kg/m}^3)(1\ \text{m}^3) = 1 \times 10^3\ \text{kg}.$$

(b) The total mass of water in the container is

$$M = \rho V = (1 \times 10^3\ \text{kg/m}^3)(5700\ \text{m}^3) = 5.70 \times 10^6\ \text{kg},$$

and the time elapsed is t = (10 h)(3600 s/h) = 3.6 × 10^4 s. Thus, the mass flow rate R is

$$R = \frac{M}{t} = \frac{5.70 \times 10^6\ \text{kg}}{3.6 \times 10^4\ \text{s}} = 158\ \text{kg/s}.$$

LEARN In terms of volume, the drain rate can be expressed as

$$R' = \frac{V}{t} = \frac{5700\ \text{m}^3}{3.6 \times 10^4\ \text{s}} = 0.158\ \text{m}^3/\text{s} \approx 42\ \text{gal/s}.$$

The greater the flow rate, the less time is required to drain a given amount of water.

1-33

THINK In this problem we are asked to differentiate between three types of tons: *displacement* ton, *freight* ton and *register* ton, all of which are units of volume.

EXPRESS The three different tons are defined in terms of *barrel bulk*, with $1\ \text{barrel bulk} = 0.1415\ \text{m}^3 = 4.0155\ \text{U.S. bushels}$ (using $1\ \text{m}^3 = 28.378\ \text{U.S. bushels}$). Thus, in terms of U.S. bushels, we have

$$1 \text{ displacement ton} = (7 \text{ barrels bulk}) \times \left(\frac{4.0155 \text{ U.S. bushels}}{1 \text{ barrel bulk}} \right) = 28.108 \text{ U.S. bushels}$$

$$1 \text{ freight ton} = (8 \text{ barrels bulk}) \times \left(\frac{4.0155 \text{ U.S. bushels}}{1 \text{ barrel bulk}} \right) = 32.124 \text{ U.S. bushels}$$

$$1 \text{ register ton} = (20 \text{ barrels bulk}) \times \left(\frac{4.0155 \text{ U.S. bushels}}{1 \text{ barrel bulk}} \right) = 80.31 \text{ U.S. bushels}$$

ANALYZE (a) The difference between 73 "freight" tons and 73 "displacement" tons is

$$\Delta V = 73(\text{freight tons} - \text{displacement tons}) = 73(32.124 \text{ U.S. bushels} - 28.108 \text{ U.S. bushels})$$
$$= 293.168 \text{ U.S. bushels} \approx 293 \text{ U.S. bushels}$$

(b) Similarly, the difference between 73 "register" tons and 73 "displacement" tons is

$$\Delta V = 73(\text{register tons} - \text{displacement tons}) = 73(80.31 \text{ U.S. bushels} - 28.108 \text{ U.S. bushels})$$
$$= 3810.746 \text{ U.S. bushels} \approx 3.81 \times 10^3 \text{ U.S. bushels}$$

LEARN With $1 \text{ register ton} > 1 \text{ freight ton} > 1 \text{ displacement ton}$, we expect the difference found in (b) to be greater than that in (a). This is indeed the case.

1-39

THINK This problem compares the U.K. gallon with U.S. gallon, two non-SI units for volume. The interpretation of the type of gallons, whether U.K. or U.S., affects the amount of gasoline one calculates for traveling a given distance.

EXPRESS If the fuel consumption rate is R (in miles/gallon), then the amount of gasoline (in gallons) needed for a trip of distance d (in miles) would be

$$V(\text{gallon}) = \frac{d \text{ (miles)}}{R \text{ (miles/gallon)}}$$

Since the car was manufactured in U.K., the fuel consumption rate is calibrated based on U.K. gallon, and the correct interpretation should be "40 miles per U.K. gallon." In U.K., one would think of gallon as U.K. gallon; however, in the U.S., the word "gallon" would naturally be interpreted as U.S. gallon. Note also that since $1 \text{ U.K. gallon} = 4.5460900 \text{ L}$ and $1 \text{ U.S. gallon} = 3.7854118 \text{ L}$, the relationship between the two is

$$1 \text{ U.K. gallon} = (4.5460900 \text{ L}) \left(\frac{1 \text{ U.S. gallon}}{3.7854118 \text{ L}} \right) = 1.20095 \text{ U.S. gallons}$$

ANALYZE (a) The amount of gasoline actually required is

$$V' = \frac{750 \text{ miles}}{40 \text{ miles/U.K. gallon}} = 18.75 \text{ U.K. gallons} \approx 18.8 \text{ U.K. gallons}$$

This means that the driver mistakenly believes that the car should need 18.8 U.S. gallons.

(b) Using the conversion factor found above, this is equivalent to

$$V' = (18.75 \text{ U.K. gallons}) \times \left(\frac{1.20095 \text{ U.S. gallons}}{1 \text{ U.K. gallon}} \right) \approx 22.5 \text{ U.S. gallons}$$

LEARN A U.K. gallon is greater than a U.S gallon by roughly a factor of 1.2 in volume. Therefore, 40 mi/U.K. gallon is less fuel-efficient than 40 mi/U.S. gallon.

1-41
THINK This problem involves converting *cord*, a non-SI unit for volume, to SI unit.

EXPRESS Using the (exact) conversion 1 in. = 2.54 cm = 0.0254 m for length, we have

$$1 \text{ ft} = 12 \text{ in} = (12 \text{ in.}) \times \left(\frac{0.0254 \text{ m}}{1 \text{ in}} \right) = 0.3048 \text{ m}.$$

Thus, $1 \text{ ft}^3 = (0.3048 \text{ m})^3 = 0.0283 \text{ m}^3$ for volume (these results also can be found in Appendix D).

ANALYZE The volume of a cord of wood is $V = (8 \text{ ft}) \times (4 \text{ ft}) \times (4 \text{ ft}) = 128 \text{ ft}^3$. Using the conversion factor found above, we obtain

$$V = 1 \text{ cord} = 128 \text{ ft}^3 = (128 \text{ ft}^3) \times \left(\frac{0.0283 \text{ m}^3}{1 \text{ ft}^3} \right) = 3.625 \text{ m}^3$$

which implies that $1 \text{ m}^3 = \left(\frac{1}{3.625} \right) \text{cord} = 0.276 \text{ cord} \approx 0.3 \text{ cord}$.

LEARN The unwanted units ft^3 all cancel out, as they should. In conversions, units obey the same algebraic rules as variables and numbers.

1-47
THINK This problem involves expressing the speed of light in astronomical units per minute.

EXPRESS We first convert meters to astronomical units (AU), and seconds to minutes, using

$$1000 \text{ m} = 1 \text{ km}, \quad 1 \text{ AU} = 1.50 \times 10^8 \text{ km}, \quad 60 \text{ s} = 1 \text{ min}.$$

ANALYZE Using the conversion factors above, the speed of light can be rewritten as

$$c = 3.0 \times 10^8 \text{ m/s} = \left(\frac{3.0 \times 10^8 \text{ m}}{\text{s}}\right)\left(\frac{1 \text{ km}}{1000 \text{ m}}\right)\left(\frac{\text{AU}}{1.50 \times 10^8 \text{ km}}\right)\left(\frac{60 \text{ s}}{\text{min}}\right) = 0.12 \text{ AU/min}.$$

LEARN When expressed the speed of light c in AU/min, we readily see that it takes about 8.3 (= 1/0.12) minutes for sunlight to reach the Earth (i.e., to travel a distance of 1 AU).

1-53

THINK The objective of this problem is to convert the Earth-Sun distance (1 AU) to parsecs and light-years.

EXPRESS To relate parsec (pc) to AU, we note that when θ is measured in radians, it is equal to the arc length s divided by the radius R. For a very large radius circle and small value of θ, the arc may be approximated as the straight line-segment of length 1 AU. Thus,

$$\theta = 1 \text{ arcsec} = (1 \text{ arcsec})\left(\frac{1 \text{ arcmin}}{60 \text{ arcsec}}\right)\left(\frac{1°}{60 \text{ arcmin}}\right)\left(\frac{2\pi \text{ radian}}{360°}\right) = 4.85 \times 10^{-6} \text{ rad}.$$

Therefore, one parsec is

$$1 \text{ pc} = \frac{s}{\theta} = \frac{1 \text{ AU}}{4.85 \times 10^{-6}} = 2.06 \times 10^5 \text{ AU}.$$

Next, we relate AU to light-year (ly). Since a year is about 3.16×10^7 s,

$$1 \text{ly} = (186{,}000 \text{ mi/s})(3.16 \times 10^7 \text{ s}) = 5.9 \times 10^{12} \text{ mi}.$$

ANALYZE (a) Since $1 \text{ pc} = 2.06 \times 10^5 \text{ AU}$, inverting the relation gives

$$1 \text{ AU} = (1 \text{ AU})\left(\frac{1 \text{ pc}}{2.06 \times 10^5 \text{ AU}}\right) = 4.9 \times 10^{-6} \text{ pc}.$$

(b) Given that $1 \text{ AU} = 92.9 \times 10^6$ mi and $1 \text{ ly} = 5.9 \times 10^{12}$ mi, the two expressions together lead to

$$1 \text{ AU} = 92.9 \times 10^6 \text{ mi} = (92.9 \times 10^6 \text{ mi})\left(\frac{1 \text{ ly}}{5.9 \times 10^{12} \text{ mi}}\right) = 1.6 \times 10^{-5} \text{ ly}.$$

LEARN Our results can be further combined to give $1 \text{ pc} = 3.2$ ly. From the above expression, we readily see that it takes 1.57×10^{-5} y, or about 8.3 min, for Sunlight to travel a distance of 1 AU to reach the Earth.

Chapter 2

2-3

THINK This one-dimensional kinematics problem consists of two parts, and we are asked to solve for the average velocity and average speed of the car.

EXPRESS Since the trip consists of two parts, let the displacements during first and second parts of the motion be Δx_1 and Δx_2, and the corresponding time intervals be Δt_1 and Δt_2, respectively. Now, because the problem is one-dimensional and both displacements are in the same direction, the total displacement is simply $\Delta x = \Delta x_1 + \Delta x_2$, and the total time for the trip is $\Delta t = \Delta t_1 + \Delta t_2$. Using the definition of average velocity given in Eq. 2-2, we have

$$v_{\text{avg}} = \frac{\Delta x}{\Delta t} = \frac{\Delta x_1 + \Delta x_2}{\Delta t_1 + \Delta t_2}.$$

To find the average speed, we note that during a time Δt if the velocity remains a positive constant, then the speed is equal to the magnitude of velocity, and the distance is equal to the magnitude of displacement, with $d = |\Delta x| = v\Delta t$.

ANALYZE

(a) During the first part of the motion, the displacement is $\Delta x_1 = 40$ km and the time taken is

$$t_1 = \frac{(40\text{ km})}{(30\text{ km/h})} = 1.33\text{ h}.$$

Similarly, during the second part of the trip the displacement is $\Delta x_2 = 40$ km and the time interval is

$$t_2 = \frac{(40\text{ km})}{(60\text{ km/h})} = 0.67\text{ h}.$$

The total displacement is $\Delta x = \Delta x_1 + \Delta x_2 = 40\text{ km} + 40\text{ km} = 80\text{ km}$, and the total time elapsed is $\Delta t = \Delta t_1 + \Delta t_2 = 2.00$ h. Consequently, the average velocity is

$$v_{\text{avg}} = \frac{\Delta x}{\Delta t} = \frac{(80\text{ km})}{(2.0\text{ h})} = 40\text{ km/h}.$$

(b) In this case, the average speed is the same as the magnitude of the average velocity: $s_{\text{avg}} = 40$ km/h.

(c) The graph of the entire trip, shown below, consists of two contiguous line segments, the first having a slope of 30 km/h and connecting the origin to $(\Delta t_1, \Delta x_1) = (1.33\text{ h}, 40\text{ km})$ and the second having a slope of 60 km/h and connecting $(\Delta t_1, \Delta x_1)$ to $(\Delta t, \Delta x) = (2.00\text{ h}, 80\text{ km})$.

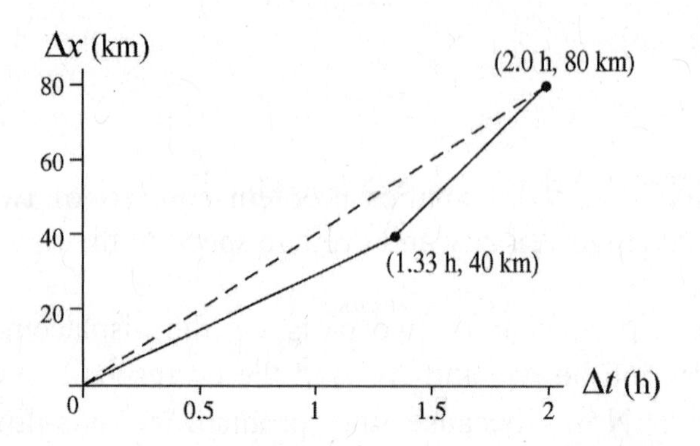

From the graphical point of view, the slope of the dashed line drawn from the origin to $(\Delta t, \Delta x)$ represents the average velocity.

LEARN The average velocity is a vector quantity that depends only on the net displacement (also a vector) between the starting and ending points.

2-5
THINK In this one-dimensional kinematics problem, we're given the position function $x(t)$, and asked to calculate the position and velocity of the object at a later time.

EXPRESS The position function is given as $x(t) = (3\text{ m/s})t - (4\text{ m/s}^2)t^2 + (1\text{ m/s}^3)t^3$. The position of the object at some instant t_0 is simply given by $x(t_0)$. For the time interval $t_1 \le t \le t_2$, the displacement is $\Delta x = x(t_2) - x(t_1)$. Similarly, using Eq. 2-2, the average velocity for this time interval is

$$v_{\text{avg}} = \frac{\Delta x}{\Delta t} = \frac{x(t_2) - x(t_1)}{t_2 - t_1}.$$

ANALYZE (a) Plugging in $t = 1$ s into $x(t)$ yields

$$x(1\text{ s}) = (3\text{ m/s})(1\text{ s}) - (4\text{ m/s}^2)(1\text{ s})^2 + (1\text{ m/s}^3)(1\text{ s})^3 = 0.$$

(b) With $t = 2$ s we get $x(2\text{ s}) = (3\text{ m/s})(2\text{ s}) - (4\text{ m/s}^2)(2\text{ s})^2 + (1\text{ m/s}^3)(2\text{ s})^3 = -2\text{ m}$.

(c) With $t = 3$ s we have $x(3\text{ s}) = (3\text{ m/s})(3\text{ s}) - (4\text{ m/s}^2)(3\text{ s})^2 + (1\text{ m/s}^3)(3\text{ s})^3 = 0\text{ m}$.

(d) Similarly, plugging in $t = 4$ s gives

$$x(4\text{ s}) = (3\text{ m/s})(4\text{ s}) - (4\text{ m/s}^2)(4\text{ s})^2 + (1\text{ m/s}^3)(4\text{ s})^3 = 12\text{ m}.$$

(e) The position at $t = 0$ is $x = 0$. Thus, the displacement between $t = 0$ and $t = 4$ s is $\Delta x = x(4\text{ s}) - x(0) = 12\text{ m} - 0 = 12\text{ m}$.

(f) The position at t = 2 s is subtracted from the position at t = 4 s to give the displacement: $\Delta x = x(4\text{ s}) - x(2\text{ s}) = 12\text{ m} - (-2\text{ m}) = 14\text{ m}$. Thus, the average velocity is

$$v_{\text{avg}} = \frac{\Delta x}{\Delta t} = \frac{14\text{ m}}{2\text{ s}} = 7\text{ m/s}.$$

(g) The position of the object for the interval $0 \le t \le 4$ is plotted below. The straight line drawn from the point at (t, x) = (2 s, –2 m) to (4 s, 12 m) would represent the average velocity, answer for part (f).

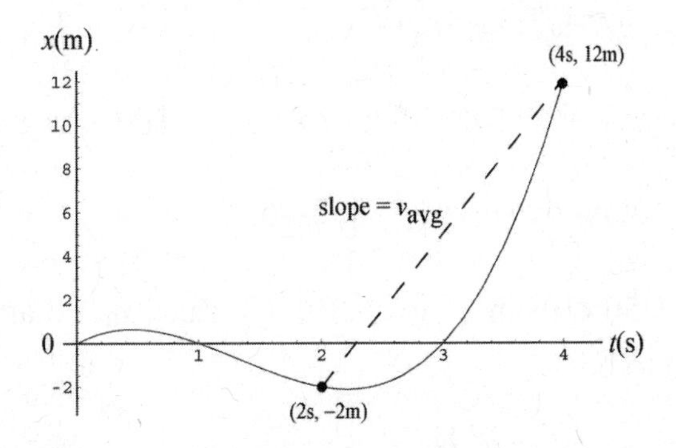

LEARN Our graphical representation illustrates once again that the average velocity for a time interval depends only on the net displacement between the starting and ending points.

2-19

THINK In this one-dimensional kinematics problem, we're given the speed of a particle at two instants and asked to calculate its average acceleration.

EXPRESS We represent the initial direction of motion as the $+x$ direction. The average acceleration over a time interval $t_1 \le t \le t_2$ is given by Eq. 2-7:

$$a_{\text{avg}} = \frac{\Delta v}{\Delta t} = \frac{v(t_2) - v(t_1)}{t_2 - t_1}.$$

ANALYZE Let v_1 = +18 m/s at $t_1 = 0$ and v_2 = –30 m/s at t_2 = 2.4 s. Using Eq. 2-7 we find

$$a_{\text{avg}} = \frac{v(t_2) - v(t_1)}{t_2 - t_1} = \frac{(-30\text{ m/s}) - (+1\text{ m/s})}{2.4\text{ s} - 0} = -20\text{ m/s}^2.$$

LEARN The average acceleration has magnitude 20 m/s^2 and is in the opposite direction to the particle's initial velocity. This makes sense because the velocity of the particle is decreasing over the time interval. With $t_1 = 0$, the velocity of the particle as a function of time can be written as $v = v_0 + at = (18\text{ m/s}) - (20\text{ m/s}^2)t$.

2-23

THINK The electron undergoes a constant acceleration. Given the final speed of the electron and the distance it has traveled, we can calculate its acceleration.

EXPRESS Since the problem involves constant acceleration, the motion of the electron can be readily analyzed using the equations given in Table 2-1:

$$v = v_0 + at \qquad (2-11)$$

$$x - x_0 = v_0 t + \frac{1}{2}at^2 \qquad (2-15)$$

$$v^2 = v_0^2 + 2a(x - x_0) \qquad (2-16)$$

The acceleration can be found by solving Eq. 2-16.

ANALYZE With $v_0 = 1.50\times10^5$ m/s, $v = 5.70\times10^6$ m/s, $x_0 = 0$ and $x = 0.010$ m, we find the average acceleration to be

$$a = \frac{v^2 - v_0^2}{2x} = \frac{(5.7\times10^6 \text{ m/s})^2 - (1.5\times10^5 \text{ m/s})^2}{2(0.010 \text{ m})} = 1.62\times10^{15} \text{ m/s}^2.$$

LEARN It is always a good idea to apply other equations in Table 2-1 not used for solving the problem as a consistency check. For example, since we now know the value of the acceleration, using Eq. 2-11, the time it takes for the electron to reach its final speed would be

$$t = \frac{v - v_0}{a} = \frac{5.70\times10^6 \text{ m/s} - 1.5\times10^5 \text{ m/s}}{1.62\times10^{15} \text{ m/s}^2} = 3.426\times10^{-9} \text{ s}$$

Substituting the value of t into Eq. 2-15, the distance the electron travels is

$$x = x_0 + v_0 t + \frac{1}{2}at^2 = 0 + (1.5\times10^5 \text{ m/s})(3.426\times10^{-9} \text{ s}) + \frac{1}{2}(1.62\times10^{15} \text{ m/s}^2)(3.426\times10^{-9} \text{ s})^2$$
$$= 0.01 \text{ m}$$

This is what was given in the problem statement. So we know the problem has been solved correctly.

2-31

THINK The rocket ship undergoes a constant acceleration from rest, and we want to know the time elapsed and the distance traveled when the rocket reaches a certain speed.

EXPRESS Since the problem involves constant acceleration, the motion of the rocket can be readily analyzed using the equations in Table 2-1:

$$v = v_0 + at \qquad (2-11)$$

$$x - x_0 = v_0 t + \frac{1}{2}at^2 \qquad (2-15)$$

$$v^2 = v_0^2 + 2a(x - x_0) \qquad (2-16)$$

ANALYZE (a) Given $a = 9.8\ \text{m/s}^2$, $v_0 = 0$ and $v = 0.1c = 3.0\times10^7\ \text{m/s}$, we solve $v = v_0 + at$ for the time:

$$t = \frac{v - v_0}{a} = \frac{3.0\times10^7\ \text{m/s} - 0}{9.8\ \text{m/s}^2} = 3.1\times10^6\ \text{s}$$

which is about 1.2 months. So it takes 1.2 months for the rocket to reach a speed of $0.1c$ starting from rest with a constant acceleration of $9.8\ \text{m/s}^2$.

(b) To calculate the distance traveled during this time, we evaluate $x = x_0 + v_0 t + \frac{1}{2}at^2$, with $x_0 = 0$ and $v_0 = 0$. The result is

$$x = \frac{1}{2}\left(9.8\ \text{m/s}^2\right)(3.1\times10^6\,\text{s})^2 = 4.6\times10^{13}\ \text{m}.$$

LEARN In solving parts (a) and (b), we did not use Eq. (2-16): $v^2 = v_0^2 + 2a(x - x_0)$. This equation can be used to check our answers. The final velocity based on this equation is

$$v = \sqrt{v_0^2 + 2a(x - x_0)} = \sqrt{0 + 2(9.8\ \text{m/s}^2)(4.6\times10^{13}\ \text{m} - 0)} = 3.0\times10^7\ \text{m/s},$$

which is what was given in the problem statement. So we know the problems have been solved correctly.

2-33

THINK The car undergoes a constant negative acceleration to avoid impacting a barrier. Given its initial speed, we want to know the distance it has traveled and the time elapsed prior to the impact.

EXPRESS Since the problem involves constant acceleration, the motion of the car can be readily analyzed using the equations in Table 2-1:

$$v = v_0 + at \qquad (2-11)$$

$$x - x_0 = v_0 t + \frac{1}{2}at^2 \qquad (2-15)$$

$$v^2 = v_0^2 + 2a(x - x_0) \qquad (2-16)$$

We take $x_0 = 0$ and $v_0 = 56.0$ km/h = 15.55 m/s to be the initial position and speed of the car. Solving Eq. 2-15 with $t = 2.00$ s gives the acceleration a. Once a is known, the speed of the car upon impact can be found by using Eq. 2-11.

ANALYZE (a) Using Eq. 2-15, we find the acceleration to be

$$a = \frac{2(x - v_0 t)}{t^2} = \frac{2[(24.0\text{ m}) - (15.55\text{ m/s})(2.00\text{ s})]}{(2.00\text{ s})^2} = -3.56\text{m/s}^2,$$

or $|a| = 3.56\text{ m/s}^2$. The negative sign indicates that the acceleration is opposite to the direction of motion of the car; the car is slowing down.

(b) The speed of the car at the instant of impact is

$$v = v_0 + at = 15.55\text{ m/s} + (-3.56\text{ m/s}^2)(2.00\text{ s}) = 8.43\text{ m/s}$$

which can also be converted to 30.3 km/h.

LEARN In solving parts (a) and (b), we did not use Eq. 1-16. This equation can be used as a consistency check. The final velocity based on this equation is

$$v = \sqrt{v_0^2 + 2a(x - x_0)} = \sqrt{(15.55\text{ m/s})^2 + 2(-3.56\text{ m/s}^2)(24\text{ m} - 0)} = 8.43\text{ m/s},$$

which is what was calculated in (b). This indicates that the problems have been solved correctly.

2-45

THINK As the ball travels vertically upward, its motion is under the influence of gravitational acceleration. The kinematics is one-dimensional.

EXPRESS We neglect air resistance for the duration of the motion (between "launching" and "landing"), so $a = -g = -9.8\text{ m/s}^2$ (we take downward to be the $-y$ direction). We use the equations in Table 2-1 (with Δy replacing Δx) because this is a = constant motion:

$$v = v_0 - gt \qquad (2-11)$$

$$y - y_0 = v_0 t - \frac{1}{2}gt^2 \qquad (2-15)$$

$$v^2 = v_0^2 - 2g(y - y_0) \qquad (2-16)$$

We set $y_0 = 0$. Upon reaching the maximum height y, the speed of the ball is momentarily zero ($v = 0$). Therefore, we can relate its initial speed v_0 to y via the equation $0 = v^2 = v_0^2 - 2gy$. The time it takes for the ball to reach maximum height is given by $v = v_0 - gt = 0$, or $t = v_0 / g$. Therefore, for the entire trip (from the time it leaves the ground until the time it returns to the ground), the total flight time is $T = 2t = 2v_0 / g$.

ANALYZE (a) At the highest point $v = 0$ and $v_0 = \sqrt{2gy}$. With $y = 50$ m, we find the initial speed of the ball to be

$$v_0 = \sqrt{2gy} = \sqrt{2(9.8\text{ m/s}^2)(50\text{ m})} = 31.3\text{ m/s}.$$

(b) Using the result from (a) for v_0, the total flight time of the ball is

$$T = \frac{2v_0}{g} = \frac{2(31.3\text{ m/s})}{9.8\text{ m/s}^2} = 6.4\text{ s}$$

(c) The plots of y, v and a as a function of time are shown below. The acceleration graph is a horizontal line at –9.8 m/s². At $t = 3.19$ s, $y = 50$ m.

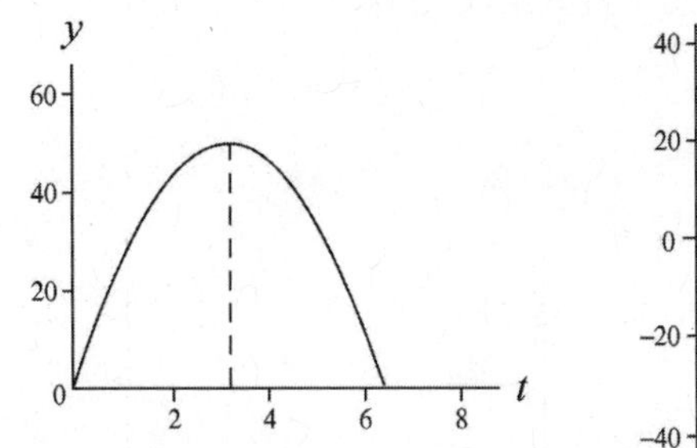

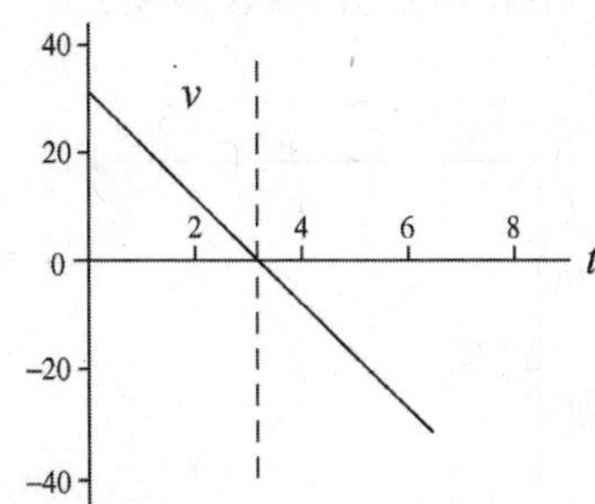

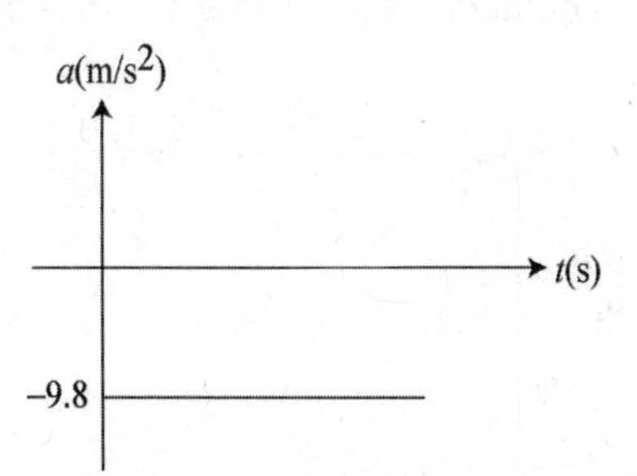

LEARN In calculating the total flight time of the ball, we could have used Eq. 2-15. At $t = T > 0$, the ball returns to its original position ($y = 0$). Therefore,

$$y = v_0 T - \frac{1}{2} g T^2 = 0 \;\Rightarrow\; T = \frac{2v_0}{g}$$

2-47

THINK The wrench is in free fall with an acceleration $a = -g = -9.8$ m/s².

EXPRESS We neglect air resistance, which justifies setting $a = -g = -9.8$ m/s² (taking *down* as the $-y$ direction) for the duration of the fall. This is constant acceleration motion, which justifies the use of Table 2-1 (with Δy replacing Δx):

$$v = v_0 - gt \qquad (2-11)$$

$$y - y_0 = v_0 t - \frac{1}{2} g t^2 \qquad (2-15)$$

$$v^2 = v_0^2 - 2g(y - y_0) \qquad (2-16)$$

Since the wrench had an initial speed $v_0 = 0$, knowing its speed of impact allows us to apply Eq. 2-16 to calculate the height from which it was dropped.

ANALYZE (a) Using $v^2 = v_0^2 + 2a\Delta y$, we find the initial height to be

$$\Delta y = \frac{v_0^2 - v^2}{2a} = \frac{0-(-24 \text{ m/s})^2}{2(-9.8 \text{ m/s}^2)} = 29.4 \text{ m}.$$

So that it fell through a height of 29.4 m.

(b) Solving $v = v_0 - gt$ for time, we obtain a flight time of

$$t = \frac{v_0 - v}{g} = \frac{0-(-24 \text{ m/s})}{9.8 \text{ m/s}^2} = 2.45 \text{ s}.$$

(c) SI units are used in the graphs, and the initial position is taken as the coordinate origin. The acceleration graph is a horizontal line at -9.8 m/s^2.

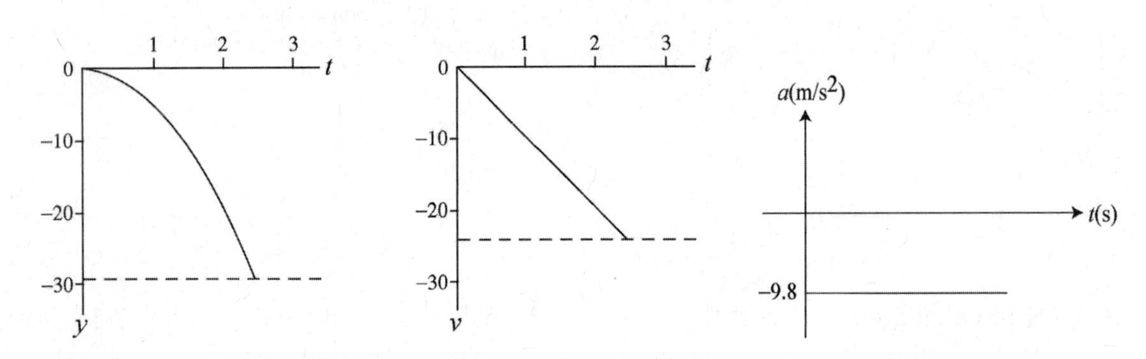

LEARN As the wrench falls, with $a = -g < 0$, its speed increases but its velocity becomes more negative, as indicated by the second graph above.

2-49

THINK In this problem a package is dropped from a hot-air balloon which is ascending vertically upward. We analyze the motion of the package under the influence of gravity.

EXPRESS We neglect air resistance, which justifies setting $a = -g = -9.8 \text{ m/s}^2$ (taking *down* as the $-y$ direction) for the duration of the motion. This allows us to use Table 2-1 (with Δy replacing Δx):

$$v = v_0 - gt \qquad (2-11)$$

$$y - y_0 = v_0 t - \frac{1}{2}gt^2 \qquad (2-15)$$

$$v^2 = v_0^2 - 2g(y - y_0) \qquad (2-16)$$

We place the coordinate origin on the ground and note that the initial velocity of the package is the same as the velocity of the balloon, $v_0 = +12$ m/s and that its initial coordinate is $y_0 = +80$ m. The time it takes for the package to hit the ground can be found by solving Eq. 2-15 with $y = 0$.

ANALYZE (a) We solve $0 = y = y_0 + v_0 t - \frac{1}{2}gt^2$ for time using the quadratic formula (choosing the positive root to yield a positive value for t):

$$t = \frac{v_0 + \sqrt{v_0^2 + 2gy_0}}{g} = \frac{12 \text{ m/s} + \sqrt{(12 \text{ m/s})^2 + 2\left(9.8 \text{ m/s}^2\right)\left(80 \text{ m}\right)}}{9.8 \text{ m/s}^2} = 5.4 \text{ s}.$$

(b) The speed of the package when it hits the ground can be calculated using Eq. 2-11. The result is

$$v = v_0 - gt = 12 \text{ m/s} - (9.8 \text{ m/s}^2)(5.447 \text{ s}) = -41.38 \text{ m/s}.$$

So its final *speed* is about 41 m/s.

LEARN Our answers can be readily verified by using Eq. 2-16 which was not used in either (a) or (b). The equation leads to

$$v = -\sqrt{v_0^2 - 2g(y - y_0)} = -\sqrt{(12 \text{ m/s})^2 - 2(9.8 \text{ m/s}^2)(0 - 80 \text{ m})} = -41.38 \text{ m/s}$$

which agrees with that calculated in (b).

2-53

THINK This problem involves two objects: a key dropped from a bridge, and a boat moving at a constant speed. We look for conditions such that the key will fall into the boat.

EXPRESS The speed of the boat is constant, given by $v_b = d/t$, where d is the distance of the boat from the bridge when the key is dropped (12 m) and t is the time the key takes in falling.

To calculate t, we take the time to be zero at the instant the key is dropped, we compute the time t when $y = 0$ using $y = y_0 + v_0 t - \frac{1}{2}gt^2$, with $y_0 = 45$ m. Once t is known, the speed of the boat can be readily calculated.

ANALYZE Since the initial velocity of the key is zero, the coordinate of the key is given by $y_0 = \frac{1}{2}gt^2$. Thus, the time it takes for the key to drop into the boat is

$$t = \sqrt{\frac{2y_0}{g}} = \sqrt{\frac{2(45 \text{ m})}{9.8 \text{ m/s}^2}} = 3.03 \text{ s}.$$

Therefore, the speed of the boat is $v_b = \dfrac{12 \text{ m}}{3.03 \text{ s}} = 4.0$ m/s.

LEARN From the general expression $v_b = \frac{d}{t} = \frac{d}{\sqrt{2y_0/g}} = d\sqrt{\frac{g}{2y_0}}$, we see that $v_b \sim 1/\sqrt{y_0}$. This agrees with our intuition that the lower the height from which the key is dropped, the greater the speed of the boat in order to catch it.

2-55

THINK The free-falling moist-clay ball strikes the ground with a non-zero speed, and it undergoes deceleration before coming to rest.

EXPRESS During contact with the ground its average acceleration is given by $a_{\text{avg}} = \frac{\Delta v}{\Delta t}$, where Δv is the change in its velocity during contact with the ground and $\Delta t = 20.0\times10^{-3}$ s is the duration of contact. Thus, we must first find the velocity of the ball just before it hits the ground ($y = 0$).

ANALYZE (a) Now, to find the velocity just *before* contact, we take $t = 0$ to be when it is dropped. Using Eq. 2-16 with $y_0 = 15.0$ m, we obtain

$$v = -\sqrt{v_0^2 - 2g(y - y_0)} = -\sqrt{0 - 2(9.8\text{ m/s}^2)(0 - 15\text{ m})} = -17.15\text{ m/s}$$

where the negative sign is chosen since the ball is traveling downward at the moment of contact. Consequently, the average acceleration during contact with the ground is

$$a_{\text{avg}} = \frac{\Delta v}{\Delta t} = \frac{0 - (-17.1\text{ m/s})}{20.0 \times 10^{-3}\text{ s}} = 857\text{ m/s}^2.$$

(b) The fact that the result is positive indicates that this acceleration vector points upward.

LEARN Since Δt is very small, it is not surprising to have a very large acceleration to stop the motion of the ball. In later chapters, we shall see that the acceleration is directly related to the magnitude and direction of the force exerted by the ground on the ball during the course of collision.

2-77

THINK The speed of the rod changes due to a nonzero acceleration.

EXPRESS Since the problem involves constant acceleration, the motion of the rod can be readily analyzed using the equations given in Table 2-1. We take $+x$ to be in the direction of motion, so

$$v = (60\text{ km/h})\left(\frac{1000\text{ m/km}}{3600\text{ s/h}}\right) = +16.7\text{ m/s}$$

and $a > 0$. The location where the rod starts from rest ($v_0 = 0$) is taken to be $x_0 = 0$.

ANALYZE (a) Using Eq. 2-7, we find the average acceleration to be

$$a_{\text{avg}} = \frac{\Delta v}{\Delta t} = \frac{v - v_0}{t - t_0} = \frac{16.7 \text{ m/s} - 0}{5.4 \text{ s} - 0} = 3.09 \text{ m/s}^2.$$

(b) Assuming constant acceleration $a = a_{\text{avg}} = 3.09 \text{ m/s}^2$, the total distance traveled during the 5.4-s time interval is

$$x = x_0 + v_0 t + \frac{1}{2}at^2 = 0 + 0 + \frac{1}{2}(3.09 \text{ m/s}^2)(5.4 \text{ s})^2 = 45 \text{ m}$$

(c) Using Eq. 2-15, the time required to travel a distance of $x = 250$ m is:

$$x = \frac{1}{2}at^2 \Rightarrow t = \sqrt{\frac{2x}{a}} = \sqrt{\frac{2(250 \text{ m})}{3.09 \text{ m/s}^2}} = 12.73 \text{ s}$$

LEARN The displacement of the rod as a function of time can be written as $x(t) = \frac{1}{2}(3.09 \text{ m/s}^2)t^2$. Note that we could have chosen Eq. 2-17 to solve for (b):

$$x = \frac{1}{2}(v_0 + v)\,t = \frac{1}{2}(16.7 \text{ m/s})(5.4 \text{ s}) = 45 \text{ m}.$$

2-81

THINK The particle undergoes a *non-constant* acceleration along the +*x*-axis. An integration is required to calculate velocity.

EXPRESS With a non-constant acceleration $a(t) = dv/dt$, the velocity of the particle at time t_1 is given by Eq. 2-22: $v_1 = v_0 + \int_{t_0}^{t_1} a(t)dt$, where v_0 is the velocity at time t_0. In our situation, we have $a = 5.0t$. In addition, we also know that $v_0 = 17$ m/s at $t_0 = 2.0$ s.

ANALYZE Integrating (from t = 2 s to variable t = 4 s) the acceleration to get the velocity and using the values given in the problem, leads to

$$v = v_0 + \int_{t_0}^{t} a\,dt = v_0 + \int_{t_0}^{t}(5.0t)dt = v_0 + \frac{1}{2}(5.0)(t^2 - t_0^2) = 17 + \tfrac{1}{2}(5.0)(4^2 - 2^2) = 47 \text{ m/s}.$$

LEARN The velocity of the particle as a function of t is

$$v(t) = v_0 + \frac{1}{2}(5.0)(t^2 - t_0^2) = 17 + \frac{1}{2}(5.0)(t^2 - 4) = 7 + 2.5t^2$$

in SI units (m/s). Since the acceleration is linear in *t*, we expect the velocity to be quadratic in *t*, and the displacement to be cubic in *t*.

2-87

THINK In this problem we're given two different speeds, and asked to find the difference in their travel times.

EXPRESS The time is takes to travel a distance *d* with a speed v_1 is $t_1 = d/v_1$. Similarly, with a speed v_2 the time would be $t_2 = d/v_2$. The two speeds in this problem are

$$v_1 = 55 \text{ mi/h} = (55 \text{ mi/h})\frac{1609 \text{ m/mi}}{3600 \text{ s/h}} = 24.58 \text{ m/s}$$

$$v_2 = 65 \text{ mi/h} = (65 \text{ mi/h})\frac{1609 \text{ m/mi}}{3600 \text{ s/h}} = 29.05 \text{ m/s}$$

ANALYZE With $d = 700 \text{ km} = 7.0\times10^5 \text{ m}$, the time difference between the two is

$$\Delta t = t_1 - t_2 = d\left(\frac{1}{v_1} - \frac{1}{v_2}\right) = (7.0\times10^5 \text{ m})\left(\frac{1}{24.58 \text{ m/s}} - \frac{1}{29.05 \text{ m/s}}\right) = 4383 \text{ s} = 73 \text{ min}$$

LEARN The travel time was reduced from 7.9 h to 6.7 h. Driving at higher speed (within the legal limit) reduces travel time.

2-89

THINK In this problem we explore the connection between the maximum height an object reaches under the influence of gravity and the total amount of time it stays in air.

EXPRESS Neglecting air resistance and setting $a = -g = -9.8 \text{ m/s}^2$ (taking *down* as the $-y$ direction) for the duration of the motion, we analyze the motion of the ball using Table 2-1 (with Δy replacing Δx). We set $y_0 = 0$. Upon reaching the maximum height *H*, the speed of the ball is momentarily zero ($v = 0$). Therefore, we can relate its initial speed v_0 to *H* via the equation

$$0 = v^2 = v_0^2 - 2gH \Rightarrow v_0 = \sqrt{2gH}.$$

The time it takes for the ball to reach maximum height is given by $v = v_0 - gt = 0$, or $t = v_0/g = \sqrt{2H/g}$.

ANALYZE If we want the ball to spend twice as much time in air as before, i.e., $t' = 2t$, then the new maximum height H' it must reach is such that $t' = \sqrt{2H'/g}$. Solving for H' we obtain

$$H' = \frac{1}{2}gt'^2 = \frac{1}{2}g(2t)^2 = 4\left(\frac{1}{2}gt^2\right) = 4H.$$

LEARN Since $H \sim t^2$, doubling t means that H must increase fourfold. Note also that for $t' = 2t$, the initial speed must be twice the original speed: $v_0' = 2v_0$.

2-95

THINK This problem involves analyzing a plot describing the position of an iceboat as function of time. The boat has a nonzero acceleration due to the wind.

EXPRESS Since we are told that the acceleration of the boat is constant, the equations of Table 2-1 can be applied. However, the challenge here is that v_0, v, and a are not explicitly given. Our strategy to deduce these values is to apply the kinematic equation $x - x_0 = v_0 t + \frac{1}{2}at^2$ to a variety of points on the graph and solve for the unknowns from the simultaneous equations.

ANALYZE (a) From the graph, we pick two points on the curve: $(t, x) = (2.0\text{ s}, 16\text{ m})$ and $(3.0\text{ s}, 27\text{ m})$. The corresponding simultaneous equations are

$$16\text{ m} - 0 = v_0(2.0\text{ s}) + \frac{1}{2}a(2.0\text{ s})^2$$

$$27\text{ m} - 0 = v_0(3.0\text{ s}) + \frac{1}{2}a(3.0\text{ s})^2$$

Solving the equations lead to the values $v_0 = 6.0$ m/s and $a = 2.0\text{ m/s}^2$.

(b) From Table 2-1,

$$x - x_0 = vt - \frac{1}{2}at^2 \Rightarrow 27\text{ m} - 0 = v(3.0\text{ s}) - \frac{1}{2}(2.0\text{ m/s}^2)(3.0\text{ s})^2$$

which leads to $v = 12$ m/s.

(c) Assuming the wind continues during $3.0 \le t \le 6.0$, we apply $x - x_0 = v_0 t + \frac{1}{2}at^2$ to this interval (where $v_0 = 12.0$ m/s from part (b)) to obtain

$$\Delta x = (12.0\text{ m/s})(3.0\text{ s}) + \frac{1}{2}(2.0\text{ m/s}^2)(3.0\text{ s})^2 = 45\text{ m}.$$

LEARN By using the results obtained in (a), the position and velocity of the iceboat as a function of time can be written as

$$x(t) = (6.0\text{ m/s})t + \frac{1}{2}(2.0\text{ m/s}^2)t^2 \text{ and } v(t) = (6.0\text{ m/s}) + (2.0\text{ m/s}^2)t.$$

One can readily verify that the same answers are obtained for (b) and (c) using the above expressions for $x(t)$ and $v(t)$.

Chapter 3

3-1

THINK In this problem we're given the magnitude and direction of a vector in two dimensions, and asked to calculate its *x*- and *y*-components.

EXPRESS The *x*- and the *y*- components of a vector $\vec{a}$ lying in the *xy* plane are given by

$$a_x = a\cos\theta, \quad a_y = a\sin\theta$$

where $a = |\vec{a}| = \sqrt{a_x^2 + a_y^2}$ is the magnitude and $\theta = \tan^{-1}(a_y / a_x)$ is the angle between $\vec{a}$ and the positive x axis. Given that $\theta = 250°$, we see that the vector is in the third quadrant, and we expect both the *x*- and the *y*-components of $\vec{a}$ to be negative.

ANALYZE (a) The x component of $\vec{a}$ is

$$a_x = a\cos\theta = (7.3\text{ m})\cos 250° = -2.5\text{ m},$$

(b) and the y component is $a_y = a\sin\theta = (7.3\text{ m})\sin 250° = -6.86\text{ m} \approx -6.9\text{ m}$. The results are depicted in the figure below:

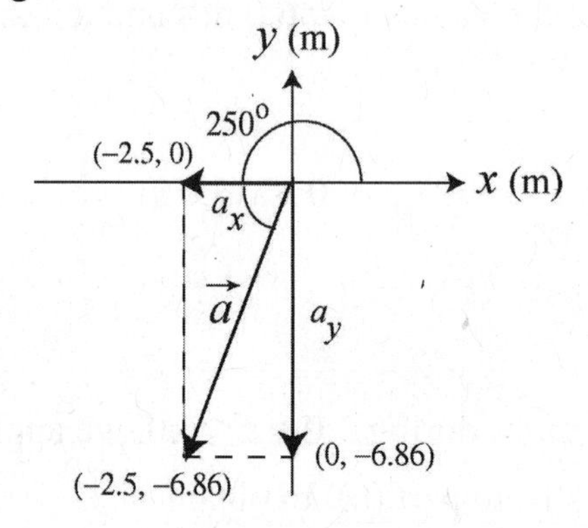

LEARN In considering the variety of ways to compute these, we note that the vector is 70° below the $-x$ axis, so the components could also have been found from

$$a_x = -(7.3\text{ m})\cos 70° = -2.50\text{ m}, \quad a_y = -(7.3\text{ m})\sin 70° = -6.86\text{ m}.$$

Similarly, we note that the vector is 20° to the left from the $-y$ axis, so one could also achieve the same results by using

$$a_x = -(7.3\text{ m})\sin 20° = -2.50\text{ m}, \quad a_y = -(7.3\text{ m})\cos 20° = -6.86\text{ m}.$$

As a consistency check, we note that $\sqrt{a_x^2+a_y^2}=\sqrt{(-2.50\text{ m})^2+(-6.86\text{ m})^2}=7.3\text{ m}$ and $\tan^{-1}(a_y/a_x)=\tan^{-1}[(-6.86\text{ m})/(-2.50\text{ m})]=250°$, which are indeed the values given in the problem statement.

3-3

THINK In this problem we're given the *x*- and *y*-components a vector $\vec{A}$ in two dimensions, and asked to calculate its magnitude and direction.

EXPRESS A vector $\vec{A}$ can be represented in the *magnitude-angle* notation (A, θ), where

$$A=\sqrt{A_x^2+A_y^2}$$

is the magnitude and

$$\theta=\tan^{-1}\left(\frac{A_y}{A_x}\right)$$

is the angle $\vec{A}$ makes with the positive *x* axis. Given that $A_x = -25.0$ m and $A_y = 40.0$ m, the above formulas can be readily used to calculate A and θ.

ANALYZE (a) The magnitude of the vector $\vec{A}$ is

$$A=\sqrt{A_x^2+A_y^2}=\sqrt{(-25.0\text{ m})^2+(40.0\text{ m})^2}=47.2\text{ m}$$

(b) Recalling that $\tan\theta = \tan(\theta + 180°)$,

$$\tan^{-1}[(40.0\text{ m})/(-25.0\text{ m})] = -58° \text{ or } 122°.$$

Noting that the vector is in the second quadrant (by the signs of its *x* and *y* components) we see that 122° is the correct answer. The results are depicted in the figure below:

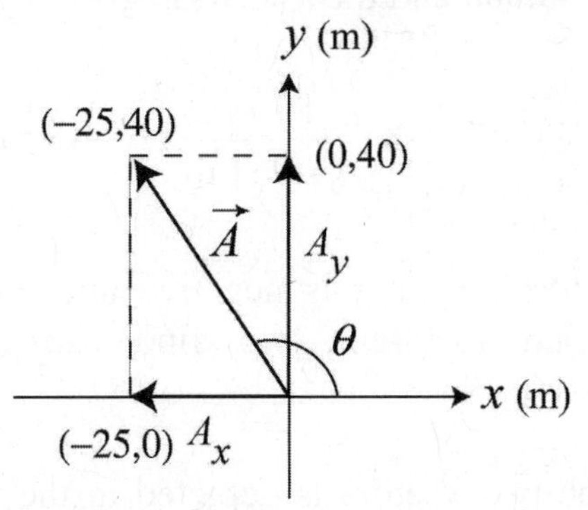

LEARN We can check our answers by noting that the *x*- and the *y*- components of $\vec{A}$ can be written as

$$A_x = A\cos\theta, \qquad A_y = A\sin\theta.$$

Substituting the results calculated above, we obtain

$$A_x = (47.2\ \text{m})\cos 122° = -25.0\ \text{m},\quad A_y = (47.2\ \text{m})\sin 122° = +40.0\ \text{m}$$

which indeed are the values given in the problem statement.

3-11

THINK This problem involves the addition of two vectors $\vec{a}$ and $\vec{b}$. We want to find the magnitude and direction of the resulting vector.

EXPRESS In two dimensions, a vector $\vec{a}$ can be written as, in unit vector notation,

$$\vec{a} = a_x\hat{\text{i}} + a_y\hat{\text{j}}.$$

Similarly, a second vector $\vec{b}$ can be expressed as $\vec{b} = b_x\hat{\text{i}} + b_y\hat{\text{j}}$. Adding the two vectors gives

$$\vec{r} = \vec{a} + \vec{b} = (a_x + b_x)\hat{\text{i}} + (a_y + b_y)\hat{\text{j}} = r_x\hat{\text{i}} + r_y\hat{\text{j}}$$

ANALYZE (a) Given that $\vec{a} = (4.0\ \text{m})\hat{\text{i}} + (3.0\ \text{m})\hat{\text{j}}$ and $\vec{b} = (-13.0\ \text{m})\hat{\text{i}} + (7.0\ \text{m})\hat{\text{j}}$, we find the x and the y components of $\vec{r}$ to be

$$r_x = a_x + b_x = (4.0\ \text{m}) + (-13\ \text{m}) = -9.0\ \text{m}$$
$$r_y = a_y + b_y = (3.0\ \text{m}) + (7.0\ \text{m}) = 10.0\ \text{m}.$$

Thus $\vec{r} = (-9.0\,\text{m})\hat{\text{i}} + (10\,\text{m})\hat{\text{j}}$.

(b) The magnitude of $\vec{r}$ is $r = |\vec{r}| = \sqrt{r_x^2 + r_y^2} = \sqrt{(-9.0\ \text{m})^2 + (10\ \text{m})^2} = 13\ \text{m}$.

(c) The angle between the resultant and the $+x$ axis is given by

$$\theta = \tan^{-1}\left(\frac{r_y}{r_x}\right) = \tan^{-1}\left(\frac{10.0\ \text{m}}{-9.0\ \text{m}}\right) = -48° \text{ or } 132°.$$

Since the x component of the resultant is negative and the y component is positive, characteristic of the second quadrant, we find the angle is 132° (measured counterclockwise from $+x$ axis).

LEARN The addition of the two vectors is depicted in the figure below (not to scale). Indeed, since $r_x < 0$ and $r_y > 0$, we expect $\vec{r}$ to be in the second quadrant.

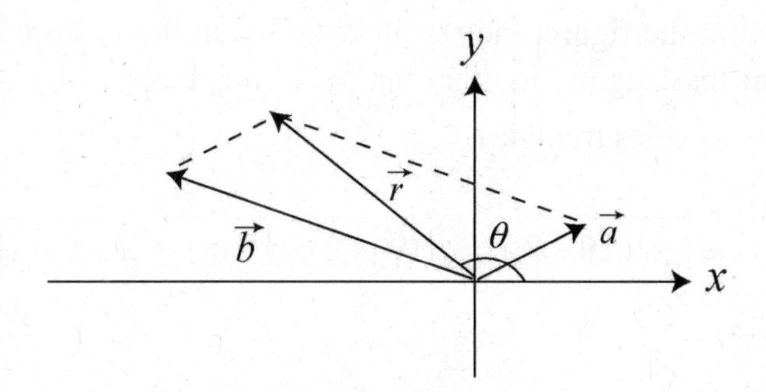

3-15

THINK This problem involves the addition of two vectors $\vec{a}$ and $\vec{b}$ in two dimensions. We're asked to find the components, magnitude and direction of the resulting vector.

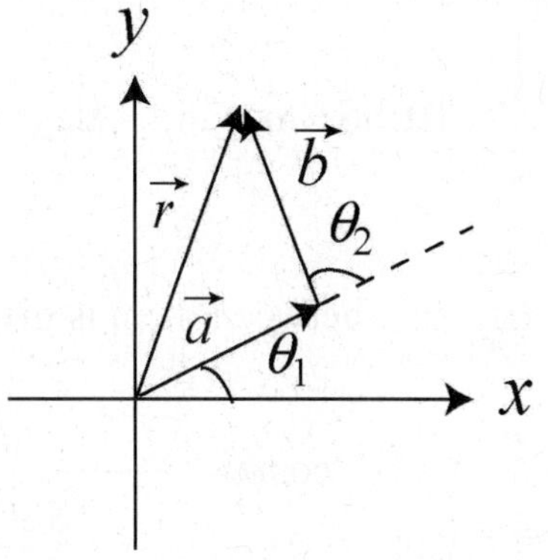

EXPRESS In two dimensions, a vector $\vec{a}$ can be written as, in unit vector notation,

$$\vec{a} = a_x\hat{\text{i}} + a_y\hat{\text{j}} = (a\cos\alpha)\hat{\text{i}} + (a\sin\alpha)\hat{\text{j}}.$$

Similarly, a second vector $\vec{b}$ can be expressed as $\vec{b} = b_x\hat{\text{i}} + b_y\hat{\text{j}} = (b\cos\beta)\hat{\text{i}} + (b\sin\beta)\hat{\text{j}}$. From the figure, we have, $\alpha = \theta_1$ and $\beta = \theta_1 + \theta_2$ (since the angles are measured from the +*x*-axis) and the resulting vector is

$$\vec{r} = \vec{a} + \vec{b} = [a\cos\theta_1 + b\cos(\theta_1 + \theta_2)]\hat{\text{i}} + [a\sin\theta_1 + b\sin(\theta_1 + \theta_2)]\hat{\text{j}} = r_x\hat{\text{i}} + r_y\hat{\text{j}}$$

ANALYZE (a) Given that $a = b = 10$ m, $\theta_1 = 30°$ and $\theta_2 = 105°$, the *x* component of $\vec{r}$ is

$$r_x = a\cos\theta_1 + b\cos(\theta_1 + \theta_2) = (10\text{ m})\cos 30° + (10\text{ m})\cos(30° + 105°) = 1.59\text{ m}$$

(b) Similarly, the *y* component of $\vec{r}$ is

$$r_y = a\sin\theta_1 + b\sin(\theta_1 + \theta_2) = (10\text{ m})\sin 30° + (10\text{ m})\sin(30° + 105°) = 12.1\text{ m}.$$

(c) The magnitude of $\vec{r}$ is $r = |\vec{r}| = \sqrt{(1.59\text{ m})^2 + (12.1\text{ m})^2} = 12.2$ m.

(d) The angle between $\vec{r}$ and the +*x*-axis is

$$\theta = \tan^{-1}\left(\frac{r_y}{r_x}\right) = \tan^{-1}\left(\frac{12.1\text{ m}}{1.59\text{ m}}\right) = 82.5°.$$

LEARN As depicted in the figure, the resultant $\vec{r}$ lies in the first quadrant. This is what we expect. Note that the magnitude of $\vec{r}$ can also be calculated by using law of cosine ($\vec{a}, \vec{b}$ and $\vec{r}$ form an isosceles triangle):

$$r=\sqrt{a^2+b^2-2ab\cos(180-\theta_2)}=\sqrt{(10\text{ m})^2+(10\text{ m})^2-2(10\text{ m})(10\text{ m})\cos 75°}$$
$$=12.2\text{ m}.$$

3-41

THINK The angle between two vectors can be calculated using the definition of scalar product.

EXPRESS Since the scalar product of two vectors $\vec{a}$ and $\vec{b}$ is

$$\vec{a}\cdot\vec{b}=ab\cos\phi=a_xb_x+a_yb_y+a_zb_z,$$

the angle between them is given by

$$\cos\phi=\frac{a_xb_x+a_yb_y+a_zb_z}{ab}\quad\Rightarrow\quad\phi=\cos^{-1}\left(\frac{a_xb_x+a_yb_y+a_zb_z}{ab}\right).$$

Once the magnitudes and components of the vectors are known, the angle ϕ can be readily calculated.

ANALYZE Given that $\vec{a}=(3.0)\hat{\text{i}}+(3.0)\hat{\text{j}}+(3.0)\hat{\text{k}}$ and $\vec{b}=(2.0)\hat{\text{i}}+(1.0)\hat{\text{j}}+(3.0)\hat{\text{k}}$, the magnitudes of the vectors are

$$a=|\vec{a}|=\sqrt{a_x^2+a_y^2+a_z^2}=\sqrt{(3.0)^2+(3.0)^2+(3.0)^2}=5.20$$

$$b=|\vec{b}|=\sqrt{b_x^2+b_y^2+b_z^2}=\sqrt{(2.0)^2+(1.0)^2+(3.0)^2}=3.74.$$

The angle between them is found to be

$$\cos\phi=\frac{(3.0)(2.0)+(3.0)(1.0)+(3.0)(3.0)}{(5.20)(3.74)}=0.926,$$

or $\phi=22°$.

LEARN As the name implies, the scalar product (or dot product) between two vectors is a scalar quantity. It can be regarded as the product between the magnitude of one of the vectors and the scalar component of the second vector along the direction of the first one, as illustrated next (see also in Fig. 3-18 of the text):

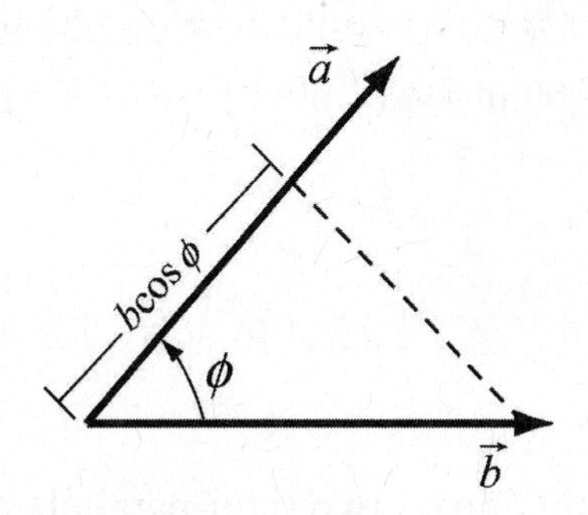

3-43

THINK In this problem we are given three vectors $\vec{a}$, $\vec{b}$ and $\vec{c}$ on the *xy*-plane, and asked to calculate their components.

EXPRESS From the figure, we note that $\vec{c} \perp \vec{b}$, which implies that the angle between $\vec{c}$ and the $+x$ axis is $\theta + 90°$. In unit-vector notation, the three vectors can be written as

$$\vec{a} = a_x\hat{\text{i}}$$
$$\vec{b} = b_x\hat{\text{i}} + b_y\hat{\text{j}} = (b\cos\theta)\hat{\text{i}} + (b\sin\theta)\hat{\text{j}}$$
$$\vec{c} = c_x\hat{\text{i}} + c_y\hat{\text{j}} = [c\cos(\theta + 90°)]\hat{\text{i}} + [c\sin(\theta + 90°)]\hat{\text{j}}.$$

The above expressions allow us to evaluate the components of the vectors.

ANALYZE (a) The *x*-component of $\vec{a}$ is $a_x = a\cos 0° = a = 3.00$ m.

(b) Similarly, the *y*-componnet of $\vec{a}$ is $a_y = a\sin 0° = 0$.

(c) The *x*-component of $\vec{b}$ is $b_x = b\cos 30° = (4.00\text{ m})\cos 30° = 3.46$ m,

(d) and the *y*-component is $b_y = b\sin 30° = (4.00\text{ m})\sin 30° = 2.00$ m.

(e) The *x*-component of $\vec{c}$ is $c_x = c\cos 120° = (10.0\text{ m})\cos 120° = -5.00$ m,

(f) and the *y*-component is $c_y = c\sin 30° = (10.0\text{ m})\sin 120° = 8.66$ m.

(g) The fact that $\vec{c} = p\vec{a} + q\vec{b}$ implies

$$\vec{c} = c_x\hat{\text{i}} + c_y\hat{\text{j}} = p(a_x\hat{\text{i}}) + q(b_x\hat{\text{i}} + b_y\hat{\text{j}}) = (pa_x + qb_x)\hat{\text{i}} + qb_y\hat{\text{j}}$$

or

$$c_x = pa_x + qb_x, \qquad c_y = qb_y.$$

Substituting the values found above, we have

$$-5.00 \text{ m} = p\,(3.00 \text{ m}) + q\,(3.46 \text{ m})$$
$$8.66 \text{ m} = q\,(2.00 \text{ m}).$$

Solving these equations, we find $p = -6.67$.

(h) Similarly, $q = 4.33$ (note that it's easiest to solve for q first). The numbers p and q have no units.

LEARN This exercise shows that given two (non-parallel) vectors in two dimensions, the third vector can always be written as a linear combination of the first two.

3-49

THINK This problem deals with the displacement of a sailboat. We want to find the displacement vector between two locations.

EXPRESS The situation is depicted in the figure below. Let $\vec{a}$ represent the first part of his actual voyage (50.0 km east) and $\vec{c}$ represent the intended voyage (90.0 km north). We look for a vector $\vec{b}$ such that $\vec{c} = \vec{a} + \vec{b}$.

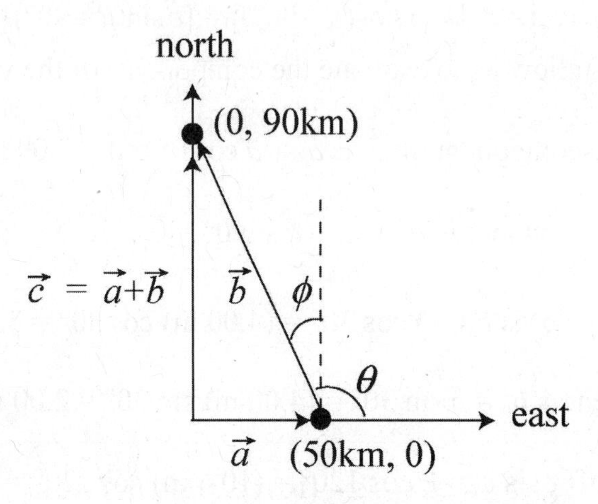

ANALYZE (a) Using the Pythagorean theorem, the distance traveled by the sailboat is

$$b = \sqrt{(50.0 \text{ km})^2 + (90.0 \text{ km})^2} = 103 \text{ km}.$$

(b) The direction is

$$\phi = \tan^{-1}\left(\frac{50.0 \text{ km}}{90.0 \text{ km}}\right) = 29.1°$$

west of north (which is equivalent to 60.9° north of due west).

LEARN This problem could also be solved by first expressing the vectors in unit-vector notation: $\vec{a} = (50.0\text{ km})\hat{\text{i}}, \vec{c} = (90.0\text{ km})\hat{\text{j}}$. This gives

$$\vec{b} = \vec{c} - \vec{a} = -(50.0\text{ km})\hat{\text{i}} + (90.0\text{ km})\hat{\text{j}}.$$

The angle between $\vec{b}$ and the $+x$-axis is

$$\theta = \tan^{-1}\left(\frac{90.0\text{ km}}{-50.0\text{ km}}\right) = 119.1°.$$

The angle θ is related to ϕ by $\theta = 90° + \phi$.

3-53

THINK This problem involves finding scalar and vector products between two vectors $\vec{a}$ and $\vec{b}$.

EXPRESS We apply Eqs. 3-20 and 3-27 to calculate the scalar and vector products between two vectors:

$$\vec{a} \cdot \vec{b} = ab\cos\phi$$
$$|\vec{a} \times \vec{b}| = ab\sin\phi.$$

ANALYZE (a) Given that $a = |\vec{a}| = 10$, $b = |\vec{b}| = 6.0$ and $\phi = 60°$, the scalar (dot) product of $\vec{a}$ and $\vec{b}$ is

$$\vec{a} \cdot \vec{b} = ab\cos\phi = (10)(6.0)\cos 60° = 30.$$

(b) Similarly, the magnitude of the vector (cross) product of the two vectors is

$$|\vec{a} \times \vec{b}| = ab\sin\phi = (10)(6.0)\sin 60° = 52.$$

LEARN When two vectors $\vec{a}$ and $\vec{b}$ are parallel ($\phi = 0$), their scalar and vector products are $\vec{a} \cdot \vec{b} = ab\cos\phi = ab$ and $|\vec{a} \times \vec{b}| = ab\sin\phi = 0$, respectively. However, when they are perpendicular ($\phi = 90°$), we have $\vec{a} \cdot \vec{b} = ab\cos\phi = 0$ and $|\vec{a} \times \vec{b}| = ab\sin\phi = ab$.

3-57

THINK This problem deals with addition and subtraction of two vectors.

EXPRESS From the problem statement, we have

$$\vec{A}+\vec{B}=(6.0)\hat{\text{i}}+(1.0)\hat{\text{j}}$$
$$\vec{A}-\vec{B}=-(4.0)\hat{\text{i}}+(7.0)\hat{\text{j}}$$

Solving the simultaneous equations gives $\vec{A}$ and $\vec{B}$.

ANALYZE Adding the above equations and dividing by 2 leads to $\vec{A}=(1.0)\hat{\text{i}}+(4.0)\hat{\text{j}}$. The magnitude of $\vec{A}$ is

$$A=|\vec{A}|=\sqrt{A_x^2+A_y^2}=\sqrt{(1.0)^2+(4.0)^2}=4.1$$

LEARN The vector $\vec{B}$ is $\vec{B}=(5.0)\hat{\text{i}}+(-3.0)\hat{\text{j}}$, and its magnitude is

$$B=|\vec{B}|=\sqrt{B_x^2+B_y^2}=\sqrt{(5.0)^2+(-3.0)^2}=5.8.$$

The results are summarized in the figure below:

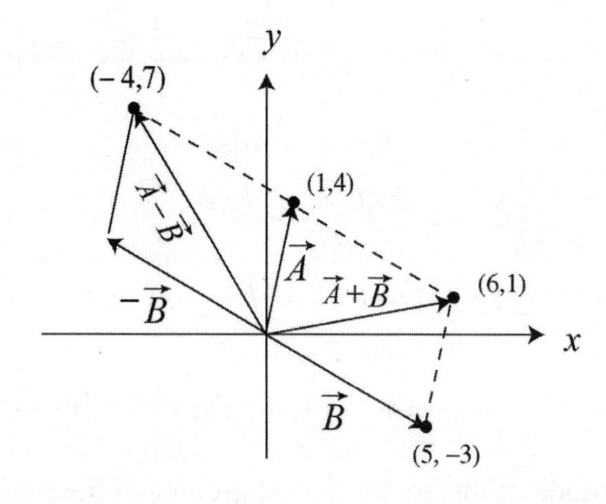

3-64

THINK This problem deals with the displacement and distance traveled by a fly from one corner of a room to the diagonally opposite corner. The displacement vector is three-dimensional.

EXPRESS The displacement of the fly is illustrated in the figure below:

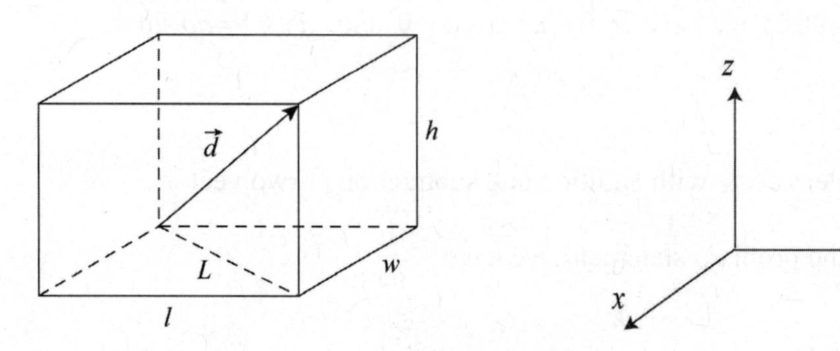

A coordinate system such as the one shown (above right) allows us to express the displacement as a three-dimensional vector.

ANALYZE (a) The magnitude of the displacement from one corner to the diagonally opposite corner is

$$d = |\vec{d}| = \sqrt{w^2 + l^2 + h^2}$$

Substituting the values given, we obtain

$$d = |\vec{d}| = \sqrt{w^2 + l^2 + h^2} = \sqrt{(3.70\text{ m})^2 + (4.30\text{ m})^2 + (3.00\text{ m})^2} = 6.42\text{ m}.$$

(b) The displacement vector is along the straight line from the beginning to the end point of the trip. Since a straight line is the shortest distance between two points, the length of the path cannot be less than *d*, the magnitude of the displacement.

(c) The length of the path of the fly can be greater than *d*, however. The fly might, for example, crawl along the edges of the room. Its displacement would be the same but the path length would be $\ell + w + h = 11.0$ m.

(d) The path length is the same as the magnitude of the displacement if the fly flies along the displacement vector.

(e) We take the x axis to be out of the page, the y axis to be to the right, and the z axis to be upward (as shown in the figure above). Then the x component of the displacement is w = 3.70 m, the y component of the displacement is 4.30 m, and the z component is 3.00 m. Thus, the displacement vector can be written as

$$\vec{d} = (3.70\text{ m})\hat{\text{i}} + (4.30\text{ m})\hat{\text{j}} + (3.00\text{ m})\hat{\text{k}}.$$

(f) Suppose the path of the fly is as shown by the dotted lines on the diagram (below left). Pretend there is a hinge where the front wall of the room joins the floor and lay the wall down as shown (above right).

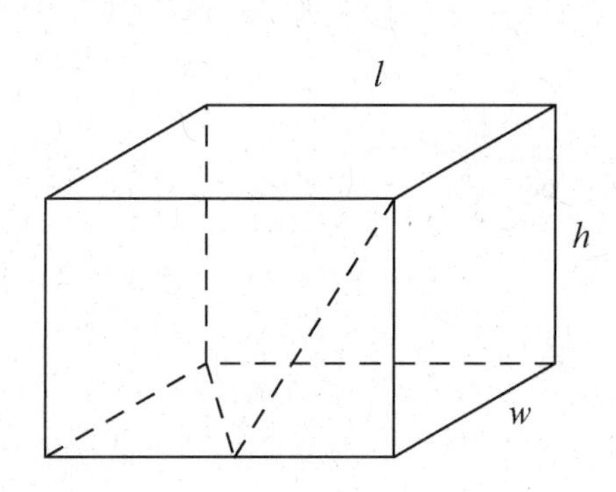

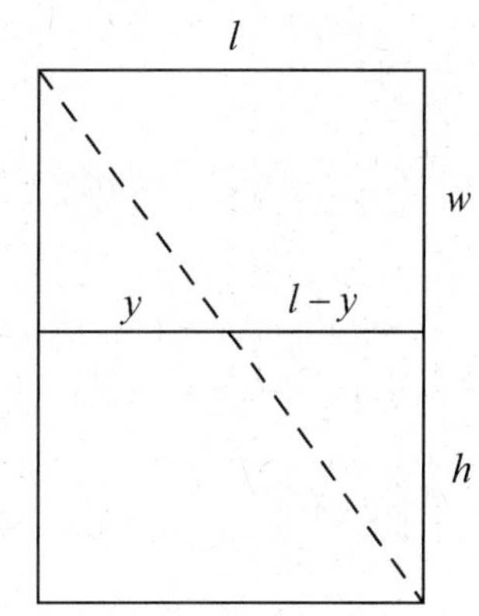

The shortest walking distance between the lower left back of the room and the upper right front corner is the dotted straight line shown on the diagram. Its length is

$$s_{\min} = \sqrt{(w + h)^2 + l^2} = \sqrt{(3.70 \text{ m} + 3.00 \text{ m})^2 + (4.30 \text{ m})^2} = 7.96 \text{ m}.$$

LEARN To show that the shortest path is indeed given by $s_{\min}$, we write the length of the path as

$$s = \sqrt{y^2 + w^2} + \sqrt{(l - y)^2 + h^2}\,.$$

The condition for minimum is given by

$$\frac{ds}{dy} = \frac{y}{\sqrt{y^2 + w^2}} - \frac{l - y}{\sqrt{(l - y)^2 + h^2}} = 0\,.$$

A little algebra shows that the condition is satisfied when $y = lw/(w + h)$, which gives

$$s_{\min} = \sqrt{w^2\left(1 + \frac{l^2}{(w + h)^2}\right)} + \sqrt{h^2\left(1 + \frac{l^2}{(w + h)^2}\right)} = \sqrt{(w + h)^2 + l^2}\,.$$

Any other path would be longer than 7.96 m.

Chapter 4

4-5

THINK This problem deals with the motion of a train in two dimensions. The entire trip consists of three parts, and we're interested in the overall average velocity.

EXPRESS The average velocity of the entire trip is given by Eq. 4-8, $\vec{v}_{avg} = \Delta\vec{r}/\Delta t$, where the total displacement $\Delta\vec{r} = \Delta\vec{r}_1 + \Delta\vec{r}_2 + \Delta\vec{r}_3$ is the sum of three displacements (each result of a constant velocity during a given time), and $\Delta t = \Delta t_1 + \Delta t_2 + \Delta t_3$ is the total amount of time for the trip. We use a coordinate system with $+x$ for East and $+y$ for North.

ANALYZE (a) In unit-vector notation, the first displacement is given by

$$\Delta\vec{r}_1 = \left(60.0\ \frac{\text{km}}{\text{h}}\right)\left(\frac{40.0\ \text{min}}{60\ \text{min/h}}\right)\hat{\mathrm{i}} = (40.0\ \text{km})\hat{\mathrm{i}}.$$

The second displacement has a magnitude of $(60.0\ \frac{\text{km}}{\text{h}})\cdot(\frac{20.0\ \text{min}}{60\ \text{min/h}}) = 20.0\ \text{km}$, and its direction is 40° north of east. Therefore,

$$\Delta\vec{r}_2 = (20.0\ \text{km})\cos(40.0°)\,\hat{\mathrm{i}} + (20.0\ \text{km})\sin(40.0°)\,\hat{\mathrm{j}} = (15.3\ \text{km})\,\hat{\mathrm{i}} + (12.9\ \text{km})\,\hat{\mathrm{j}}.$$

Similarly, the third displacement is

$$\Delta\vec{r}_3 = -\left(60.0\ \frac{\text{km}}{\text{h}}\right)\left(\frac{50.0\ \text{min}}{60\ \text{min/h}}\right)\hat{\mathrm{i}} = (-50.0\ \text{km})\,\hat{\mathrm{i}}.$$

Thus, the total displacement is

$$\begin{aligned}\Delta\vec{r} &= \Delta\vec{r}_1 + \Delta\vec{r}_2 + \Delta\vec{r}_3 = (40.0\ \text{km})\hat{\mathrm{i}} + (15.3\ \text{km})\,\hat{\mathrm{i}} + (12.9\ \text{km})\,\hat{\mathrm{j}} - (50.0\ \text{km})\,\hat{\mathrm{i}} \\ &= (5.30\ \text{km})\,\hat{\mathrm{i}} + (12.9\ \text{km})\,\hat{\mathrm{j}}.\end{aligned}$$

The time for the trip is $\Delta t = (40.0 + 20.0 + 50.0)$ min = 110 min, which is equivalent to 1.83 h. Eq. 4-8 then yields

$$\vec{v}_{avg} = \frac{(5.30\ \text{km})\,\hat{\mathrm{i}} + (12.9\ \text{km})\,\hat{\mathrm{j}}}{1.83\ \text{h}} = (2.90\ \text{km/h})\,\hat{\mathrm{i}} + (7.01\ \text{km/h})\,\hat{\mathrm{j}}.$$

The magnitude of $\vec{v}_{avg}$ is $|\vec{v}_{avg}| = \sqrt{(2.90\ \text{km/h})^2 + (7.01\ \text{km/h})^2} = 7.59\ \text{km/h}$.

(b) The angle is given by

$$\theta = \tan^{-1}\left(\frac{v_{\text{avg},y}}{v_{\text{avg},x}}\right) = \tan^{-1}\left(\frac{7.01 \text{ km/h}}{2.90 \text{ km/h}}\right) = 67.5° \text{ (north of east)},$$

or 22.5° east of due north.

LEARN The displacement of the train is depicted in the figure below:

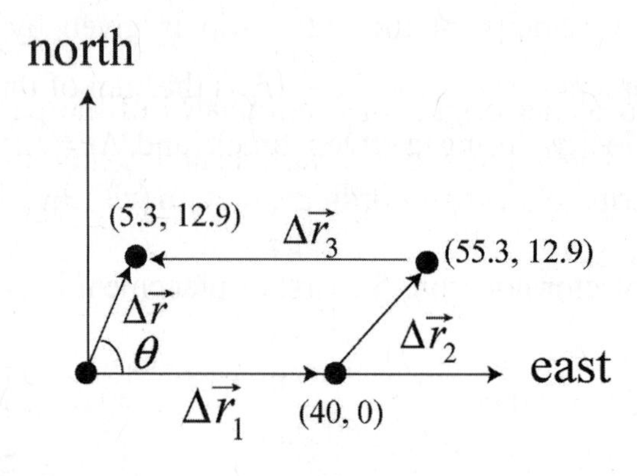

Note that the net displacement $\Delta\vec{r}$ is found by adding $\Delta\vec{r}_1$, $\Delta\vec{r}_2$ and $\Delta\vec{r}_3$ vectorially.

4-13
THINK Knowing the position of a particle as function of time allows us to calculate its corresponding velocity and acceleration by taking time derivatives.

EXPRESS From the position vector $\vec{r}(t)$, the velocity and acceleration of the particle can be found by differentiating $\vec{r}(t)$ with respect to time:

$$\vec{v} = \frac{d\vec{r}}{dt}, \quad \vec{a} = \frac{d\vec{v}}{dt} = \frac{d^2\vec{r}}{dt^2}.$$

ANALYZE (a) Taking the derivative of the position vector $\vec{r}(t) = \hat{\text{i}} + (4t^2)\hat{\text{j}} + t\hat{\text{k}}$ with respect to time, we have, in SI units (m/s),

$$\vec{v} = \frac{d}{dt}(\hat{\text{i}} + 4t^2\,\hat{\text{j}} + t\,\hat{\text{k}}) = 8t\,\hat{\text{j}} + \hat{\text{k}}.$$

(b) Taking another derivative with respect to time leads to, in SI units (m/s^2),

$$\vec{a} = \frac{d}{dt}(8t\,\hat{\text{j}} + \hat{\text{k}}) = 8\,\hat{\text{j}}.$$

LEARN The particle undergoes constant acceleration in the +*y*-direction. This can be seen by noting that the *y* component of $\vec{r}(t)$ is $4t^2$, which is quadratic in *t*.

4-15

THINK Given the initial velocity and acceleration of a particle, we're interested in finding its velocity and position at a later time.

EXPRESS Since the acceleration, $\vec{a} = a_x\hat{\text{i}} + a_y\hat{\text{j}} = (-1.0\text{ m/s}^2)\hat{\text{i}} + (-0.50\text{ m/s}^2)\hat{\text{j}}$, is constant in both x and y directions, we may use Table 2-1 for the motion along each direction. This can be handled individually (for x and y) or together with the unit-vector notation (for $\Delta\vec{r}$).

Since the particle started at the origin, the coordinates of the particle at any time t are given by $\vec{r} = \vec{v}_0 t + \frac{1}{2}\vec{a}t^2$. The velocity of the particle at any time t is given by $\vec{v} = \vec{v}_0 + \vec{a}t$, where $\vec{v}_0$ is the initial velocity and $\vec{a}$ is the (constant) acceleration. Along the x-direction, we have

$$x(t) = v_{0x}t + \frac{1}{2}a_x t^2, \qquad v_x(t) = v_{0x} + a_x t$$

Similarly, along the y-direction, we get

$$y(t) = v_{0y}t + \frac{1}{2}a_y t^2, \qquad v_y(t) = v_{0y} + a_y t\,.$$

Known: $v_{0x} = 3.0\text{ m/s},\ v_{0y} = 0,\ a_x = -1.0\text{ m/s}^2,\ a_y = -0.5\text{ m/s}^2$.

ANALYZE (a) Substituting the values given, the components of the velocity are

$$v_x(t) = v_{0x} + a_x t = (3.0\text{ m/s}) - (1.0\text{ m/s}^2)t$$
$$v_y(t) = v_{0y} + a_y t = -(0.50\text{ m/s}^2)t$$

When the particle reaches its maximum x coordinate at $t = t_m$, we must have $v_x = 0$. Therefore, $3.0 - 1.0t_m = 0$ or $t_m = 3.0$ s. The y component of the velocity at this time is

$$v_y(t = 3.0\text{ s}) = -(0.50\text{ m/s}^2)(3.0) = -1.5\text{ m/s}$$

Thus, $\vec{v}_m = (-1.5\text{ m/s})\hat{\text{j}}$.

(b) At $t = 3.0$ s, the components of the position are

$$x(t = 3.0\text{ s}) = v_{0x}t + \frac{1}{2}a_x t^2 = (3.0\text{ m/s})(3.0\text{ s}) + \frac{1}{2}(-1.0\text{ m/s}^2)(3.0\text{ s})^2 = 4.5\text{ m}$$
$$y(t = 3.0\text{ s}) = v_{0y}t + \frac{1}{2}a_y t^2 = 0 + \frac{1}{2}(-0.5\text{ m/s}^2)(3.0\text{ s})^2 = -2.25\text{ m}$$

Using unit-vector notation, the results can be written as $\vec{r}_m = (4.50\text{ m})\hat{\text{i}} - (2.25\text{ m})\hat{\text{j}}$.

LEARN The motion of the particle in this problem is two-dimensional, and the kinematics in the x- and y-directions can be analyzed separately.

4-33

THINK This problem deals with projectile motion. We're interested in the horizontal displacement and velocity of the projectile before it strikes the ground.

EXPRESS We adopt the positive direction choices used in the textbook so that equations such as Eq. 4-22 are directly applicable. The coordinate origin is at ground level directly below the release point. We write $\theta_0 = -37.0°$ for the angle measured from $+x$, since the angle $\phi_0 = 53.0°$ given in the problem is measured from the $-y$ direction. The initial setup of the problem is shown in the figure below (not to scale).

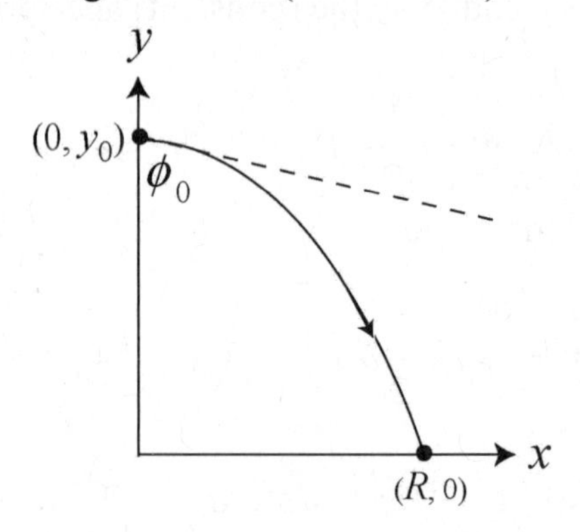

ANALYZE (a) The initial speed of the projectile is the plane's speed at the moment of release. Given that $y_0 = 730$ m and $y = 0$ at $t = 5.00$ s, we use Eq. 4-22 to find v_0:

$$y - y_0 = (v_0 \sin\theta_0)\,t - \frac{1}{2}gt^2 \;\Rightarrow\; 0 - 730 \text{ m} = v_0 \sin(-37.0°)(5.00 \text{ s}) - \frac{1}{2}(9.80 \text{ m/s}^2)(5.00 \text{ s})^2$$

which yields $v_0 = 202$ m/s.

(b) The horizontal distance traveled is

$$R = v_x t = (v_0 \cos\theta_0)t = [(202 \text{ m/s})\cos(-37.0°)](5.00 \text{ s}) = 806 \text{ m}.$$

(c) The x component of the velocity (just before impact) is

$$v_x = v_0 \cos\theta_0 = (202 \text{ m/s})\cos(-37.0°) = 161 \text{ m/s}.$$

(d) The y component of the velocity (just before impact) is

$$v_y = v_0 \sin\theta_0 - gt = (202 \text{ m/s})\sin(-37.0°) - (9.80 \text{ m/s}^2)(5.00 \text{ s}) = -171 \text{ m/s}.$$

LEARN In this projectile problem we analyzed the kinematics in the vertical and horizontal directions separately since they do not affect each other. The x-component of the velocity, $v_x = v_0 \cos\theta_0$, remains unchanged throughout since there's no horizontal acceleration.

4-35

THINK This problem deals with projectile motion of a bullet. We're interested in the firing angle that allows the bullet to strike a target at some distance away.

EXPRESS We adopt the positive direction choices used in the textbook so that equations such as Eq. 4-22 are directly applicable. The coordinate origin is at the end of the rifle (the initial point for the bullet as it begins projectile motion in the sense of § 4-5), and we let θ_0 be the firing angle. If the target is a distance d away, then its coordinates are $x = d$, $y = 0$.

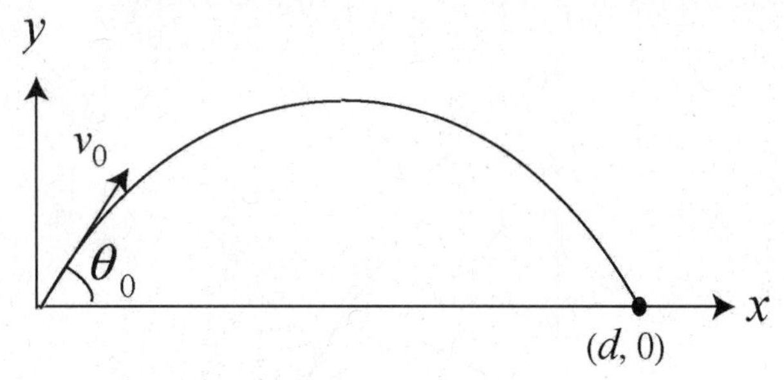

The projectile motion equations lead to

$$d = (v_0 \cos\theta_0)t, \quad 0 = v_0 t \sin\theta_0 - \frac{1}{2}gt^2$$

where θ_0 is the firing angle. The setup of the problem is shown in the figure above (scale exaggerated).

ANALYZE The time at which the bullet strikes the target is given by $t = d/(v_0 \cos\theta_0)$. Eliminating t leads to $2v_0^2 \sin\theta_0 \cos\theta_0 - gd = 0$. Using $\sin\theta_0 \cos\theta_0 = \frac{1}{2}\sin(2\theta_0)$, we obtain

$$v_0^2 \sin\,(2\theta_0) = gd \Rightarrow \sin(2\theta_0) = \frac{gd}{v_0^2} = \frac{(9.80\text{ m/s}^2)(45.7\text{ m})}{(460\text{ m/s})^2}$$

which yields $\sin(2\theta_0) = 2.11\times10^{-3}$, or $\theta_0 = 0.0606°$. If the gun is aimed at a point a distance ℓ above the target, then $\tan\theta_0 = \ell/d$ so that

$$\ell = d\tan\theta_0 = (45.7\text{ m})\tan(0.0606°) = 0.0484\text{ m} = 4.84\text{ cm}.$$

LEARN Due to the downward gravitational acceleration, in order for the bullet to strike the target, the gun must be aimed at a point slightly above the target.

4-37

THINK The trajectory of the diver is a projectile motion. We are interested in the displacement of the diver at a later time.

EXPRESS The initial velocity has no vertical component ($\theta_0 = 0$), but only an x component. Eqs. 4-21 and 4-22 can be simplified to

$$x - x_0 = v_{0x}t$$
$$y - y_0 = v_{0y}t - \frac{1}{2}gt^2 = -\frac{1}{2}gt^2.$$

where $x_0 = 0$, $v_{0x} = v_0 = +2.0$ m/s and y_0 = +10.0 m (taking the water surface to be at $y = 0$). The setup of the problem is shown in the figure below.

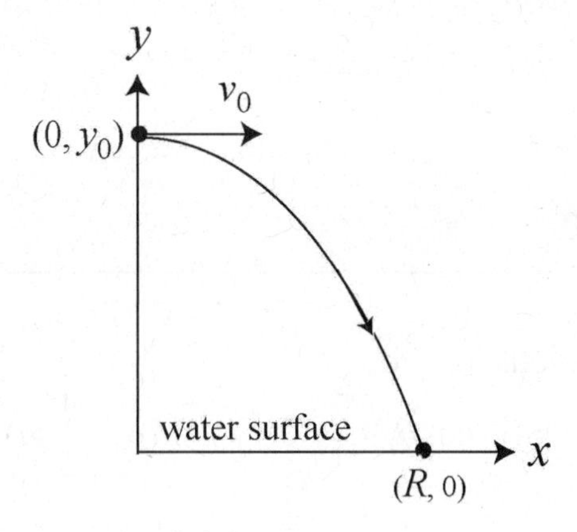

ANALYZE (a) At $t = 0.80$ s, the horizontal distance of the diver from the edge is

$$x = x_0 + v_{0x}t = 0 + (2.0 \text{ m/s})(0.80 \text{ s}) = 1.60 \text{ m}.$$

(b) Similarly, using the second equation for the vertical motion, we obtain

$$y = y_0 - \frac{1}{2}gt^2 = 10.0 \text{ m} - \frac{1}{2}(9.80 \text{ m/s}^2)(0.80 \text{ s})^2 = 6.86 \text{ m}.$$

(c) At the instant the diver strikes the water surface, y = 0. Solving for t using the equation $y = y_0 - \frac{1}{2}gt^2 = 0$ leads to

$$t = \sqrt{\frac{2y_0}{g}} = \sqrt{\frac{2(10.0 \text{ m})}{9.80 \text{ m/s}^2}} = 1.43 \text{ s}.$$

During this time, the x-displacement of the diver is $R = x$ = (2.00 m/s)(1.43 s) = 2.86 m.

LEARN Using Eq. 4-25 with $\theta_0 = 0$, the trajectory of the diver can also be written as

$$y = y_0 - \frac{gx^2}{2v_0^2}.$$

Part (c) can also be solved by using this equation:

$$y = y_0 - \frac{gx^2}{2v_0^2} = 0 \Rightarrow x = R = \sqrt{\frac{2v_0^2 y_0}{g}} = \sqrt{\frac{2(2.0\text{ m/s})^2(10.0\text{ m})}{9.8\text{ m/s}^2}} = 2.86\text{ m}.$$

4-47

THINK The baseball undergoes projectile motion after being hit by the batter. We'd like to know if the ball clears a high fence at some distance away.

EXPRESS We adopt the positive direction choices used in the textbook so that equations such as Eq. 4-22 are directly applicable. The coordinate origin is at ground level directly below impact point between bat and ball. In the absence of a fence, with $\theta_0 = 45°$, the horizontal range (same launch level) is $R = 107$ m. We want to know how high the ball is from the ground when it is at $x' = 97.5$ m, which requires knowing the initial velocity. The trajectory of the baseball can be described by Eq. 4-25:

$$y - y_0 = (\tan\theta_0)x - \frac{gx^2}{2(v_0\cos\theta_0)^2}.$$

The setup of the problem is shown in the figure below (not to scale).

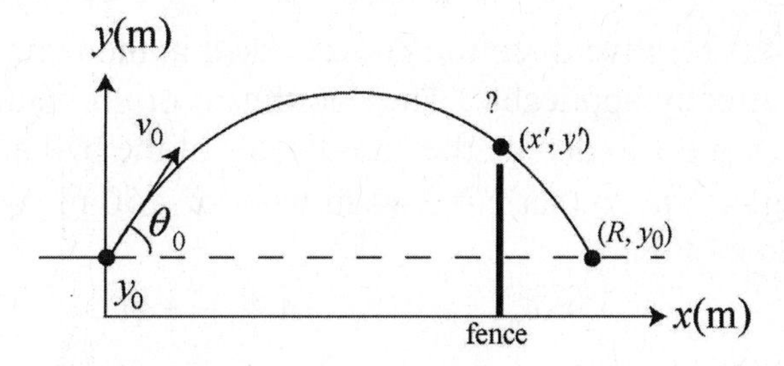

ANALYZE (a) We first solve for the initial speed v_0. Using the range information ($y = y_0$ when $x = R$) and $\theta_0 = 45°$, Eq. 4-25 gives

$$v_0 = \sqrt{\frac{gR}{\sin 2\theta_0}} = \sqrt{\frac{(9.8\text{ m/s}^2)(107\text{ m})}{\sin(2\cdot 45°)}} = 32.4\text{ m/s}.$$

Thus, the time at which the ball flies over the fence is:

$$x' = (v_0\cos\theta_0)t' \Rightarrow t' = \frac{x'}{v_0\cos\theta_0} = \frac{97.5\text{ m}}{(32.4\text{ m/s})\cos 45°} = 4.26\text{ s}.$$

At this moment, the ball is at a height (above the ground) of

$$\begin{aligned} y' &= y_0 + (v_0 \sin\theta_0)t' - \frac{1}{2}gt'^2 \\ &= 1.22\ \text{m} + [(32.4\ \text{m/s})\sin 45°](4.26\ \text{s}) - \frac{1}{2}(9.8\ \text{m/s}^2)(4.26\ \text{s})^2 \\ &= 9.88\ \text{m} \end{aligned}$$

which implies that it does clear the 7.32 m high fence.

(b) At $t' = 4.26\ \text{s}$, the center of the ball is 9.88 m – 7.32 m = 2.56 m above the fence.

LEARN Using the trajectory equation above, one can show that the minimum initial velocity required to clear the fence is given by

$$y' - y_0 = (\tan\theta_0)x' - \frac{gx'^2}{2(v_0\cos\theta_0)^2},$$

or about 31.9 m/s.

4-49

THINK In this problem a football is given an initial speed and it undergoes projectile motion. We'd like to know the smallest and greatest angles at which a field goal can be scored.

EXPRESS We adopt the positive direction choices used in the textbook so that equations such as Eq. 4-22 are directly applicable. The coordinate origin is at the point where the ball is kicked. We use x and y to denote the coordinates of the ball at the goalpost, and try to find the kicking angle(s) θ_0 so that $y = 3.44$ m when $x = 50$ m. Writing the kinematics equations for projectile motion:

$$x = v_0\cos\theta_0, \quad y = v_0 t\sin\theta_0 - \tfrac{1}{2}gt^2,$$

we see the first equation gives $t = x/v_0 \cos\theta_0$, and when this is substituted into the second the result is

$$y = x\tan\theta_0 - \frac{gx^2}{2v_0^2\cos^2\theta_0}.$$

ANALYZE One may solve the above equation by trial and error: systematically trying values of θ_0 until you find the two that satisfy the equation. A little manipulation, however, will give an algebraic solution: Using the trigonometric identity

$$1/\cos^2\theta_0 = 1 + \tan^2\theta_0,$$

we obtain

$$\frac{1}{2}\frac{gx^2}{v_0^2}\tan^2\theta_0 - x\tan\theta_0 + y + \frac{1}{2}\frac{gx^2}{v_0^2} = 0$$

which is a second-order equation for tan θ_0. To simplify writing the solution, we denote

$$c = \frac{1}{2}gx^2/v_0^2 = \frac{1}{2}\left(9.80\ \text{m/s}^2\right)\left(50\ \text{m}\right)^2/\left(25\ \text{m/s}\right)^2 = 19.6\ \text{m}.$$

Then the second-order equation becomes $c\tan^2\theta_0 - x\tan\theta_0 + y + c = 0$. Using the quadratic formula, we obtain its solution(s).

$$\tan\theta_0 = \frac{x \pm \sqrt{x^2 - 4(y+c)c}}{2c} = \frac{50\ \text{m} \pm \sqrt{(50\ \text{m})^2 - 4(3.44\ \text{m} + 19.6\ \text{m})(19.6\ \text{m})}}{2(19.6\ \text{m})}.$$

The two solutions are given by $\tan\theta_0 = 1.95$ and $\tan\theta_0 = 0.605$. The corresponding (first-quadrant) angles are $\theta_0 = 63°$ and $\theta_0 = 31°$. Thus,

(a) The smallest elevation angle is $\theta_0 = 31°$, and

(b) The greatest elevation angle is $\theta_0 = 63°$.

LEARN If kicked at any angle between 31° and 63°, the ball will travel above the cross bar on the goalposts.

4-55

THINK In this problem a ball rolls off the top of a stairway with an initial speed, and we'd like to know on which step it lands first.

EXPRESS We denote h as the height of a step and w as the width. To hit step n, the ball must fall a distance nh and travel horizontally a distance between $(n-1)w$ and nw. We take the origin of a coordinate system to be at the point where the ball leaves the top of the stairway, and we choose the y axis to be positive in the upward direction, as shown in the figure.

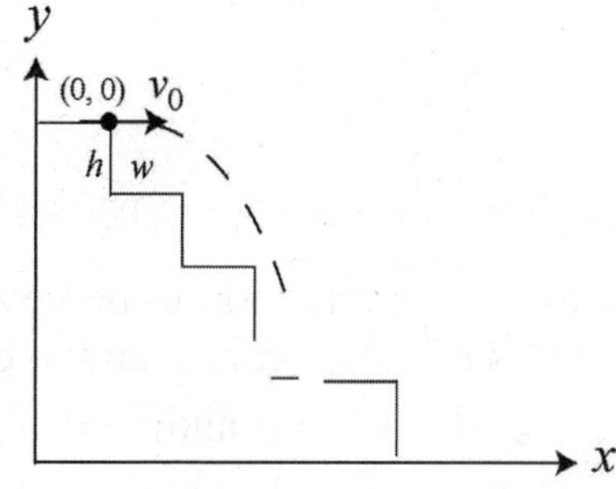

The coordinates of the ball at time t are given by $x = v_{0x}t$ and $y = -\frac{1}{2}gt^2$ (since $v_{0y} = 0$).

ANALYZE We equate y to $-nh$ and solve for the time to reach the level of step n:

$$t = \sqrt{\frac{2nh}{g}}.$$

The x coordinate then is

$$x = v_{0x}\sqrt{\frac{2nh}{g}} = (1.52\text{ m/s})\sqrt{\frac{2n(0.203\text{ m})}{9.8\text{ m/s}^2}} = (0.309\text{ m})\sqrt{n}.$$

The method is to try values of n until we find one for which x/w is less than n but greater than $n-1$. For $n = 1$, $x = 0.309$ m and $x/w = 1.52$, which is greater than n. For $n = 2$, $x = 0.437$ m and $x/w = 2.15$, which is also greater than n. For $n = 3$, $x = 0.535$ m and $x/w = 2.64$. Now, this is less than n and greater than $n-1$, so the ball hits the third step.

LEARN To check the consistency of our calculation, we can substitute $n = 3$ into the above equations. The results are $t = 0.353$ s, $y = 0.609$ m and $x = 0.535$ m. This indeed corresponds to the third step.

4-67

THINK In this problem we have a stone whirled in a horizontal circle. After the string breaks, the stone undergoes projectile motion.

EXPRESS The stone moves in a circular path (top view shown below left) initially, but undergoes projectile motion after the string breaks (side view shown below right). Since $a = v^2/R$, to calculate the centripetal acceleration of the stone, we need to know its speed during its circular motion (this is also its initial speed when it flies off). We use the kinematics equations of projectile motion (discussed in §4-6) to find that speed.

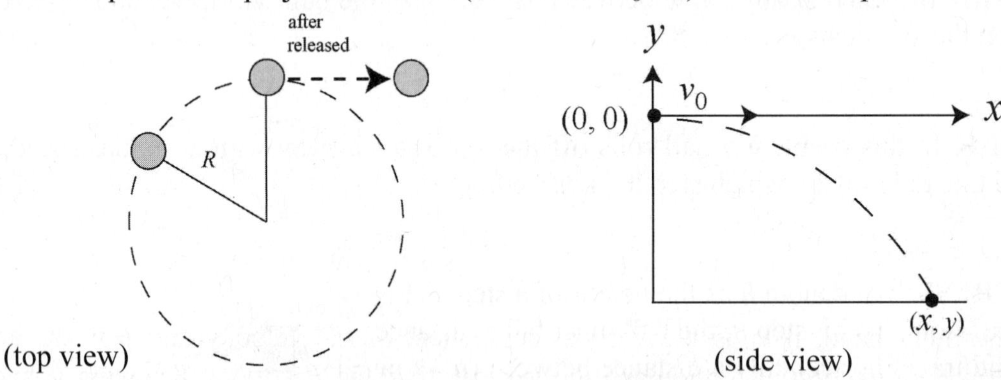

Taking the $+y$ direction to be upward and placing the origin at the point where the stone leaves its circular orbit, then the coordinates of the stone during its motion as a projectile are given by $x = v_0 t$ and $y = -\frac{1}{2}gt^2$ (since $v_{0y} = 0$). It hits the ground at $x = 10$ m and $y = -2.0$ m.

ANALYZE Formally solving the y-component equation for the time, we obtain $t = \sqrt{-2y/g}$, which we substitute into the first equation:

$$v_0 = x\sqrt{-\frac{g}{2y}} = (10\text{ m})\sqrt{-\frac{9.8\text{ m/s}^2}{2(-2.0\text{ m})}} = 15.7\text{ m/s}.$$

Therefore, the magnitude of the centripetal acceleration is

$$a = \frac{v_0^2}{R} = \frac{(15.7\text{ m/s})^2}{1.5\text{ m}} = 160\text{ m/s}^2.$$

LEARN The above equations can be combined to give $a = \frac{gx^2}{-2yR}$. The equation implies that the greater the centripetal acceleration, the greater the initial speed of the projectile, and the greater the distance traveled by the stone. This is precisely what we expect.

4-75

THINK This problem deals with relative motion in two dimensions. Raindrops appear to fall vertically by an observer on a moving train.

EXPRESS Since the raindrops fall vertically relative to the train, the horizontal component of the velocity of a raindrop, $v_h = 30$ m/s, must be the same as the speed of the train, i.e., $v_h = v_{\text{train}}$ (see figure below).

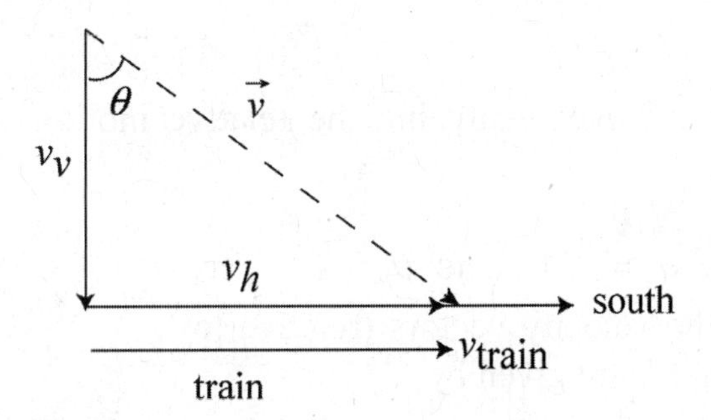

On the other hand, if v_v is the vertical component of the velocity and θ is the angle between the direction of motion and the vertical, then $\tan\theta = v_h/v_v$. Knowing v_v and v_h allows us to determine the speed of the raindrops.

ANALYZE With $\theta = 70°$, we find the vertical component of the velocity to be

$$v_v = v_h/\tan\theta = (30\text{ m/s})/\tan 70° = 10.9\text{ m/s}.$$

Therefore, the speed of a raindrop is

$$v = \sqrt{v_h^2 + v_v^2} = \sqrt{(30\text{ m/s})^2 + (10.9\text{ m/s})^2} = 32\text{ m/s}.$$

LEARN As long as the horizontal component of the velocity of the raindrops coincides with the speed of the train, the passenger on board will see the rain falling perfectly vertically.

4-77

THINK This problem deals with relative motion in two dimensions. Snowflakes falling vertically downward are seen to fall at an angle by a moving observer.

EXPRESS Relative to the car the velocity of the snowflakes has a vertical component of $v_v = 8.0\text{ m/s}$ and a horizontal component of $v_h = 50\text{ km/h} = 13.9\text{ m/s}$.

ANALYZE The angle θ from the vertical is found from

$$\tan\theta = \frac{v_h}{v_v} = \frac{13.9\text{ m/s}}{8.0\text{ m/s}} = 1.74$$

which yields $\theta = 60°$.

LEARN The problem can also be solved by expressing the velocity relation in vector notation: $\vec{v}_{\text{rel}} = \vec{v}_{\text{car}} + \vec{v}_{\text{snow}}$, as shown in the figure.

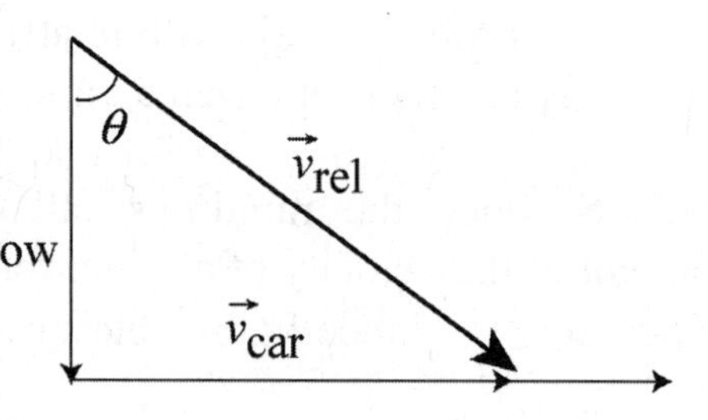

4-79

THINK This problem involves analyzing the relative motion of two ships sailing in different directions.

EXPRESS Given that $\theta_A = 45°$, and $\theta_B = 40°$, as defined in the figure, the velocity vectors (relative to the shore) for ships A and B are given by

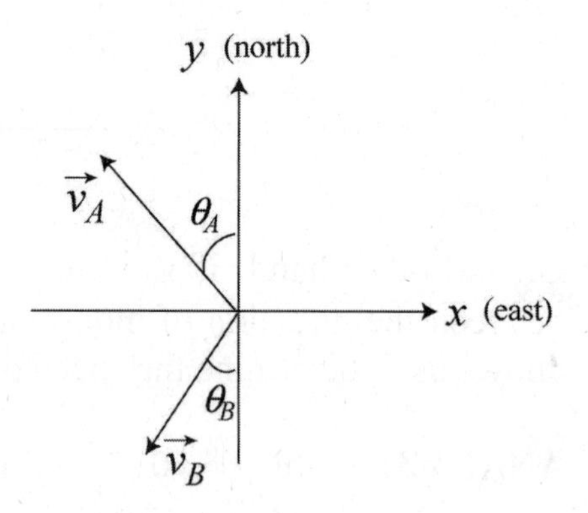

$$\vec{v}_A = -(v_A \cos 45°)\,\hat{\text{i}} + (v_A \sin 45°)\,\hat{\text{j}}$$
$$\vec{v}_B = -(v_B \sin 40°)\,\hat{\text{i}} - (v_B \cos 40°)\,\hat{\text{j}},$$

with v_A = 24 knots and v_B = 28 knots. We take east as $+\hat{\text{i}}$ and north as $\hat{\text{j}}$.

The velocity of ship A relative to ship B is simply given by $\vec{v}_{AB} = \vec{v}_A - \vec{v}_B$.

ANALYZE (a) The relative velocity is

$$\vec{v}_{AB} = \vec{v}_A - \vec{v}_B = (v_B \sin 40° - v_A \cos 45°)\hat{\text{i}} + (v_B \cos 40° + v_A \sin 45°)\,\hat{\text{j}}$$
$$= (1.03\text{ knots})\hat{\text{i}} + (38.4\text{ knots})\hat{\text{j}}$$

the magnitude of which is $|\vec{v}_{AB}| = \sqrt{(1.03 \text{ knots})^2 + (38.4 \text{ knots})^2} \approx 38.4$ knots.

(b) The angle θ_{AB} which $\vec{v}_{AB}$ makes with north is given by

$$\theta_{AB} = \tan^{-1}\left(\frac{v_{AB,x}}{v_{AB,y}}\right) = \tan^{-1}\left(\frac{1.03 \text{ knots}}{38.4 \text{ knots}}\right) = 1.5°$$

which is to say that $\vec{v}_{AB}$ points 1.5° east of north.

(c) Since the two ships started at the same time, their relative velocity describes at what rate the distance between them is increasing. Because the rate is steady, we have

$$t = \frac{|\Delta r_{AB}|}{|\vec{v}_{AB}|} = \frac{160 \text{ nautical miles}}{38.4 \text{ knots}} = 4.2 \text{ h}.$$

(d) The velocity $\vec{v}_{AB}$ does not change with time in this problem, and $\vec{r}_{AB}$ is in the same direction as $\vec{v}_{AB}$ since they started at the same time. Reversing the points of view, we have $\vec{v}_{AB} = -\vec{v}_{BA}$ so that $\vec{r}_{AB} = -\vec{r}_{BA}$ (i.e., they are 180° opposite to each other). Hence, we conclude that *B* stays at a bearing of 1.5° west of south relative to *A* during the journey (neglecting the curvature of Earth).

LEARN The relative velocity is depicted in the figure on the right. When analyzing relative motion in two dimensions, a vector diagram such as the one shown can be very helpful.

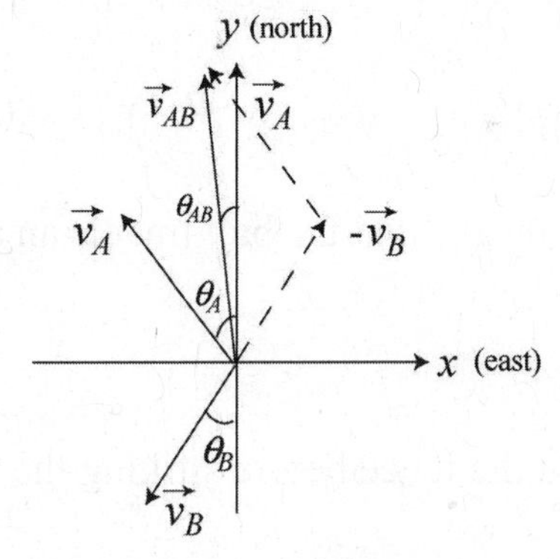

4-87

THINK This problem deals with the projectile motion of a baseball. Given the information on the position of the ball at two instants, we are asked to analyze its trajectory.

EXPRESS The trajectory of the baseball is shown in the figure on the right. According to the problem statement, at $t_1 = 3.0$ s, the ball reaches it maximum height y_{max}, and at $t_2 = t_1 + 2.5 \text{ s} = 5.5$ s, it barely clears a fence at $x_2 = 97.5$ m.

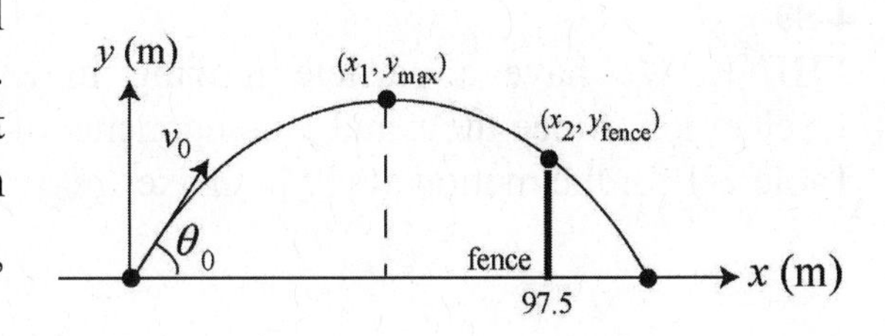

Eq. 2-15 can be applied to the vertical (y axis) motion related to reaching the maximum height (when $t_1 = 3.0$ s and $v_y = 0$):

$$y_{\max} - y_0 = v_y t - \frac{1}{2}gt^2 .$$

ANALYZE (a) With ground level chosen so $y_0 = 0$, this equation gives the result

$$y_{\max} = \frac{1}{2}gt_1^2 = \frac{1}{2}(9.8 \text{ m/s}^2)(3.0 \text{ s})^2 = 44.1 \text{ m}$$

(b) After the moment it reached maximum height, it is falling; at $t_2 = t_1 + 2.5 \text{ s} = 5.5 \text{ s}$, it will have fallen an amount given by Eq. 2-18:

$$y_{\text{fence}} - y_{\max} = 0 - \frac{1}{2}g(t_2 - t_1)^2 .$$

Thus, the height of the fence is

$$y_{\text{fence}} = y_{\max} - \frac{1}{2}g(t_2 - t_1)^2 = 44.1 \text{ m} - \frac{1}{2}(9.8 \text{ m/s}^2)(2.5 \text{ s})^2 = 13.48 \text{ m} .$$

(c) Since the horizontal component of velocity in a projectile-motion problem is constant (neglecting air friction), we find from 97.5 m = v_{0x}(5.5 s) that v_{0x} = 17.7 m/s. The total flight time of the ball is $T = 2t_1 = 2(3.0 \text{ s}) = 6.0 \text{ s}$. Thus, the range of the baseball is

$$R = v_{0x}T = (17.7 \text{ m/s})(6.0 \text{ s}) = 106.4 \text{ m}$$

which means that the ball travels an additional distance

$$\Delta x = R - x_2 = 106.4 \text{ m} - 97.5 \text{ m} = 8.86 \text{ m}$$

beyond the fence before striking the ground.

LEARN Part (c) can also be solved by noting that after passing the fence, the ball will strike the ground in 0.5 s (so that the total "fall-time" equals the "rise-time"). With v_{0x} = 17.7 m/s, we have $\Delta x = (17.7 \text{ m/s})(0.5 \text{ s}) = 8.86 \text{ m}$.

4-89

THINK We have a particle moving in a two-dimensional plane with a constant acceleration. Since the x and y components of the acceleration are constants, we can use Table 2-1 for the motion along both axes.

EXPRESS Using vector notation with $\vec{r}_0 = 0$, the position and velocity of the particle as a function of time are given by $\vec{r}(t) = \vec{v}_0 t + \frac{1}{2}\vec{a}t^2$ and $\vec{v}(t) = \vec{v}_0 + \vec{a}t$, respectively. Where units are not shown, SI units are to be understood.

ANALYZE (a) Given the initial velocity $\vec{v}_0 = (8.0\text{ m/s})\hat{\text{j}}$ and the acceleration $\vec{a} = (4.0\text{ m/s}^2)\hat{\text{i}} + (2.0\text{ m/s}^2)\hat{\text{j}}$, the position vector of the particle is

$$\vec{r} = \vec{v}_0 t + \frac{1}{2}\vec{a}t^2 = \left(8.0\hat{\text{j}}\right)t + \frac{1}{2}\left(4.0\hat{\text{i}} + 2.0\hat{\text{j}}\right)t^2 = \left(2.0t^2\right)\hat{\text{i}} + \left(8.0t + 1.0t^2\right)\hat{\text{j}}.$$

Therefore, the time that corresponds to $x = 29$ m can be found by solving the equation $2.0t^2 = 29$, which leads to $t = 3.8$ s. The y coordinate at that time is

$$y = (8.0\text{ m/s})(3.8\text{ s}) + (1.0\text{ m/s}^2)(3.8\text{ s})^2 = 45\text{ m}.$$

(b) The velocity of the particle is given by $\vec{v} = \vec{v}_0 + \vec{a}t$. Thus, at $t = 3.8$ s, the velocity is

$$\vec{v} = (8.0\text{ m/s})\hat{\text{j}} + \left((4.0\text{ m/s}^2)\hat{\text{i}} + (2.0\text{ m/s}^2)\hat{\text{j}}\right)(3.8\text{ s}) = (15.2\text{ m/s})\hat{\text{i}} + (15.6\text{ m/s})\hat{\text{j}}$$

which has a magnitude of $v = \sqrt{v_x^2 + v_y^2} = \sqrt{(15.2\text{ m/s})^2 + (15.6\text{ m/s})^2} = 22\text{ m/s}$.

LEARN Instead of using the vector notation, we can also deal with the x- and the y-components individually.

4-93

THINK This problem deals with the two-dimensional kinematics of a desert camel moving from oasis A to oasis B.

EXPRESS The journey of the camel is illustrated in the figure on the right. We use a 'standard' coordinate system with $+x$ East and $+y$ North. Lengths are in kilometers and times are in hours. Using vector notation, we write the displacements for the first two segments of the trip as:

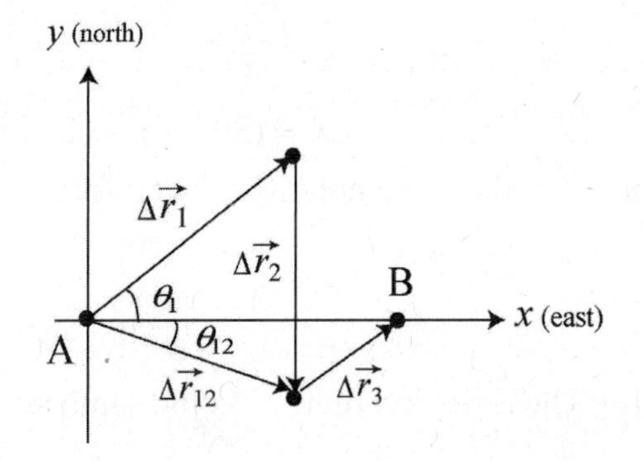

$$\Delta\vec{r}_1 = (75\text{ km})\cos(37°)\hat{\text{i}} + (75\text{ km})\sin(37°)\hat{\text{j}}$$

$$\Delta\vec{r}_2 = (-65\text{ km})\hat{\text{j}}$$

The net displacement is $\Delta\vec{r}_{12} = \Delta\vec{r}_1 + \Delta\vec{r}_2$. As can be seen from the figure, to reach oasis B requires an additional displacement $\Delta\vec{r}_3$.

ANALYZE (a) We perform the vector addition of individual displacements to find the net displacement of the camel: $\Delta\vec{r}_{12} = \Delta\vec{r}_1 + \Delta\vec{r}_2 = (60\text{ km})\hat{\mathrm{i}} - (20\text{ km})\hat{\mathrm{j}}$. Its corresponding magnitude is

$$|\Delta\vec{r}_{12}| = \sqrt{(60\text{ km})^2 + (-20\text{ km})^2} = 63\text{ km}.$$

(b) The direction of $\Delta\vec{r}_{12}$ is $\theta_{12} = \tan^{-1}[(-20\text{ km})/(60\text{ km})] = -18°$, or $18°$ south of east.

(c) To calculate the average velocity for the first two segments of the journey (including rest), we use the result from part (a) in Eq. 4-8 along with the fact that

$$\Delta t_{12} = \Delta t_1 + \Delta t_2 + \Delta t_{\text{rest}} = 50\text{ h} + 35\text{ h} + 5.0\text{ h} = 90\text{ h}.$$

In unit vector notation, we have $\vec{v}_{12,\text{avg}} = \dfrac{(60\hat{\mathrm{i}} - 20\hat{\mathrm{j}})\text{ km}}{90\text{ h}} = (0.67\hat{\mathrm{i}} - 0.22\hat{\mathrm{j}})\text{ km/h}.$

This leads to $|\vec{v}_{12,\text{avg}}| = 0.70\text{ km/h}$.

(d) The direction of $\vec{v}_{12,\text{avg}}$ is

$$\theta_{12} = \tan^{-1}[(-0.22\text{ km/h})/(0.67\text{ km/h})] = -18°,$$

or $18°$ south of east.

(e) The average speed is distinguished from the magnitude of average velocity in that it depends on the total distance as opposed to the net displacement. Since the camel travels 140 km, we obtain (140 km)/(90 h) = 1.56 km/h $\approx$ 1.6 km/h.

(f) The net displacement is required to be the 90 km East from *A* to *B*. The displacement from the resting place to *B* is denoted $\Delta\vec{r}_3$. Thus, we must have

$$\Delta\vec{r}_1 + \Delta\vec{r}_2 + \Delta\vec{r}_3 = (90\text{ km})\hat{\mathrm{i}}$$

which produces $\Delta\vec{r}_3 = (30\text{ km})\hat{\mathrm{i}} + (20\text{ km})\hat{\mathrm{j}}$ in unit-vector notation, or $(36 \angle 33°)$ in magnitude-angle notation. Therefore, using Eq. 4-8 we obtain

$$|\vec{v}_{3,\text{avg}}| = \frac{36\text{ km}}{(120-90)\text{ h}} = 1.2\text{ km/h}.$$

(g) The direction of $\vec{v}_{3,\text{avg}}$ is the same as $\Delta\vec{r}_3$ (that is, 33° north of east).

LEARN With a vector-capable calculator in polar mode, we could perform the vector addition of the displacements as $(75 \angle 37°) + (65 \angle -90°) = (63 \angle -18°)$. Note the distinction between average velocity and average speed.

4-97

THINK A bullet fired horizontally from a rifle strikes the target at some distance below its aiming point. We're asked to find its total flight time and speed.

EXPRESS The trajectory of the bullet is shown in the figure on the right (not to scale). Note that the origin is chosen to be at the firing point. With this convention, the y coordinate of the bullet is given by $y = -\frac{1}{2}gt^2$. Knowing the coordinates (x, y) at the target allows us to calculate the total flight time and speed of the bullet.

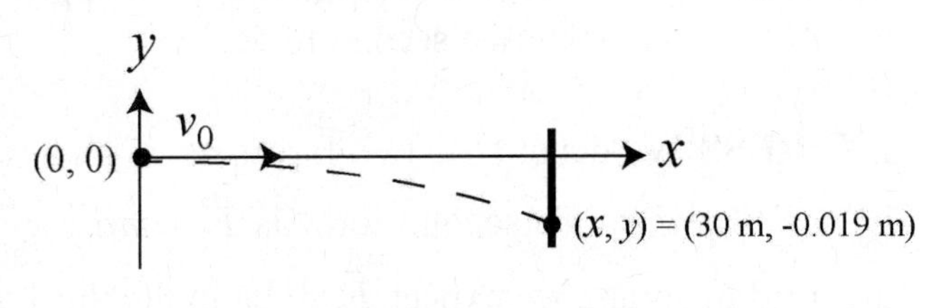

ANALYZE (a) If t is the time of flight and $y = -0.019$ m indicates where the bullet hits the target, then

$$t = \sqrt{\frac{-2y}{g}} = \sqrt{\frac{-2(-0.019\text{ m})}{9.8\text{ m/s}^2}} = 6.2 \times 10^{-2}\text{ s}.$$

(b) The muzzle velocity is the initial (horizontal) velocity of the bullet. Since $x = 30$ m is the horizontal position of the target, we have $x = v_0t$. Thus,

$$v_0 = \frac{x}{t} = \frac{30\text{ m}}{6.3 \times 10^{-2}\text{ s}} = 4.8 \times 10^2\text{ m/s}.$$

LEARN Alternatively, we may use Eq. 4-25 to solve for the initial velocity. With $\theta_0 = 0$ and $y_0 = 0$, the equation simplifies to $y = -\frac{gx^2}{2v_0^2}$, from which we find

$$v_0 = \sqrt{-\frac{gx^2}{2y}} = \sqrt{-\frac{(9.8\text{ m/s}^2)(30\text{ m})^2}{2(-0.019\text{ m})}} = 4.8 \times 10^2\text{ m/s},$$

in agreement with what we calculated in part (b).

Chapter 5

5-7

THINK A box is under acceleration by two applied forces. We use Newton's second law to solve for the unknown second force.

EXPRESS We denote the two forces as $\vec{F}_1$ and $\vec{F}_2$. According to Newton's second law, $\vec{F}_1 + \vec{F}_2 = m\vec{a}$, so the second force is $\vec{F}_2 = m\vec{a} - \vec{F}_1$. Note that since the acceleration is in the third quadrant, we expect $\vec{F}_2$ to be in the third quadrant as well.

ANALYZE (a) In unit vector notation $\vec{F}_1 = (20.0\text{ N})\hat{\mathrm{i}}$ and

$$\vec{a} = -\left(12.0\sin 30.0°\,\text{m/s}^2\right)\hat{\mathrm{i}} - \left(12.0\cos 30.0°\,\text{m/s}^2\right)\hat{\mathrm{j}} = -\left(6.00\text{ m/s}^2\right)\hat{\mathrm{i}} - \left(10.4\text{ m/s}^2\right)\hat{\mathrm{j}}.$$

Therefore, we find the second force to be

$$\begin{aligned}\vec{F}_2 &= m\vec{a} - \vec{F}_1 \\ &= (2.00\text{ kg})(-6.00\text{ m/s}^2)\hat{\mathrm{i}} + (2.00\text{ kg})(-10.4\text{ m/s}^2)\hat{\mathrm{j}} - (20.0\text{ N})\hat{\mathrm{i}} \\ &= (-32.0\text{ N})\hat{\mathrm{i}} - (20.8\text{ N})\hat{\mathrm{j}}.\end{aligned}$$

(b) The magnitude of $\vec{F}_2$ is $|\vec{F}_2| = \sqrt{F_{2x}^2 + F_{2y}^2} = \sqrt{(-32.0\text{ N})^2 + (-20.8\text{ N})^2} = 38.2\text{ N}.$

(c) The angle that $\vec{F}_2$ makes with the positive x-axis is found from

$$\tan\phi = \left(\frac{F_{2y}}{F_{2x}}\right) = \frac{-20.8\text{ N}}{-32.0\text{ N}} = 0.656.$$

Consequently, the angle is either 33.0° or 33.0° + 180° = 213°. Since both the x and y components are negative, the correct result is ϕ = 213° from the +x-axis. An alternative answer is $213° - 360° = -147°$.

LEARN The result is shown in the figure on the right. The calculation confirms our expectation that $\vec{F}_2$ lies in the third quadrant (same as $\vec{a}$). The net force is

$$\begin{aligned}\vec{F}_{\text{net}} &= \vec{F}_1 + \vec{F}_2 = (20.0\text{ N})\hat{\mathrm{i}} + \left[(-32.0\text{ N})\hat{\mathrm{i}} - (20.8\text{ N})\hat{\mathrm{j}}\right] \\ &= (-12.0\text{ N})\hat{\mathrm{i}} - (20.8\text{ N})\hat{\mathrm{j}}\end{aligned}$$

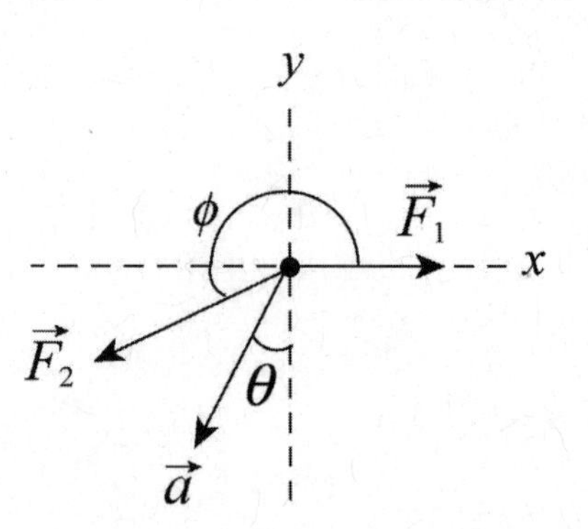

which points in the same direction as $\vec{a}$.

5-15

THINK We have a piece of salami hung to a spring scale in various ways. The problem is to explore the concept of weight.

EXPRESS We first note that the reading on the spring scale is proportional to the weight of the salami. In all three cases (a) – (c) depicted in Fig. 5-34, the scale is not accelerating, which means that the two cords exert forces of equal magnitude on it. The scale reads the magnitude of either of these forces. In each case the tension force of the cord attached to the salami must be the same in magnitude as the weight of the salami because the salami is not accelerating. Thus the scale reading is mg, where m is the mass of the salami.

ANALYZE In all three cases (a) – (c), the reading on the scale is

$$w = mg = (11.0 \text{ kg})(9.8 \text{ m/s}^2) = 108 \text{ N}.$$

LEARN The weight of an object is measured when the object is not accelerating vertically relative to the ground. If it is, then the weight measured is called the apparent weight.

5-17

THINK A block attached to a cord is resting on an incline plane. We apply Newton's second law to solve for the tension in the cord and the normal force on the block.

EXPRESS The free-body diagram of the problem is shown to the right. Since the acceleration of the block is zero, the components of Newton's second law equation yield

$$T - mg \sin\theta = 0$$
$$F_N - mg \cos\theta = 0,$$

where T is the tension in the cord, and F_N is the normal force on the block.

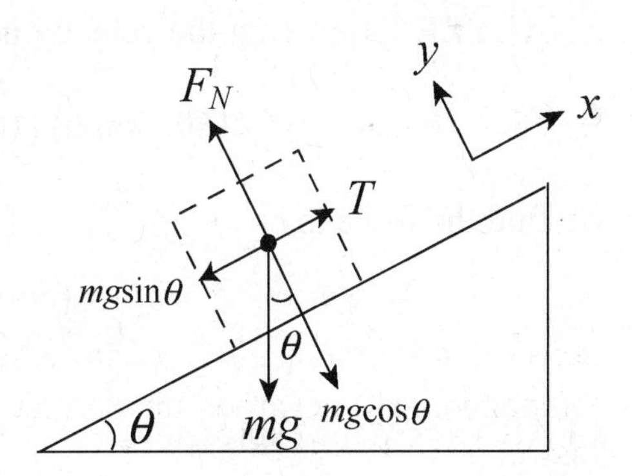

ANALYZE (a) Solving the first equation for the tension in the string, we find

$$T = mg \sin\theta = (8.5 \text{ kg})(9.8 \text{ m/s}^2) \sin 30° = 42 \text{ N}.$$

(b) We solve the second equation above for the normal force F_N:

$$F_N = mg \cos\theta = (8.5 \text{ kg})(9.8 \text{ m/s}^2) \cos 30° = 72 \text{ N}.$$

(c) When the cord is cut, it no longer exerts a force on the block and the block accelerates. The x component of the second law becomes $-mg\sin\theta = ma$, so the acceleration becomes

$$a = -g\sin\theta = -(9.8\ \text{m/s}^2)\sin 30° = -4.9\ \text{m/s}^2.$$

The negative sign indicates the acceleration is down the plane. The magnitude of the acceleration is 4.9 m/s^2.

LEARN The normal force F_N on the block must be equal to $mg\cos\theta$ so that the block is in contact with the surface of the incline at all time. When the cord is cut, the block has an acceleration $a = -g\sin\theta$, which in the limit $\theta \to 90°$ becomes $-g$, as in the case of a free fall.

5-19

THINK In this problem we're interested in the force applied to a rocket sled to accelerate it from rest to a given speed in a given time interval.

EXPRESS In terms of magnitudes, Newton's second law is $F = ma$, where $F = |\vec{F}_{\text{net}}|$, $a = |\vec{a}|$, and m is the (always positive) mass. The magnitude of the acceleration can be found using constant acceleration kinematics (Table 2-1). Solving $v = v_0 + at$ for the case where it starts from rest, we have $a = v/t$ (which we interpret in terms of magnitudes, making specification of coordinate directions unnecessary). Thus, the required force is $F = ma = mv/t$.

ANALYZE Expressing the velocity in SI units as

$$v = (1600\ \text{km/h})\ (1000\ \text{m/km})/(3600\ \text{s/h}) = 444\ \text{m/s},$$

we find the force to be

$$F = m\frac{v}{t} = (500\,\text{kg})\,\frac{444\,\text{m/s}}{1.8\,\text{s}} = 1.2 \times 10^5\ \text{N}.$$

LEARN From the expression $F = mv/t$, we see that the shorter the time to attain a given speed, the greater the force required.

5-27

THINK An electron moving horizontally is under the influence of a vertical force. Its path will be deflected toward the direction of the applied force.

EXPRESS The setup is shown in the figure below. The acceleration of the electron is vertical and for all practical purposes the only force acting on it is the electric force. The force of gravity is negligible. We take the $+x$ axis to be in the direction of the initial velocity v_0 and the $+y$ axis to be in the direction of the electrical force, and place the origin at the initial position of the electron.

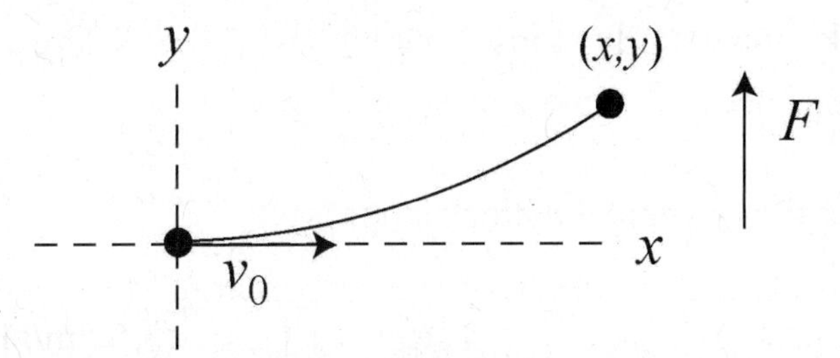

Since the force and acceleration are constant, we use the equations from Table 2-1: $x = v_0 t$ and

$$y = \frac{1}{2}at^2 = \frac{1}{2}\left(\frac{F}{m}\right)t^2.$$

ANALYZE The time taken by the electron to travel a distance x (= 30 mm) horizontally is $t = x/v_0$ and its deflection in the direction of the force is

$$y = \frac{1}{2}\frac{F}{m}\left(\frac{x}{v_0}\right)^2 = \frac{1}{2}\left(\frac{4.5\times10^{-16}\ \text{N}}{9.11\times10^{-31}\ \text{kg}}\right)\left(\frac{30\times10^{-3}\ \text{m}}{1.2\times10^{7}\ \text{m/s}}\right)^2 = 1.5\times10^{-3}\ \text{m}.$$

LEARN Since the applied force is constant, the acceleration in the y-direction is also constant and the path is parabolic with $y \propto x^2$.

5-31

THINK In this problem we analyze the motion of a block sliding up an inclined plane and back down.

EXPRESS The free-body diagram is shown below. $\vec{F}_N$ is the normal force of the plane on the block and $m\vec{g}$ is the force of gravity on the block. We take the $+x$ direction to be up the incline, and the $+y$ direction to be in the direction of the normal force exerted by the incline on the block.

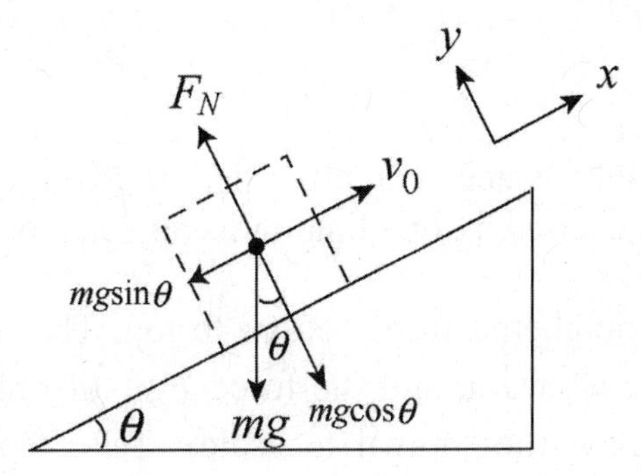

The x component of Newton's second law is then $mg\sin\theta = -ma$; thus, the acceleration is $a = -g\sin\theta$. Placing the origin at the bottom of the plane, the kinematic equations (Table 2-1) for motion along the x axis which we will use are $v^2 = v_0^2 + 2ax$ and $v = v_0 + at$. The block momentarily stops at its highest point, where $v = 0$; according to the second equation, this occurs at time $t = -v_0/a$.

ANALYZE (a) The position where the block stops is

$$x = v_0 t + \frac{1}{2}at^2 = v_0\left(\frac{-v_0}{a}\right) + \frac{1}{2}a\left(\frac{-v_0}{a}\right)^2 = -\frac{1}{2}\frac{v_0^2}{a} = -\frac{1}{2}\left(\frac{(3.50\text{ m/s})^2}{-(9.8\text{ m/s}^2)\sin 32.0°}\right) = 1.18\text{ m}.$$

(b) The time it takes for the block to get there is

$$t = \frac{v_0}{a} = -\frac{v_0}{-g\sin\theta} = -\frac{3.50\text{m/s}}{-(9.8\text{m/s}^2)\sin 32.0°} = 0.674\text{ s}.$$

(c) That the return speed is identical to the initial speed is to be expected since there are no dissipative forces in this problem. In order to prove this, one approach is to set $x = 0$ and solve $x = v_0 t + \frac{1}{2}at^2$ for the total time (up and back down) t. The result is

$$t = -\frac{2v_0}{a} = -\frac{2v_0}{-g\sin\theta} = -\frac{2(3.50\text{ m/s})}{-(9.8\text{ m/s}^2)\sin 32.0°} = 1.35\text{ s}.$$

The velocity when it returns is therefore

$$v = v_0 + at = v_0 - gt\sin\theta = 3.50\text{ m/s} - (9.8\text{ m/s}^2)(1.35\text{ s})\sin 32° = -3.50\text{ m/s}.$$

The negative sign indicates the direction is down the plane.

LEARN As expected, the speed of the block when it gets back to the bottom of the incline is the same as its initial speed. As we shall see in Chapter 8, this is a consequence of energy conservation. If friction is present, then the return speed will be smaller than the initial speed.

5-43

THINK A chain of five links is accelerated vertically upward by an external force. We are interested in the forces exerted by one link on its adjacent one.

EXPRESS The links are numbered from bottom to top. The forces on the first link are the force of gravity $m\vec{g}$, downward, and the force $\vec{F}_{2on1}$ of link 2, upward, as shown in the free-body diagram below (not drawn to scale). Take the positive direction to be upward. Then Newton's second law for the first link is $F_{2on1} - m_1 g = m_1 a$. The equations for the other links can be written in a similar manner (see below).

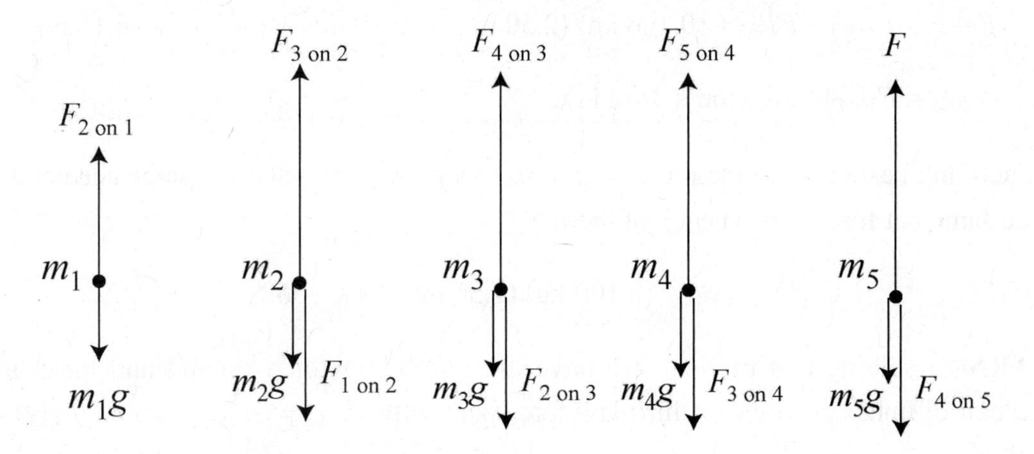

ANALYZE (a) Given that $a = 2.50\text{ m/s}^2$, from $F_{2on1} - m_1 g = m_1 a$, the force exerted by link 2 on link 1 is

$$F_{2on1} = m_1(a + g) = (0.100\text{ kg})(2.5\text{ m/s}^2 + 9.80\text{ m/s}^2) = 1.23\text{ N}.$$

(b) From the free-body diagram above, we see that the forces on the second link are the force of gravity $m_2\vec{g}$, downward, the force $\vec{F}_{1on2}$ of link 1, downward, and the force $\vec{F}_{3on2}$ of link 3, upward. According to Newton's third law $\vec{F}_{1on2}$ has the same magnitude as $\vec{F}_{2on1}$. Newton's second law for the second link is

$$F_{3on2} - F_{1on2} - m_2 g = m_2 a$$

so

$$F_{3on2} = m_2(a + g) + F_{1on2} = (0.100\text{ kg})\,(2.50\text{ m/s}^2 + 9.80\text{ m/s}^2) + 1.23\text{ N} = 2.46\text{ N}.$$

(c) Newton's second law equation for link 3 is $F_{4on3} - F_{2on3} - m_3 g = m_3 a$, so

$$F_{4on3} = m_3(a + g) + F_{2on3} = (0.100\text{ N})\,(2.50\text{ m/s}^2 + 9.80\text{ m/s}^2) + 2.46\text{ N} = 3.69\text{ N},$$

where Newton's third law implies $F_{2on3} = F_{3on2}$ (since these are magnitudes of the force vectors).

(d) Newton's second law for link 4 is

$$F_{5on4} - F_{3on4} - m_4 g = m_4 a,$$

so

$$F_{5on4} = m_4(a + g) + F_{3on4} = (0.100\text{ kg})\,(2.50\text{ m/s}^2 + 9.80\text{ m/s}^2) + 3.69\text{ N} = 4.92\text{ N},$$

where Newton's third law implies $F_{3on4} = F_{4on3}$.

(e) Newton's second law for the top link is $F - F_{4on5} - m_5 g = m_5 a$, so

$$F = m_5(a + g) + F_{4on5} = (0.100 \text{ kg})\,(2.50 \text{ m/s}^2 + 9.80 \text{ m/s}^2) + 4.92 \text{ N} = 6.15 \text{ N},$$

where $F_{4on5} = F_{5on4}$ by Newton's third law.

(f) Each link has the same mass ($m_1 = m_2 = m_3 = m_4 = m_5 = m$) and the same acceleration, so the same net force acts on each of them:

$$F_{net} = ma = (0.100 \text{ kg})\,(2.50 \text{ m/s}^2) = 0.250 \text{ N}.$$

LEARN In solving this problem we have used both Newton's second and third laws. Each pair of links constitutes a third-law force pair, with $\vec{F}_{i\text{ on }j} = -\vec{F}_{j\text{ on }i}$.

5-55

THINK In this problem a horizontal force is applied to block 1 which then pushes against block 2. Both blocks move together as a rigid connected system.

EXPRESS The free-body diagrams for the two blocks in (a) are shown below. $\vec{F}$ is the applied force and $\vec{F}_{1on2}$ is the force exerted by block 1 on block 2. We note that $\vec{F}$ is applied directly to block 1 and that block 2 exerts a force $\vec{F}_{2on1} = -\vec{F}_{1on2}$ on block 1 (taking Newton's third law into account).

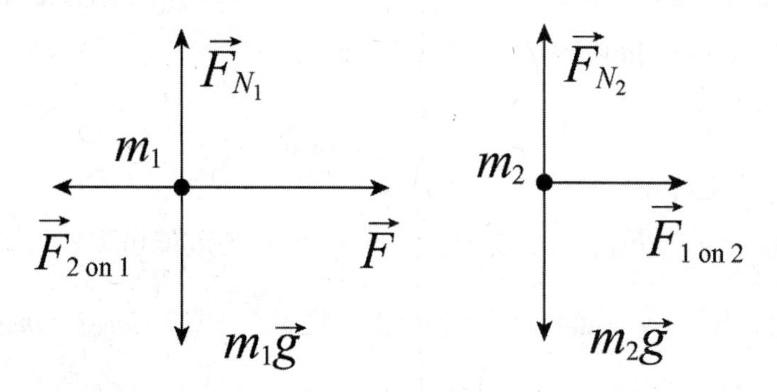

Newton's second law for block 1 is $F - F_{2on1} = m_1 a$, where a is the acceleration. The second law for block 2 is $F_{1on2} = m_2 a$. Since the blocks move together they have the same acceleration and the same symbol is used in both equations.

ANALYZE (a) From the second equation we obtain the expression $a = F_{1on2}/m_2$, which we substitute into the first equation to get $F - F_{2on1} = m_1 F_{1on2}/m_2$. Since $F_{2on1} = F_{1on2}$ (same magnitude for third-law force pair), we obtain

$$F_{2on1} = F_{1on2} = \frac{m_2}{m_1 + m_2}F = \frac{1.2 \text{ kg}}{2.3 \text{ kg} + 1.2 \text{ kg}}(3.2 \text{ N}) = 1.1 \text{ N}.$$

(b) If $\vec{F}$ is applied to block 2 instead of block 1 (and in the opposite direction), the free-body diagrams would look like the following:

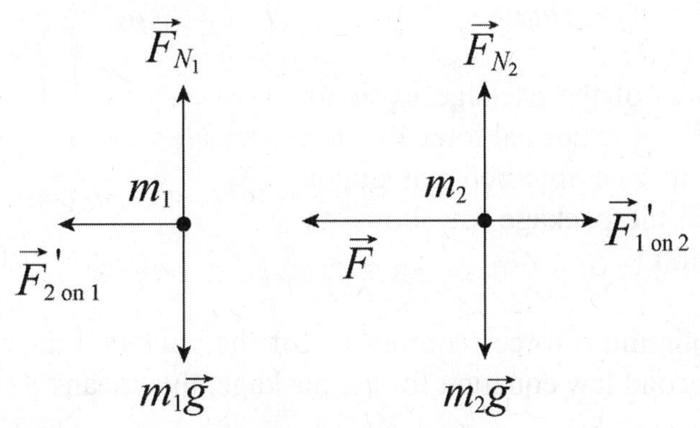

The corresponding force of contact between the blocks would be

$$F'_{2on1} = F'_{1on2} = \frac{m_1}{m_1 + m_2}F = \frac{2.3 \text{ kg}}{2.3 \text{ kg} + 1.2 \text{ kg}}(3.2 \text{ N}) = 2.1 \text{ N}.$$

(c) We note that the acceleration of the blocks is the same in the two cases. In part (a), the force F_{1on2} is the only horizontal force on the block of mass m_2 and in part (b) F'_{2on1} is the only horizontal force on the block with $m_1 > m_2$. Since $F_{1on2} = m_2 a$ in part (a) and $F'_{2on1} = m_1 a$ in part (b), then for the accelerations to be the same, $F'_{2on1} > F_{1on2}$, i.e., force between blocks must be larger in part (b).

LEARN This problem demonstrates that when two blocks are being accelerated together under an external force, the contact force between the two blocks is greater if the smaller mass is pushing against the bigger one, as in part (b). In the special case where the two masses are equal, $m_1 = m_2 = m$, $F'_{2on1} = F_{2on1} = F/2$.

5-59

THINK This problem involves the application of Newton's third law. As the monkey climbs up a tree, it pulls downward on the rope, but the rope pulls upward on the monkey.

EXPRESS We take $+y$ to be up for both the monkey and the package. The force the monkey pulls downward on the rope has magnitude F. According to Newton's third law, the rope pulls upward on the monkey with a force of the same magnitude, so Newton's second law for forces acting on the monkey leads to

$$F - m_m g = m_m a_m,$$

where m_m is the mass of the monkey and a_m is its acceleration. Since the rope is massless $F = T$ is the tension in the rope.

The rope pulls upward on the package with a force of magnitude F, so Newton's second law for the package is

$$F + F_N - m_p g = m_p a_p,$$

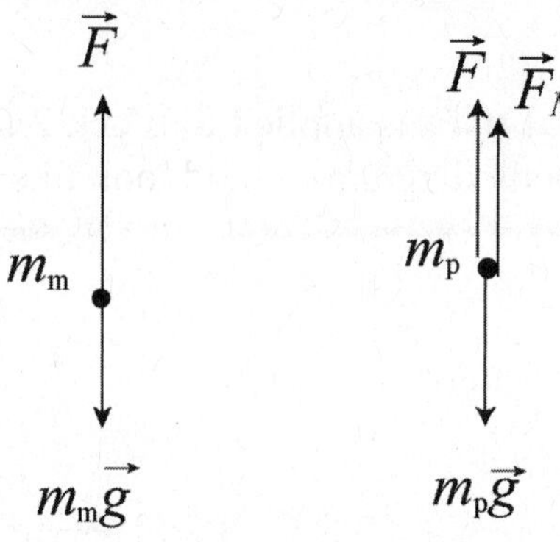

where m_p is the mass of the package, a_p is its acceleration, and F_N is the normal force exerted by the ground on it. The free-body diagrams for the monkey and the package are shown to the right (not to scale).

Now, if F is the minimum force required to lift the package, then $F_N = 0$ and $a_p = 0$. According to the second law equation for the package, this means $F = m_p g$.

ANALYZE (a) Substituting $m_p g$ for F in the equation for the monkey, we solve for a_m:

$$a_m = \frac{F - m_m g}{m_m} = \frac{(m_p - m_m)g}{m_m} = \frac{(15\text{ kg} - 10\text{ kg})(9.8\text{ m/s}^2)}{10\text{ kg}} = 4.9\text{ m/s}^2.$$

(b) As discussed, Newton's second law leads to $F - m_p g = m_p a'_p$ for the package and $F - m_m g = m_m a'_m$ for the monkey. If the acceleration of the package is downward, then the acceleration of the monkey is upward, so $a'_m = -a'_p$. Solving the first equation for F

$$F = m_p(g + a'_p) = m_p(g - a'_m)$$

and substituting this result into the second equation:

$$m_p(g - a'_m) - m_m g = m_m a'_m,$$

we solve for a'_m:

$$a'_m = \frac{(m_p - m_m)g}{m_p + m_m} = \frac{(15\text{ kg} - 10\text{ kg})(9.8\text{ m/s}^2)}{15\text{ kg} + 10\text{ kg}} = 2.0\text{ m/s}^2.$$

(c) The result is positive, indicating that the acceleration of the monkey is upward.

(d) Solving the second law equation for the package, the tension in the rope is

$$F = m_p(g - a'_m) = (15\text{ kg})(9.8\text{ m/s}^2 - 2.0\text{ m/s}^2) = 120\text{N}.$$

LEARN The situations described in (b)-(d) are similar to that of an Atwood machine. With $m_p > m_m$, the package accelerates downward while the monkey accelerates upward.

5-61

THINK As more mass is thrown out of the hot-air balloon, its upward acceleration increases.

EXPRESS The forces on the balloon are the force of gravity $m\vec{g}$ (down) and the force of the air $\vec{F}_a$ (up). We take the $+y$ to be up, and use a to mean the *magnitude* of the acceleration. When the mass is M (before the ballast is thrown out) the acceleration is downward and Newton's second law is

$$Mg - F_a = Ma$$

After the ballast is thrown out, the mass is $M - m$ (where m is the mass of the ballast) and the acceleration is now upward. Newton's second law leads to

$$F_a - (M - m)g = (M - m)a.$$

Combing the two equations allows us to solve for m.

ANALYZE The first equation gives $F_a = M(g - a)$, and this plugs into the new equation to give

$$M(g - a) - (M - m)g = (M - m)a \Rightarrow m = \frac{2\,Ma}{g + a}.$$

LEARN More generally, if a ballast mass m' is tossed, the resulting acceleration is a' which is related to m' via:

$$m' = M\frac{a' + a}{g + a},$$

showing that the more mass thrown out, the greater is the upward acceleration. For $a' = a$, we get $m' = 2Ma/(g + a)$, which agrees with what was found above.

5-71

THINK We have two boxes connected together by a cord and placed on a wedge. The system accelerates together and we'd like to know the tension in the cord.

EXPRESS The $+x$ axis is "uphill" for $m_1 = 3.0$ kg and "downhill" for $m_2 = 2.0$ kg (so they both accelerate with the same sign). The x components of the two masses along the x axis are given by $m_1 g\sin\theta_1$ and $m_2 g\sin\theta_2$, respectively. The free-body diagram is shown below. Applying Newton's second law, we obtain

$$T - m_1 g \sin\theta_1 = m_1 a$$
$$m_2 g \sin\theta_2 - T = m_2 a$$

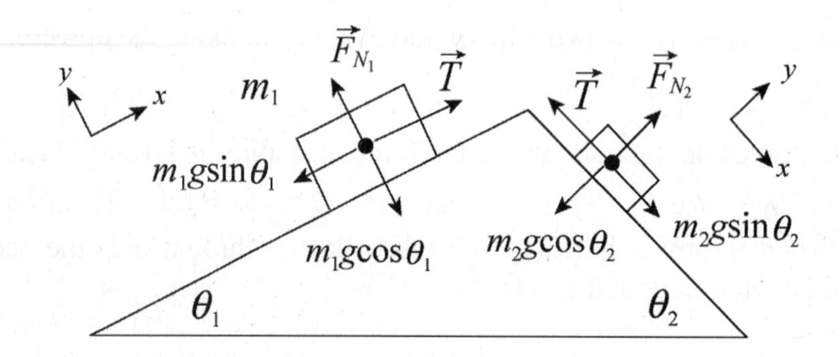

Adding the two equations allows us to solve for the acceleration:

$$a = \left(\frac{m_2 \sin\theta_2 - m_1 \sin\theta_1}{m_2 + m_1}\right) g$$

ANALYZE With $\theta_1 = 30°$ and $\theta_2 = 60°$, we have a = 0.45 m/s^2. This value is plugged back into either of the two equations to yield the tension

$$T = \frac{m_1 m_2 g}{m_2 + m_1}(\sin\theta_2 + \sin\theta_1) = 16.1\ \text{N}$$

LEARN In this problem we find $m_2 \sin\theta_2 > m_1 \sin\theta_1$, so that $a > 0$, indicating that m_2 slides down and m_1 slides up. The situation would reverse if $m_2 \sin\theta_2 < m_1 \sin\theta_1$. When $m_2 \sin\theta_2 = m_1 \sin\theta_1$, the acceleration is $a = 0$ and the two masses hang in balance. Notice also the symmetry between the two masses in the expression for *T*.

5-73

THINK We have two masses connected together by a cord. A force is applied to the second mass and the system accelerates together. We apply Newton's second law to solve the problem.

EXPRESS The free-body diagrams for the two masses are shown below (not to scale). We first analyze the forces on m_1=1.0 kg. The +*x* direction is "downhill" (parallel to $\vec{T}$). With an acceleration a = 5.5 m/s^2 in the positive *x* direction for m_1, Newton's second law applied to the *x*-axis gives

$$T + m_1 g \sin\beta = m_1 a\,.$$

On the other hand, for the second mass m_2=2.0 kg, we have $m_2 g - F - T = m_2 a$, where the tension comes in as an upward force (the cord can pull, not push). The two equations can be combined to solve for T and β.

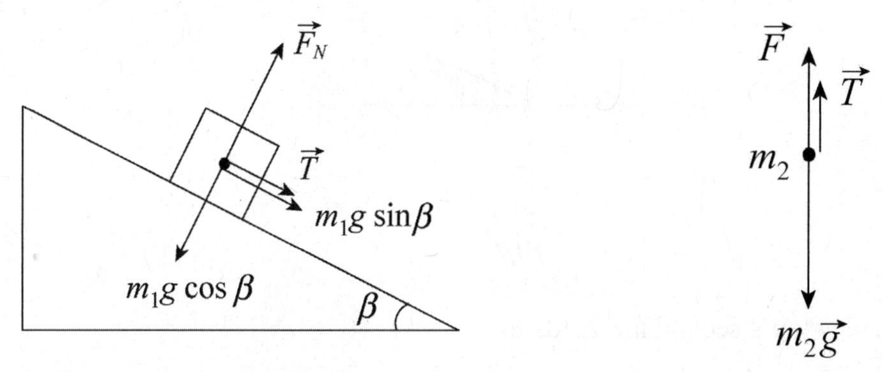

ANALYZE We solve (b) first. By combining the two equations above, we obtain

$$\sin\beta = \frac{(m_1+m_2)a + F - m_2 g}{m_1 g} = \frac{(1.0\text{ kg}+2.0\text{ kg})(5.5\text{ m/s}^2)+6.0\text{ N}-(2.0\text{ kg})(9.8\text{ m/s}^2)}{(1.0\text{ kg})(9.8\text{ m/s}^2)}$$
$$= 0.296$$

which gives $\beta = 17.2°$.

(a) Substituting the value for β found in (a) into the first equation, we have

$$T = m_1(a - g\sin\beta) = (1.0\text{ kg})\left[5.5\text{ m/s}^2 - (9.8\text{ m/s}^2)\sin 17.2°\right] = 2.60\text{ N}.$$

LEARN For $\beta = 0$, the problem becomes the same as that discussed in Sample Problem "Block on table, block hanging." In this case, our results reduce to the familiar expressions: $a = m_2 g/(m_1+m_2)$, and $T = m_1 m_2 g/(m_1+m_2)$.

5-77

THINK We have a crate that is being pulled at an angle. We apply Newton's second law to analyze the motion.

EXPRESS Although the full specification of $\vec{F}_{\text{net}} = m\vec{a}$ in this situation involves both x and y axes, only the x-application is needed to find what this particular problem asks for. We note that $a_y = 0$ so that there is no ambiguity denoting a_x simply as a. We choose $+x$ to the right and $+y$ up. The free-body diagram (not to scale) is shown next. The x component of the rope's tension (acting on the crate) is

$$F_x = F\cos\theta = (450\text{ N})\cos 38° = 355\text{ N},$$

and the resistive force (pointing in the $-x$ direction) has magnitude f = 125 N.

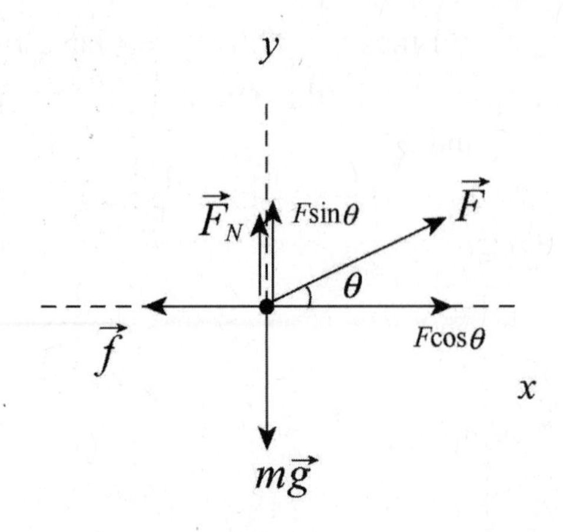

ANALYZE (a) Newton's second law leads to

$$F_x - f = ma \Rightarrow a = \frac{F\cos\theta - f}{m} = \frac{355\text{ N} - 125\text{ N}}{310\text{ kg}} = 0.74\text{m/s}^2.$$

(b) In this case, we use Eq. 5-12 to find the mass: $m' = W/g = 31.6\text{ kg}$. Newton's second law then leads to

$$F_x - f = m'a' \Rightarrow a' = \frac{F_x - f}{m'} = \frac{355\text{ N} - 125\text{ N}}{31.6\text{ kg}} = 7.3\text{ m/s}^2.$$

LEARN The resistive force opposing the motion is due to the friction between the crate and the floor. This topic is discussed in greater detail in Chapter 6.

5-83

THINK This problem deals with the relationship between the three quantities: force, mass and acceleration in Newton's second law $F = ma$.

EXPRESS The "certain force," denoted as F, is assumed to be the net force on the object when it gives m_1 an acceleration a_1 = 12 m/s^2 and when it gives m_2 an acceleration a_2 = 3.3 m/s^2, i.e., $F = m_1a_1 = m_2a_2$. The accelerations for $m_2 + m_1$ and $m_2 - m_1$ can be solved by substituting $m_1 = F/a_1$ and $m_2 = F/a_2$.

ANALYZE (a) Now we seek the acceleration a of an object of mass $m_2 - m_1$ when F is the net force on it. The result is

$$a = \frac{F}{m_2 - m_1} = \frac{F}{(F/a_2)-(F/a_1)} = \frac{a_1a_2}{a_1 - a_2} = \frac{(12.0\text{ m/s}^2)(3.30\text{ m/s}^2)}{12.0\text{ m/s}^2 - 3.30\text{ m/s}^2} = 4.55\text{ m/s}^2.$$

(b) Similarly for an object of mass $m_2 + m_1$, we have:

$$a' = \frac{F}{m_2 + m_1} = \frac{F}{(F/a_2) + (F/a_1)} = \frac{a_1 a_2}{a_1 + a_2} = \frac{(12.0 \text{ m/s}^2)(3.30 \text{ m/s}^2)}{12.0 \text{ m/s}^2 + 3.30 \text{ m/s}^2} = 2.59 \text{ m/s}^2.$$

LEARN With the same applied force, the greater the mass the smaller the acceleration. In this problem, we have $a_1 > a > a_2 > a'$. This implies $m_1 < m_2 - m_1 < m_2 < m_2 + m_1$.

5-91

THINK We have a motorcycle going up a ramp at a constant acceleration. We apply Newton's second law to calculate the net force on the rider and the force on the rider from the motorcycle.

EXPRESS The free-body diagram is shown to the right (not to scale). Note that F_{m,r_y} and F_{m,r_x}, respectively, denote the y and x components of the force $\vec{F}_{m,r}$ exerted by the motorcycle on the rider. The net force on the rider is

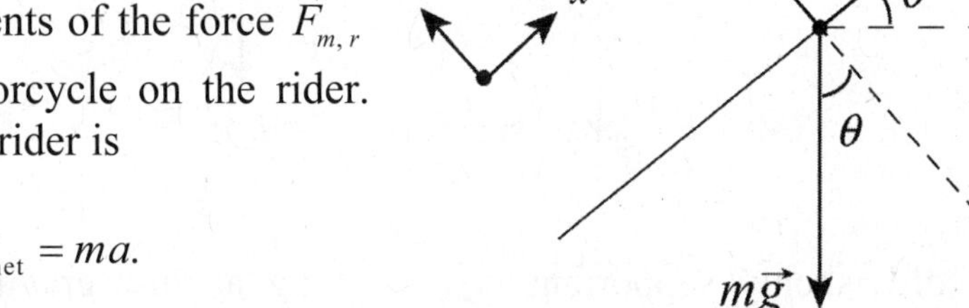

$$F_{\text{net}} = ma.$$

ANALYZE (a) Since the net force equals ma, then the magnitude of the net force on the rider is

$$F_{\text{net}} = ma = (60.0 \text{ kg})\,(3.0 \text{ m/s}^2) = 1.8 \times 10^2 \text{ N}.$$

(b) To calculate the force by the motorcycle on the rider, we apply Newton's second law to the x- and the y-axes separately. For the x-axis, we have:

$$F_{m,r_x} - mg \sin \theta = ma$$

where m = 60.0 kg, a = 3.0 m/s^2, and θ = 10°. Thus, $F_{m,r_x} = 282$ N. Applying it to the y-axis (where there is no acceleration), we have

$$F_{m,r_y} - mg \cos \theta = 0$$

which gives $F_{m,r_y} = 579$ N. Using the Pythagorean theorem, we find

$$F_{m,r} = \sqrt{F_{m,r_x}^2 + F_{m,r_y}^2} = \sqrt{(282 \text{ N})^2 + (579 \text{ N})^2} = 644 \text{ N}.$$

Now, the magnitude of the force exerted on the rider by the motorcycle is the same magnitude of force exerted by the rider on the motorcycle, so the answer is 644 N.

LEARN The force exerted by the motorcycle on the rider keeps the rider accelerating in the $+x$-direction, while maintaining contact with the inclines surface ($a_y = 0$).

5-93
THINK In this problem we have mobiles consisting of masses connected by cords. We apply Newton's second law to calculate the tensions in the cords.

EXPRESS The free-body diagrams for m_1 and m_2 for part (a) are shown to the right.

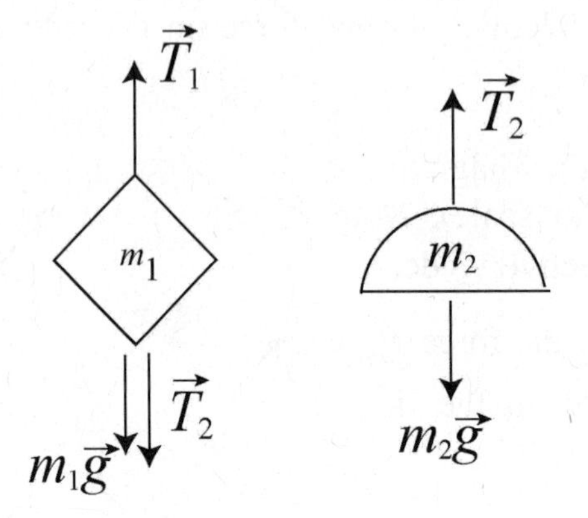

The bottom cord is only supporting m_2 = 4.5 kg against gravity, so its tension is $T_2 = m_2 g$. On the other hand, the top cord is supporting a total mass of $m_1 + m_2$ = (3.5 kg + 4.5 kg) = 8.0 kg against gravity. Applying Newton's second law gives

$$T_1 - T_2 - m_1 g = 0$$

so the tension is

$$T_1 = m_1 g + T_2 = (m_1 + m_2)g.$$

ANALYZE (a) From the equations above, we find the tension in the bottom cord to be

$$T_2 = m_2 g = (4.5 \text{ kg})(9.8 \text{ m/s}^2) = 44 \text{ N}.$$

(b) Similarly, the tension in the top cord is $T_1 = (m_1 + m_2)g$ = (8.0 kg)(9.8 m/s^2) = 78 N.

(c) The free-body diagrams for m_3, m_4 and m_5 for part (b) are shown below (not to scale).

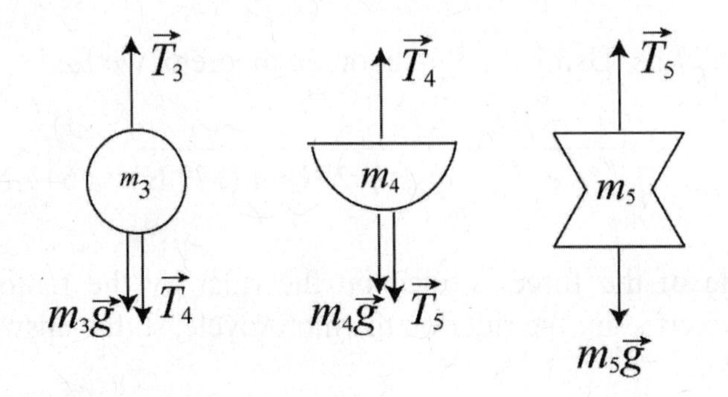

From the diagram, we see that the lowest cord supports a mass of m_5 = 5.5 kg against gravity and consequently has a tension of

$$T_5 = m_5 g = (5.5 \text{ kg})(9.8 \text{ m/s}^2) = 54 \text{ N}.$$

(d) The top cord, as we are told, has a tension T_3 =199 N which supports a total of (199 N)/(9.80 m/s^2) = 20.3 kg, 10.3 kg of which is already accounted for in the figure. Thus, the unknown mass in the middle must be m_4 = 20.3 kg – 10.3 kg = 10.0 kg, and the tension in the cord above it must be enough to support

$$m_4 + m_5 = (10.0 \text{ kg} + 5.50 \text{ kg}) = 15.5 \text{ kg},$$

so T_4 = (15.5 kg)(9.80 m/s^2) = 152 N.

LEARN Another way to calculate T_4 is to examine the forces on m_3 – one of the downward forces on it is T_4. From Newton's second law, we have $T_3 - m_3 g - T_4 = 0$, which can be solved to give

$$T_4 = T_3 - m_3 g = 199 \text{ N} - (4.8 \text{ kg})(9.8 \text{ m/s}^2) = 152 \text{ N}.$$

Chapter 6

6-3

THINK In the presence of friction between the floor and the bureau, a minimum horizontal force must be applied before the bureau would begin to move.

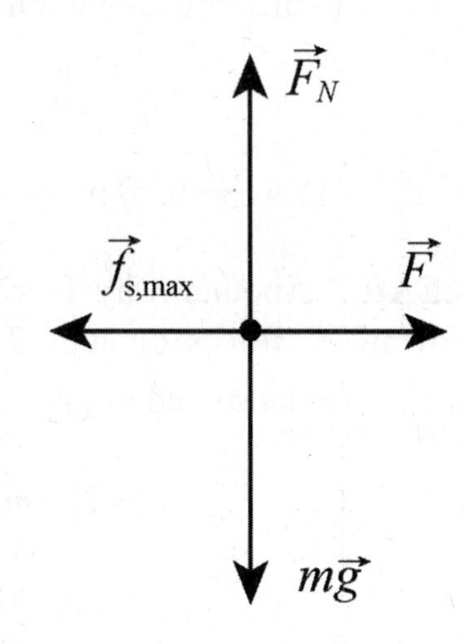

EXPRESS The free-body diagram for the bureau is shown to the right. We denote $\vec{F}$ as the horizontal force of the person, $\vec{f}_s$ is the force of static friction (in the $-x$ direction), F_N is the vertical normal force exerted by the floor (in the $+y$ direction), and $m\vec{g}$ is the force of gravity. We do not consider the possibility that the bureau might tip, and treat this as a purely horizontal motion problem (with the person's push $\vec{F}$ in the $+x$ direction). Applying Newton's second law to the x and y axes, we obtain

$$F - f_{s,\,\max} = ma$$
$$F_N - mg = 0$$

respectively. The second equation yields the normal force $F_N = mg$, whereupon the maximum static friction is found to be (from Eq. 6-1) $f_{s,\max} = \mu_s mg$.

Thus, the first equation becomes

$$F - \mu_s mg = ma = 0$$

where we have set $a = 0$ to be consistent with the fact that the static friction is still (just barely) able to prevent the bureau from moving.

ANALYZE (a) With $\mu_s = 0.45$ and $m = 45$ kg, the equation above leads to

$$F = \mu_s mg = (0.45)(45\text{ kg})(9.8\text{ m/s}^2) = 198\text{ N}.$$

To bring the bureau into a state of motion, the person should push with any force greater than this value. Rounding to two significant figures, we can therefore say the minimum required push is $F = 2.0 \times 10^2$ N.

(b) Replacing $m = 45$ kg with $m = 28$ kg, the reasoning above leads to roughly $F = 1.2\times10^2$ N.

LEARN The values found above represent the minimum force required to overcome the friction. Applying a force greater than $\mu_s mg$ results in a net force in the $+x$-direction, and hence, non-zero acceleration.

6-7

THINK A force is being applied to accelerate a crate in the presence of friction. We apply Newton's second law to solve for the acceleration.

EXPRESS The free-body diagram for the crate is shown below.

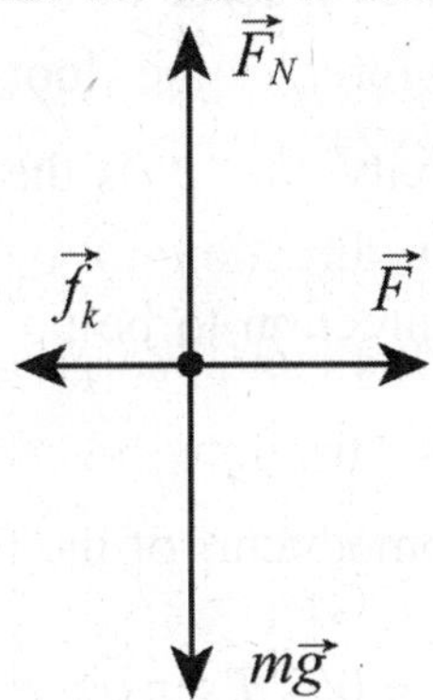

We denote $\vec{F}$ as the horizontal force of the person exerted on the crate (in the $+x$ direction), $\vec{f}_k$ is the force of kinetic friction (in the $-x$ direction), F_N is the vertical normal force exerted by the floor (in the $+y$ direction), and $m\vec{g}$ is the force of gravity. The magnitude of the force of friction is given by Eq. 6-2: $f_k = \mu_k F_N$. Applying Newton's second law to the x and y axes, we obtain

$$F - f_k = ma$$
$$F_N - mg = 0$$

respectively.

ANALYZE (a) The second equation above yields the normal force $F_N = mg$, so that the friction is

$$f_k = \mu_k F_N = \mu_k mg = (0.35)(55\text{ kg})(9.8\text{ m/s}^2) = 1.9 \times 10^2\text{ N}.$$

(b) The first equation becomes

$$F - \mu_k mg = ma$$

which (with $F = 220$ N) we solve to find

$$a = \frac{F}{m} - \mu_k g = \frac{220\text{ N}}{55\text{ kg}} - (0.35)(9.8\text{ m/s}^2) = 0.56\text{ m/s}^2.$$

LEARN For the crate to accelerate, the condition $F > f_k = \mu_k mg$ must be met. As can be seen from the equation above, the greater the value of μ_k, the smaller the acceleration under the same applied force.

6-11

THINK Since the crate is being pulled by a rope at an angle with the horizontal, we need to analyze the force components in both the x and y-directions.

EXPRESS The free-body diagram for the crate is shown to the right. Here $\vec{T}$ is the tension force of the rope on the crate, $\vec{F}_N$ is the normal force of the floor on the crate, $m\vec{g}$ is the force of gravity, and $\vec{f}$ is the force of friction. We take the $+x$ direction to be horizontal to the right and the $+y$ direction to be up. We assume the crate is motionless.

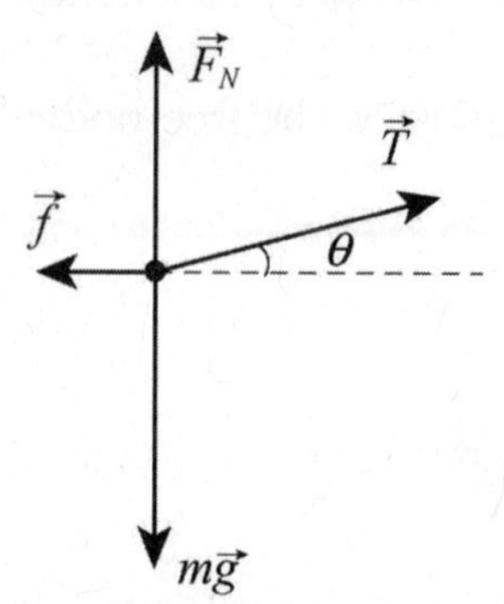

The equations for the x and the y components of the force according to Newton's second law are:

$$T\cos\theta - f = 0, \quad T\sin\theta + F_N - mg = 0$$

where $\theta = 15°$ is the angle between the rope and the horizontal. The first equation gives $f = T\cos\theta$ and the second gives $F_N = mg - T\sin\theta$. If the crate is to remain at rest, f must be less than $\mu_s F_N$, or $T\cos\theta < \mu_s (mg - T\sin\theta)$. When the tension force is sufficient to just start the crate moving, we must have $T\cos\theta = \mu_s (mg - T\sin\theta)$.

ANALYZE (a) We solve for the tension:

$$T = \frac{\mu_s mg}{\cos\theta + \mu_s \sin\theta} = \frac{(0.50)(68\text{ kg})(9.8\text{ m/s}^2)}{\cos 15° + 0.50\sin 15°} = 304\text{ N} \approx 3.0\times10^2\text{ N}.$$

(b) The second law equations for the moving crate are

$$T\cos\theta - f = ma, \quad T\sin\theta + F_N - mg = 0.$$

Now $f = \mu_k F_N$, and the second equation above gives $F_N = mg - T\sin\theta$, which then yields $f = \mu_k (mg - T\sin\theta)$. This expression is substituted for f in the first equation to obtain

$$T\cos\theta - \mu_k (mg - T\sin\theta) = ma,$$

so the acceleration is

$$\begin{aligned} a &= \frac{T(\cos\theta + \mu_k \sin\theta)}{m} - \mu_k g \\ &= \frac{(304\text{ N})(\cos 15° + 0.35\sin 15°)}{68\text{ kg}} - (0.35)(9.8\text{ m/s}^2) = 1.3\text{ m/s}^2. \end{aligned}$$

LEARN Let's check the limit where $\theta = 0$. In this case, we recover the familiar expressions: $T = \mu_s mg$ and $a = (T - \mu_k mg)/m$.

6-25

THINK In order that the two blocks remain in equilibrium, friction must be present between block *B* and the surface.

EXPRESS The free-body diagrams for block *B* and for the knot just above block *A* are shown below. $\vec{T}_1$ is the tension force of the rope pulling on block *B* or pulling on the knot (as the case may be), $\vec{T}_2$ is the tension force exerted by the second rope (at angle $\theta = 30°$) on the knot, $\vec{f}$ is the force of static friction exerted by the horizontal surface on block *B*, $\vec{F}_N$ is normal force exerted by the surface on block *B*, W_A is the weight of block *A* (W_A is the magnitude of $m_A\vec{g}$), and W_B is the weight of block *B* ($W_B = 711$ N is the magnitude of $m_B\vec{g}$).

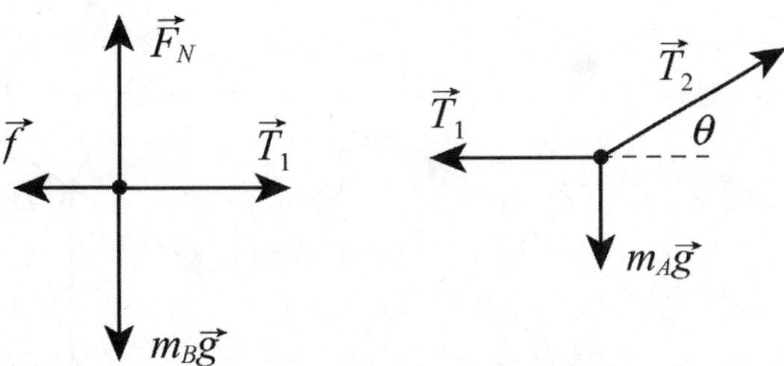

For each object we take +*x* horizontally rightward and +*y* upward. Applying Newton's second law in the *x* and *y* directions for block *B* and then doing the same for the knot results in four equations:

$$
\begin{aligned}
T_1 - f_{s,\max} &= 0 \\
F_N - W_B &= 0 \\
T_2 \cos\theta - T_1 &= 0 \\
T_2 \sin\theta - W_A &= 0
\end{aligned}
$$

where we assume the static friction to be at its maximum value (permitting us to use Eq. 6-1). The above equations yield

$$T_1 = \mu_s F_N,\ F_N = W_B \text{ and } T_1 = T_2 \cos\theta .$$

ANALYZE Solving these equations with $\mu_s = 0.25$, we obtain

$$
\begin{aligned}
W_A &= T_2 \sin\theta = T_1 \tan\theta = \mu_s F_N \tan\theta = \mu_s W_B \tan\theta \\
&= (0.25)(711\text{ N})\tan 30° = 1.0\times10^2\text{ N}
\end{aligned}
$$

LEARN As expected, the maximum weight of *A* is proportional to the weight of *B*, as well as the coefficient of static friction. In addition, we see that W_A is proportional to $\tan\theta$ (the larger the angle, the greater the vertical component of T_2 that supports its weight).

6-31

THINK In this problem we have two blocks connected by a string sliding down an inclined plane; the blocks have different coefficient of kinetic friction.

EXPRESS The free-body diagrams for the two blocks are shown below. T is the magnitude of the tension force of the string, $\vec{F}_{NA}$ is the normal force on block A (the leading block), $\vec{F}_{NB}$ is the normal force on block B, $\vec{f}_A$ is kinetic friction force on block A, $\vec{f}_B$ is kinetic friction force on block B. Also, m_A is the mass of block A (where $m_A = W_A/g$ and $W_A = 3.6$ N), and m_B is the mass of block B (where $m_B = W_B/g$ and $W_B = 7.2$ N). The angle of the incline is $\theta = 30°$.

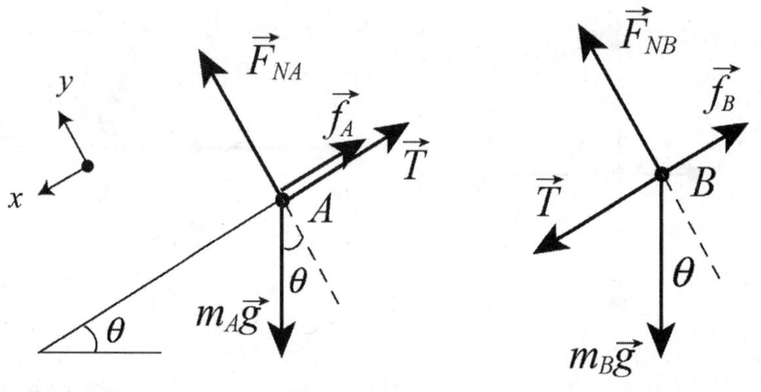

For each block we take $+x$ downhill (which is toward the lower-left in these diagrams) and $+y$ in the direction of the normal force. Applying Newton's second law to the x and y directions of both blocks A and B, we arrive at four equations:

$$W_A \sin\theta - f_A - T = m_A a$$
$$F_{NA} - W_A \cos\theta = 0$$
$$W_B \sin\theta - f_B + T = m_B a$$
$$F_{NB} - W_B \cos\theta = 0$$

which, when combined with Eq. 6-2 ($f_A = \mu_{kA} F_{NA}$ where $\mu_{k\,A} = 0.10$ and $f_B = \mu_{kB} F_{NB} f_B$ where $\mu_{k\,B} = 0.20$), fully describe the dynamics of the system so long as the blocks have the same acceleration and $T > 0$.

ANALYZE (a) From these equations, we find the acceleration to be

$$a = g\left(\sin\theta - \left(\frac{\mu_{kA}W_A + \mu_{kB}W_B}{W_A + W_B}\right)\cos\theta\right) = 3.5 \text{ m/s}^2.$$

(b) We solve the above equations for the tension and obtain

$$T = \left(\frac{W_A W_B}{W_A + W_B}\right)(\mu_{kB} - \mu_{kA})\cos\theta = \frac{(3.6\text{ N})(7.2\text{ N})}{3.6\text{ N} + 7.2\text{ N}}(0.20 - 0.10)\cos 30° = 0.21\text{ N}.$$

LEARN The tension in the string is proportional to $\mu_{kB} - \mu_{kA}$, the difference in coefficients of kinetic friction for the two blocks. When the coefficients are equal ($\mu_{kB} = \mu_{kA}$), the two blocks can be viewed as moving independent of one another and the tension is zero. Similarly, when $\mu_{kB} < \mu_{kA}$ (the leading block *A* has larger coefficient than the *B*), the string is slack, so the tension is also zero.

6-33

THINK In this problem, the frictional force is not a constant, but instead proportional to the speed of the boat. An integration is required to solve for the speed.

EXPRESS We denote the magnitude of the frictional force as αv, where $\alpha = 70\text{ N}\cdot\text{s/m}$. We take the direction of the boat's motion to be positive. Newton's second law gives

$$-\alpha v = m\frac{dv}{dt} \quad\Rightarrow\quad \frac{dv}{v} = -\frac{\alpha}{m}dt.$$

Integrating the equation gives

$$\int_{v_0}^{v}\frac{dv}{v} = -\frac{\alpha}{m}\int_0^t dt$$

where v_0 is the velocity at time zero and v is the velocity at time t. Solving the integral allows us to calculate the time it takes for the boat to slow down to 45 km/h, or $v = v_0/2$, where $v_0 = 90\text{ km/h}$.

ANALYZE The integrals are evaluated with the result

$$\ln\left(\frac{v}{v_0}\right) = -\frac{\alpha t}{m}$$

With $v = v_0/2$ and $m = 1000$ kg, we find the time to be

$$t = -\frac{m}{\alpha}\ln\left(\frac{v}{v_0}\right) = -\frac{m}{\alpha}\ln\left(\frac{1}{2}\right) = -\frac{1000\text{ kg}}{70\text{ N}\cdot\text{s/m}}\ln\left(\frac{1}{2}\right) = 9.9\text{ s}.$$

LEARN The speed of the boat is given by $v(t) = v_0 e^{-\alpha t/m}$, showing exponential decay with time. The greater the value of α, the more rapidly the boat slows down.

6-45

THINK Ferris wheel ride is a vertical circular motion. The apparent weight of the rider varies with his position.

EXPRESS The free-body diagrams of the student at the top and bottom of the Ferris wheel are shown below:

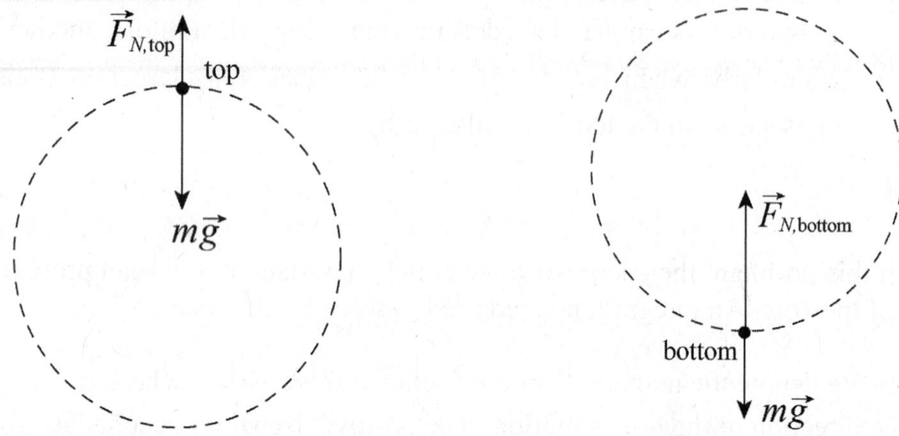

At the top (the highest point in the circular motion) the seat pushes up on the student with a force of magnitude $F_{N,\text{top}}$, while the Earth pulls down with a force of magnitude mg. Newton's second law for the radial direction gives

$$mg - F_{N,\text{top}} = \frac{mv^2}{R}.$$

At the bottom of the ride, $F_{N,\text{bottom}}$ is the magnitude of the upward force exerted by the seat. The net force toward the center of the circle is (choosing upward as the positive direction):

$$F_{N,\text{bottom}} - mg = \frac{mv^2}{R}.$$

The Ferris wheel is "steadily rotating" so the value $F_c = mv^2/R$ is the same everywhere. The apparent weight of the student is given by F_N.

ANALYZE (a) At the top, we are told that $F_{N,\text{top}} = 556$ N and $mg = 667$ N. This means that the seat is pushing up with a force that is smaller than the student's weight, and we say the student experiences a decrease in his "apparent weight" at the highest point. Thus, he feels "light."

(b) From (a), we find the centripetal force to be

$$F_c = \frac{mv^2}{R} = mg - F_{N,\text{top}} = 667\text{ N} - 556\text{ N} = 111\text{ N}.$$

Thus, the normal force at the bottom is

$$F_{N,\text{bottom}} = \frac{mv^2}{R} + mg = F_c + mg = 111\text{ N} + 667\text{ N} = 778\text{ N}.$$

(c) If the speed is doubled, $F_c' = \frac{m(2v)^2}{R} = 4(111\text{ N}) = 444\text{ N}$. Therefore, at the highest point we have

$$F'_{N,\text{top}} = mg - F_c' = 667\text{ N} - 444\text{ N} = 223\text{ N}.$$

(d) Similarly, the normal force at the lowest point is now found to be

$$F'_{N,\text{bottom}} = F_c' + mg = 444\text{ N} + 667\text{ N} = 1111\text{ N}.$$

LEARN The apparent weight of the student is the greatest at the bottom and smallest at the top of the ride. The speed $v = \sqrt{gR}$ would result in $F_{N,\text{top}} = 0$, giving the student a sudden sensation of "weightlessness" at the top of the ride.

6-51
THINK An airplane with its wings tilted at an angle is in a circular motion. Centripetal force is involved in this problem.

EXPRESS The free-body diagram for the airplane of mass m is shown to the right. We note that $\vec{F}_l$ is the force of aerodynamic lift and $\vec{a}$ points rightwards in the figure. We also note that $|\vec{a}| = v^2/R$. Applying Newton's law to the axes of the problem (+x rightward and +y upward) we obtain

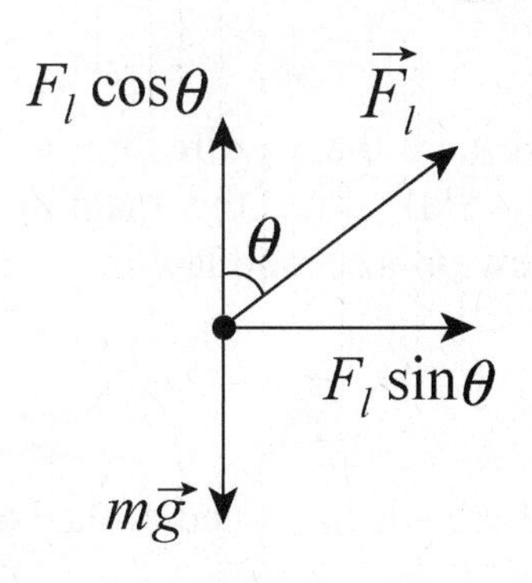

$$F_l \sin\theta = m\frac{v^2}{R}$$
$$F_l \cos\theta = mg$$

Eliminating mass from these equations leads to $\tan\theta = \frac{v^2}{gR}$. The equation allows us to solve for the radius R.

ANALYZE With v = 480 km/h = 133 m/s and θ = 40°, we find

$$R = \frac{v^2}{g\tan\theta} = \frac{(133\text{ m/s})^2}{(9.8\text{ m/s}^2)\tan 40°} = 2151\text{ m} \approx 2.2\times10^3\text{ m}.$$

LEARN Our approach to solving this problem is identical to that discussed in the Sample Problem 6.06 – "Car in banked circular turn." Do you see the similarities?

6-59

THINK As illustrated in Fig. 6-45, our system consists of a ball connected by two strings to a rotating rod. The tensions in the strings provide the source of centripetal force.

EXPRESS The free-body diagram for the ball is shown below. $\vec{T}_u$ is the tension exerted by the upper string on the ball, $\vec{T}_\ell$ is the tension in the lower string, and m is the mass of the ball. Note that the tension in the upper string is greater than the tension in the lower string. It must balance the downward pull of gravity and the force of the lower string.

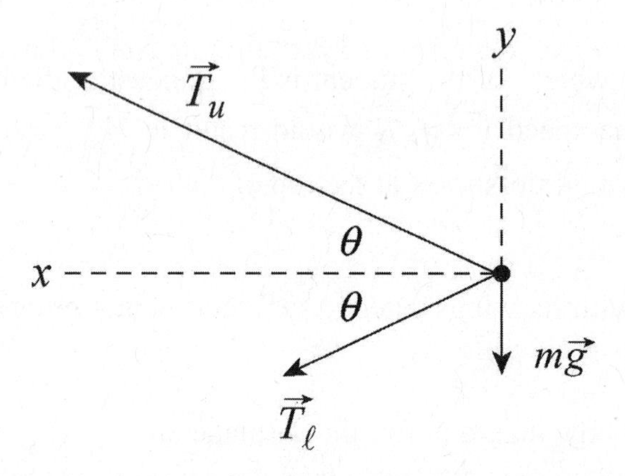

We take the $+x$ direction to be leftward (toward the center of the circular orbit) and $+y$ upward. Since the magnitude of the acceleration is $a = v^2/R$, the x component of Newton's second law is

$$T_u \cos\theta + T_\ell \cos\theta = \frac{mv^2}{R},$$

where v is the speed of the ball and R is the radius of its orbit. The y component is

$$T_u \sin\theta - T_\ell \sin\theta - mg = 0.$$

The second equation gives the tension in the lower string: $T_\ell = T_u - mg/\sin\theta$.

ANALYZE (a) Since the triangle is equilateral, the angle is $\theta = 30.0°$. Thus

$$T_\ell = T_u - \frac{mg}{\sin\theta} = 35.0\text{ N} - \frac{(1.34\text{ kg})(9.80\text{ m/s}^2)}{\sin 30.0°} = 8.74\text{ N}.$$

(b) The net force in the y-direction is zero. In the x-direction, the net force has magnitude

$$F_{\text{net,str}} = (T_u + T_\ell)\cos\theta = (35.0\text{ N} + 8.74\text{ N})\cos 30.0° = 37.9\text{ N}.$$

(c) The radius of the path is

$$R = L\cos\theta = (1.70\ \text{m})\cos 30° = 1.47\ \text{m}.$$

Using $F_{\text{net,str}} = mv^2/R$, we find the speed of the ball to be

$$v = \sqrt{\frac{RF_{\text{net,str}}}{m}} = \sqrt{\frac{(1.47\ \text{m})(37.9\ \text{N})}{1.34\ \text{kg}}} = 6.45\ \text{m/s}.$$

(d) The direction of $\vec{F}_{\text{net,str}}$ is leftward ("radially inward").

LEARN The upper string, with a tension about 4 times that in the lower string ($T_u \approx 4T_\ell$), will break more easily than the lower one.

5-61

THINK Our system consists of two blocks, one on top of the other. If we pull the bottom block too hard, the top block will slip on the bottom one. We're interested in the maximum force that can be applied such that the two will move together.

EXPRESS The free-body diagrams for the two blocks are shown below.

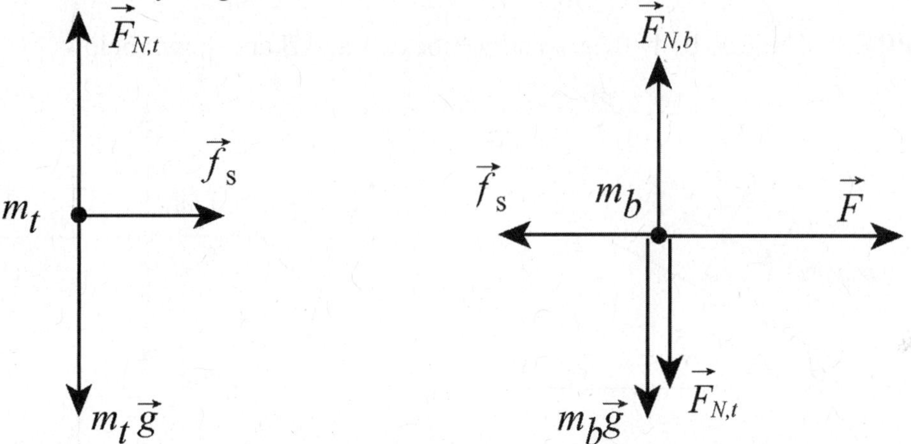

We first calculate the coefficient of static friction for the surface between the two blocks. When the force applied is at a maximum, the frictional force between the two blocks must also be a maximum. Since $F_t = 12$ N of force has to be applied to the top block for slipping to take place, using $F_t = f_{s,\max} = \mu_s F_{N,t} = \mu_s m_t g$, we have

$$\mu_s = \frac{F_t}{m_t g} = \frac{12\ \text{N}}{(4.0\ \text{kg})(9.8\ \text{m/s}^2)} = 0.31.$$

Using the same reasoning, for the two masses to move together, the maximum applied force would be

$$F = \mu_s (m_t + m_b) g$$

ANALYZE (a) Substituting the value of μ_s found above, the maximum horizontal force has a magnitude

$$F = \mu_s (m_t + m_b) g = (0.31)(4.0\,\text{kg} + 5.0\,\text{kg})(9.8\,\text{m/s}^2) = 27\,\text{N}$$

(b) The maximum acceleration is

$$a_{\max} = \frac{F}{m_t + m_b} = \mu_s g = (0.31)(9.8\,\text{m/s}^2) = 3.0\,\text{m/s}^2.$$

LEARN Slipping will occur if the applied force exceeds 27.3 N. In the absence of friction ($\mu_s = 0$) between the two blocks, any amount of force would cause the top block to slip.

6-79

THINK We have two blocks connected by a cord, as shown in Fig. 6-56. As block *A* slides down the frictionless inclined plane, it pulls block *B*, so there's a tension in the cord.

EXPRESS The free-body diagrams for blocks *A* and *B* are shown below:

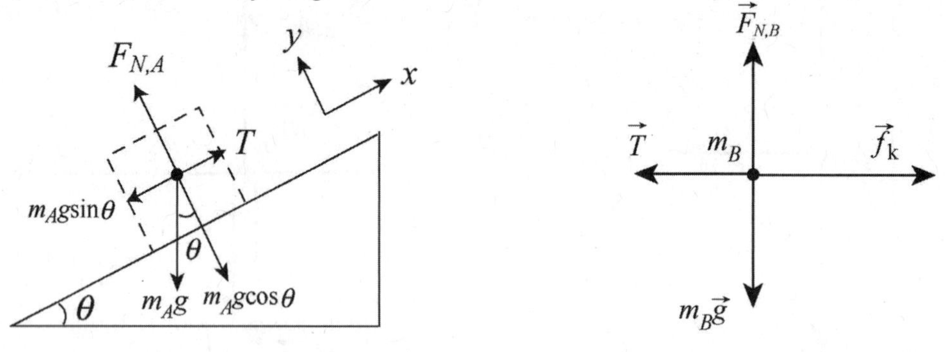

Newton's law gives

$$m_A g \sin\theta - T = m_A a$$

for block *A* (where θ= 30°). For block *B*, we have

$$T - f_k = m_B a$$

Now the frictional force is given by $f_k = \mu_k F_{N,B} = \mu_k m_B g$. The equations allow us to solve for the tension *T* and the acceleration *a*.

ANALYZE (a) Combining the above equations to solve for *T*, we obtain

$$T = \frac{m_A m_B}{m_A + m_B}(\sin\theta + \mu_k) g = \frac{(4.0\,\text{kg})(2.0\,\text{kg})}{4.0\,\text{kg} + 2.0\,\text{kg}}(\sin 30° + 0.50)(9.80\,\text{m/s}^2) = 13\,\text{N}.$$

(b) Similarly, the acceleration of the two-block system is

$$a = \left(\frac{m_A \sin\theta - \mu_k m_B}{m_A + m_B} \right) g = \frac{(4.0\,\text{kg})\sin 30° - (0.50)(2.0\,\text{kg})}{4.0\,\text{kg} + 2.0\,\text{kg}} (9.80\,\text{m/s}^2) = 1.6\,\text{m/s}^2 .$$

LEARN In the case where $\theta = 90°$ and $\mu_k = 0$, we have

$$T = \frac{m_A m_B}{m_A + m_B} g, \quad a = \frac{m_A}{m_A + m_B} g$$

which correspond to the Sample Problem – "Block on table, block hanging," discussed in Chapter 5.

6-81

THINK How can a cyclist move in a circle? It is the force of friction that provides the centripetal force required for the circular motion.

EXPRESS The free-body diagram is shown below. The magnitude of the acceleration of the cyclist as it moves along the horizontal circular path is given by v^2/R, where v is the speed of the cyclist and R is the radius of the curve.

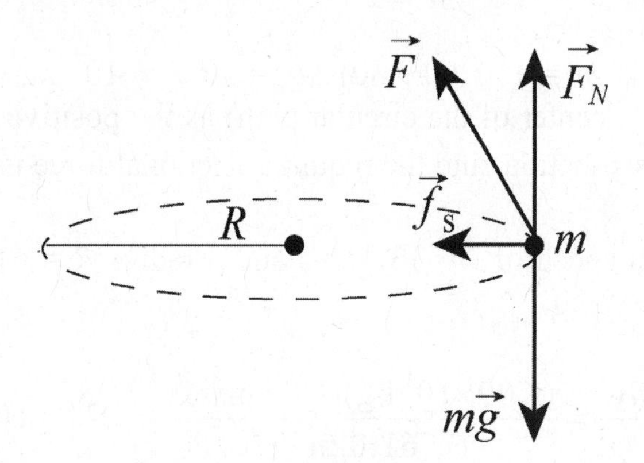

The horizontal component of Newton's second law is $f_s = mv^2/R$, where f_s is the static friction exerted horizontally by the ground on the tires. Similarly, if F_N is the vertical force of the ground on the bicycle and m is the mass of the bicycle and rider, the vertical component of Newton's second law leads to $F_N = mg = 833\,\text{N}$.

ANALYZE (a) The frictional force is $f_s = \dfrac{mv^2}{R} = \dfrac{(85.0\,\text{kg})(9.00\,\text{m/s})^2}{25.0\,\text{m}} = 275\ \text{N}.$

(b) Since the frictional force $\vec{f}_s$ and $\vec{F}_N$, the normal force exerted by the road, are perpendicular to each other, the magnitude of the force exerted by the ground on the bicycle is

$$F = \sqrt{f_s^2 + F_N^2} = \sqrt{(275\text{ N})^2 + (833\text{ N})^2} = 877\text{ N}.$$

LEARN The force exerted by the ground on the bicycle is at an angle $\theta = \tan^{-1}(275\text{ N}/833\text{ N}) = 18.3°$ with respect to the *vertical* axis.

6-87

THINK A car is making a turn on an unbanked curve. Friction is what provides the centripetal force needed for this circular motion.

EXPRESS The free-body diagram is shown to the below.

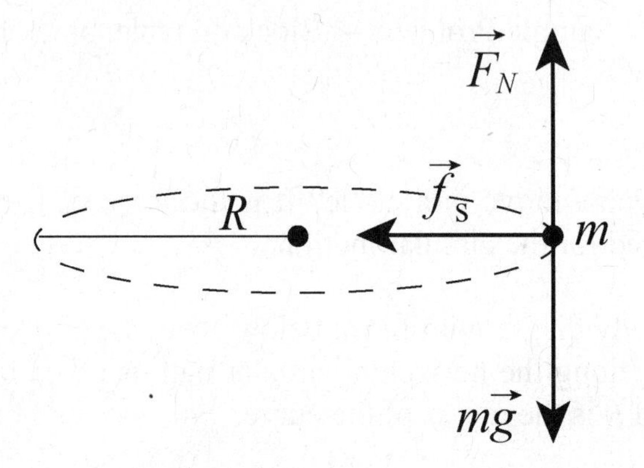

The mass of the car is m = (10700/9.80) kg = 1.09×10^3 kg. We choose "inward" (horizontally toward the center of the circular path) as the positive direction. The normal force is $F_N = mg$ in this situation, and the required frictional force is $f_s = mv^2/R$.

ANALYZE (a) With a speed of v = 13.4 m/s and a radius R = 61 m, Newton's second law (using Eq. 6-18) leads to

$$f_s = \frac{mv^2}{R} = \frac{(1.09\times10^3\text{ kg})(13.4\text{ m/s})^2}{61.0\text{ m}} = 3.21\times10^3\text{ N}.$$

(b) The maximum possible static friction is found to be

$$f_{s,\max} = \mu_s mg = (0.35)(10700\text{ N}) = 3.75\times10^3\text{ N}$$

using Eq. 6-1. We see that the static friction found in part (a) is less than this, so the car rolls (no skidding) and successfully negotiates the curve.

LEARN From the above expressions, we see that with a coefficient of friction μ_s, the maximum speed of the car negotiating a curve of radius R is $v_{\max} = \sqrt{\mu_s gR}$. So in this case, the car can go up to a maximum speed of

$$v_{\max} = \sqrt{(0.35)(9.8 \text{ m/s}^2)(61 \text{ m})} = 14.5 \text{ m/s}$$

without skidding.

6-89

THINK In order to move a filing cabinet, the force applied must be able to overcome the frictional force.

EXPRESS We apply Newton's second law (as $F_{\text{push}} - f = ma$). If we find the applied force F_{push} to be less than $f_{s,\max}$, the maximum static frictional force, our conclusion would then be "no, the cabinet does not move" (which means a is actually 0 and the frictional force is simply $f = F_{\text{push}}$). On the other hand, if we obtain $a > 0$ then the cabinet moves (so $f = f_k$). For $f_{s,\max}$ and f_k we use Eq. 6-1 and Eq. 6-2 (respectively), and in those formulas we set the magnitude of the normal force to the weight of the cabinet: $F_N = mg = 556 \text{ N}$. Thus, the maximum static frictional force is

$$f_{s,\max} = \mu_s F_N = \mu_s mg = (0.68)(556 \text{ N}) = 378 \text{ N}.$$

and the kinetic frictional force is

$$f_k = \mu_k F_N = \mu_k mg = (0.56)(556 \text{ N}) = 311 \text{ N}.$$

ANALYZE (a) Here we find $F_{\text{push}} < f_{s,\max}$ which leads to $f = F_{\text{push}} = 222$ N. The cabinet does not move.

(b) Again we find $F_{\text{push}} < f_{s,\max}$ which leads to $f = F_{\text{push}} = 334$ N. The cabinet does not move.

(c) Now we have $F_{\text{push}} > f_{s,\max}$ which means the cabinet moves and $f = f_k = 311$ N.

(d) Again we have $F_{\text{push}} > f_{s,\max}$ which means the cabinet moves and $f = f_k = 311$ N.

(e) The cabinet moves in (c) and (d).

LEARN In summary, in order to make the cabinet move, the minimum applied force is equal to the maximum static frictional force $f_{s,\max}$.

6-91

THINK Whether the block is sliding down or up the incline, there is a frictional force in the opposite direction of the motion.

EXPRESS The free-body diagram for the first part of this problem (when the block is sliding downhill with zero acceleration) is shown to the right. Newton's second law gives

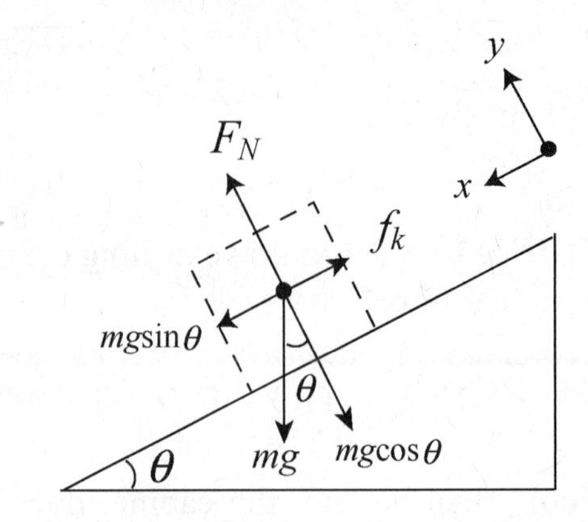

$$mg\sin\theta - f_k = mg\sin\theta - \mu_k F_N = ma_x = 0$$
$$mg\cos\theta - F_N = ma_y = 0$$

The two equations can be combined to give

$$\mu_k = \tan\theta.$$

Now (for the second part of the problem, with the block projected uphill) the friction direction is reversed (see figure to the right). Newton's second law for the uphill motion (and Eq. 6-12) leads to

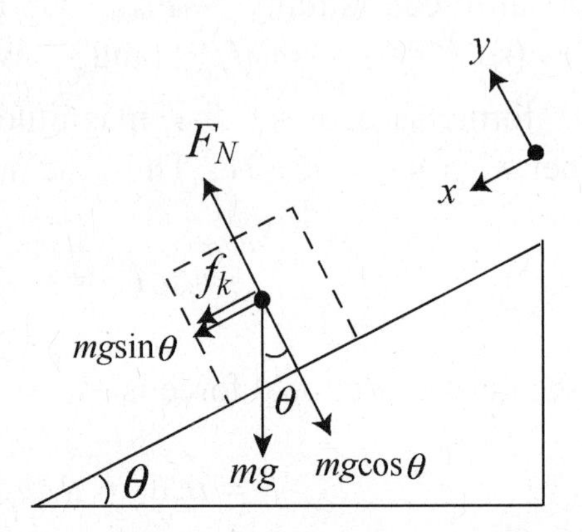

$$mg\sin\theta + f_k = mg\sin\theta + \mu_k F_N = ma_x$$
$$mg\cos\theta - F_N = ma_y = 0$$

Note that by our convention, $a_x > 0$ means that the acceleration is downhill, and therefore, the speed of the block will decrease as it moves up the incline.

ANALYZE (a) Using $\mu_k = \tan\theta$ and $F_N = mg\cos\theta$, we find the *x*-component of the acceleration to be

$$a_x = g\sin\theta + \frac{\mu_k F_N}{m} = g\sin\theta + \frac{(\tan\theta)(mg\cos\theta)}{m} = 2g\sin\theta.$$

The distance the block travels before coming to a stop can be found by using Eq. 2-16: $v_f^2 = v_0^2 - 2a_x\Delta x$, which yields

$$\Delta x = \frac{v_0^2}{2a_x} = \frac{v_0^2}{2(2g\sin\theta)} = \frac{v_0^2}{4g\sin\theta}.$$

(b) We usually expect $\mu_s > \mu_k$ (see the discussion in Section 6-1). The "angle of repose" (the minimum angle necessary for a stationary block to start sliding downhill) is $\mu_s = \tan(\theta_{repose})$. Therefore, we expect $\theta_{repose} > \theta$ found in part (a). Consequently, when the block comes to rest, the incline is not steep enough to cause it to start slipping down the incline again.

LEARN An alternative way to see that the block will not slide down again is to note that the downward force of gravitation is not large enough to overcome the force of friction, i.e., $mg\sin\theta = f_k < f_{s,\max}$.

6-97

THINK In this problem a force is applied to accelerate a box. From the distance traveled and the speed at that instant, we can calculate the coefficient of kinetic friction between the box and the floor.

EXPRESS The free-body diagram is shown below.

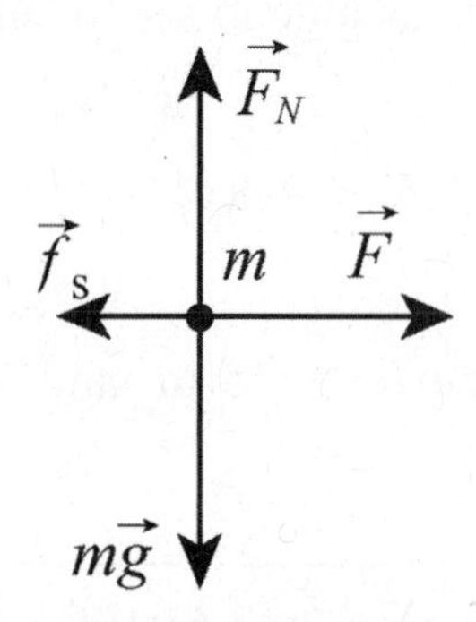

We adopt the familiar axes with $+x$ rightward and $+y$ upward, and refer to the 85 N horizontal push of the worker as F (and assume it to be rightward). Applying Newton's second law to the x axis and y axis, respectively, produces

$$F - f_k = ma_x, \qquad F_N - mg = 0.$$

On the other hand, using Eq. 2-16 ($v^2 = v_0^2 + 2a_x\Delta x$), we find the acceleration to be

$$a_x = \frac{v^2 - v_0^2}{2\Delta x} = \frac{(1.0\text{ m/s})^2 - 0}{2(1.4\text{ m})} = 0.357\text{ m/s}^2.$$

The above equations can be combined to give μ_k.

ANALYZE Using $f_k = \mu_k F_N$, we find the coefficient of kinetic friction between the box and the floor to be

$$\mu_k = \frac{f_k}{F_N} = \frac{F - ma_x}{mg} = \frac{85\text{ N} - (40\text{ kg})(0.357\text{ m/s}^2)}{(40\text{ kg})(9.8\text{ m/s}^2)} = 0.18.$$

LEARN In general, the acceleration can be written as $a_x = (F/m) - \mu_k g$. We see that the smaller the value of μ_k, the greater the acceleration. In the limit $\mu_k = 0$, we simply have $a_x = F/m$.

Chapter 7

7-1

THINK As the proton is being accelerated, its speed increases, and so does its kinetic energy.

EXPRESS To calculate the speed of the proton at a later time, we use the equation $v^2 = v_0^2 + 2a\Delta x$ from Table 2-1. The change in kinetic energy is then equal to

$$\Delta K = \frac{1}{2}m(v_f^2 - v_i^2).$$

ANALYZE (a) With $\Delta x = 3.5\ \text{cm} = 0.035\ \text{m}$ and $a = 3.6\times10^{15}\ \text{m/s}^2$, we find the proton speed to be

$$v = \sqrt{v_0^2 + 2a\Delta x} = \sqrt{\left(2.4\times10^7\ \text{m/s}\right)^2 + 2\ \left(3.6\times10^{15}\ \text{m/s}^2\right)\left(0.035\ \text{m}\right)} = 2.9\times10^7\ \text{m/s}.$$

(b) The initial kinetic energy is

$$K_i = \frac{1}{2}mv_0^2 = \frac{1}{2}\ \left(1.67\times10^{-27}\ \text{kg}\right)\left(2.4\times10^7\ \text{m/s}\right)^2 = 4.8\times10^{-13}\ \text{J},$$

and the final kinetic energy is

$$K_f = \frac{1}{2}mv^2 = \frac{1}{2}\ \left(1.67\times10^{-27}\ \text{kg}\right)\left(2.9\times10^7\ \text{m/s}\right)^2 = 6.9\times10^{-13}\ \text{J}.$$

Thus, the change in kinetic energy is

$$\Delta K = K_f - K_i = 6.9\times10^{-13}\ \text{J} - 4.8\times10^{-13}\ \text{J} = 2.1\times10^{-13}\ \text{J}.$$

LEARN The change in kinetic energy can be rewritten as

$$\Delta K = \frac{1}{2}m(v_f^2 - v_i^2) = \frac{1}{2}m(2a\Delta x) = ma\Delta x = F\Delta x = W$$

which, according to the work-kinetic energy theorem, is simply the work done on the particle.

7-17

THINK The helicopter does work to lift the astronaut upward against gravity. The work done on the astronaut is converted to the kinetic energy of the astronaut.

EXPRESS We use $\vec{F}$ to denote the upward force exerted by the cable on the astronaut. The force of the cable is upward and the force of gravity is mg downward. Furthermore, the acceleration of the astronaut is $a = g/10$ upward. According to Newton's second law, the force is given by

$$F - mg = ma \;\Rightarrow\; F = m(g+a) = \frac{11}{10}mg,$$

in the same direction as the displacement. On the other hand, the force of gravity has magnitude $F_g = mg$ and is opposite in direction to the displacement.

ANALYZE (a) Since the force of the cable $\vec{F}$ and the displacement $\vec{d}$ are in the same direction, the work done by $\vec{F}$ is

$$W_F = Fd = \frac{11mgd}{10} = \frac{11\,(72\text{ kg})(9.8\text{ m/s}^2)(15\text{ m})}{10} = 1.164\times10^4\text{ J} \approx 1.2\times10^4\text{ J}.$$

(b) Using Eq. 7-7, the work done by gravity is

$$W_g = -F_g d = -mgd = -\,(72\text{ kg})(9.8\text{ m/s}^2)(15\text{ m}) = -1.058\times10^4\text{ J} \approx -1.1\times10^4\text{ J}.$$

(c) The total work done is the sum of the two works:

$$W_\text{net} = W_F + W_g = 1.164\times10^4\text{ J} - 1.058\times10^4\text{ J} = 1.06\times10^3\text{ J} \approx 1.1\times10^3\text{ J}.$$

Since the astronaut started from rest, the work-kinetic energy theorem tells us that this is her final kinetic energy.

(d) Since $K = \frac{1}{2}mv^2$, her final speed is $v = \sqrt{\dfrac{2K}{m}} = \sqrt{\dfrac{2(1.06\times10^3\text{J})}{72\text{ kg}}} = 5.4\text{ m/s}.$

LEARN For a general upward acceleration a, the net work done is

$$W_\text{net} = W_F + W_g = Fd - F_g d = m(g+a)d - mgd = mad.$$

Since $W_\text{net} = \Delta K = mv^2/2$ by the work-kinetic energy theorem, the speed of the astronaut would be $v = \sqrt{2ad}$, which is independent of the mass of the astronaut. In our case, $v = \sqrt{2(9.8\text{ m/s}^2/10)(15\text{ m})} = 5.4\text{ m/s}$, which agrees with that calculated in (d).

7-21

THINK In this problem the cord is doing work on the block so that it does not undergo free fall.

EXPRESS We use F to denote the magnitude of the force of the cord on the block. This force is upward, opposite to the force of gravity (which has magnitude $F_g = Mg$), to prevent the block from undergoing free fall. The acceleration is $\vec{a} = g/4$ downward. Taking the downward direction to be positive, then Newton's second law yields

$$\vec{F}_{\text{net}} = m\vec{a} \Rightarrow Mg - F = M\left(\frac{g}{4}\right),$$

so $F = 3Mg/4$, in the opposite direction of the displacement. On the other hand, the force of gravity $F_g = mg$ is in the same direction to the displacement.

ANALYZE (a) Since the displacement is downward, the work done by the cord's force is, using Eq. 7-7,

$$W_F = -Fd = -\frac{3}{4}Mgd.$$

(b) Similarly, the work done by the force of gravity is $W_g = F_g d = Mgd$.

(c) The total work done on the block is simply the sum of the two works:

$$W_{\text{net}} = W_F + W_g = -\frac{3}{4}Mgd + Mgd = \frac{1}{4}Mgd\,.$$

Since the block starts from rest, we use Eq. 7-15 to conclude that this $(Mgd/4)$ is the block's kinetic energy K at the moment it has descended the distance d.

(d) With $K = \frac{1}{2}Mv^2$, the speed is

$$v = \sqrt{\frac{2K}{M}} = \sqrt{\frac{2(Mgd/4)}{M}} = \sqrt{\frac{gd}{2}}$$

at the moment the block has descended the distance d.

LEARN For a general downward acceleration a, the force exerted by the cord is $F = m(g-a)$, so that the net work done on the block is $W_{\text{net}} = F_{\text{net}}d = mad$. The speed of the block after falling a distance d is $v = \sqrt{2ad}$. In the special case where the block hangs still, $a = 0$, $F = mg$ and $v = 0$. In our case, $a = g/4$, and $v = \sqrt{2(g/4)d} = \sqrt{gd/2}$, which agrees with that calculated in (d).

7-31

THINK The applied force varies with x, so an integration is required to calculate the work done on the body.

EXPRESS As the body moves along the x axis from x_i = 3.0 m to x_f = 4.0 m the work done by the force is

$$W=\int_{x_i}^{x_f} F_x\,dx=\int_{x_i}^{x_f} -6x\,dx=-3(x_f^2-x_i^2)=-3\,(4.0^2-3.0^2)=-21\text{ J}.$$

According to the work-kinetic energy theorem, this gives the change in the kinetic energy:

$$W=\Delta K=\frac{1}{2}m\left(v_f^2-v_i^2\right)$$

where v_i is the initial velocity (at x_i) and v_f is the final velocity (at x_f). Given v_i, we can readily calculate v_f.

ANALYZE (a) The work-kinetic theorem yields

$$v_f=\sqrt{\frac{2W}{m}+v_i^2}=\sqrt{\frac{2(-21\text{ J})}{2.0\text{ kg}}+(8.0\text{ m/s})^2}=6.6\text{ m/s}.$$

(b) The velocity of the particle is v_f = 5.0 m/s when it is at $x = x_f$. The work-kinetic energy theorem is used to solve for x_f. The net work done on the particle is $W=-3\left(x_f^2-x_i^2\right)$, so the theorem leads to

$$W=\Delta K \quad\Rightarrow\quad -3\left(x_f^2-x_i^2\right)=\frac{1}{2}m\,\left(v_f^2-v_i^2\right).$$

Thus,

$$x_f=\sqrt{-\frac{m}{6}\left(v_f^2-v_i^2\right)+x_i^2}=\sqrt{-\frac{2.0\text{ kg}}{6\text{ N/m}}\left((5.0\text{ m/s})^2-(8.0\text{ m/s})^2\right)+(3.0\text{ m})^2}=4.7\text{ m}.$$

LEARN Since $x_f > x_i$, $W=-3(x_f^2-x_i^2)<0$, i.e., the work done by the force is negative. From the work-kinetic energy theorem, this implies $\Delta K<0$. Hence, the speed of the particle will continue to decrease as it moves in the $+x$-direction.

7-35

THINK We have an applied force that varies with x. An integration is required to calculate the work done on the particle.

EXPRESS Given a one-dimensional force $F(x)$, the work done is simply equal to the area under the curve: $W=\int_{x_i}^{x_f} F(x)\,dx$.

ANALYZE (a) The plot of $F(x)$ is shown to the right. Here we take x_0 to be positive. The work is negative as the object moves from $x=0$ to $x=x_0$ and positive as it moves from $x=x_0$ to $x=2x_0$.

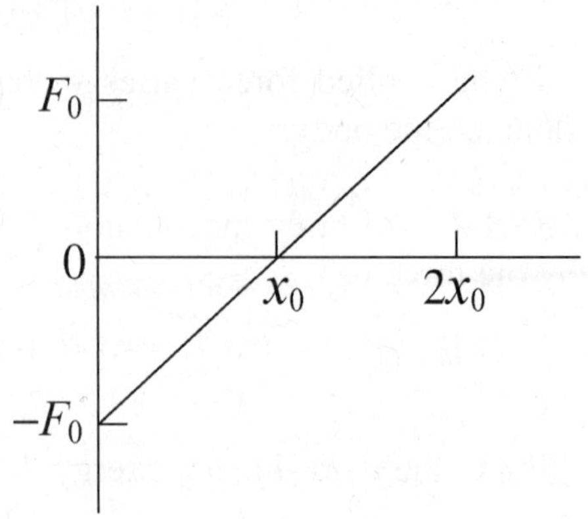

Since the area of a triangle is (base)(altitude)/2, the work done from $x=0$ to $x=x_0$ is $W_1=-(x_0)(F_0)/2$ and the work done from $x=x_0$ to $x=2x_0$ is

$$W_2=(2x_0-x_0)(F_0)/2=(x_0)(F_0)/2$$

The total work is the sum of the two:

$$W=W_1+W_2=-\frac{1}{2}F_0x_0+\frac{1}{2}F_0x_0=0.$$

(b) The integral for the work is

$$W=\int_0^{2x_0}F_0\left(\frac{x}{x_0}-1\right)dx=F_0\left(\frac{x^2}{2x_0}-x\right)\Bigg|_0^{2x_0}=0.$$

LEARN If the particle starts out at $x = 0$ with an initial speed v_i, with a negative work $W_1=-F_0x_0/2<0$, its speed at $x=x_0$ will decrease to

$$v=\sqrt{v_i^2+\frac{2W_1}{m}}=\sqrt{v_i^2-\frac{F_0x_0}{m}}<v_i,$$

but return to v_i again at $x=2x_0$ with a positive work $W_2=F_0x_0/2>0$.

7-43

THINK This problem deals with the power and work done by a constant force.

EXPRESS The power done by a constant force F is given by $P = Fv$ and the work done by F from time t_1 to time t_2 is

$$W=\int_{t_1}^{t_2}P\,dt=\int_{t_1}^{t_2}Fv\,dt$$

Since F is the magnitude of the net force, the magnitude of the acceleration is $a = F/m$. Thus, if the initial velocity is $v_0=0$, then the velocity of the body as a function of time is given by $v=v_0+at=(F/m)t$. Substituting the expression for v into the equation above, the work done during the time interval (t_1,t_2) becomes

$$W=\int_{t_1}^{t_2}(F^2/m)t\,dt=\frac{F^2}{2m}\left(t_2^2-t_1^2\right).$$

ANALYZE (a) For $t_1=0$ and $t_2=1.0$ s, $W=\frac{1}{2}\left(\frac{(5.0\text{ N})^2}{15\text{ kg}}\right)[(1.0\text{ s})^2-0]=0.83$ J.

(b) For $t_1=1.0\text{s}$, and $t_2=2.0$ s, $W=\frac{1}{2}\left(\frac{(5.0\text{ N})^2}{15\text{ kg}}\right)[(2.0\text{ s})^2-(1.0\text{ s})^2]=2.5$ J.

(c) For $t_1=2.0$ s and $t_2=3.0$ s, $W=\frac{1}{2}\left(\frac{(5.0\text{ N})^2}{15\text{ kg}}\right)[(3.0\text{ s})^2-(2.0\text{ s})^2]=4.2$ J.

(d) Substituting $v=(F/m)t$ into $P=Fv$ we obtain $P=F^2t/m$ for the power at any time t. At the end of the third second, the instantaneous power is

$$P=\left(\frac{(5.0\text{ N})^2\ (3.0\text{ s})}{15\text{ kg}}\right)=5.0\text{ W}.$$

LEARN The work done here is quadratic in t. Therefore, from the definition $P=dW/dt$ for the instantaneous power, we see that P increases linearly with t.

7-45

THINK A block is pulled at a constant speed by a force directed at some angle with respect to the direction of motion. The quantity we're interested in is the power, or the time rate at which work is done by the applied force.

EXPRESS The power associated with force $\vec{F}$ is given by $P=\vec{F}\cdot\vec{v}=Fv\cos\phi$, where $\vec{v}$ is the velocity of the object on which the force acts, and ϕ is the angle between $\vec{F}$ and $\vec{v}$.

ANALYZE With $F=122\text{ N}$, $v=5.0\text{ m/s}$ and $\phi=37.0°$, we find the power to be

$$P=Fv\cos\phi=(122\text{ N})(5.0\text{ m/s})\cos37.0°=4.9\times10^2\text{ W}.$$

LEARN From the expression $P=Fv\cos\phi$, we see that the power is at a maximum when $\vec{F}$ and $\vec{v}$ are in the same direction ($\phi=0$), and is zero when they are perpendicular of each other. In addition, we're told that the block moves at a constant speed, so $\Delta K=0$, and the net work done on it must also be zero by the work-kinetic energy theorem. Thus, the applied force here must be compensating another force (e.g., friction) for the net rate to be zero.

7-49

THINK We have a loaded elevator moving upward at a constant speed. The forces involved are: gravitational force on the elevator, gravitational force on the counterweight, and the force by the motor via cable.

EXPRESS The total work is the sum of the work done by gravity on the elevator, the work done by gravity on the counterweight, and the work done by the motor on the system:

$$W = W_e + W_c + W_m.$$

Since the elevator moves at constant velocity, its kinetic energy does not change and according to the work-kinetic energy theorem the total work done is zero, i.e., $W = \Delta K = 0$.

ANALYZE The elevator moves *upward* through 54 m, so the work done by gravity on it is

$$W_e = -m_e g d = -(1200 \text{ kg})(9.80 \text{ m/s}^2)(54 \text{ m}) = -6.35 \times 10^5 \text{ J}.$$

The counterweight moves *downward* the same distance, so the work done by gravity on it is

$$W_c = m_c g d = (950 \text{ kg})(9.80 \text{ m/s}^2)(54 \text{ m}) = 5.03 \times 10^5 \text{ J}.$$

Since $W = 0$, the work done by the motor on the system is

$$W_m = -W_e - W_c = 6.35 \times 10^5 \text{ J} - 5.03 \times 10^5 \text{ J} = 1.32 \times 10^5 \text{ J}.$$

This work is done in a time interval of $\Delta t = 3.0 \text{ min} = 180 \text{ s}$, so the power supplied by the motor to lift the elevator is

$$P = \frac{W_m}{\Delta t} = \frac{1.32 \times 10^5 \text{ J}}{180 \text{ s}} = 7.4 \times 10^2 \text{ W}.$$

LEARN In general, the work done by the motor is $W_m = (m_e - m_c) g d$. So when the counterweight mass balances the total mass, $m_c = m_e$, no work is required by the motor.

7-55

THINK A horse is doing work to pull a cart at a constant speed. We'd like to know the work done during a time interval and the corresponding average power.

EXPRESS The horse pulls with a force $\vec{F}$. As the cart moves through a displacement $\vec{d}$, the work done by $\vec{F}$ is $W = \vec{F} \cdot \vec{d} = Fd\cos\phi$, where ϕ is the angle between $\vec{F}$ and $\vec{d}$.

ANALYZE (a) In 10 min the cart moves a distance

$$d = v\Delta t = \left(6.0\ \frac{\text{mi}}{\text{h}}\right)\left(\frac{5280\ \text{ft/mi}}{60\ \text{min/h}}\right)(10\ \text{min}) = 5280\ \text{ft}$$

so that Eq. 7-7 yields

$$W = Fd\cos\phi = (40\ \text{lb})(5280\ \text{ft})\cos 30° = 1.8\times10^5\ \text{ft}\cdot\text{lb}.$$

(b) The average power is given by Eq. 7-42. With $\Delta t = 10\ \text{min} = 600\,\text{s}$, we obtain

$$P_{\text{avg}} = \frac{W}{\Delta t} = \frac{1.8\times10^5\ \text{ft}\cdot\text{lb}}{600\ \text{s}} = 305\ \text{ft}\cdot\text{lb/s},$$

which (using the conversion factor $1\ \text{hp} = 550\ \text{ft}\cdot\text{lb/s}$) converts to $P_{\text{avg}} = 0.55$ hp.

LEARN The average power can also be calculate by using Eq. 7-48: $P_{\text{avg}} = Fv\cos\phi$. Converting the speed to $v = (6.0\ \text{mi/h})\left(\frac{5280\ \text{ft/mi}}{3600\ \text{s/h}}\right) = 8.8\ \text{ft/s}$, we get

$$P_{\text{avg}} = Fv\cos\phi = (40\ \text{lb})(8.8\ \text{ft/s})\cos 30° = 305\ \text{ft}\cdot\text{lb} = 0.55\ \text{hp}$$

which agrees with that found in (b).

7-63

THINK A crate is being pushed up a frictionless inclined plane. The forces involved are: gravitational force on the crate, normal force on the crate, and the force applied by the worker.

EXPRESS The work done by a force $\vec{F}$ on an object as it moves through a displacement $\vec{d}$ is $W = \vec{F}\cdot\vec{d} = Fd\cos\phi$, where ϕ is the angle between $\vec{F}$ and $\vec{d}$.

ANALYZE (a) The applied force is parallel to the inclined plane. Thus, using Eq. 7-6, the work done on the crate by the worker's applied force is

$$W_a = Fd\cos 0° = (209\ \text{N})(1.50\ \text{m}) \approx 314\ \text{J}.$$

(b) Using Eq. 7-12, we find the work done by the gravitational force to be

$$\begin{aligned} W_g &= F_g d\cos(90° + 25°) = mgd\cos 115° \\ &= (25.0\,\text{kg})(9.8\,\text{m/s}^2)(1.50\ \text{m})\cos 115° \\ &\approx -155\ \text{J}. \end{aligned}$$

(c) The angle between the normal force and the direction of motion remains 90º at all times, so the work it does is zero:

$$W_N = F_N d\cos 90° = 0.$$

(d) The total work done on the crate is the sum of all three works:

$$W = W_a + W_g + W_N = 314\text{ J} + (-155\text{ J}) + 0\text{ J} = 158\text{ J}.$$

LEARN By work-kinetic energy theorem, if the crate is initially at rest ($K_i = 0$), then its kinetic energy after having moved 1.50 m up the incline would be $K_f = W = 158\text{ J}$, and the speed of the crate at that instant is

$$v = \sqrt{2K_f / m} = \sqrt{2(158\text{ J})/25.0\text{ kg}} = 3.56\text{ m/s}.$$

7-67

THINK In this problem we have packages hung from the spring. The extent of stretching can be determined from Hooke's law.

EXPRESS According to Hooke's law, the spring force is given by

$$F_x = -k(x - x_0) = -k\Delta x,$$

where Δx is the displacement from the equilibrium position. Thus, the first two situations in Fig. 7-46 can be written as

$$-110\text{ N} = -k(40\text{ mm} - x_0)$$
$$-240\text{ N} = -k(60\text{ mm} - x_0)$$

The two equations allow us to solve for *k*, the spring constant, as well as x_0, the relaxed position when no mass is hung.

ANALYZE (a) The two equations can be added to give

$$240\text{ N} - 110\text{ N} = k(60\text{ mm} - 40\text{ mm})$$

which yields $k = 6.5$ N/mm. Substituting the result into the first equation, we find

$$x_0 = 40\text{ mm} - \frac{110\text{ N}}{k} = 40\text{ mm} - \frac{110\text{ N}}{6.5\text{ N/mm}} = 23\text{ mm}.$$

(b) Using the results from part (a) to analyze that last picture, we find the weight to be

$$W = k(30\text{mm} - x_o) = (6.5\text{ N/mm})(30\text{ mm} - 23\text{ mm}) = 45\text{ N}.$$

LEARN An alternative method to calculate W in the third picture is to note that since the amount of stretching is proportional to the weight hung, we have $\frac{W}{W'}=\frac{\Delta x}{\Delta x'}$. Applying this relation to the second and the third pictures, the weight W is

$$W=\left(\frac{\Delta x_3}{\Delta x_2}\right)W_2=\left(\frac{30\text{ mm}-23\text{ mm}}{60\text{ mm}-23\text{ mm}}\right)(240\text{ N})=45\text{ N},$$

in agreement with the result shown in (b).

7-75

THINK Power must be supplied in order to lift the elevator with load upward at a constant speed.

EXPRESS For the elevator-load system to move upward at a constant speed (zero acceleration), the applied force F must exactly balance the gravitational force on the system, i.e., $F=F_g=(m_{\text{elev}}+m_{\text{load}})g$. The power required can then be calculated using Eq. 17-48: $P=Fv$.

ANALYZE With $m_{\text{elev}}=4500\text{ kg}$, $m_{\text{load}}=1800\text{ kg}$ and $v=3.80\text{ m/s}$, we find the power to be

$$P=Fv=(m_{\text{elev}}+m_{\text{load}})gv=(4500\text{ kg}+1800\text{ kg})(9.8\text{ m/s}^2)(3.80\text{ m/s})=235\text{ kW}.$$

LEARN The power required is proportional to the speed at which the system moves; the greater the speed, the greater the power that must be supplied.

7-79

THINK A box sliding in the +*x*-direction is slowed down by a steady wind in the –*x*-direction. The problem involves graphical analysis.

EXPRESS Fig. 7-49 represents $x(t)$, the position of the lunch box as a function of time. It is convenient to fit the curve to a concave-downward parabola:

$$x(t)=\frac{1}{10}t(10-t)=t-\frac{1}{10}t^2.$$

By taking one and two derivatives, we find the velocity and acceleration to be

$$v(t)=\frac{dx}{dt}=1-\frac{t}{5}\ \text{(in m/s)},\quad a=\frac{d^2x}{dt^2}=-\frac{1}{5}=-0.2\ \text{(in m/s}^2\text{)}.$$

The equations imply that the initial speed of the box is $v_i = v(0) = 1.0\text{ m/s}$, and the constant force by the wind is

$$F = ma = (2.0\text{ kg})(-0.2\text{ m/s}^2) = -0.40\text{ N}.$$

The corresponding work is given by (SI units understood)

$$W(t) = F \cdot x(t) = -0.04t(10-t).$$

The initial kinetic energy of the lunch box is

$$K_i = \frac{1}{2}mv_i^2 = \frac{1}{2}(2.0\text{ kg})(1.0\text{ m/s})^2 = 1.0\text{ J}.$$

With $\Delta K = K_f - K_i = W$, the kinetic energy at a later time is given by (in SI units)

$$K(t) = K_i + W = 1 - 0.04t(10-t)$$

ANALYZE (a) When t = 1.0 s, the above expression gives

$$K(1\text{ s}) = 1 - 0.04(1)(10-1) = 1 - 0.36 = 0.64 \approx 0.6\text{ J}$$

where the second significant figure is not to be taken too seriously.

(b) At t = 5.0 s, the above method gives $K(5.0\text{ s}) = 1 - 0.04(5)(10-5) = 1 - 1 = 0.$

(c) The work done by the force from the wind from t = 1.0 s to t = 5.0 s is

$$W = K(5.0) - K(1.0\text{ s}) = 0 - 0.6 \approx -0.6\text{ J}.$$

LEARN The result in (c) can also be obtained by evaluating $W(t) = -0.04t(10-t)$ directly at t = 5.0 s and t = 1.0 s, and subtracting:

$$W(5) - W(1) = -0.04(5)(10-5) - [-0.04(1)(10-1)] = -1 - (-0.36) = -0.64 \approx -0.6\text{ J}.$$

Note that at t = 5.0 s, K = 0, the box comes to a stop and then reverses its direction subsequently (with x decreasing).

Chapter 8

8-1

THINK A compressed spring stores potential energy. This exercise explores the relationship between the energy stored and the spring constant.

EXPRESS The potential energy stored by the spring is given by $U = kx^2/2$, where k is the spring constant and x is the displacement of the end of the spring from its position when the spring is in equilibrium. Thus, the spring constant is $k = 2U/x^2$.

ANALYZE Substituting $U = 25$ J and $x = 7.5$ m $= 0.075$ cm into the above expression, we find the spring constant to be

$$k = \frac{2U}{x^2} = \frac{2(25\text{ J})}{(0.075\text{ m})^2} = 8.9 \times 10^3\text{ N/m}.$$

LEARN The spring constant k has units N/m. The quantity provides a measure of stiffness of the spring, for a given x, the greater the value of k, the greater the potential energy U.

8-5

THINK As the ice flake slides down the frictionless bowl, its potential energy changes due to the work done by the gravitational force.

EXPRESS The force of gravity is constant, so the work it does is given by $W = \vec{F} \cdot \vec{d}$, where $\vec{F}$ is the force and $\vec{d}$ is the displacement. The force is vertically downward and has magnitude mg, where m is the mass of the flake, so this reduces to $W = mgh$, where h is the height from which the flake falls. The force of gravity is conservative, so the change in gravitational potential energy of the flake-Earth system is the negative of the work done: $\Delta U = -W$.

ANALYZE (a) The ice flake falls a distance $h = r = 22.0$ cm $= 0.22$ m. Therefore, the work done by gravity is

$$W = mgr = (2.00 \times 10^{-3}\text{ kg})(9.8\text{ m/s}^2)(22.0 \times 10^{-2}\text{ m}) = 4.31 \times 10^{-3}\text{ J}.$$

(b) The change in gravitational potential energy is $\Delta U = -W = -4.31 \times 10^{-3}$ J.

(c) The potential energy when the flake is at the top is greater than when it is at the bottom by $|\Delta U|$. If $U = 0$ at the bottom, then $U = +4.31 \times 10^{-3}$ J at the top.

(d) If $U = 0$ at the top, then $U = -4.31 \times 10^{-3}$ J at the bottom.

(e) All the answers are proportional to the mass of the flake. If the mass is doubled, all answers are doubled.

LEARN While the potential energy depends on the reference point (location where $U = 0$), the change in potential energy, ΔU, does not. In both (c) and (d), we find $\Delta U = -4.31\times10^{-3}$ J.

8-11

THINK As the ice flake slides down the frictionless bowl, its potential energy decreases (discussed in Problem 8-5). By conservation of mechanical energy, its kinetic energy must increase.

EXPRESS If K_i is the kinetic energy of the flake at the edge of the bowl, K_f is its kinetic energy at the bottom, U_i is the gravitational potential energy of the flake-Earth system with the flake at the top, and U_f is the gravitational potential energy with it at the bottom, then

$$K_f + U_f = K_i + U_i .$$

Taking the potential energy to be zero at the bottom of the bowl, then the potential energy at the top is $U_i = mgr$ where $r = 0.220$ m is the radius of the bowl and m is the mass of the flake. $K_i = 0$ since the flake starts from rest. Since the problem asks for the speed at the bottom, we write $K_f = mv^2/2$.

ANALYZE (a) Energy conservation leads to

$$K_f + U_f = K_i + U_i \;\Rightarrow\; \frac{1}{2}mv^2 + 0 = 0 + mgr .$$

The speed is $v = \sqrt{2gr} = 2.08$ m/s.

(b) Since the expression for speed is $v = \sqrt{2gr}$, which does not contain the mass of the flake, the speed would be the same, 2.08 m/s, regardless of the mass of the flake.

(c) The final kinetic energy is given by $K_f = K_i + U_i - U_f$. If K_i is greater than before, then K_f will also be greater. This means the final speed of the flake is greater.

LEARN The mechanical energy conservation principle can also be expressed as $\Delta E_{\text{mech}} = \Delta K + \Delta U = 0$, which implies $\Delta K = -\Delta U$, i.e., the increase in kinetic energy is equal to the negative of the change in potential energy.

8-13

THINK As the marble moves vertically upward, its gravitational potential energy increases. This energy comes from the release of elastic potential energy stored in the spring.

EXPRESS We take the reference point for gravitational potential energy to be at the position of the marble when the spring is compressed. The gravitational potential energy when the marble is at the top of its motion is $U_g = mgh$. On the other had, the energy stored in the spring is $U_s = kx^2/2$. Applying mechanical energy conservation principle allows us to solve the problem.

ANALYZE (a) The height of the highest point is h = 20 m. With initial gravitational potential energy set to zero, we find

$$\Delta U_g = mgh = (5.0\times10^{-3}\text{ kg})(9.8\text{ m/s}^2)(20\text{ m}) = 0.98\text{ J}.$$

(b) Since the kinetic energy is zero at the release point and at the highest point, then conservation of mechanical energy implies $\Delta U_g + \Delta U_s = 0$, where ΔU_s is the change in the spring's elastic potential energy. Therefore, $\Delta U_s = -\Delta U_g = -0.98$ J.

(c) We take the spring potential energy to be zero when the spring is relaxed. Then, our result in the previous part implies that its initial potential energy is U_s = 0.98 J. This must be $\frac{1}{2}kx^2$, where k is the spring constant and x is the initial compression. Consequently,

$$k = \frac{2U_s}{x^2} = \frac{0.98\text{ J}}{(0.080\text{ m})^2} = 3.1\times10^2\text{ N/m} = 3.1\text{ N/cm}.$$

LEARN In general, the marble has both kinetic and potential energies:

$$\frac{1}{2}kx^2 = \frac{1}{2}mv^2 + mgy$$

At the maximum height $y_{\text{max}} = h,\ v = 0$ and $mgh = kx^2/2$, or $h = \dfrac{kx^2}{2mg}$.

8-15

THINK The truck with failed brakes is moving up an escape ramp. In order for it to come to a complete stop, all of its kinetic energy must be converted into gravitational potential energy.

EXPRESS We ignore any work done by friction. In SI units, the initial speed of the truck just before entering the escape ramp is v_i = 130(1000/3600) = 36.1 m/s. When the truck comes to a stop, its kinetic and potential energies are K_f = 0 and $U_f = mgh$. We apply mechanical energy conservation to solve the problem.

ANALYZE (a) Energy conservation implies $K_f + U_f = K_i + U_i$. With U_i = 0, and $K_i = \frac{1}{2}mv_i^2$, we obtain

$$\frac{1}{2}mv_i^2 + 0 = 0 + mgh \quad \Rightarrow \quad h = \frac{v_i^2}{2g} = \frac{(36.1\text{ m/s})^2}{2(9.8\text{ m/s}^2)} = 66.5\text{ m}.$$

If L is the minimum length of the ramp, then $L\sin\theta = h$, or L sin 15° = 66.5 m so that $L = (66.5\text{ m})/\sin 15° = 257\text{ m}$. That is, the ramp must be about 2.6×10^2 m long if friction is negligible.

(b) The minimum length is

$$L = \frac{h}{\sin\theta} = \frac{v_i^2}{2g\sin\theta}$$

which does not depend on the mass of the truck. Thus, the answer remains the same if the mass is reduced.

(c) If the speed is decreased, then h and L both decrease (note that h is proportional to the square of the speed and that L is proportional to h).

LEARN The greater the speed of the truck, the longer the ramp required. This length can be shortened considerably if the friction between the tires and the ramp surface is factored in.

8-29

THINK As the block slides down the inclined plane, it compresses the spring, then stops momentarily before sliding back up again.

EXPRESS We refer to its starting point as A, the point where it first comes into contact with the spring as B, and the point where the spring is compressed by $x_0 = 0.055$ m as C (see the figure below). Point C is our reference point for computing gravitational potential energy. Elastic potential energy (of the spring) is zero when the spring is relaxed.

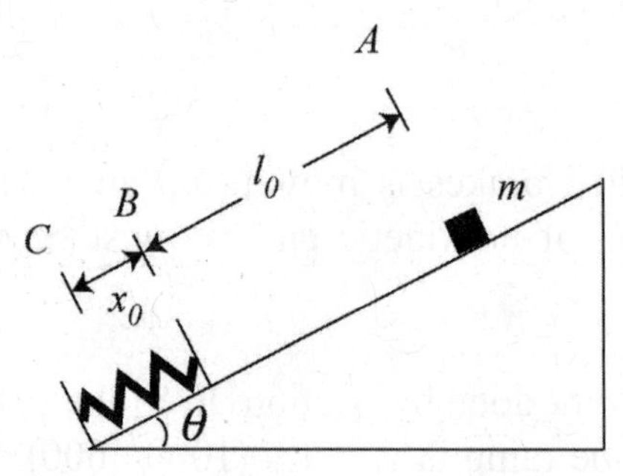

Information given in the second sentence allows us to compute the spring constant. From Hooke's law, we find

$$k = \frac{F}{x} = \frac{270 \text{ N}}{0.02 \text{ m}} = 1.35 \times 10^4 \text{ N/m}.$$

The distance between points A and B is l_0 and we note that the total sliding distance $l_0 + x_0$ is related to the initial height h_A of the block (measured relative to C) by $\sin\theta = \frac{h_A}{l_0 + x_0}$, where the incline angle θ is 30°.

ANALYZE (a) Mechanical energy conservation leads to

$$K_A + U_A = K_C + U_C \Rightarrow 0 + mgh_A = \frac{1}{2}kx_0^2$$

which yields

$$h_A = \frac{kx_0^2}{2mg} = \frac{(1.35\times10^4 \text{ N/m})(0.055 \text{ m})^2}{2(12 \text{ kg})(9.8 \text{ m/s}^2)} = 0.174 \text{ m}.$$

Therefore, the total distance traveled by the block before coming to a stop is

$$l_0 + x_0 = \frac{h_A}{\sin 30°} = \frac{0.174 \text{ m}}{\sin 30°} = 0.347 \text{ m} \approx 0.35 \text{ m}.$$

(b) From this result, we find $l_0 = x_0 = 0.347 \text{ m} - 0.055 \text{ m} = 0.292 \text{ m}$, which means that the block has descended a vertical distance

$$|\Delta y| = h_A - h_B = l_0 \sin\theta = (0.292 \text{ m})\sin 30° = 0.146 \text{ m}$$

in sliding from point A to point B. Thus, using Eq. 8-18, we have

$$0 + mgh_A = \frac{1}{2}mv_B^2 + mgh_B \Rightarrow \frac{1}{2}mv_B^2 = mg|\Delta y|$$

which yields $v_B = \sqrt{2g|\Delta y|} = \sqrt{2(9.8 \text{ m/s}^2)(0.146 \text{ m})} = 1.69 \text{ m/s} \approx 1.7 \text{ m/s}$.

LEARN Energy is conserved in the process. The total energy of the block at position B is

$$E_B = \frac{1}{2}mv_B^2 + mgh_B = \frac{1}{2}(12 \text{ kg})(1.69 \text{ m/s})^2 + (12 \text{ kg})(9.8 \text{ m/s}^2)(0.028 \text{ m}) = 20.4 \text{ J},$$

which is equal to the elastic potential energy in the spring:

$$\frac{1}{2}kx_0^2 = \frac{1}{2}(1.35\times10^4 \text{ N/m})(0.055 \text{ m})^2 = 20.4 \text{ J}.$$

8-45

THINK Work is done against friction while pulling a block along the floor at a constant speed.

EXPRESS Place the x-axis along the path of the block and the y-axis normal to the floor. The free-body diagram is shown below. The x and the y component of Newton's second law are

$$x: \quad F\cos\theta - f = 0$$
$$y: \quad F_N + F\sin\theta - mg = 0,$$

where m is the mass of the block, F is the force exerted by the rope, f is the magnitude of the kinetic friction force, and θ is the angle between that force and the horizontal.

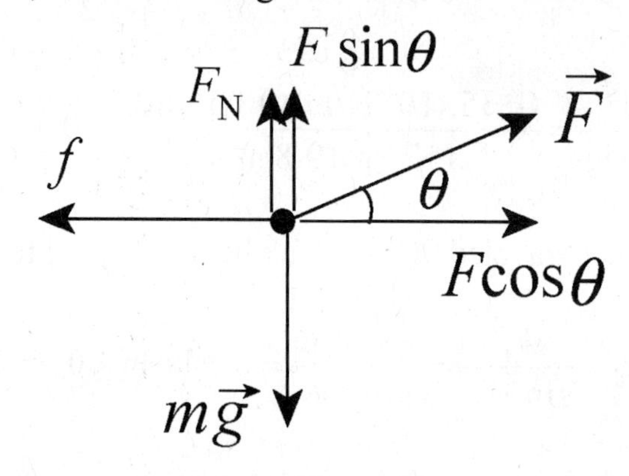

The work done on the block by the force in the rope is $W = Fd\cos\theta$. Similarly, the increase in thermal energy of the block-floor system due to the frictional force is given by Eq. 8-29, $\Delta E_{\text{th}} = fd$.

ANALYZE (a) Substituting the values given, we find the work done on the block by the rope's force to be

$$W = Fd\cos\theta = (7.68\,\text{N})(4.06\,\text{m})\cos 15.0° = 30.1\ \text{J}.$$

(b) The increase in thermal energy is $\Delta E_{\text{th}} = fd = (7.42\,\text{N})(4.06\,\text{m}) = 30.1\ \text{J}$.

(c) We can use Newton's second law of motion to obtain the frictional and normal forces, then use $\mu_k = f/F_N$ to obtain the coefficient of friction. The x-component of Newton's law gives

$$f = F\cos\theta = (7.68\ \text{N})\cos 15.0° = 7.42\ \text{N}.$$

Similarly, the y-component yields

$$F_N = mg - F\sin\theta = (3.57\ \text{kg})(9.8\ \text{m/s}^2) - (7.68\ \text{N})\sin 15.0° = 33.0\ \text{N}.$$

Thus, the coefficient of kinetic friction is

$$\mu_k = \frac{f}{F_N} = \frac{7.42\text{ N}}{33.0\text{ N}} = 0.225.$$

LEARN In this problem, the block moves at a constant speed so that $\Delta K = 0$, i.e., no change in kinetic energy. The work done by the external force is converted into thermal energy of the system, $W = \Delta E_{\text{th}}$.

8-49

THINK As the bear slides down the tree, its gravitational potential energy is converted into both kinetic energy and thermal energy.

EXPRESS We take the initial gravitational potential energy to be $U_i = mgL$, where L is the length of the tree, and final gravitational potential energy at the bottom to be $U_f = 0$. To solve this problem, we note that the changes in the mechanical and thermal energies must sum to zero.

ANALYZE (a) Substituting the values given, the change in gravitational potential energy is

$$\Delta U = U_f - U_i = -mgL = -(25\text{ kg})(9.8\text{ m/s}^2)(12\text{ m}) = -2.9\times 10^3\text{ J}.$$

(b) The final speed is $v_f = 5.6\text{ m/s}$. Therefore, the kinetic energy is

$$K_f = \frac{1}{2}mv_f^2 = \frac{1}{2}(25\text{ kg})(5.6\text{ m/s})^2 = 3.9\times 10^2\text{ J}.$$

(c) The change in thermal energy is $\Delta E_{\text{th}} = fL$, where f is the magnitude of the average frictional force; therefore, from $\Delta E_{\text{th}} + \Delta K + \Delta U = 0$, we find f to be

$$f = -\frac{\Delta K + \Delta U}{L} = -\frac{3.9\times 10^2\text{ J} - 2.9\times 10^3\text{ J}}{12\text{ m}} = 2.1\times 10^2\text{ N}.$$

LEARN In this problem, no external work is done to the bear. Therefore,

$$W = \Delta E_{\text{th}} + \Delta E_{\text{mech}} = \Delta E_{\text{th}} + \Delta K + \Delta U = 0,$$

which implies $\Delta K = -\Delta U - \Delta E_{\text{th}} = -\Delta U - fL$. Thus, $\Delta E_{\text{th}} = fL$ can be interpreted as the additional change (decrease) in kinetic energy due to frictional force.

8-67

THINK As the block is projected up the inclined plane, its kinetic energy is converted into gravitational potential energy and elastic potential energy of the spring. The block compresses the spring, stopping momentarily before sliding back down again.

EXPRESS Let *A* be the starting point and the reference point for computing gravitational potential energy ($U_A = 0$). The block first comes into contact with the spring at *B*. The spring is compressed by an additional amount x at *C*, as shown in the figure below.

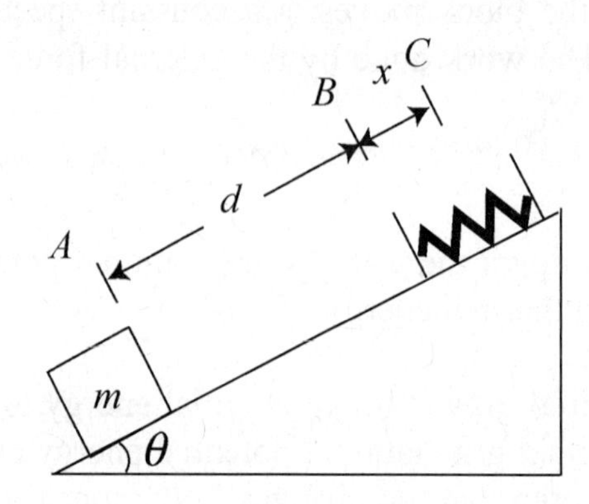

By energy conservation, $K_A + U_A = K_B + U_B = K_C + U_C$. Note that

$$U = U_g + U_s = mgy + \frac{1}{2}kx^2,$$

i.e., the total potential energy is the sum of gravitational potential energy and elastic potential energy of the spring.

ANALYZE (a) At the instant when $x_C = 0.20 \text{ m}$, the vertical height is

$$y_C = (d + x_C)\sin\theta = (0.60 \text{ m} + 0.20 \text{ m})\sin 40° = 0.514 \text{ m}.$$

Applying energy conservation principle gives

$$K_A + U_A = K_C + U_C \Rightarrow \quad 16 \text{ J} + 0 = K_C + mgy_C + \frac{1}{2}kx_C^2$$

from which we obtain

$$\begin{aligned} K_C &= K_A - mgy_C - \frac{1}{2}kx_C^2 \\ &= 16 \text{ J} - (1.0 \text{ kg})(9.8 \text{ m/s}^2)(0.514 \text{ m}) - \frac{1}{2}(200 \text{ N/m})(0.20 \text{ m})^2 = 6.96 \text{ J} \approx 7.0 \text{ J}. \end{aligned}$$

(b) At the instant when $x'_C = 0.40 \text{ m}$, the vertical height is

$$y'_C = (d + x'_C)\sin\theta = (0.60 \text{ m} + 0.40 \text{ m})\sin 40° = 0.64 \text{ m}.$$

Applying energy conservation principle, we have $K'_A + U'_A = K'_C + U'_C$. Since $U'_A = 0$, the initial kinetic energy that gives $K'_C = 0$ is

$$K'_A = U'_C = mgy'_C + \frac{1}{2}kx'^2_C$$
$$= (1.0\,\text{kg})(9.8\,\text{m/s}^2)(0.64\,\text{m}) + \frac{1}{2}(200\,\text{N/m})(0.40\,\text{m})^2$$
$$= 22\,\text{J}.$$

LEARN Comparing the results found in (a) and (b), we see that more kinetic energy is required to move the block higher in the inclined plane to achieve a greater spring compression.

8-69

THINK The two blocks are connected by a cord. As block *B* falls, block *A* moves up the incline.

EXPRESS If the larger mass (block *B*, m_B = 2.0 kg) falls a vertical distance $d = 0.25$ m, then the smaller mass (blocks *A*, m_A = 1.0 kg) must increase its height by $h = d\sin 30°$. The change in gravitational potential energy is

$$\Delta U = -m_B g d + m_A g h.$$

By mechanical energy conservation, $\Delta E_{\text{mech}} = \Delta K + \Delta U = 0$, the change in kinetic energy of the system is $\Delta K = -\Delta U$.

ANALYZE Since the initial kinetic energy is zero, the final kinetic energy is

$$K_f = \Delta K = m_B g d - m_A g h = m_B g d - m_A g d \sin\theta$$
$$= (m_B - m_A \sin\theta) g d = [2.0\,\text{kg} - (1.0\,\text{kg})\sin 30°](9.8\,\text{m/s}^2)(0.25\,\text{m})$$
$$= 3.7\,\text{J}.$$

LEARN From the above expression, we see that in the special case where $m_B = m_A \sin\theta$, the two-block system would remain stationary. On the other hand, if $m_A \sin\theta > m_B$, block *A* will slide down the incline, with block *B* moving vertically upward.

8-71

THINK As the block slides down the frictionless incline, its gravitational potential energy is converted to kinetic energy, so the speed of the block increases.

EXPRESS By energy conservation, $K_A + U_A = K_B + U_B$. Thus, the change in kinetic energy as the block moves from points *A* to *B* is

$$\Delta K = K_B - K_A = -\Delta U = -(U_B - U_A).$$

In both circumstances, we have the same potential energy change. Thus, $\Delta K_1 = \Delta K_2$.

ANALYZE With $\Delta K_1 = \Delta K_2$, the speed of the block at B the second time is given by

$$\frac{1}{2}mv_{B,1}^2 - \frac{1}{2}mv_{A,1}^2 = \frac{1}{2}mv_{B,2}^2 \frac{1}{2}mv_{A,2}^2$$

or

$$v_{B,2} = \sqrt{v_{B,1}^2 - v_{A,1}^2 + v_{A,2}^2} = \sqrt{(2.60\ \text{m/s})^2 - (2.00\ \text{m/s})^2 + (4.00\ \text{m/s})^2} = 4.33\ \text{m/s}.$$

LEARN The speed of the block at A is greater the second time, $v_{A,2} > v_{A,1}$. This can happen if the block slides down from a higher position with greater initial gravitational potential energy.

8-73

THINK As the cube is pushed across the floor, both the thermal energies of floor and the cube increase because of friction.

EXPRESS By law of conservation of energy, we have $W = \Delta E_{\text{mech}} + \Delta E_{\text{th}}$ for the floor-cube system. Since the speed is constant, $\Delta K = 0$, Eq. 8-33 (an application of the energy conservation concept) implies

$$W = \Delta E_{\text{mech}} + \Delta E_{\text{th}} = \Delta E_{\text{th}} = \Delta E_{\text{th (cube)}} + \Delta E_{\text{th (floor)}}.$$

ANALYZE With W = (15 N)(3.0 m) = 45 J, and we are told that $\Delta E_{\text{th (cube)}}$ = 20 J, then we conclude that $\Delta E_{\text{th (floor)}}$ = 25 J.

LEARN The applied work here has all been converted into thermal energies of the floor and the cube. The amount of thermal energy transferred to a material depends on its thermal properties, as we shall discuss in Chapter 18.

8-75

THINK This problem deals with pendulum motion. The kinetic and potential energies of the ball attached to the rod change with position, but the mechanical energy remains conserved throughout the process.

EXPRESS Let L be the length of the pendulum. The connection between angle θ (measured from vertical) and height h (measured from the lowest point, which is our choice of reference position in computing the gravitational potential energy mgh) is given by $h = L(1 - \cos\theta)$. The free-body diagram is shown below. The initial height is at $h_1 = 2L$, and at the lowest point, we have $h_2 = 0$. The total mechanical energy is conserved throughout.

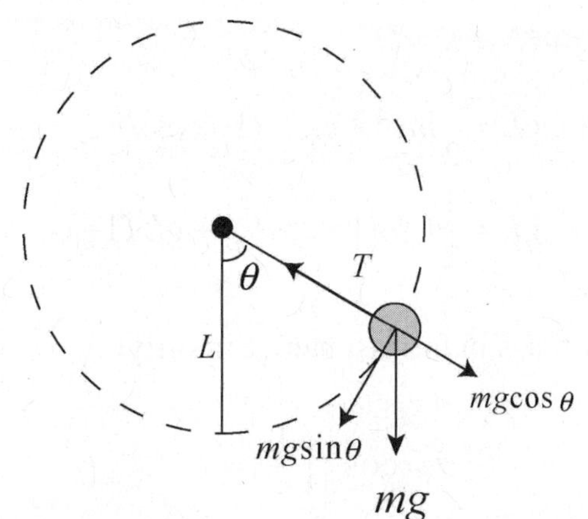

ANALYZE (a) Initially the ball is at $h_1 = 2L$ with $K_1 = 0$ and $U_1 = mgh_1 = mg(2L)$. At the lowest point $h_2 = 0$, we have $K_2 = \frac{1}{2}mv_2^2$ and $U_2 = 0$. Using energy conservation in the form of Eq. 8-17 leads to

$$K_1 + U_1 = K_2 + U_2 \quad \Rightarrow \quad 0 + 2mgL = \frac{1}{2}mv_2^2 + 0$$

This leads to $v_2 = 2\sqrt{gL}$. With $L = 0.62$ m, we have

$$v_2 = 2\sqrt{(9.8 \text{ m/s}^2)(0.62 \text{ m})} = 4.9 \text{ m/s}.$$

(b) At the lowest point, the ball is in circular motion with the center of the circle above it, so $\vec{a} = v^2 / r$ upward, where $r = L$. Newton's second law leads to

$$T - mg = m\frac{v^2}{r} \Rightarrow T = m\left(g + \frac{4gL}{L}\right) = 5\ mg.$$

With $m = 0.092$ kg, the tension is $T = 4.5$ N.

(c) The pendulum is now started (with zero speed) at $\theta_i = 90°$ (that is, $h_i = L$), and we look for an angle θ such that $T = mg$. When the ball is moving through a point at angle θ, as can be seen from the free-body diagram shown above, Newton's second law applied to the axis along the rod yields

$$\frac{mv^2}{r} = T - mg\cos\theta = mg(1 - \cos\theta)$$

which (since $r = L$) implies $v^2 = gL(1 - \cos\theta)$ at the position we are looking for. Energy conservation leads to

$$K_i + U_i = K + U$$

$$0 + mgL = \frac{1}{2}mv^2 + mgL\ (1 - \cos\theta)$$

$$gL = \frac{1}{2}(gL(1 - \cos\theta)) + gL\ (1 - \cos\theta)$$

where we have divided by mass in the last step. Simplifying, we obtain

$$\theta = \cos^{-1}\left(\frac{1}{3}\right) = 71°.$$

(d) Since the angle found in (c) is independent of the mass, the result remains the same if the mass of the ball is changed.

LEARN At a given angle θ with respect to the vertical, the tension in the rod is

$$T = m\left(\frac{v^2}{r} + g\cos\theta\right)$$

The tangential acceleration, $a_t = g\sin\theta$, is what causes the speed and, therefore, the kinetic energy to change with time. Nonetheless, mechanical energy is conserved.

8-77

THINK This problem involves graphical analyses. From the graph of potential energy as a function of position, the conservative force can de deduced.

EXPRESS The connection between the potential energy function $U(x)$ and the conservative force $F(x)$ is given by Eq. 8-22: $F(x) = -dU/dx$. A positive slope of $U(x)$ at a point means that $F(x)$ is negative, and vice versa.

ANALYZE (a) The force at $x = 2.0$ m is

$$F = -\frac{dU}{dx} \approx -\frac{\Delta U}{\Delta x} = -\frac{U(x=4\text{ m}) - U(x=1\text{ m})}{4.0\text{ m} - 1.0\text{ m}} = -\frac{-(17.5\text{ J}) - (-2.8\text{ J})}{4.0\text{ m} - 1.0\text{ m}} = 4.9\text{ N}.$$

(b) Since the slope of $U(x)$ at $x = 2.0$ m is negative, the force points in the $+x$ direction (but there is some uncertainty in reading the graph which makes the last digit not very significant).

(c) At $x = 2.0$ m, we estimate the potential energy to be

$$U(x = 2.0\text{ m}) \approx U(x = 1.0\text{ m}) + (-4.9\text{ J/m})(1.0\text{ m}) = -7.7\text{ J}$$

Thus, the total mechanical energy is

$$E = K + U = \frac{1}{2}mv^2 + U = \frac{1}{2}(2.0\text{ kg})(-1.5\text{ m/s})^2 + (-7.7\text{ J}) = -5.5\text{ J}.$$

Again, there is some uncertainty in reading the graph which makes the last digit not very significant. At that level (–5.5 J) on the graph, we find two points where the potential energy curve has that value — at $x \approx 1.5$ m and $x \approx 13.5$ m. Therefore, the particle remains in the region $1.5 < x < 13.5$ m. The left boundary is at $x = 1.5$ m.

(d) From the above results, the right boundary is at $x = 13.5$ m.

(e) At $x = 7.0$ m, we read $U \approx -17.5$ J. Thus, if its total energy (calculated in the previous part) is $E \approx -5.5$ J, then we find

$$\frac{1}{2}mv^2 = E - U \approx 12\text{ J} \Rightarrow v = \sqrt{\frac{2}{m}(E - U)} \approx 3.5\text{ m/s}$$

where there is certainly room for disagreement on that last digit for the reasons cited above.

LEARN Since the total mechanical energy is negative, the particle is bounded by the potential, with its motion confined to the region $1.5\text{ m} < x < 13.5$ m. At the turning points (1.5 m and 13.5 m), kinetic energy is zero and the particle is momentarily at rest.

8-79

THINK As the car slides down the incline, due to the presence of frictional force, some of its mechanical energy is converted into thermal energy.

EXPRESS The incline angle is $\theta = 5.0°$. Thus, the change in height between the car's highest and lowest points is $\Delta y = -(50\text{ m})\sin\theta = -4.4$ m. We take the lowest point (the car's final reported location) to correspond to the $y = 0$ reference level. The change in potential energy is given by $\Delta U = mg\Delta y$.

As for the kinetic energy, we first convert the speeds to SI units, $v_0 = 8.3$ m/s and $v = 11.1$ m/s. The change in kinetic energy is $\Delta K = \frac{1}{2}m(v_f^2 - v_i^2)$. The total change in mechanical energy is $\Delta E_{\text{mech}} = \Delta K + \Delta U$.

ANALYZE (a) Substituting the values given, we find ΔE_{mech} to be

$$\Delta E_{\text{mech}} = \Delta K + \Delta U = \frac{1}{2}m(v_f^2 - v_i^2) + mg\Delta y$$
$$= \frac{1}{2}(1500\text{ kg})\left[(11.1\text{ m/s})^2 - (8.3\text{ m/s})^2\right] + (1500\text{ kg})(9.8\text{ m/s}^2)(-4.4\text{ m})$$
$$= -23940\text{ J} \approx -2.4\times10^4\text{ J}$$

That is, the mechanical energy decreases (due to friction) by 2.4×10^4 J.

(b) Using Eq. 8-31 and Eq. 8-33, we find $\Delta E_{\text{th}} = f_k d = -\Delta E_{\text{mech}}$. With d = 50 m, we solve for f_k and obtain

$$f_k = \frac{-\Delta E_{\text{mech}}}{d} = \frac{-(-2.4\times10^4\text{ J})}{50\text{ m}} = 4.8\times10^2\text{ N}.$$

LEARN The amount of mechanical energy lost is proportional to the frictional force; in the absence of friction, mechanical energy would have been conserved.

8-83

THINK Energy is transferred from an external agent to the block so that its speed continues to increase.

EXPRESS According to Eq. 8-25, the work done by the external force is $W = \Delta E_{\text{mech}} = \Delta K + \Delta U$. When there is no change in potential energy, $\Delta U = 0$, the expression simplifies to

$$W = \Delta E_{\text{mech}} = \Delta K = \frac{1}{2}m(v_f^2 - v_i^2).$$

The average power, or average rate of work done, is given by $P_{\text{avg}} = W/\Delta t$.

ANALYZE (a) Substituting the values given, the change in mechanical energy is

$$\Delta E_{\text{mech}} = \Delta K = \frac{1}{2}m(v_f^2 - v_i^2) = \frac{1}{2}(15\text{ kg})[(30\text{ m/s})^2 - (10\text{ m/s})^2] = 6000\text{ J} = 6.0\times10^3\text{ J}$$

(b) From the above, we have W = 6.0 × 10^3 J. Also, from Chapter 2, we know that $\Delta t = \Delta v/a = 10$ s. Thus, using Eq. 7-42, the average rate at which energy is transferred to the block is

$$P_{\text{avg}} = \frac{W}{\Delta t} = \frac{6.0\times10^3\text{ J}}{10.0\text{ s}} = 600\text{ W}.$$

(c) and (d) The constant applied force is $F = ma$ = 30 N and clearly in the direction of motion, so Eq. 7-48 provides the results for instantaneous power:

$$P = \vec{F} \cdot \vec{v} = \begin{cases} 300 \text{ W} & \text{for } v = 10 \text{ m/s} \\ 900 \text{ W} & \text{for } v = 30 \text{ m/s} \end{cases}$$

LEARN The average of these two values found in (c) and (d) agrees with the result in part (b). Note that the expression for the instantaneous rate used above can be derived from:

$$P = \frac{dW}{dt} = \frac{d}{dt}\left(\frac{1}{2}mv^2\right) = m\vec{v} \cdot \frac{d\vec{v}}{dt} = m\vec{v} \cdot \vec{a} = \vec{F} \cdot \vec{v}$$

8-85

THINK This problem deals with the concept of hydroelectric generator – kinetic energy of water can be converted into electrical energy.

EXPRESS By energy conservation, the change in kinetic energy of water in one second is

$$\Delta K = -\Delta U = mgh = \rho Vgh = (10^3 \text{ kg/m}^3)(1200 \text{ m}^3)(9.8 \text{ m/s}^2)(100 \text{ m}) = 1.176 \times 10^9 \text{ J}$$

Only 3/4 of this amount is transferred to electrical energy.

ANALYZE The power generation (assumed constant, so average power is the same as instantaneous power) is

$$P_{\text{avg}} = \frac{(3/4)\Delta K}{t} = \frac{(3/4)(1.176 \times 10^9 \text{ J})}{1.0 \text{ s}} = 8.82 \times 10^8 \text{ W}.$$

LEARN Hydroelectricity is the most widely used renewable energy; it accounts for almost 20% of the world's electricity supply.

8-87

THINK We have a ball attached to a rod that moves in a vertical circle. The total mechanical energy of the system is conserved.

EXPRESS Let position A be the reference point for potential energy, $U_A = 0$. The total mechanical energies at A, B and C are:

$$E_A = \frac{1}{2}mv_A^2 + U_A = \frac{1}{2}mv_0^2$$

$$E_B = \frac{1}{2}mv_B^2 + U_B = \frac{1}{2}mv_B^2 - mgL$$

$$E_D = \frac{1}{2}mv_D^2 + U_D = mgL$$

where $v_D = 0$. The problem can be analyzed by applying energy conservation: $E_A = E_B = E_D$.

ANALYZE (a) The condition $E_A = E_D$ gives

$$\frac{1}{2}mv_0^2 = mgL \Rightarrow v_0 = \sqrt{2gL}$$

(b) To find the tension in the rod when the ball passes through *B*, we first calculate the speed at *B*. Using $E_B = E_D$, we find

$$\frac{1}{2}mv_B^2 - mgL = mgL$$

or $v_B = \sqrt{4gL}$. The direction of the centripetal acceleration is upward (at that moment), as is the tension force. Thus, Newton's second law gives

$$T - mg = \frac{mv_B^2}{r} = \frac{m(4gL)}{L} = 4mg$$

or $T = 5mg$.

(c) The difference in height between *C* and *D* is *L*, so the "loss" of mechanical energy (which goes into thermal energy) is $-mgL$.

(d) The difference in height between *B* and *D* is 2*L*, so the total "loss" of mechanical energy (which all goes into thermal energy) is $-2mgL$.

LEARN An alternative way to calculate the energy loss in (d) is to note that

$$E_B' = \frac{1}{2}mv_B'^2 + U_B = 0 - mgL = -mgL$$

which gives

$$\Delta E = E_B' - E_A = -mgL - mgL = -2mgL.$$

8-119

THINK We apply energy method to analyze the projectile motion of a ball.

EXPRESS We choose the initial position at the window to be our reference point for calculating the potential energy. The initial energy of the ball is $E_0 = \frac{1}{2}mv_0^2$. At the top of its flight, the vertical component of the velocity is zero, and the horizontal component (neglecting air friction) is the same as it was when it was thrown: $v_x = v_0 \cos\theta$. At a position *h* below the window, the energy of the ball is

$$E = K + U = \frac{1}{2}mv^2 - mgh$$

where v is the speed of the ball.

ANALYZE (a) The kinetic energy of the ball at the top of the flight is

$$K_{\text{top}} = \frac{1}{2}mv_x^2 = \frac{1}{2}m(v_0\cos\theta)^2 = \frac{1}{2}(0.050\text{ kg})[(8.0\text{ m/s})\cos 30°]^2 = 1.2\text{ J}.$$

(b) When the ball is $h = 3.0$ m below the window, by energy conservation, we have

$$\frac{1}{2}mv_0^2 = \frac{1}{2}mv^2 - mgh$$

or

$$v = \sqrt{v_0^2 + 2gh} = \sqrt{(8.0\text{ m/s})^2 + 2(9.8\text{ m/s}^2)(3.0\text{ m})} = 11.1\text{ m/s}.$$

(c) As can be seen from our expression above, $v = \sqrt{v_0^2 + 2gh}$, which is independent of the mass m.

(d) Similarly, the speed v is independent of the initial angle θ.

LEARN Our results demonstrate that the quantity v in the kinetic energy formula is the magnitude of the velocity vector; it does not depend on direction. In addition, mass cancels out in the energy conservation equation, so that v is independent of m.

8-122

THINK A shuffleboard disk is accelerated over some distance by an external force, but it eventually comes to rest due to the frictional force.

EXPRESS In the presence of frictional force, the work done on a system is $W = \Delta E_{\text{mech}} + \Delta E_{\text{th}}$, where $\Delta E_{\text{mech}} = \Delta K + \Delta U$ and $\Delta E_{\text{th}} = f_k d$. In our situation, work has been done by the cue only to the first 2.0 m, and not to the subsequent 12 m of distance traveled.

ANALYZE (a) During the final $d = 12$ m of motion, $W = 0$ and we use

$$K_1 + U_1 = K_2 + U_2 + f_k d$$
$$\frac{1}{2}mv_i^2 + 0 = 0 + 0 + f_k d$$

where $m = 0.42$ kg and $v = 4.2$ m/s. This gives $f_k = 0.31$ N. Therefore, the thermal energy change is $\Delta E_{\text{th}} = f_k d = 3.7$ J.

(b) Using $f_k = 0.31$ N for the entire distance $d_{\text{total}} = 14$ m, we obtain

$$\Delta E_{\text{th,total}} = f_k d_{\text{total}} = (0.31\ \text{N})(14\ \text{m}) = 4.3\ \text{J}$$

for the thermal energy generated by friction.

(c) During the initial $d' = 2$ m of motion, we have

$$W = \Delta E_{\text{mech}} + \Delta E'_{\text{th}} = \Delta K + \Delta U + f_k d' = \frac{1}{2}mv^2 + 0 + f_k d'$$

which essentially combines Eq. 8-31 and Eq. 8-33. Thus, the work done on the disk by the cue is

$$W = \frac{1}{2}mv^2 + f_k d' = \frac{1}{2}(0.42\ \text{kg})(4.2\ \text{m/s})^2 + (0.31\ \text{N})(2.0\ \text{m}) = 4.3\ \text{J}.$$

LEARN Our answer in (c) is the same as that in (b). This is expected because all the work done becomes thermal energy at the end.

Chapter 9

9-13

THINK A shell explodes into two segments at the top of its trajectory. Knowing the motion of one segment allows us to analyze the motion of the other using the momentum conservation principle.

EXPRESS We need to find the coordinates of the point where the shell explodes and the velocity of the fragment that does not fall straight down. The coordinate origin is at the firing point, the $+x$ axis is rightward, and the $+y$ direction is upward. The y component of the velocity is given by $v = v_{0y} - gt$ and this is zero at time $t = v_{0y}/g = (v_0/g)\sin\theta_0$, where v_0 is the initial speed and θ_0 is the firing angle. The coordinates of the highest point on the trajectory are

$$x = v_{0x}t = v_0 t\cos\theta_0 = \frac{v_0^2}{g}\sin\theta_0\cos\theta_0 = \frac{(20\ \text{m/s})^2}{9.8\ \text{m/s}^2}\sin 60°\cos 60° = 17.7\ \text{m}$$

and

$$y = v_{0y}t - \frac{1}{2}gt^2 = \frac{1}{2}\frac{v_0^2}{g}\sin^2\theta_0 = \frac{1}{2}\frac{(20\ \text{m/s})^2}{9.8\ \text{m/s}^2}\sin^2 60° = 15.3\ \text{m}.$$

Since no horizontal forces act, the horizontal component of the momentum is conserved. In addition, since one fragment has a velocity of zero after the explosion, the momentum of the other equals the momentum of the shell before the explosion. At the highest point the velocity of the shell is $v_0\cos\theta_0$, in the positive x direction. Let M be the mass of the shell and let V_0 be the velocity of the fragment. Then

$$Mv_0\cos\theta_0 = MV_0/2,$$

since the mass of the fragment is $M/2$. This means

$$V_0 = 2v_0\cos\theta_0 = 2(20\ \text{m/s})\cos 60° = 20\ \text{m/s}.$$

This information is used in the form of initial conditions for a projectile motion problem to determine where the fragment lands.

ANALYZE Resetting our clock, we now analyze a projectile launched horizontally at time $t = 0$ with a speed of 20 m/s from a location having coordinates $x_0 = 17.7$ m, $y_0 = 15.3$ m. Its y coordinate is given by $y = y_0 - \frac{1}{2}gt^2$, and when it lands this is zero. The time of landing is $t = \sqrt{2y_0/g}$ and the x coordinate of the landing point is

$$x = x_0 + V_0 t = x_0 + V_0 \sqrt{\frac{2y_0}{g}} = 17.7\text{ m} + (20\text{ m/s})\sqrt{\frac{2(15.3\text{ m})}{9.8\text{ m/s}^2}} = 53\text{ m}.$$

LEARN In the absence of explosion, the shell with a mass M would have landed at

$$R = 2x_0 = \frac{v_0^2}{g}\sin 2\theta_0 = \frac{(20\text{ m/s})^2}{9.8\text{ m/s}^2}\sin[2(60°)] = 35.3\text{ m}$$

which is shorter than $x = 53$ m found above. This makes sense because the broken fragment, having a smaller mass but greater horizontal speed, can travel much farther than the original shell.

9-27

THINK The velocity of the ball is changing because of the external force applied. Impulse-linear momentum theorem is involved.

EXPRESS The initial direction of motion is in the +x direction. The magnitude of the average force F_{avg} is given by

$$F_{\text{avg}} = \frac{J}{\Delta t} = \frac{32.4\text{ N}\cdot\text{s}}{2.70\times10^{-2}\text{ s}} = 1.20\times10^3\text{ N}.$$

The force is in the negative direction. Using the linear momentum-impulse theorem stated in Eq. 9-31, we have

$$-F_{\text{avg}}\Delta t = J = \Delta p = m(v_f - v_i).$$

where m is the mass, v_i the initial velocity, and v_f the final velocity of the ball. The equation can be used to solve for v_f.

ANALYZE (a) Using the above expression, we find

$$v_f = \frac{mv_i - F_{\text{avg}}\Delta t}{m} = \frac{(0.40\text{ kg})(14\text{ m/s}) - (1200\text{ N})(27\times10^{-3}\text{ s})}{0.40\text{ kg}} = -67\text{ m/s}.$$

The final speed of the ball is $|v_f| = 67$ m/s.

(b) The negative sign in v_f indicates that the velocity is in the $-x$ direction, which is opposite to the initial direction of travel.

(c) From the above, the average magnitude of the force is $F_{\text{avg}} = 1.20\times10^3$ N.

(d) The direction of the impulse on the ball is $-x$, same as the applied force.

LEARN In vector notation, $\vec{F}_{\text{avg}}\Delta t = \vec{J} = \Delta\vec{p} = m(\vec{v}_f - \vec{v}_i)$, which gives

$$\vec{v}_f = \vec{v}_i + \frac{\vec{J}}{m} = \vec{v}_i + \frac{\vec{F}_{\text{avg}}\Delta t}{m}$$

Since $\vec{J}$ or $\vec{F}_{\text{avg}}$ is in the opposite direction of $\vec{v}_i$, the velocity of the ball will decrease under the applied force. The ball first moves in the $+x$-direction, but then slows down and comes to a stop, and then reverses its direction of travel.

9-37

THINK We're given the force as a function of time, and asked to calculate the corresponding impulse, the average force and the maximum force.

EXPRESS Since the motion is one-dimensional, we work with the magnitudes of the vector quantities. The impulse J due to a force $F(t)$ exerted on a body is

$$J = \int_{t_i}^{t_f} F(t)dt = F_{\text{avg}}\Delta t,$$

where F_{avg} is the average force and $\Delta t = t_f - t_i$. To find the time at which the maximum force occurs, we set the derivative of F with respect to time equal to zero, and solve for t.

ANALYZE (a) We take the force to be in the positive direction, at least for earlier times. Then the impulse is

$$J = \int_0^{3.0\times10^{-3}} F dt = \int_0^{3.0\times10^{-3}} \left[(6.0\times10^6)t - (2.0\times10^9)t^2\right]dt$$

$$= \left[\frac{1}{2}(6.0\times10^6)t^2 - \frac{1}{3}(2.0\times10^9)t^3\right]\Bigg|_0^{3.0\times10^{-3}} = 9.0\,\text{N}\cdot\text{s}.$$

(b) Using $J = F_{\text{avg}}\,\Delta t$, we find the average force to be

$$F_{\text{avg}} \frac{J}{\Delta t} = \frac{9.0\text{ N}\cdot\text{s}}{3.0\times10^{-3}\text{ s}} = 3.0\times10^3\text{ N}.$$

(c) Differentiating $F(t)$ with respect to t and setting it to zero, we have

$$\frac{dF}{dt} = \frac{d}{dt}\left[(6.0\times10^6)t - (2.0\times10^9)t^2\right] = (6.0\times10^6) - (4.0\times10^9)t = 0,$$

which can be solved to give $t = 1.5\times10^{-3}$ s. At that time the force is

$$F_{max} = (6.0\times10^6)(1.5\times10^{-3}) - (2.0\times10^9)(1.5\times10^{-3})^2 = 4.5\times10^3\ \text{N}.$$

(d) Since it starts from rest, the ball acquires momentum equal to the impulse from the kick. Let m be the mass of the ball and v its speed as it leaves the foot. The speed of the ball immediately after it loses contact with the player's foot is

$$v = \frac{p}{m} = \frac{J}{m} = \frac{9.0\ \text{N}\cdot\text{s}}{0.45\ \text{kg}} = 20\ \text{m/s}.$$

LEARN The force as function of time is shown below. The area under the curve is the impulse J. From the plot, we readily see that $F(t)$ is a maximum at $t = 0.0015\,\text{s}$, with $F_{max} = 4500\ \text{N}$.

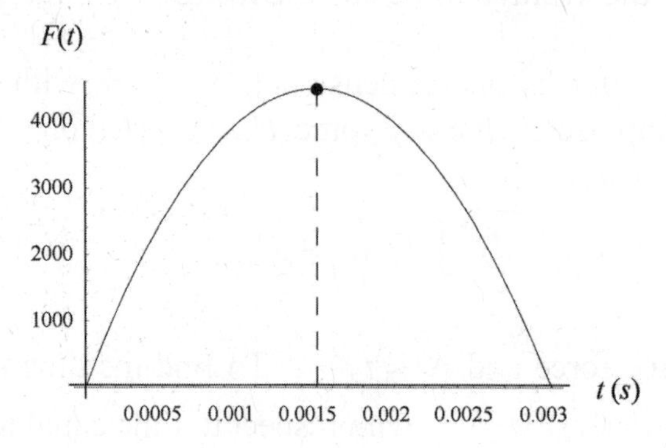

9-39

THINK This problem deals with momentum conservation. Since no external forces with horizontal components act on the man-stone system and the vertical forces sum to zero, the total momentum of the system is conserved.

EXPRESS Since the man and the stone are initially at rest, the total momentum is zero both before and after the stone is kicked. Let m_s be the mass of the stone and v_s be its velocity after it is kicked. Also, let m_m be the mass of the man and v_m be his velocity after he kicks the stone. Then, by momentum conservation,

$$m_s v_s + m_m v_m = 0 \Rightarrow v_m = -\frac{m_s}{m_m} v_s .$$

ANALYZE We take the axis to be positive in the direction of motion of the stone. Then

$$v_m = -\frac{m_s}{m_m} v_s = -\frac{0.068\ \text{kg}}{91\ \text{kg}}(4.0\ \text{m/s}) = -3.0\times10^{-3}\ \text{m/s}$$

or $|v_m| = 3.0\times10^{-3}\ \text{m/s}$.

LEARN The negative sign in v_m indicates that the man moves in the direction opposite to the motion of the stone. Note that his speed is much smaller (by a factor of m_s/m_m) compared to the speed of the stone.

9-45

THINK The moving body is an isolated system with no external force acting on it. When it breaks up into three pieces, momentum remains conserved, both in the *x*- and the *y*-directions.

EXPRESS Our notation is as follows: the mass of the original body is $M = 20.0$ kg; its initial velocity is $\vec{v}_0 = (200\text{ m/s})\hat{\text{i}}$; the mass of one fragment is $m_1 = 10.0$ kg; its velocity is $\vec{v}_1 = (100\text{ m/s})\hat{\text{j}}$; the mass of the second fragment is $m_2 = 4.0$ kg; its velocity is $\vec{v}_2 = (-500\text{ m/s})\hat{\text{i}}$; and, the mass of the third fragment is $m_3 = 6.00$ kg. Conservation of linear momentum requires

$$M\vec{v}_0 = m_1\vec{v}_1 + m_2\vec{v}_2 + m_3\vec{v}_3 .$$

The energy released in the explosion is equal to ΔK, the change in kinetic energy.

ANALYZE (a) The above momentum-conservation equation leads to

$$\begin{aligned}\vec{v}_3 &= \frac{M\vec{v}_0 - m_1\vec{v}_1 - m_2\vec{v}_2}{m_3} \\ &= \frac{(20.0\text{ kg})(200\text{ m/s})\hat{\text{i}} - (10.0\text{ kg})(100\text{ m/s})\hat{\text{j}} - (4.0\text{ kg})(-500\text{ m/s})\hat{\text{i}}}{6.00\text{ kg}} . \\ &= (1.00\times10^3\text{ m/s})\hat{\text{i}} - (0.167\times10^3\text{ m/s})\hat{\text{j}}\end{aligned}$$

The magnitude of $\vec{v}_3$ is $v_3 = \sqrt{(1000\text{ m/s})^2 + (-167\text{ m/s})^2} = 1.01\times10^3\text{ m/s}$. It points at $\theta = \tan^{-1}(-167/1000) = -9.48°$ (that is, at 9.5° measured clockwise from the $+x$ axis).

(b) The energy released is ΔK:

$$\Delta K = K_f - K_i = \left(\frac{1}{2}m_1v_1^2 + \frac{1}{2}m_2v_2^2 + \frac{1}{2}m_3v_3^2\right) - \frac{1}{2}Mv_0^2 = 3.23\times10^6\text{ J}.$$

LEARN The energy released in the explosion, of chemical nature, is converted into the kinetic energy of the fragments.

9-61

THINK We have a moving cart colliding with a stationary cart. Since the collision is elastic, the total kinetic energy of the system remains unchanged.

EXPRESS Let m_1 be the mass of the cart that is originally moving, v_{1i} be its velocity before the collision, and v_{1f} be its velocity after the collision. Let m_2 be the mass of the cart that is originally at rest and v_{2f} be its velocity after the collision. Conservation of linear momentum gives $m_1v_{1i} = m_1v_{1f} + m_2v_{2f}$. Similarly, the total kinetic energy is conserved and we have

$$\frac{1}{2}m_1v_{1i}^2 = \frac{1}{2}m_1v_{1f}^2 + \frac{1}{2}m_2v_{2f}^2.$$

Solving for v_{1f} and v_{2f}, we obtain:

$$v_{1f} = \frac{m_1 - m_2}{m_1 + m_2}v_{1i}, \quad v_{2f} = \frac{2m_1}{m_1 + m_2}v_{1i}$$

The speed of the center of mass is $v_{com} = \dfrac{m_1v_{1i} + m_2v_{2i}}{m_1 + m_2}$.

ANALYZE (a) With $m_1 = 0.34$ kg, $v_{1i} = 1.2$ m/s and $v_{1f} = 0.66$ m/s, we obtain

$$m_2 = \frac{v_{1i} - v_{1f}}{v_{1i} + v_{1f}}m_1 = \left(\frac{1.2\text{ m/s} - 0.66\text{ m/s}}{1.2\text{ m/s} + 0.66\text{ m/s}}\right)(0.34\text{ kg}) = 0.0987\text{ kg} \approx 0.099\text{ kg}.$$

(b) The velocity of the second cart is:

$$v_{2f} = \frac{2m_1}{m_1 + m_2}v_{1i} = \left(\frac{2(0.34\text{ kg})}{0.34\text{ kg} + 0.099\text{ kg}}\right)(1.2\text{ m/s}) = 1.9\text{ m/s}.$$

(c) From the above, we find the speed of the center of mass to be

$$v_{com} = \frac{m_1v_{1i} + m_2v_{2i}}{m_1 + m_2} = \frac{(0.34\text{ kg})(1.2\text{ m/s}) + 0}{0.34\text{ kg} + 0.099\text{ kg}} = 0.93\text{ m/s}.$$

LEARN In solving for v_{com}, values for the initial velocities were used. Since the system is isolated with no external force acting on it, v_{com} remains the same after the collision, so the same result is obtained if values for the final velocities are used. That is,

$$v_{com} = \frac{m_1v_{1f} + m_2v_{2f}}{m_1 + m_2} = \frac{(0.34\text{ kg})(0.66\text{ m/s}) + (0.099\text{ kg})(1.9\text{ m/s})}{0.34\text{ kg} + 0.099\text{ kg}} = 0.93\text{ m/s}.$$

9-65

THINK We have a mass colliding with another stationary mass. Since the collision is elastic, the total kinetic energy of the system remains unchanged.

EXPRESS Let m_1 be the mass of the body that is originally moving, v_{1i} be its velocity before the collision, and v_{1f} be its velocity after the collision. Let m_2 be the mass of the body that is originally at rest and v_{2f} be its velocity after the collision. Conservation of linear momentum gives

$$m_1 v_{1i} = m_1 v_{1f} + m_2 v_{2f}.$$

Similarly, the total kinetic energy is conserved and we have

$$\frac{1}{2} m_1 v_{1i}^2 = \frac{1}{2} m_1 v_{1f}^2 + \frac{1}{2} m_2 v_{2f}^2.$$

The solution to v_{1f} is given by Eq. 9-67: $v_{1f} = \dfrac{m_1 - m_2}{m_1 + m_2} v_{1i}$. We solve for m_2 to obtain

$$m_2 = \frac{v_{1i} - v_{1f}}{v_{1i} + v_{1f}} m_1.$$

The speed of the center of mass is

$$v_{com} = \frac{m_1 v_{1i} + m_2 v_{2i}}{m_1 + m_2}.$$

ANALYZE (a) given that $v_{1f} = v_{1i}/4$, we find the second mass to be

$$m_2 = \frac{v_{1i} - v_{1f}}{v_{1i} + v_{1f}} m_1 = \left(\frac{v_{1i} - v_{1i}/4}{v_{1i} + v_{1i}/4} \right) m_1 = \frac{3}{5} m_1 = \frac{3}{5}(2.0\text{ kg}) = 1.2\text{ kg}.$$

(b) The speed of the center of mass is $v_{com} = \dfrac{m_1 v_{1i} + m_2 v_{2i}}{m_1 + m_2} = \dfrac{(2.0\text{ kg})(4.0\text{ m/s})}{2.0\text{ kg} + 1.2\text{ kg}} = 2.5\text{ m/s}.$

LEARN The final speed of the second mass is

$$v_{2f} = \frac{2m_1}{m_1 + m_2} v_{1i} = \left(\frac{2(2.0\text{ kg})}{2.0\text{ kg} + 1.2\text{ kg}} \right)(4.0\text{ m/s}) = 5.0\text{ m/s}.$$

Since the system is isolated with no external force acting on it, v_{com} remains the same after the collision, so the same result is obtained if values for the final velocities are used:

$$v_{com} = \frac{m_1 v_{1f} + m_2 v_{2f}}{m_1 + m_2} = \frac{(2.0\text{ kg})(1.0\text{ m/s}) + (1.2\text{ kg})(5.0\text{ kg})}{2.0\text{ kg} + 1.2\text{ kg}} = 2.5\text{ m/s}.$$

9-77

THINK The mass of the faster barge is increasing at a constant rate. Additional force must be provided in order to maintain a constant speed.

EXPRESS We consider what must happen to the coal that lands on the faster barge during a time interval Δt. In that time, a total of Δm of coal must experience a change of velocity (from slow to fast) $\Delta v = v_{\text{fast}} - v_{\text{slow}}$, where rightwards is considered the positive direction. The rate of change in momentum for the coal is therefore

$$\frac{\Delta p}{\Delta t} = \frac{(\Delta m)}{\Delta t}\Delta v = \left(\frac{\Delta m}{\Delta t}\right)(v_{\text{fast}} - v_{\text{slow}})$$

which, by Eq. 9-23, must equal the force exerted by the (faster) barge on the coal. The processes (the shoveling, the barge motions) are constant, so there is no ambiguity in equating $\frac{\Delta p}{\Delta t}$ with $\frac{dp}{dt}$. Note that we ignore the transverse speed of the coal as it is shoveled from the slower barge to the faster one.

ANALYZE (a) With $v_{\text{fast}} = 20\,\text{km/h} = 5.56\,\text{m/s}$, $v_{\text{slow}} = 10\,\text{km/h} = 2.78\,\text{m/s}$ and the rate of mass change $(\Delta m/\Delta t) = 1000\,\text{kg/min} = (16.67\,\text{kg/s})$, the force that must be applied to the faster barge is

$$F_{\text{fast}} = \left(\frac{\Delta m}{\Delta t}\right)(v_{\text{fast}} - v_{\text{slow}}) = (16.67\text{ kg/s})(5.56\text{ m/s} - 2.78\text{ m/s}) = 46.3\,\text{N}$$

(b) The problem states that the frictional forces acting on the barges does not depend on mass, so the loss of mass from the slower barge does not affect its motion (so no extra force is required as a result of the shoveling).

LEARN The force that must be applied to the faster barge in order to maintain a constant speed is equal to the rate of change of momentum of the coal.

9-79

THINK As fuel is consumed, both the mass and the speed of the rocket will change.

EXPRESS The thrust of the rocket is given by $T = Rv_{\text{rel}}$ where R is the rate of fuel consumption and v_{rel} is the speed of the exhaust gas relative to the rocket. On the other hand, the mass of fuel ejected is given by $M_{\text{fuel}} = R\Delta t$, where Δt is the time interval of the burn. Thus, the mass of the rocket after the burn is

$$M_f = M_i - M_{\text{fuel}}\,.$$

ANALYZE (a) Given that $R = 480$ kg/s and $v_{\text{rel}} = 3.27 \times 10^3$ m/s, we find the thrust to be

$$T = Rv_{\text{rel}} = (480\,\text{kg/s})(3.27\times10^3\text{ m/s}) = 1.57\times10^6\text{ N}.$$

(b) With the mass of fuel ejected given by $M_{\text{fuel}} = R\Delta t = (480\text{ kg/s})(250\text{ s}) = 1.20\times10^5$ kg, the final mass of the rocket is

$$M_f = M_i - M_{\text{fuel}} = (2.55\times10^5\text{ kg}) - (1.20\times10^5\text{ kg}) = 1.35\times10^5\text{ kg}.$$

(c) Since the initial speed is zero, the final speed of the rocket is

$$v_f = v_{\text{rel}} \ln \frac{M_i}{M_f} = (3.27\times10^3 \text{ m/s}) \ln\left(\frac{2.55\times10^5 \text{ kg}}{1.35\times10^5 \text{ kg}}\right) = 2.08\times10^3 \text{ m/s}.$$

LEARN The speed of the rocket continues to rise as the fuel is consumed. From the first rocket equation given in Eq. 9-87, the thrust of the rocket is related to the acceleration by $T = Ma$. Using this equation, we find the initial acceleration to be

$$a_i = \frac{T}{M_i} = \frac{1.57\times10^6 \text{ N}}{2.55\times10^5 \text{ kg}} = 6.16 \text{ m/s}^2.$$

9-89

THINK The momentum of the car changes as it turns and collides with a tree.

EXPRESS Let the initial and final momenta of the car be $\vec{p}_i = m\vec{v}_i$ and $\vec{p}_f = m\vec{v}_f$, respectively. The impulse on it equals the change in its momentum:

$$\vec{J} = \Delta\vec{p} = \vec{p}_f - \vec{p}_i = m(\vec{v}_f - \vec{v}_i).$$

The average force over the duration Δt is given by $\vec{F}_{\text{avg}} = \vec{J}/\Delta t$.

ANALYZE (a) The initial momentum of the car is

$$\vec{p}_i = m\vec{v}_i = (1400\,\text{kg})(5.3\,\text{m/s})\hat{\text{j}} = (7400\,\text{kg}\cdot\text{m/s})\hat{\text{j}}$$

and the final momentum after making the turn is $\vec{p}_f = (7400\,\text{kg}\cdot\text{m/s})\hat{\text{i}}$ (note that the magnitude remains the same, only the direction is changed). Thus, the impulse is

$$\vec{J} = \vec{p}_f - \vec{p}_i = (7.4\times10^3\,\text{N}\cdot\text{s})(\hat{\text{i}} - \hat{\text{j}}).$$

(b) The initial momentum of the car after the turn is $\vec{p}_i' = (7400\,\text{kg}\cdot\text{m/s})\hat{\text{i}}$ and the final momentum after colliding with a tree is $\vec{p}_f' = 0$. The impulse acting on it is

$$\vec{J}' = \vec{p}_f' - \vec{p}_i' = (-7.4\times10^3\,\text{N}\cdot\text{s})\hat{\text{i}}.$$

(c) The average force on the car during the turn is

$$\vec{F}_{\text{avg}} = \frac{\Delta\vec{p}}{\Delta t} = \frac{\vec{J}}{\Delta t} = \frac{(7400\,\text{kg}\cdot\text{m/s})(\hat{\text{i}} - \hat{\text{j}})}{4.6\ \text{s}} = (1600\,\text{N})(\hat{\text{i}} - \hat{\text{j}})$$

and its magnitude is $F_{\text{avg}} = (1600\,\text{N})\sqrt{2} = 2.3\times10^3\,\text{N}$.

(d) The average force during the collision with the tree is

$$\vec{F}'_{\text{avg}} = \frac{\vec{J}'}{\Delta t} = \frac{(-7400\,\text{kg}\cdot\text{m/s})\hat{\text{i}}}{350\times10^{-3}\,\text{s}} = (-2.1\times10^4\,\text{N})\hat{\text{i}}$$

and its magnitude is $F'_{\text{avg}} = 2.1\times10^4\,\text{N}$.

(e) As shown in (c), the average force during the turn, in unit vector notation, is $\vec{F}_{\text{avg}} = (1600\,\text{N})(\hat{\text{i}} - \hat{\text{j}})$. Its direction is 45° below the positive x axis.

LEARN During the turn, the average force $\vec{F}_{\text{avg}}$ is in the same direction as $\vec{J}$, or $\Delta\vec{p}$. Its x and y components have equal magnitudes. The x component is positive and the y component is negative, so the force is 45° below the positive x axis.

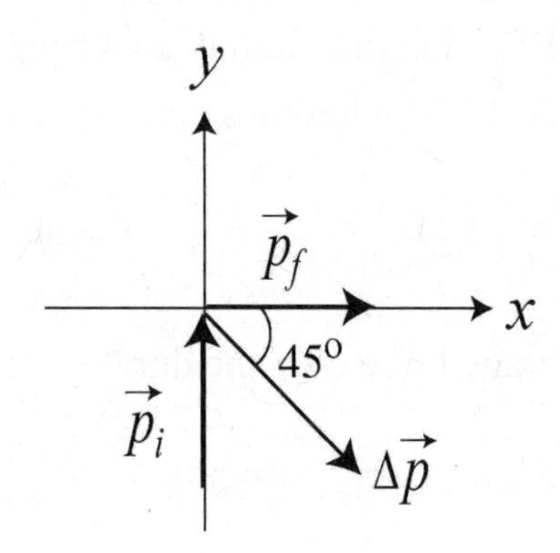

9-93

THINK A completely inelastic collision means that the railroad freight car and the caboose car move together after the collision. The motion is one-dimensional.

EXPRESS Let m_F be the mass of the freight car and v_F be its initial velocity. Let m_C be the mass of the caboose and v be the common final velocity of the two when they are coupled. Conservation of the total momentum of the two-car system leads to

$$m_F v_F = (m_F + m_C)v \;\Rightarrow\; v = \frac{m_F v_F}{m_F + m_C}.$$

The initial kinetic energy of the system is $K_i = \frac{1}{2}m_F v_F^2$ and the final kinetic energy is

$$K_f = \frac{1}{2}(m_F + m_C)v^2 = \frac{1}{2}(m_F + m_C)\frac{m_F^2 v_F^2}{(m_F + m_C)^2} = \frac{1}{2}\frac{m_F^2 v_F^2}{(m_F + m_C)}.$$

Since 27% of the original kinetic energy is lost, we have $K_f = 0.73K_i$. Combining with the two equations above allows us to solve for m_C, the mass of the caboose.

ANALYZE With $K_f = 0.73K_i$, or

$$\frac{1}{2}\frac{m_F^2 v_F^2}{(m_F + m_C)} = (0.73)\left(\frac{1}{2}m_F v_F^2\right)$$

we obtain $m_F/(m_F + m_C) = 0.73,$ which we use in solving for the mass of the caboose:

$$m_C = \frac{0.27}{0.73}m_F = 0.37m_F = (0.37)(3.18\times10^4\ \text{kg}) = 1.18\times10^4\ \text{kg}.$$

LEARN Energy is lost during an inelastic collision, but momentum is still conserved because there's no external force acting on the two-car system.

9-95

THINK A billiard ball undergoes glancing collision with another identical billiard ball. The collision is two-dimensional.

EXPRESS The mass of each ball is m, and the initial speed of one of the balls is $v_{1i} = 2.2\,\text{m/s}$. We apply the conservation of linear momentum to the x and y axes respectively:

$$mv_{1i} = mv_{1f}\cos\theta_1 + mv_{2f}\cos\theta_2$$
$$0 = mv_{1f}\sin\theta_1 - mv_{2f}\sin\theta_2$$

The mass m cancels out of these equations, and we are left with two unknowns and two equations, which is sufficient to solve.

ANALYZE (a) Solving the simultaneous equations leads to

$$v_{1f} = \frac{\sin\theta_2}{\sin(\theta_1+\theta_2)}v_{1i},\quad v_{2f} = \frac{\sin\theta_1}{\sin(\theta_1+\theta_2)}v_{1i}$$

Since $v_{2f} = v_{1i}/2 = 1.1\ \text{m/s}$ and $\theta_2 = 60°$, we have

$$\frac{\sin\theta_1}{\sin(\theta_1+60°)} = \frac{1}{2} \quad\Rightarrow\quad \tan\theta_1 = \frac{1}{\sqrt{3}}$$

or $\theta_1 = 30°$. Thus, the speed of ball 1 after collision is

$$v_{1f} = \frac{\sin\theta_2}{\sin(\theta_1+\theta_2)}v_{1i} = \frac{\sin 60°}{\sin(30°+60°)}v_{1i} = \frac{\sqrt{3}}{2}v_{1i} = \frac{\sqrt{3}}{2}(2.2\ \text{m/s}) = 1.9\ \text{m/s}.$$

(b) From the above, we have θ_1 = 30°, measured *clockwise* from the +*x*-axis, or equivalently, –30°, measured *counterclockwise* from the +*x*-axis.

(c) The kinetic energy before collision is $K_i = \frac{1}{2}mv_{1i}^2$. After the collision, we have

$$K_f = \frac{1}{2}m\left(v_{1f}^2 + v_{2f}^2\right)$$

Substituting the expressions for v_{1f} and v_{2f} found above gives

$$K_f = \frac{1}{2}m\left[\frac{\sin^2\theta_2}{\sin^2(\theta_1+\theta_2)} + \frac{\sin^2\theta_1}{\sin^2(\theta_1+\theta_2)}\right]v_{1i}^2$$

Since $\theta_1 = 30°$ and $\theta_2 = 60°$, $\sin(\theta_1+\theta_2) = 1$ and $\sin^2\theta_1 + \sin^2\theta_2 = \sin^2\theta_1 + \cos^2\theta_1 = 1$, and indeed, we have $K_f = \frac{1}{2}mv_{1i}^2 = K_i$, which means that energy is conserved.

LEARN One may verify that when two identical masses collide elastically, they will move off perpendicularly to each other with $\theta_1 + \theta_2 = 90°$.

9-105

THINK Both momentum and energy are conserved during an elastic collision.

EXPRESS Let m_1 be the mass of the object that is originally moving, v_{1i} be its velocity before the collision, and v_{1f} be its velocity after the collision. Let $m_2 = M$ be the mass of the object that is originally at rest and v_{2f} be its velocity after the collision. Conservation of linear momentum gives $m_1v_{1i} = m_1v_{1f} + m_2v_{2f}$. Similarly, the total kinetic energy is conserved and we have

$$\frac{1}{2}m_1v_{1i}^2 = \frac{1}{2}m_1v_{1f}^2 + \frac{1}{2}m_2v_{2f}^2.$$

Solving for v_{1f} and v_{2f}, we obtain:

$$v_{1f} = \frac{m_1 - m_2}{m_1 + m_2}v_{1i}, \quad v_{2f} = \frac{2m_1}{m_1 + m_2}v_{1i}$$

The second equation can be inverted to give $m_2 = m_1\left(\frac{2v_{1i}}{v_{2f}} - 1\right)$.

ANALYZE With m_1 = 3.0 kg, v_{1i} = 8.0 m/s and v_{2f} = 6.0 m/s, the above expression leads to

$$m_2 = M = m_1\left(\frac{2v_{1i}}{v_{2f}} - 1\right) = (3.0\text{ kg})\left(\frac{2(8.0\text{ m/s})}{6.0\text{ m/s}} - 1\right) = 5.0\text{ kg}$$

LEARN Our analytic expression for m_2 shows that if the two masses are equal, then $v_{2f} = v_{1i}$, and the pool player's result is recovered.

9-107

THINK To successfully launch a rocket from the ground, fuel is consumed at a rate that results in a thrust big enough to overcome the gravitational force.

EXPRESS The thrust of the rocket is given by $T = Rv_{\text{rel}}$ where R is the rate of fuel consumption and v_{rel} is the speed of the exhaust gas relative to the rocket.

ANALYZE (a) The exhaust speed is $v_{\text{rel}} = 1200$ m/s. For the thrust to equal the weight Mg where $M = 6100$ kg, we must have

$$T = Rv_{\text{rel}} = Mg \quad \Rightarrow \quad R = \frac{Mg}{v_{\text{rel}}} = \frac{(6100\text{ kg})(9.8\text{ m/s}^2)}{1200\text{ m/s}} = 49.8\text{ kg/s} \approx 50\text{ kg/s}.$$

(b) Using Eq. 9-42 with the additional effect due to gravity, we have

$$Rv_{\text{rel}} - Mg = Ma$$

so that requiring $a = 21$ m/s^2 leads to

$$R = \frac{M(g+a)}{v_{\text{rel}}} = \frac{(6100\text{ kg})(9.8\text{ m/s}^2 + 21\text{ m/s}^2)}{1200\text{ m/s}} = 156.6\text{ kg/s} \approx 1.6\times10^2\text{ kg/s}.$$

LEARN A greater upward acceleration requires a greater fuel consumption rate. To be launched from Earth's surface, the initial acceleration of the rocket must exceed $g = 9.8\text{ m/s}^2$. This means that the rate R must be greater than 50 kg/s.

9-109

THINK In this problem, we are asked to locate the center of mass of the Earth-Moon system.

EXPRESS We locate the coordinate origin at the center of Earth. Then the distance r_{com} of the center of mass of the Earth-Moon system is given by

$$r_{\text{com}} = \frac{m_M r_{ME}}{m_M + m_E}$$

where m_M is the mass of the Moon, m_E is the mass of Earth, and r_{ME} is their separation.

ANALYZE (a) With $m_E = 5.98\times10^{24}$ kg , $m_M = 7.36\times10^{22}$ kg and $r_{ME} = 3.82\times10^8$ m (these values are given in Appendix C), we find the center of mass to be at

$$r_{\text{com}} = \frac{(7.36\times10^{22}\ \text{kg})(3.82\times10^{8}\ \text{m})}{7.36\times10^{22}\ \text{kg}+5.98\times10^{24}\ \text{kg}} = 4.64\times10^{6}\ \text{m} \approx 4.6\times10^{3}\ \text{km}.$$

(b) The radius of Earth is $R_E = 6.37 \times 10^6$ m, so $r_{\text{com}} / R_E = 0.73 = 73\%$.

LEARN The center of mass of the Earth-Moon system is located inside the Earth!

9-111

THINK The water added to the sled will move at the same speed as the sled.

EXPRESS Let the mass of the sled be m_s and its initial speed be v_i. If the total mass of water being scooped up is m_w, then by momentum conservation, $m_s v_i = (m_s + m_w) v_f$, where v_f is the final speed of the sled-water system.

ANALYZE With $m_s = 2900\ \text{kg}$, $m_w = 920\ \text{kg}$ and $v_i = 250\ \text{m/s}$, we obtain

$$v_f = \frac{m_s v_i}{m_s + m_w} = \frac{(2900\ \text{kg})(250\ \text{m/s})}{2900\ \text{kg}+920\ \text{kg}} = 189.8\ \text{m/s} \approx 190\ \text{m/s}.$$

LEARN The water added to the sled can be regarded as undergoing completely inelastic collision with the sled. Some kinetic energy is converted into other forms of energy (thermal, sound, etc.) and the final speed of the sled-water system is smaller than the initial speed of the sled alone.

9-112

THINK The pellets that were fired carry both kinetic energy and momentum. Force is exerted by the rigid wall in stopping the pellets.

EXPRESS Let m be the mass of a pellet and v be its velocity as it hits the wall, then its momentum is $p = mv$, toward the wall. The kinetic energy of a pellet is $K = mv^2/2$. The force on the wall is given by the rate at which momentum is transferred from the pellets to the wall. Since the pellets do not rebound, each pellet that hits transfers p. If ΔN pellets hit in time Δt, then the average rate at which momentum is transferred would be $F_{\text{avg}} = p(\Delta N / \Delta t)$.

ANALYZE (a) With $m = 2.0 \times 10^{-3}$ kg and $v = 500$ m/s, the momentum of a pellet is

$$p = mv = (2.0 \times 10^{-3}\ \text{kg})(500\ \text{m/s}) = 1.0\ \text{kg} \cdot \text{m/s}.$$

(b) The kinetic energy of a pellet is $K = \frac{1}{2}mv^2 = \frac{1}{2}(2.0\times10^{-3}\ \text{kg})(500\ \text{m/s})^2 = 2.5\times10^{2}\ \text{J}$.

(c) With $(\Delta N/\Delta t)=10/\text{s}$, the average force on the wall from the stream of pellets is

$$F_{\text{avg}}=p\left(\frac{\Delta N}{\Delta t}\right)=(1.0\,\text{kg}\cdot\text{m/s})(10\,\text{s}^{-1})=10\,\text{N}.$$

The force on the wall is in the direction of the initial velocity of the pellets.

(d) If $\Delta t'$ is the time interval for a pellet to be brought to rest by the wall, then the average force exerted on the wall by a pellet is

$$F'_{\text{avg}}=\frac{p}{\Delta t'}=\frac{1.0\,\text{kg}\cdot\text{m/s}}{0.6\times10^{-3}\,\text{s}}=1.7\times10^{3}\,\text{N}.$$

The force is in the direction of the initial velocity of the pellet.

(e) In part (d) the force is averaged over the time a pellet is in contact with the wall, while in part (c) it is averaged over the time for many pellets to hit the wall. Hence, $F'_{\text{avg}}\neq F_{\text{avg}}$.

LEARN During the majority of this time, no pellet is in contact with the wall, so the average force in part (c) is much less than the average force in part (d).

9-115

THINK We have two forces acting on two masses separately. The masses will move according to Newton's second law.

EXPRESS Let $\vec{F}_1$ be the force acting on m_1, and $\vec{F}_2$ the force acting on m_2. According to Newton's second law, their displacements are

$$\vec{d}_1=\frac{1}{2}\vec{a}_1t^2=\frac{1}{2}\left(\frac{\vec{F}_1}{m_1}\right)t^2,\quad \vec{d}_2=\frac{1}{2}\vec{a}_2t^2=\frac{1}{2}\left(\frac{\vec{F}_2}{m_2}\right)t^2$$

The corresponding displacement of the center of mass is

$$\vec{d}_{\text{cm}}=\frac{m_1\vec{d}_1+m_2\vec{d}_2}{m_1+m_2}=\frac{1}{2}\frac{m_1}{m_1+m_2}\left(\frac{\vec{F}_1}{m_1}\right)t^2+\frac{1}{2}\frac{m_2}{m_1+m_2}\left(\frac{\vec{F}_2}{m_2}\right)t^2=\frac{1}{2}\left(\frac{\vec{F}_1+\vec{F}_2}{m_1+m_2}\right)t^2.$$

ANALYZE (a) The two masses are $m_1=2.00\times10^{-3}$ kg and $m_2=4.00\times10^{-3}$ kg. With the forces given by $\vec{F}_1=(-4.00\text{ N})\hat{\text{i}}+(5.00\text{ N})\hat{\text{j}}$ and $\vec{F}_2=(2.00\text{ N})\hat{\text{i}}-(4.00\text{ N})\hat{\text{j}}$, and $t=2.00\times10^{-3}$ s, we obtain

$$\vec{d}_{cm} = \frac{1}{2}\left(\frac{\vec{F}_1+\vec{F}_2}{m_1+m_2}\right)t^2 = \frac{1}{2}\frac{(-4.00\text{ N}+2.00\text{ N})\hat{\mathrm{i}}+(5.00\text{ N}-4.00\text{ N})\hat{\mathrm{j}}}{2.00\times10^{-3}\text{ kg}+4.00\times10^{-3}\text{ kg}}(2.00\times10^{-3}\text{ s})^2$$
$$= (-6.67\times10^{-4}\text{ m})\hat{\mathrm{i}}+(3.33\times10^{-4}\text{ m})\hat{\mathrm{j}}.$$

The magnitude of $\vec{d}_{cm}$ is

$$d_{cm} = \sqrt{(-6.67\times10^{-4}\text{ m})^2+(3.33\times10^{-4}\text{ m})^2} = 7.45\times10^{-4}\text{ m}$$

or 0.745 mm.

(b) The angle of $\vec{d}_{cm}$ is given by $\theta = \tan^{-1}\left(\frac{3.33\times10^{-4}\text{ m}}{-6.67\times10^{-4}\text{ m}}\right) = \tan^{-1}\left(-\frac{1}{2}\right) = 153°$, measured counterclockwise from $+x$-axis.

(c) The velocities of the two masses are

$$\vec{v}_1 = \vec{a}_1 t = \frac{\vec{F}_1 t}{m_1}, \quad \vec{v}_2 = \vec{a}_2 t = \frac{\vec{F}_2 t}{m_2},$$

and the velocity of the center of mass is

$$\vec{v}_{cm} = \frac{m_1\vec{v}_1+m_2\vec{v}_2}{m_1+m_2} = \frac{m_1}{m_1+m_2}\left(\frac{\vec{F}_1 t}{m_1}\right)+\frac{m_2}{m_1+m_2}\left(\frac{\vec{F}_2 t}{m_2}\right) = \left(\frac{\vec{F}_1+\vec{F}_2}{m_1+m_2}\right)t.$$

The corresponding kinetic energy of the center of mass is

$$K_{cm} = \frac{1}{2}(m_1+m_2)v_{cm}^2 = \frac{1}{2}\frac{|\vec{F}_1+\vec{F}_2|^2}{m_1+m_2}t^2$$

With $|\vec{F}_1+\vec{F}_2| = |(-2.00\text{ N})\hat{\mathrm{i}}+(1.00\text{ N})\hat{\mathrm{j}}| = \sqrt{5}\text{ N}$, we get

$$K_{cm} = \frac{1}{2}\frac{|\vec{F}_1+\vec{F}_2|^2}{m_1+m_2}t^2 = \frac{1}{2}\frac{(\sqrt{5}\text{ N})^2}{2.00\times10^{-3}\text{ kg}+4.00\times10^{-3}\text{ kg}}(2.00\times10^{-3}\text{ s})^2 = 1.67\times10^{-3}\text{ J}.$$

LEARN The motion of the center of the mass could be analyzed as though a force $\vec{F} = \vec{F}_1+\vec{F}_2$ is acting on a mass $M = m_1+m_2$. Thus, the acceleration of the center of the mass is $\vec{a}_{cm} = \dfrac{\vec{F}_1+\vec{F}_2}{m_1+m_2}$.

Chapter 10

10-15

THINK We have a wheel rotating at a constant angular acceleration. We use the equations given in Table 10-1 to analyze its motion.

EXPRESS Since the wheel starts from rest, its angular displacement as a function of time is given by $\theta = \frac{1}{2}\alpha t^2$. We take t_1 to be the start time of the interval so that $t_2 = t_1 + 4.0\text{ s}$. The corresponding angular displacements at these times are

$$\theta_1 = \frac{1}{2}\alpha t_1^2, \quad \theta_2 = \frac{1}{2}\alpha t_2^2$$

Given $\Delta\theta = \theta_2 - \theta_1$, we can solve for t_1, which tells us how long the wheel has been in motion up to the beginning of the 4.0-s interval.

ANALYZE The above expressions can be combined to give

$$\Delta\theta = \theta_2 - \theta_1 = \frac{1}{2}\alpha\left(t_2^2 - t_1^2\right) = \frac{1}{2}\alpha(t_2 + t_1)(t_2 - t_1)$$

With $\Delta\theta = 120\text{ rad}$, $\alpha = 3.0\text{ rad/s}^2$, and $t_2 - t_1 = 4.0\text{ s}$, we obtain

$$t_2 + t_1 = \frac{2(\Delta\theta)}{\alpha(t_2 - t_1)} = \frac{2(120\text{ rad})}{(3.0\text{ rad/s}^2)(4.0\text{ s})} = 20\text{ s}$$

which can be further solved to give $t_2 = 12.0\text{ s}$ and $t_1 = 8.0\text{ s}$. So, the wheel started from rest 8.0 s before the start of the described 4.0-s interval.

LEARN We can readily verify the results by calculating θ_1 and θ_2 explicitly:

$$\theta_1 = \frac{1}{2}\alpha t_1^2 = \frac{1}{2}(3.0\text{ rad/s}^2)(8.0\text{ s})^2 = 96\text{ rad}$$
$$\theta_2 = \frac{1}{2}\alpha t_2^2 = \frac{1}{2}(3.0\text{ rad/s}^2)(12.0\text{ s})^2 = 216\text{ rad}$$

Indeed the difference is $\Delta\theta = \theta_2 - \theta_1 = 120\text{ rad}$.

10-23

THINK A positive angular acceleration is required in order to increase the angular speed of the flywheel.

EXPRESS The linear speed of the flywheel is related to its angular speed by $v = \omega r$, where r is the radius of the wheel. As the wheel is accelerated, its angular speed at a later time is $\omega = \omega_0 + \alpha t$.

ANALYZE (a) The angular speed of the wheel, expressed in rad/s, is

$$\omega_0 = \frac{(200 \text{ rev/min})(2\pi \text{ rad/rev})}{60 \text{ s/min}} = 20.9 \text{ rad/s}.$$

(b) With $r = (1.20 \text{ m})/2 = 0.60$ m, using Eq. 10-18, we find the linear speed to be

$$v = r\omega_0 = (0.60 \text{ m})(20.9 \text{ rad/s}) = 12.5 \text{ m/s}.$$

(c) With $t = 1$ min, $\omega = 1000$ rev/min and $\omega_0 = 200$ rev/min, Eq. 10-12 gives the required acceleration:

$$\alpha = \frac{\omega - \omega_0}{t} = 800 \text{ rev/min}^2.$$

(d) With the same values used in part (c), Eq. 10-15 becomes

$$\theta = \frac{1}{2}(\omega_0 + \omega)t = \frac{1}{2}(200 \text{ rev/min} + 1000 \text{ rev/min})(1.0 \text{ min}) = 600 \text{ rev}.$$

LEARN An alternative way to solve for (d) is to use Eq. 10-13:

$$\theta = \theta_0 + \omega_0 t + \frac{1}{2}\alpha t^2 = 0 + (200 \text{ rev/min})(1.0 \text{ min}) + \frac{1}{2}(800 \text{ rev/min}^2)(1.0 \text{ min})^2 = 600 \text{ rev}.$$

10-25

THINK The linear speed of a point on Earth's surface depends on its distance from the Earth's axis of rotation.

EXPRESS To solve for the linear speed, we use $v = \omega r$, where r is the radius of its orbit. A point on Earth at a latitude of 40° moves along a circular path of radius $r = R \cos 40°$, where R is the radius of Earth (6.4×10^6 m). On the other hand, $r = R$ at the equator.

ANALYZE (a) Earth makes one rotation per day and 1 d is (24 h) (3600 s/h) = 8.64×10^4 s, so the angular speed of Earth is

$$\omega = \frac{2\pi \text{ rad}}{8.64 \times 10^4 \text{ s}} = 7.3 \times 10^{-5} \text{ rad/s}.$$

(b) At latitude of 40°, the linear speed is

$$v = \omega(R \cos 40°) = (7.3 \times 10^{-5} \text{ rad/s})(6.4 \times 10^6 \text{ m})\cos 40° = 3.5 \times 10^2 \text{ m/s}.$$

(c) At the equator (and all other points on Earth) the value of ω is the same (7.3×10^{-5} rad/s).

(d) The latitude at the equator is 0° and the speed is

$$v = \omega R = (7.3 \times 10^{-5} \text{ rad/s})(6.4 \times 10^6 \text{ m}) = 4.6 \times 10^2 \text{ m/s}.$$

LEARN The linear speed at the poles is zero since $r = R\cos 90° = 0$.

10-33

THINK We want to calculate the rotational inertia of a wheel, given its rotational energy and rotational speed.

EXPRESS The kinetic energy (in J) is given by $K = \frac{1}{2} I\omega^2$, where I is the rotational inertia (in $\text{kg} \cdot \text{m}^2$) and ω is the angular velocity (in rad/s).

ANALYZE Expressing the angular speed as

$$\omega = \frac{(602 \text{ rev/min})(2\pi \text{ rad/rev})}{60 \text{ s/min}} = 63.0 \text{ rad/s},$$

we find the rotational inertia to be $I = \dfrac{2K}{\omega^2} = \dfrac{2(24400 \text{ J})}{(63.0 \text{ rad / s})^2} = 12.3 \text{ kg} \cdot \text{m}^2$.

LEARN Note the analogy between rotational kinetic energy $\frac{1}{2} I\omega^2$ and $\frac{1}{2} mv^2$, the kinetic energy associated with linear motion.

10-35

THINK The rotational inertia of a rigid body depends on how its mass is distributed.

EXPRESS Since the rotational inertia of a cylinder is $I = \frac{1}{2} MR^2$ (Table 10-2(c)), its rotational kinetic energy is

$$K = \frac{1}{2} I\omega^2 = \frac{1}{4} MR^2\omega^2.$$

ANALYZE (a) For the smaller cylinder, we have

$$K_1 = \frac{1}{4}(1.25\,\text{kg})(0.25\ \text{m})^2(235\,\text{rad/s})^2 = 1.08\times10^3\ \text{J}.$$

(b) For the larger cylinder, we obtain

$$K_2 = \frac{1}{4}(1.25\,\text{kg})(0.75\ \text{m})^2(235\,\text{rad/s})^2 = 9.71\times10^3\ \text{J}.$$

LEARN The ratio of the rotational kinetic energies of the two cylinders having the same mass and angular speed is

$$\frac{K_2}{K_1} = \left(\frac{R_2}{R_1}\right)^2 = \left(\frac{0.75\ \text{m}}{0.25\ \text{m}}\right)^2 = (3)^2 = 9.$$

10-37

THINK We want to calculate the rotational inertia of a meter stick about an axis perpendicular to the stick but not through its center.

EXPRESS We use the parallel-axis theorem: $I = I_{com} + Mh^2$, where I_{com} is the rotational inertia about the center of mass (see Table 10-2(d)), M is the mass, and h is the distance between the center of mass and the chosen rotation axis. The center of mass is at the center of the meter stick, which implies $h = 0.50\ \text{m} - 0.20\ \text{m} = 0.30\ \text{m}$.

ANALYZE With $M = 0.56\,\text{kg}$ and $L = 1.0\ \text{m},$ we have

$$I_{com} = \frac{1}{12}ML^2 = \frac{1}{12}(0.56\ \text{kg})(1.0\ \text{m})^2 = 4.67\times10^{-2}\ \text{kg}\cdot\text{m}^2.$$

Consequently, the parallel-axis theorem yields

$$I = 4.67\times10^{-2}\ \text{kg}\cdot\text{m}^2 + (0.56\ \text{kg})(0.30\ \text{m})^2 = 9.7\times10^{-2}\ \text{kg}\cdot\text{m}^2.$$

LEARN A greater moment of inertia $I > I_{com}$ means that it is more difficult to rotate the meter stick about this axis than the case where the axis passes through the center.

10-43

THINK Since the rotation axis does not pass through the center of the block, we use the parallel-axis theorem to calculate the rotational inertia.

EXPRESS According to Table 10-2(i), the rotational inertia of a uniform slab about an axis through the center and perpendicular to the large faces is given by $I_{com} = \frac{M}{12}(a^2 + b^2)$. A parallel axis through the corner is a distance $h = \sqrt{(a/2)^2 + (b/2)^2}$ from the center. Therefore,

$$I = I_{\text{com}} + Mh^2 = \frac{M}{12}\left(a^2+b^2\right) + \frac{M}{4}\left(a^2+b^2\right) = \frac{M}{3}\left(a^2+b^2\right).$$

ANALYZE With $M = 0.172\text{ kg}$, $a = 3.5\text{ cm}$ and $b = 8.4\text{ cm}$, we have

$$I = \frac{M}{3}\left(a^2+b^2\right) = \frac{0.172\text{ kg}}{3}[(0.035\text{ m})^2 + (0.084\text{ m})^2] = 4.7\times10^{-4}\text{ kg}\cdot\text{m}^2.$$

LEARN A greater moment of inertia $I > I_{\text{com}}$ means that it is more difficult to rotate the block about the axis through the corner than the case where the axis passes through the center.

10-45

THINK Torque is the product of the force applied and the moment arm. When two torques act on a body, the net torque is their vector sum.

EXPRESS We take a torque that tends to cause a counterclockwise rotation from rest to be positive and a torque tending to cause a clockwise rotation to be negative. Thus, a positive torque of magnitude $r_1 F_1 \sin\theta_1$ is associated with $\vec{F}_1$ and a negative torque of magnitude $r_2F_2 \sin\theta_2$ is associated with $\vec{F}_2$. The net torque is consequently

$$\tau = r_1F_1\sin\theta_1 - r_2F_2\sin\theta_2.$$

ANALYZE Substituting the given values, we obtain

$$\tau = r_1F_1\sin\theta_1 - r_2F_2\sin\theta_2 = (1.30\text{ m})(4.20\text{ N})\sin75° - (2.15\text{ m})(4.90\text{ N})\sin60° = -3.85\text{ N}\cdot\text{m}.$$

LEARN Since $\tau < 0$, the body will rotate clockwise about the pivot point.

10-47

THINK In this problem we have a pendulum made up of a ball attached to a massless rod. There are two forces acting on the ball, the force of the rod and the force of gravity.

EXPRESS No torque about the pivot point is associated with the force of the rod since that force is along the line from the pivot point to the ball. As can be seen from the diagram, the component of the force of gravity that is perpendicular to the rod is $mg\sin\theta$. If ℓ is the length of the rod, then the torque associated with this force has magnitude

$$\tau = mg\ell\sin\theta.$$

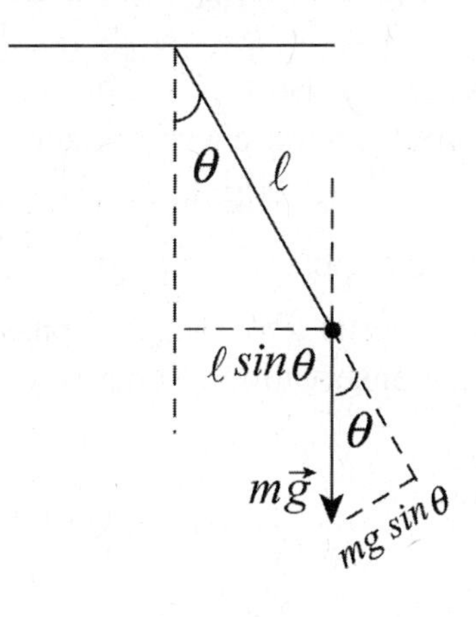

ANALYZE With $m = 0.75\,\text{kg}$, $\ell = 1.25$ m and $\theta = 30°$, we find the torque to be

$$\tau = mg\ell\sin\theta = (0.75)(9.8)(1.25)\sin 30° = 4.6\ \text{N}\cdot\text{m}.$$

LEARN The moment arm of the gravitational force mg is $\ell\sin\theta$. Alternatively, we may say that ℓ is the moment arm of $mg\sin\theta$, the tangential component of the gravitational force. Both interpretations lead to the same result: $\tau = (mg)(\ell\sin\theta) = (mg\sin\theta)(\ell)$.

10-49

THINK Since the angular velocity of the diver changes with time, there must be a non-vanishing angular acceleration.

EXPRESS To calculate the angular acceleration α, we use the kinematic equation $\omega = \omega_0 + \alpha t$, where ω_0 is the initial angular velocity, ω is the final angular velocity and t is the time. If I is the rotational inertia of the diver, then the magnitude of the torque acting on her is $\tau = I\alpha$.

ANALYZE (a) Using the values given, the angular acceleration is

$$\alpha = \frac{\omega - \omega_0}{t} = \frac{6.20\ \text{rad/s}}{220 \times 10^{-3}\ \text{s}} = 28.2\ \text{rad/s}^2.$$

(b) Similarly, we find the magnitude of the torque on the diver to be

$$\tau = I\alpha = \left(12.0\ \text{kg}\cdot\text{m}^2\right)\left(28.2\ \text{rad/s}^2\right) = 3.38 \times 10^2\ \text{N}\cdot\text{m}.$$

LEARN A net toque results in an angular acceleration that changes angular velocity. The equation $\tau = I\alpha$ implies that the greater the rotational inertia I, the greater the torque required for a given angular acceleration α.

10-63

THINK As the meter stick falls by rotating about the axis passing through one end of the stick, its potential energy is converted into rotational kinetic energy.

EXPRESS We use ℓ to denote the length of the stick. The meter stick is initially at rest so its initial kinetic energy is zero. Since its center of mass is $\ell/2$ from either end, its initial potential energy is $U_g = \frac{1}{2}mg\ell$, where m is its mass. Just before the stick hits the floor, its final potential energy is zero, and its final kinetic energy is $\frac{1}{2}I\omega^2$, where I is its rotational inertia about an axis passing through one end of the stick and ω is the angular velocity. Conservation of energy yields

$$\frac{1}{2}mg\ell = \frac{1}{2}I\omega^2 \Rightarrow \omega = \sqrt{\frac{mg\ell}{I}}.$$

The free end of the stick is a distance ℓ from the rotation axis, so its speed as it hits the floor is (from Eq. 10-18)

$$v = \omega\ell = \sqrt{\frac{mg\ell^3}{I}}.$$

ANALYZE Using Table 10-2 and the parallel-axis theorem, the rotational inertial is $I = \frac{1}{3}m\ell^2$, so

$$v = \sqrt{3g\ell} = \sqrt{3\left(9.8\ \text{m/s}^2\right)\left(1.00\ \text{m}\right)} = 5.42\ \text{m/s}.$$

LEARN The linear speed of a point on the meter stick depends on its distance from the axis of rotation. One may show that the speed of the center of mass is

$$v_{\text{cm}} = \omega(\ell/2) = \frac{1}{2}\sqrt{3g\ell}.$$

10-71

THINK Since the string that connects the two blocks does not slip, the pulley rotates about its axel as the blocks move.

EXPRESS We choose positive coordinate directions (different choices for each item) so that each is accelerating positively, which will allow us to set $a_2 = a_1 = R\alpha$ (for simplicity, we denote this as a). Thus, we choose rightward positive for $m_2 = M$ (the block on the table), downward positive for $m_1 = M$ (the block at the end of the string) and (somewhat unconventionally) clockwise for positive sense of disk rotation. This means that we interpret θ given in the problem as a positive-valued quantity. Applying Newton's second law to m_1, m_2 and (in the form of Eq. 10-45) to M, respectively, we arrive at the following three equations (where we allow for the possibility of friction f_2 acting on m_2):

$$\begin{aligned} m_1 g - T_1 &= m_1 a_1 \\ T_2 - f_2 &= m_2 a_2 \\ T_1 R - T_2 R &= I\alpha \end{aligned}$$

ANALYZE (a) From Eq. 10-13 (with $\omega_0 = 0$) we find the magnitude of the pulley's angular acceleration to be

$$\theta = \omega_0 t + \frac{1}{2}\alpha t^2 \Rightarrow \alpha = \frac{2\theta}{t^2} = \frac{2(0.130\ \text{rad})}{(0.0910\ \text{s})^2} = 31.4\ \text{rad/s}^2.$$

(b) From the fact that $a = R\alpha$ (noted above), the acceleration of the blocks is

$$a = \frac{2R\theta}{t^2} = \frac{2(0.024\ \text{m})(0.130\ \text{rad})}{(0.0910\ \text{s})^2} = 0.754\ \text{m/s}^2.$$

(c) From the first of the above equations, we find the string tension T_1 to be

$$T_1 = m_1(g - a_1) = M\left(g - \frac{2R\theta}{t^2}\right) = (6.20 \text{ kg})\left(9.80 \text{ m/s}^2 - \frac{2(0.024 \text{ m})(0.130 \text{ rad})}{(0.0910 \text{ s})^2}\right) = 56.1 \text{ N}.$$

(d) From the last of the above equations, we obtain the second tension:

$$T_2 = T_1 - \frac{I\alpha}{R} = 56.1 \text{ N} - \frac{(7.40\times 10^{-4}\text{kg}\cdot\text{m}^2)(31.4 \text{ rad/s}^2)}{0.024 \text{ m}} = 55.1 \text{ N}.$$

LEARN The torque acting on the pulley is $\tau = I\alpha = (T_1 - T_2)R$. If the pulley becomes massless, then $I = 0$ and we recover the expected result: $T_1 = T_2$.

10-77

THINK The record turntable comes to a stop due to a constant angular acceleration. We apply equations given in Table 10-1 to analyze the rotational motion.

EXPRESS We take the sense of initial rotation to be positive. Then, with $\omega_0 > 0$ and $\omega = 0$ (since it stops at time t), our angular acceleration is negative-valued. The angular acceleration is constant, so we can apply Eq. 10-12 ($\omega = \omega_0 + \alpha t$), which gives $\alpha = (\omega - \omega_0)/t$. Similarly, the angular displacement can be found by using Eq. 10-13: $\theta = \theta_0 + \omega_0 t + \frac{1}{2}\alpha t^2$.

ANALYZE (a) To obtain the requested units, we use t = 30 s = 0.50 min. With $\omega_0 = 33.33$ rev/min, we find the angular acceleration to be

$$\alpha = -\frac{33.33 \text{ rev/min}}{0.50 \text{min}} = -66.7 \text{ rev/min}^2 \approx -67 \text{ rev/min}^2.$$

(b) Substituting the value of α obtained above into Eq. 10-13, we get

$$\theta = \omega_0 t + \frac{1}{2}\alpha t^2 = (33.33 \text{ rev/min})(0.50 \text{ min}) + \frac{1}{2}(-66.7\text{rev/min}^2)(0.50 \text{ min})^2 = 8.33 \text{ rev}.$$

LEARN To solve for the angular displacement in (b), we may also use Eq. 10-15:

$$\theta = \frac{1}{2}(\omega_0 + \omega)t = \frac{1}{2}(33.33 \text{ rev/min} + 0)(0.50 \text{ min}) = 8.33 \text{ rev}.$$

10-79

THINK In this problem we compare the rotational inertia between a solid cylinder and a hoop.

EXPRESS According to Table 10-2, the rotational inertia formulas for a cylinder of radius *R* and mass *M*, and a hoop of radius *r* and mass *M* are

$$I_C = \frac{1}{2}MR^2, \quad I_H = Mr^2.$$

Equating $I_C = I_H$ allows us to deduce the relationship between *r* and *R*.

ANALYZE (a) Since both the cylinder and the hoop have the same mass, then they will have the same rotational inertia ($I_C = I_H$) if $R^2/2 = r^2 \rightarrow r = R/\sqrt{2}$.

(b) We require the rotational inertia of any given body to be written as $I = Mk^2$, where *M* is the mass of the given body and *k* is the radius of the "equivalent hoop." It follows directly that $k = \sqrt{I/M}$.

LEARN Listed below are some examples of equivalent hoop and their radii:

$$I_C = \frac{1}{2}MR^2 = M(R/\sqrt{2})^2 \Rightarrow k_C = R/\sqrt{2}$$

$$I_S = \frac{2}{5}MR^2 = M\left(\sqrt{\frac{2}{5}}R\right)^2 \Rightarrow k_S = \sqrt{\frac{2}{5}}R$$

10-91

THINK As the box falls, gravitational force gives rise to a torque that causes the wheel to rotate.

EXPRESS We employ energy methods to solve this problem; thus, considerations of positive versus negative sense (regarding the rotation of the wheel) are not relevant.

(a) The speed of the box is related to the angular speed of the wheel by $v = R\omega$, where $K_{\text{box}} = m_{\text{box}}v^2/2$. The rotational kinetic energy of the wheel is $K_{\text{rot}} = I\omega^2/2$.

ANALYZE (a) With $K_{\text{box}} = 0.60$ J, we find the speed of the box to be

$$K_{\text{box}} = \frac{1}{2}m_{\text{box}}v^2 \Rightarrow v = \sqrt{\frac{2K_{\text{box}}}{m_{\text{box}}}} = \sqrt{\frac{2(6.0\text{ J})}{6.0\text{ kg}}} = 1.41\text{ m/s},$$

implying that the angular speed is ω = (1.41 m/s)/(0.20 m) = 7.07 rad/s. Thus, the kinetic energy of rotation is

$$K_{\text{rot}} = \frac{1}{2}I\omega^2 = \frac{1}{2}(0.40\ \text{kg}\cdot\text{m}^2)(7.07\ \text{rad/s})^2 = 10.0\ \text{J}.$$

(b) Since it was released from rest, we will take the initial position to be our reference point for gravitational potential. Energy conservation requires

$$K_0 + U_0 = K + U \;\Rightarrow\; 0 + 0 = (6.0\ \text{J} + 10.0\ \text{J}) + m_{\text{box}}g(-h).$$

Therefore,

$$h = \frac{K}{m_{\text{box}}g} = \frac{6.0\ \text{J} + 10.0\ \text{J}}{(6.0\ \text{kg})(9.8\ \text{m/s}^2)} = 0.27\ \text{m}.$$

LEARN As the box falls, its gravitational potential energy gets converted into kinetic energy of the box as well as rotational kinetic energy of the wheel; the total energy remains conserved.

10-93

THINK The applied force *P* accelerates the block. In addition, it gives rise to a torque that causes the wheel to undergo angular acceleration.

EXPRESS We take rightward to be positive for the block and clockwise negative for the wheel (as is conventional). With this convention, we note that the tangential acceleration of the wheel is of opposite sign from the block's acceleration (which we simply denote as a); that is, $a_t = -a$. Applying Newton's second law to the block leads to $P - T = ma$, where T is the tension in the cord. Similarly, applying Newton's second law (for rotation) to the wheel leads to $-TR = I\alpha$. Noting that $R\alpha = a_t = -a$, we multiply this equation by R and obtain

$$-TR^2 = -Ia \Rightarrow T = a\frac{I}{R^2}.$$

Adding this to the above equation (for the block) leads to $P = (m + I/R^2)a$. Thus, the angular acceleration is

$$\alpha = -\frac{a}{R} = -\frac{P}{(m + I/R^2)R}$$

ANALYZE With $m = 2.0$ kg, $I = 0.050\ \text{kg}\cdot\text{m}^2$, $P = 3.0$ N and $R = 0.20$ m, we find

$$\alpha = -\frac{P}{(m + I/R^2)R} = -\frac{3.0\ \text{N}}{[2.0\ \text{kg} + (0.050\ \text{kg}\cdot\text{m}^2)/(0.20\ \text{m})^2](0.20\ \text{m})} = -4.62\ \text{rad/s}^2 .$$

LEARN The greater the applied force *P*, the greater the (magnitude of) angular acceleration. Note that the negative sign in α should not be mistaken for a deceleration; it simply indicates the clockwise sense to the motion.

Chapter 11

11-3

THINK The work required to stop the hoop is the negative of the initial kinetic energy of the hoop.

EXPRESS From Eq. 11-5, the initial kinetic energy of the hoop is $K_i = \frac{1}{2}I\omega^2 + \frac{1}{2}mv^2$, where $I = mR^2$ is its rotational inertia about the center of mass. Eq. 11-2 relates the angular speed to the speed of the center of mass: $\omega = v/R$. Thus,

$$K_i = \frac{1}{2}I\omega^2 + \frac{1}{2}mv^2 = \frac{1}{2}(mR^2)\left(\frac{v}{R}\right)^2 + \frac{1}{2}mv^2 = mv^2$$

ANALYZE With m = 140 kg, and the speed of its center of mass v = 0.150 m/s, we find the initial kinetic energy to be

$$K_i = mv^2 = (140\text{ kg})(0.150\text{ m/s})^2 = 3.15\text{ J}$$

which implies that the work required is $W = \Delta K = K_f - K_i = -K_i = -3.15\text{ J}$.

LEARN By the work-kinetic energy theorem, the work done is negative since it decreases the kinetic energy. A rolling body has two types of kinetic energy: rotational and translational.

11-17

THINK The yo-yo has both translational and rotational types of motion.

EXPRESS The derivation of the acceleration is given by Eq. 11-13:

$$a_{\text{com}} = -\frac{g}{1 + I_{\text{com}}/MR_0^2}$$

where M is the mass of the yo-yo, I_{cm} is the rotational inertia and R_0 is the radius of the axel. The positive direction is upward. The time it takes for the yo-yo to reach the end of the string can be found by solving the kinematic equation $y_{\text{com}} = \frac{1}{2}a_{\text{com}}t^2$.

ANALYZE (a) With $I_{\text{com}} = 950\ \text{g}\cdot\text{cm}^2$, M=120 g, R_0 = 0.320 cm and g = 980 cm/s², we obtain

$$|a_{com}| = \frac{980 \text{ cm/s}^2}{1+(950 \text{ g}\cdot\text{cm}^2)/(120 \text{ g})(0.32 \text{ cm})^2} = 12.5 \text{ cm/s}^2 \approx 13 \text{ cm/s}^2.$$

(b) Taking the coordinate origin at the initial position, Eq. 2-15 leads to $y_{com} = \frac{1}{2}a_{com}t^2$. Thus, we set $y_{com} = -\ 120$ cm and find

$$t = \sqrt{\frac{2y_{com}}{a_{com}}} = \sqrt{\frac{2(-120\text{cm})}{-12.5 \text{ cm/s}^2}} = 4.38 \text{ s} \approx 4.4 \text{ s}.$$

(c) As the yo-yo reaches the end of the string, its center of mass velocity is given by Eq. 2-11:

$$v_{com} = a_{com}t = (-12.5 \text{ cm/s}^2)(4.38\text{s}) = -54.8 \text{ cm/s},$$

so its linear speed then is approximately $|v_{com}| = 55$ cm/s.

(d) The translational kinetic energy of the yo-yo is

$$K_{trans} = \frac{1}{2}mv_{com}^2 = \frac{1}{2}(0.120 \text{ kg})(0.548 \text{ m/s})^2 = 1.8\times10^{-2} \text{ J}.$$

(e) The angular velocity is $\omega = -\ v_{com}/R_0$, so the rotational kinetic energy is

$$K_{rot} = \frac{1}{2}I_{com}\omega^2 = \frac{1}{2}I_{com}\left(\frac{v_{com}}{R_0}\right)^2 = \frac{1}{2}(9.50\times10^{-5} \text{ kg}\cdot\text{m}^2)\left(\frac{0.548 \text{ m/s}}{3.2\times10^{-3} \text{ m}}\right)^2$$
$$= 1.393 \text{ J} \approx 1.4 \text{ J}$$

(f) The angular speed is $\omega = \frac{|v_{com}|}{R_0} = \frac{0.548 \text{ m/s}}{3.2\times10^{-3} \text{ m}} = 1.7\times10^2 \text{ rad/s} = 27 \text{ rev/s}$.

LEARN As the yo-yo rolls down, its gravitational potential energy gets converted into both translational kinetic energy as well as rotational kinetic energy of the wheel. To show that the total energy remains conserved, we note that the initial energy is

$$U_i = Mgy_i = (0.120 \text{ kg})(9.80 \text{ m/s}^2)(1.20 \text{ m}) = 1.411 \text{ J}$$

which is equal to the sum of K_{trans} (= 0.018 J) and K_{rot} (= 1.393 J).

11-25

THINK We take the cross product of $\vec{r}$ and $\vec{F}$ to find the torque $\vec{\tau}$ on a particle.

EXPRESS If we write $\vec{r} = x\hat{\mathrm{i}} + y\hat{\mathrm{j}} + z\hat{\mathrm{k}}$ and $\vec{F} = F_x\hat{\mathrm{i}} + F_y\hat{\mathrm{j}} + F_z\hat{\mathrm{k}}$, then (using Eq. 3-30) the general expression for torque can be written as

$$\vec{\tau} = \vec{r} \times \vec{F} = \left(yF_z - zF_y\right)\hat{\mathrm{i}} + \left(zF_x - xF_z\right)\hat{\mathrm{j}} + \left(xF_y - yF_x\right)\hat{\mathrm{k}}.$$

ANALYZE (a) With $\vec{r} = (3.0 \text{ m})\hat{\mathrm{i}} + (4.0 \text{ m})\hat{\mathrm{j}}$ and $\vec{F} = (-8.0 \text{ N})\hat{\mathrm{i}} + (6.0 \text{ N})\hat{\mathrm{j}}$, we have

$$\vec{\tau} = \left[(3.0\text{m})(6.0\text{N}) - (4.0\text{m})(-8.0\text{N})\right]\hat{\mathrm{k}} = (50\,\text{N}\cdot\text{m})\,\hat{\mathrm{k}}.$$

(b) To find the angle ϕ between $\vec{r}$ and $\vec{F}$, we use Eq. 3-27: $|\vec{r} \times \vec{F}| = rF\sin\phi$. Now $r = \sqrt{x^2 + y^2} = 5.0$ m and $F = \sqrt{F_x^2 + F_y^2} = 10$ N. Thus,

$$rF = (5.0 \text{ m})(10 \text{ N}) = 50 \text{ N}\cdot\text{m},$$

the same as the magnitude of the vector product calculated in part (a). This implies $\sin\phi = 1$ and $\phi = 90°$.

LEARN Our result ($\phi = 90°$) implies that $\vec{r}$ and $\vec{F}$ are perpendicular to each other. A useful check is to show that their dot product is zero. This is indeed the case:

$$\begin{aligned}\vec{r}\cdot\vec{F} &= [(3.0 \text{ m})\hat{\mathrm{i}} + (4.0 \text{ m})\hat{\mathrm{j}}]\cdot[(-8.0 \text{ N})\hat{\mathrm{i}} + (6.0 \text{ N})\hat{\mathrm{j}}] \\ &= (3.0 \text{ m})(-8.0 \text{ N}) + (4.0 \text{ m})(6.0 \text{ N}) = 0.\end{aligned}$$

11-27

THINK We evaluate the cross product $\vec{\ell} = m\vec{r} \times \vec{v}$ to find the angular momentum $\vec{\ell}$ on the object, and the cross product of $\vec{r} \times \vec{F}$ for the torque $\vec{\tau}$.

EXPRESS Let $\vec{r} = x\hat{\mathrm{i}} + y\hat{\mathrm{j}} + z\hat{\mathrm{k}}$ be the position vector of the object, $\vec{v} = v_x\hat{\mathrm{i}} + v_y\hat{\mathrm{j}} + v_z\hat{\mathrm{k}}$ its velocity vector, and m its mass. The cross product of $\vec{r}$ and $\vec{v}$ is (using Eq. 3-30)

$$\vec{r} \times \vec{v} = \left(yv_z - zv_y\right)\hat{\mathrm{i}} + \left(zv_x - xv_z\right)\hat{\mathrm{j}} + \left(xv_y - yv_x\right)\hat{\mathrm{k}}.$$

Since only the x and z components of the position and velocity vectors are nonzero (i.e., $y = 0$ and $v_y = 0$), the above expression becomes $\vec{r} \times \vec{v} = \left(-xv_z + zv_z\right)\hat{\mathrm{j}}$. As for the torque, writing $\vec{F} = F_x\hat{\mathrm{i}} + F_y\hat{\mathrm{j}} + F_z\hat{\mathrm{k}}$, we find $\vec{r} \times \vec{F}$ to be

$$\vec{\tau} = \vec{r} \times \vec{F} = \left(yF_z - zF_y\right)\hat{\mathrm{i}} + \left(zF_x - xF_z\right)\hat{\mathrm{j}} + \left(xF_y - yF_x\right)\hat{\mathrm{k}}.$$

ANALYZE (a) With $\vec{r} = (2.0\text{ m})\hat{\text{i}} - (2.0\text{ m})\hat{\text{k}}$ and $\vec{v} = (-5.0\text{ m/s})\hat{\text{i}} + (5.0\text{ m/s})\hat{\text{k}}$, in unit-vector notation, the angular momentum of the object is

$$\vec{\ell} = m(-xv_z + zv_x)\hat{\text{j}} = (0.25\text{ kg})\big(-(2.0\text{ m})(5.0\text{ m/s}) + (-2.0\text{ m})(-5.0\text{ m/s})\big)\hat{\text{j}} = 0.$$

(b) With $x = 2.0$ m, $z = -2.0$ m, $F_y = 4.0$ N and all other components zero, the expression above yields

$$\vec{\tau} = \vec{r} \times \vec{F} = (8.0\text{ N}\cdot\text{m})\hat{\text{i}} + (8.0\text{ N}\cdot\text{m})\hat{\text{k}}.$$

LEARN The fact that $\vec{\ell} = 0$ implies that $\vec{r}$ and $\vec{v}$ are parallel to each other ($\vec{r} \times \vec{v} = 0$). Using $\tau = |\vec{r} \times \vec{F}| = rF\sin\phi$, we find the angle between $\vec{r}$ and $\vec{F}$ to be

$$\sin\phi = \frac{\tau}{rF} = \frac{8\sqrt{2}\text{ N}\cdot\text{m}}{(2\sqrt{2}\text{ m})(4.0\text{ N})} = 1 \Rightarrow \phi = 90°$$

That is, $\vec{r}$ and $\vec{F}$ are perpendicular to each other.

11-33

THINK We evaluate the cross product $\vec{\ell} = m\vec{r} \times \vec{v}$ to find the angular momentum $\vec{\ell}$ on the particle, and the cross product of $\vec{r} \times \vec{F}$ for the torque $\vec{\tau}$.

EXPRESS Let $\vec{r} = x\hat{\text{i}} + y\hat{\text{j}} + z\hat{\text{k}}$ be the position vector of the object, $\vec{v} = v_x\hat{\text{i}} + v_y\hat{\text{j}} + v_z\hat{\text{k}}$ its velocity vector, and m its mass. The cross product of $\vec{r}$ and $\vec{v}$ is

$$\vec{r} \times \vec{v} = (yv_z - zv_y)\hat{\text{i}} + (zv_x - xv_z)\hat{\text{j}} + (xv_y - yv_x)\hat{\text{k}}.$$

The angular momentum is given by the vector product $\vec{\ell} = m\vec{r} \times \vec{v}$. As for the torque, writing $\vec{F} = F_x\hat{\text{i}} + F_y\hat{\text{j}} + F_z\hat{\text{k}}$, then we find $\vec{r} \times \vec{F}$ to be

$$\vec{\tau} = \vec{r} \times \vec{F} = (yF_z - zF_y)\hat{\text{i}} + (zF_x - xF_z)\hat{\text{j}} + (xF_y - yF_x)\hat{\text{k}}.$$

ANALYZE (a) Substituting $m = 3.0$ kg, $x = 3.0$ m, $y = 8.0$ m, $z = 0$, $v_x = 5.0$ m/s, $v_y = -6.0$ m/s and $v_z = 0$ into the above expression, we obtain

$$\vec{\ell} = (3.0\text{ kg})[(3.0\text{ m})(-6.0\text{ m/s}) - (8.0\text{ m})(5.0\text{ m/s})]\hat{\text{k}} = (-174\text{ kg}\cdot\text{m}^2/\text{s})\hat{\text{k}}.$$

(b) Given that $\vec{r} = x\hat{\text{i}} + y\hat{\text{j}}$ and $\vec{F} = F_x\hat{\text{i}}$, the corresponding torque is

$$\vec{\tau} = \left(x\hat{\text{i}} + y\hat{\text{j}}\right) \times \left(F_x\hat{\text{i}}\right) = -yF_x\hat{\text{k}}.$$

Substituting the values given, we find $\vec{\tau} = -(8.0\,\text{m})(-7.0\,\text{N})\,\hat{\text{k}} = (56\,\text{N}\cdot\text{m})\hat{\text{k}}$.

(c) According to Newton's second law $\vec{\tau} = d\vec{\ell}/dt$, so the rate of change of the angular momentum is 56 kg·m²/s², in the positive z direction.

LEARN The direction of $\vec{\ell}$ is in the –z-direction, which is perpendicular to both $\vec{r}$ and $\vec{v}$. Similarly, the torque $\vec{\tau}$ is perpendicular to both $\vec{r}$ and $\vec{F}$ (i.e, $\vec{\tau}$ is in the direction normal to the plane formed by $\vec{r}$ and $\vec{F}$).

11-39

THINK A non-zero torque is required to change the angular momentum of the flywheel. We analyze the rotational motion of the wheel using the equations given in Table 10-1.

EXPRESS Since the torque is equal to the rate of change of angular momentum, $\tau = dL/dt$, the average torque acting during any interval Δt is simply given by $\tau_{\text{avg}} = \left(L_f - L_i\right)/\Delta t$, where L_i is the initial angular momentum and L_f is the final angular momentum. For uniform angular acceleration, the angle turned is $\theta = \omega_0 t + \alpha t^2/2$, and the work done on the wheel is $W = \tau\theta$.

ANALYZE (a) Substituting the values given, the average torque is

$$\tau_{\text{avg}} = \frac{L_f - L_i}{\Delta t} = \frac{(0.800\,\text{kg}\cdot\text{m}^2/\text{s}) - (3.00\,\text{kg}\cdot\text{m}^2/\text{s})}{1.50\ \text{s}} = -1.47\ \text{N}\cdot\text{m},$$

or $|\tau_{\text{avg}}| = 1.47\ \text{N}\cdot\text{m}$. In this case the negative sign indicates that the direction of the torque is opposite the direction of the initial angular momentum, implicitly taken to be positive.

(b) If the angular acceleration α is uniform, so is the torque and $\alpha = \tau/I$. Furthermore, $\omega_0 = L_i/I$, and we obtain

$$\theta = \frac{L_i t + \tau t^2/2}{I} = \frac{\left(3.00\,\text{kg}\cdot\text{m}^2/\text{s}\right)\left(1.50\ \text{s}\right) + \left(-1.467\,\text{N}\cdot\text{m}\right)\left(1.50\ \text{s}\right)^2/2}{0.140\,\text{kg}\cdot\text{m}^2} = 20.4\,\text{rad}.$$

(c) Using the values of τ and θ found above, we find the work done on the wheel to be

$$W = \tau\theta = \left(-1.47\,\text{N}\cdot\text{m}\right)\left(20.4\,\text{rad}\right) = -29.9\ \text{J}.$$

(d) The average power is the work done by the flywheel (the negative of the work done on the flywheel) divided by the time interval:

$$P_{\text{avg}} = -\frac{W}{\Delta t} = -\frac{-29.9 \text{ J}}{1.50 \text{ s}} = 19.9 \text{ W}.$$

LEARN An alternative way to calculate the work done on the wheel is to apply the work-kinetic energy theorem:

$$W = \Delta K = K_f - K_i = \frac{1}{2} I(\omega_f^2 - \omega_i^2) = \frac{1}{2} I\left[\left(\frac{L_f}{I}\right)^2 - \left(\frac{L_i}{I}\right)^2\right] = \frac{L_f^2 - L_i^2}{2I}$$

Substituting the values given, we have

$$W = \frac{L_f^2 - L_i^2}{2I} = \frac{(0.800 \text{ kg}\cdot\text{m}^2/\text{s})^2 - (3.00 \text{ kg}\cdot\text{m}^2/\text{s})^2}{2(0.140 \text{ kg}\cdot\text{m}^2)} = -29.9 \text{ J}$$

which agrees with that calculated in part (c).

11-45

THINK No external torque acts on the system consisting of the man, bricks, and platform, so the total angular momentum of the system is conserved.

EXPRESS Let I_i be the initial rotational inertia of the system and let I_f be the final rotational inertia. Then $I_i\omega_i = I_f\omega_f$ by angular momentum conservation. The kinetic energy (of rotational nature) is given by $K = I\omega^2/2$.

ANALYZE (a) The final angular momentum of the system is

$$\omega_f = \left(\frac{I_i}{I_f}\right)\omega_i = \left(\frac{6.0 \text{ kg}\cdot\text{m}^2}{2.0 \text{ kg}\cdot\text{m}^2}\right)(1.2 \text{ rev/s}) = 3.6 \text{ rev/s}.$$

(b) The initial kinetic energy is $K_i = \frac{1}{2} I_i\omega_i^2$, and the final kinetic energy is $K_f = \frac{1}{2} I_f\omega_f^2$, so that their ratio is

$$\frac{K_f}{K_i} = \frac{I_f\omega_f^2/2}{I_i\omega_i^2/2} = \frac{(2.0 \text{ kg}\cdot\text{m}^2)(3.6 \text{ rev/s})^2/2}{(6.0 \text{ kg}\cdot\text{m}^2)(1.2 \text{ rev/s})^2/2} = 3.0.$$

(c) The man did work in decreasing the rotational inertia by pulling the bricks closer to his body. This energy came from the man's internal energy.

LEARN The work done by the person is equal to the change in kinetic energy:

$$W = K_f - K_i = 3K_i - K_i = 2K_i = I_i\omega_i^2 = (6.0\,\text{kg}\cdot\text{m}^2)(2\pi\cdot 1.2\,\text{rad/s})^2 = 341\ \text{J}.$$

11-47

THINK No external torque acts on the system consisting of the train and wheel, so the total angular momentum of the system (which is initially zero) remains zero.

EXPRESS Let $I = MR^2$ be the rotational inertia of the wheel (which we treat as a hoop). Its angular momentum is

$$\vec{L}_{\text{wheel}} = (I\omega)\hat{\text{k}} = -MR^2|\omega|\hat{\text{k}},$$

where $\hat{\text{k}}$ is *up* in Fig. 11-48 and that last step (with the minus sign) is done in recognition that the wheel's clockwise rotation implies a negative value for ω. The linear speed of a point on the track is $-|\omega|R$ and the speed of the train (going counterclockwise in Fig. 11-48 with speed v' relative to an outside observer) is therefore $v' = v - |\omega|R$ where v is its speed relative to the tracks. Consequently, the angular momentum of the train is $\vec{L}_{\text{train}} = m\left(v - |\omega|R\right)R\hat{\text{k}}$. Conservation of angular momentum yields

$$0 = \vec{L}_{\text{wheel}} + \vec{L}_{\text{train}} = -MR^2|\omega|\hat{\text{k}} + m\left(v - |\omega|R\right)R\hat{\text{k}}$$

which we can use to solve for $|\omega|$.

ANALYZE Solving for the angular speed, the result is

$$|\omega| = \frac{mvR}{(M+m)R^2} = \frac{v}{(M/m+1)R} = \frac{0.15\ \text{m/s}}{(1.1+1)(0.43\ \text{m})} = 0.17\ \text{rad/s}.$$

LEARN By angular momentum conservation, we must have $\vec{L}_{\text{wheel}} = -\vec{L}_{\text{train}}$, which means that train and the wheel must have opposite senses of rotation.

11-51

THINK No external torques act on the system consisting of the two wheels, so its total angular momentum is conserved.

EXPRESS Let I_1 be the rotational inertia of the wheel that is originally spinning $(\text{at } \omega_i)$ and I_2 be the rotational inertia of the wheel that is initially at rest. Then by angular momentum conservation, $L_i = L_f$, or $I_1\,\omega_i = (I_1 + I_2)\omega_f$ and

$$\omega_f = \frac{I_1}{I_1 + I_2}\omega_i$$

where ω_f is the common final angular velocity of the wheels.

ANALYZE (a) Substituting $I_2 = 2I_1$ and $\omega_i = 800 \text{ rev/min}$, we obtain

$$\omega_f = \frac{I_1}{I_1 + I_2}\omega_i = \frac{I_1}{I_1 + 2(I_1)}(800 \text{ rev/min}) = \frac{1}{3}(800 \text{ rev/min}) = 267 \text{ rev/min}.$$

(b) The initial kinetic energy is $K_i = \frac{1}{2}I_1\omega_i^2$ and the final kinetic energy is $K_f = \frac{1}{2}(I_1 + I_2)\omega_f^2$. We rewrite this as

$$K_f = \frac{1}{2}(I_1 + 2I_1)\left(\frac{I_1\omega_i}{I_1 + 2I_1}\right)^2 = \frac{1}{6}I\omega_i^2.$$

Therefore, the fraction lost is

$$\frac{\Delta K}{K_i} = \frac{K_i - K_f}{K_i} = 1 - \frac{K_f}{K_i} = 1 - \frac{I\omega_i^2/6}{I\omega_i^2/2} = \frac{2}{3} = 0.667.$$

LEARN The situation here is analogous to the case of completely inelastic collision, in which some energy is lost but momentum remains conserved.

11-65

THINK If we consider a short time interval from just before the wad hits to just after it hits and sticks, we may use the principle of conservation of angular momentum. The initial angular momentum is the angular momentum of the falling putty wad.

EXPRESS The wad initially moves along a line that is $d/2$ distant from the axis of rotation, where d is the length of the rod. The angular momentum of the wad is $mvd/2$ where m and v are the mass and initial speed of the wad. After the wad sticks, the rod has angular velocity ω and angular momentum $I\omega$, where I is the rotational inertia of the system consisting of the rod with the two balls (each having a mass M) and the wad at its end. Conservation of angular momentum yields $mvd/2 = I\omega$ where

$$I = (2M + m)(d/2)^2 .$$

The equation allows us to solve for ω.

ANALYZE (a) With $M = 2.00$ kg, $d = 0.500$ m, $m = 0.0500$ kg and $v = 3.00$ m/s, we find the angular speed to be

$$\omega = \frac{mvd}{2I} = \frac{2mv}{(2M + m)d} = \frac{2(0.0500 \text{ kg})(3.00 \text{ m/s})}{(2(2.00 \text{ kg}) + 0.0500 \text{ kg})(0.500 \text{ m})} = 0.148 \text{ rad/s}.$$

(b) The initial kinetic energy is $K_i = \frac{1}{2}mv^2$, the final kinetic energy is $K_f = \frac{1}{2}I\omega^2$, and their ratio is $K_f/K_i = I\omega^2/mv^2$. When $I = (2M + m)d^2/4$ and $\omega = 2mv/(2M + m)d$ are substituted, the ratio becomes

$$\frac{K_f}{K_i} = \frac{m}{2M + m} = \frac{0.0500 \text{ kg}}{2(2.00 \text{ kg}) + 0.0500 \text{ kg}} = 0.0123.$$

(c) As the rod rotates, the sum of its kinetic and potential energies is conserved. If one of the balls is lowered a distance h, the other is raised the same distance and the sum of the potential energies of the balls does not change. We need consider only the potential energy of the putty wad. It moves through a 90° arc to reach the lowest point on its path, gaining kinetic energy and losing gravitational potential energy as it goes. It then swings up through an angle θ, losing kinetic energy and gaining potential energy, until it momentarily comes to rest. Take the lowest point on the path to be the zero of potential energy. It starts a distance $d/2$ above this point, so its initial potential energy is $U_i = mg(d/2)$. If it swings up to the angular position θ, as measured from its lowest point, then its final height is $(d/2)(1 - \cos\theta)$ above the lowest point and its final potential energy is

$$U_f = mg(d/2)(1 - \cos\theta).$$

The initial kinetic energy is the sum of that of the balls and wad:

$$K_i = \frac{1}{2}I\omega^2 = \frac{1}{2}(2M + m)(d/2)^2\omega^2.$$

At its final position, we have $K_f = 0$. Conservation of energy provides the relation:

$$U_i + K_i = U_f + K_f \quad \Rightarrow \quad mg\frac{d}{2} + \frac{1}{2}(2M + m)\left(\frac{d}{2}\right)^2\omega^2 = mg\frac{d}{2}(1 - \cos\theta).$$

When this equation is solved for $\cos\theta$, the result is

$$\cos\theta = -\frac{1}{2}\left(\frac{2M + m}{mg}\right)\left(\frac{d}{2}\right)\omega^2 = -\frac{1}{2}\left(\frac{2(2.00 \text{ kg}) + 0.0500 \text{ kg}}{(0.0500 \text{ kg})(9.8 \text{ m/s}^2)}\right)\left(\frac{0.500 \text{ m}}{2}\right)(0.148 \text{ rad/s})^2$$

$$= -0.0226.$$

Consequently, the result for θ is 91.3°. The total angle through which it has swung is 90° + 91.3° = 181°.

LEARN This problem is rather involved. To summarize, we calculated ω using angular momentum conservation. Some energy is lost due to the inelastic collision between the putty wad and one of the balls. However, in the subsequent motion, energy is conserved, and we apply energy conservation to find the angle at which the system comes to rest momentarily.

11-71

THINK The applied force gives rise to a torque that causes the cylinder to rotate to the right at a constant angular acceleration.

EXPRESS We make the unconventional choice of *clockwise* sense as positive, so that the angular acceleration is positive (as is the linear acceleration of the center of mass, since we take rightwards as positive). We approach this in the manner of Eq. 11-3 (*pure rotation* about point *P*) but use torques instead of energy. The torque (relative to point *P*) is $\tau = I_P\alpha$, where

$$I_P = \frac{1}{2}MR^2 + MR^2 = \frac{3}{2}MR^2$$

with the use of the parallel-axis theorem and Table 10-2(c). The torque is due to the F_{app} force and can be written as $\tau = F_{\text{app}}(2R)$. In this way, we find

$$\tau = I_P\alpha = \left(\frac{3}{2}MR^2\right)\alpha = 2RF_{\text{app}}.$$

The equation allows us to solve for the angular acceleration α, which is related to the acceleration of the center of mass as $\alpha = a_{\text{com}}/R$.

ANALYZE (a) With $M = 10\text{ kg}$, $R = 0.10\text{ m}$ and $F_{\text{app}} = 12\text{ N}$, we obtain

$$a_{\text{com}} = \alpha R = \frac{2R^2F_{\text{app}}}{3MR^2/2} = \frac{4F_{\text{app}}}{3M} = \frac{4(12\text{ N})}{3(10\text{ kg})} = 1.6\text{ m/s}^2.$$

(b) The magnitude of the angular acceleration is

$$\alpha = a_{\text{com}}/R = (1.6\text{ m/s}^2)/(0.10\text{ m}) = 16\text{ rad/s}^2.$$

(c) Applying Newton's second law in its linear form yields $(12\text{N}) - f = Ma_{\text{com}}$. Therefore, $f = -4.0$ N. Contradicting what we assumed in setting up our force equation, the friction force is found to point *rightward* with magnitude 4.0 N, i.e., $\vec{f} = (4.0\text{ N})\hat{\text{i}}$.

LEARN As the cylinder rolls to the right, the frictional force also points to the right to oppose the tendency to slip.

11-73

THINK We evaluate the cross product $\vec{\ell} = m\vec{r}\times\vec{v}$ to find the angular momentum $\vec{\ell}$ on the object, and the cross product of $\vec{r}\times\vec{F}$ for the torque.

EXPRESS This problem involves the vector cross product of vectors lying in the *xy* plane. For such vectors, if we write $\vec{r}\,' = x'\hat{\text{i}} + y'\hat{\text{j}}$, then (using Eq. 3-30) we find

$$\vec{r}\,'\times\vec{v} = \left(x'v_y - y'v_x\right)\hat{\text{k}}.$$

ANALYZE (a) Here, $\vec{r}\,'$ points in either the $+\hat{\text{i}}$ or the $-\hat{\text{i}}$ direction (since the particle moves along the *x* axis). It has no y' or z' components, and neither does $\vec{v}$, so it is clear from the above expression (or, more simply, from the fact that $\hat{\text{i}}\times\hat{\text{i}} = 0$) that $\vec{\ell} = m(\vec{r}\,'\times\vec{v}) = 0$ in this case.

(b) The net force is in the $-\hat{\text{i}}$ direction (as one finds from differentiating the velocity expression, yielding the acceleration), so, similar to what we found in part (a), we obtain $\tau = \vec{r}\,'\times\vec{F} = 0$.

(c) Now, $\vec{r}\,' = \vec{r} - \vec{r}_{\text{o}}$ where $\vec{r}_{\text{o}} = 2.0\hat{\text{i}} + 5.0\hat{\text{j}}$ (with SI units understood) and points from (2.0, 5.0, 0) to the instantaneous position of the car (indicated by $\vec{r}$, which points in either the +*x* or –*x* directions, or nowhere (if the car is passing through the origin)). Since $\vec{r}\times\vec{v} = 0$ we have (plugging into our general expression above)

$$\vec{\ell} = m(\vec{r}\,'\times\vec{v}) = -m(\vec{r}_{\text{o}}\times\vec{v}) = -(3.0)\big((2.0)(0) - (5.0)(-2.0t^3)\big)\hat{\text{k}}$$

which gives $\vec{\ell} = (-30t^3\hat{\text{k}})\ \text{kg}\cdot\text{m/s}^2$.

(d) The acceleration vector is given by $\vec{a} = \frac{d\vec{v}}{dt} = -6.0t^2\hat{\text{i}}$ in SI units, and the net force on the car is $m\vec{a}$. In a similar argument to that given in the previous part, we have

$$\vec{\tau} = m(\vec{r}\,'\times\vec{a}) = -m(\vec{r}_{\text{o}}\times\vec{a}) = -(3.0)\big((2.0)(0) - (5.0)(-6.0t^2)\big)\hat{\text{k}}$$

which yields $\vec{\tau} = (-90t^2\hat{\text{k}})\ \text{N}\cdot\text{m}$.

(e) In this situation, $\vec{r}\,' = \vec{r} - \vec{r}_{\text{o}}$ where $\vec{r}_{\text{o}} = 2.0\hat{\text{i}} - 5.0\hat{\text{j}}$ (with SI units understood) and points from (2.0, –5.0, 0) to the instantaneous position of the car (indicated by $\vec{r}$, which points in either the $+x$ or $-x$ directions, or nowhere (if the car is passing through the origin)). Since $\vec{r}\times\vec{v}=0$ we have (plugging into our general expression above)

$$\vec{\ell}=m(\vec{r}\,'\times\vec{v})=-m(\vec{r}_o\times\vec{v})=-(3.0)\big((2.0)(0)-(-5.0)(-2.0t^3)\big)\hat{\text{k}}$$

which yields $\vec{\ell}=(30t^3\hat{\text{k}})\ \text{kg}\cdot\text{m}^2/\text{s}$.

(f) Again, the acceleration vector is given by $\vec{a}=-6.0t^2\hat{\text{i}}$ in SI units, and the net force on the car is $m\vec{a}$. In a similar argument to that given in the previous part, we have

$$\vec{\tau}=m(\vec{r}\,'\times\vec{a})=-m(\vec{r}_o\times\vec{a})=-(3.0)\big((2.0)(0)-(-5.0)(-6.0t^2)\big)\hat{\text{k}}$$

which yields $\vec{\tau}=(90t^2\hat{\text{k}})\ \text{N}\cdot\text{m}$.

LEARN Both $\vec{\ell}=m\vec{r}\times\vec{v}$ and $\vec{\tau}=\vec{r}\times\vec{F}$ depend on $\vec{r}$, the position vector of the toy car with respect to some reference point which we take to be the origin.

11-75

THINK No external torque acts on the system consisting of the child and the merry-go-round, so the total angular momentum of the system is conserved.

EXPRESS An object moving along a straight line has angular momentum about any point that is not on the line. The magnitude of the angular momentum of the child about the center of the merry-go-round is given by Eq. 11-21, mvR, where R is the radius of the merry-go-round.

ANALYZE (a) In terms of the radius of gyration k, the rotational inertia of the merry-go-round is $I = Mk^2$. With $M = 180$ kg and $k = 0.91$ m, we obtain

$$I = (180\ \text{kg})\ (0.910\ \text{m})^2 = 149\ \text{kg}\cdot\text{m}^2.$$

(b) The magnitude of angular momentum of the running child about the axis of rotation of the merry-go-round is

$$L_{\text{child}}=mvR=(44.0\ \text{kg})(3.00\ \text{m/s})(1.20\ \text{m})=158\ \text{kg}\cdot\text{m}^2/\text{s}.$$

(c) The initial angular momentum is given by $L_i=L_{\text{child}}=mvR$; the final angular momentum is given by $L_f = (I + mR^2)\,\omega$, where ω is the final common angular velocity of the merry-go-round and child. Thus $mvR=(I+mR^2)\omega$ and

$$\omega=\frac{mvR}{I+mR^2}=\frac{158\ \text{kg}\cdot\text{m}^2/\text{s}}{149\ \text{kg}\cdot\text{m}^2+(44.0\ \text{kg})(1.20\ \text{m})^2}=0.744\ \text{rad/s}.$$

LEARN The child initially had an angular velocity of

$$\omega_0 = \frac{v}{R} = \frac{3.00\text{ m/s}}{1.20\text{ m}} = 2.5\text{ rad/s}.$$

After he jumped onto the merry-go-round, the rotational inertia of the system (merry-go-round + child) increases, so the angular velocity decreases by angular momentum conservation.

11-77

THINK Our system consists of two particles moving in opposite directions along parallel lines. The angular momentum of the system about a point is the vector sum of the two individual angular momenta.

EXPRESS The diagram below shows the particles and their lines of motion. The origin is marked *O* and may be anywhere. We set up our coordinate system in such a way that +*x* is to the right, +*y* up and +*z* out of the page. The angular momentum of the system about *O* is

$$\vec{\ell} = \vec{\ell}_1 + \vec{\ell}_2 = \vec{r}_1 \times \vec{p}_1 + \vec{r}_2 \times \vec{p}_2 = m(\vec{r}_1 \times \vec{v}_1 + \vec{r}_2 \times \vec{v}_2)$$

since $m_1 = m_2 = m$.

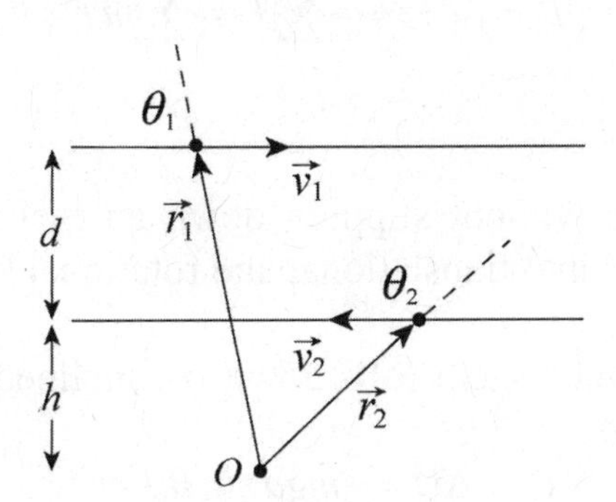

ANALYZE (a) With $\vec{v}_1 = v_1\hat{\text{i}}$, the angular momentum of particle 1 has magnitude

$$\ell_1 = mvr_1 \sin\theta_1 = mv(d+h)$$

and is in the –*z*-direction, or into the page. On the other hand, with $\vec{v}_2 = -v_2\hat{\text{i}}$, the angular momentum of particle 2 has magnitude

$$\ell_2 = mvr_2 \sin\theta_2 = mvh$$

and is in the +*z*-direction, or out of the page. The net angular momentum has magnitude

$$\ell = mv(d+h) - mvh = mvd$$

which depends only on the separation between the two lines and not on the location of the origin. Thus, if O is midway between the two lines, the total angular momentum is

$$\ell = mvd = (2.90\times10^{-4}\ \text{kg})(5.46\ \text{m/s})(0.042\ \text{m}) = 6.65\times10^{-5}\ \text{kg}\cdot\text{m}^2/\text{s}$$

and is into the page.

(b) As indicated above, the expression does not change.

(c) Suppose particle 2 is traveling to the right. Then

$$\ell = mv(d+h) + mvh = mv(d+2h).$$

This result now depends on h, the distance from the origin to one of the lines of motion. If the origin is midway between the lines of motion, then $h = -d/2$ and $\ell = 0$.

(d) As we have seen in part (c), the result depends on the choice of origin.

LEARN Angular momentum is a vector quantity. For a system of many particles, the total angular momentum about a point is

$$\vec{\ell} = \vec{\ell}_1 + \vec{\ell}_2 + \cdots = \sum_i \vec{\ell}_i = \sum_i m_i \vec{r}_i \times \vec{v}_i .$$

11-81

THINK As the wheel rolls without slipping down an inclined plane, its gravitational potential energy is converted into translational and rotational kinetic energies.

EXPRESS As the wheel-axel system rolls down the inclined plane by a distance d, the change in potential energy is

$$\Delta U = -mgd\sin\theta .$$

By energy conservation, the total kinetic energy gained is

$$-\Delta U = \Delta K = \Delta K_{\text{trans}} + \Delta K_{\text{rot}} \quad \Rightarrow \quad mgd\sin\theta = \frac{1}{2}mv^2 + \frac{1}{2}I\omega^2 .$$

Since the axel rolls without slipping, the angular speed is given by $\omega = v/r$, where r is the radius of the axel. The above equation then becomes

$$mgd\sin\theta = \frac{1}{2}I\omega^2\left(\frac{mr^2}{I}+1\right) = \Delta K_{\text{rot}}\left(\frac{mr^2}{I}+1\right).$$

ANALYZE (a) With m=10.0 kg, d = 2.00 m, r = 0.200 m, and $I = 0.600\ \text{kg}\cdot\text{m}^2$, the rotational kinetic energy may be obtained as

$$\Delta K_{\text{rot}} = \frac{mgd\sin\theta}{\dfrac{mr^2}{I}+1} = \frac{(10.0\ \text{kg})(9.80\ \text{m/s}^2)(2.00\ \text{m})\sin 30.0°}{\dfrac{(10.0\ \text{kg})(0.200\ \text{m})^2}{0.600\ \text{kg}\cdot\text{m}^2}+1} = 58.8\ \text{J}.$$

(b) The translational kinetic energy is

$$\Delta K_{\text{trans}} = \Delta K - \Delta K_{\text{rot}} = 98\ \text{J} - 58.8\ \text{J} = 39.2\ \text{J}.$$

LEARN One may show that $mr^2/I = 2/3$, which implies that $\Delta K_{\text{trans}}/\Delta K_{\text{rot}} = 2/3$. Equivalently, we may write $\Delta K_{\text{trans}}/\Delta K = 2/5$ and $\Delta K_{\text{rot}}/\Delta K = 3/5$. So as the wheel rolls down, 40% of the kinetic energy is translational while the other 60% is rotational.

Chapter 12

12-3

THINK Three forces act on the sphere: the tension force $\vec{T}$ of the rope, the force of the wall $\vec{F}_N$, and the force of gravity $m\vec{g}$.

EXPRESS The free-body diagram is shown to the right. The tension force $\vec{T}$ acts along the rope, the force of the wall $\vec{F}_N$ acts horizontally away from the wall, and the force of gravity $m\vec{g}$ acts downward. Since the sphere is in equilibrium they sum to zero. Let θ be the angle between the rope and the vertical. Then Newton's second law gives

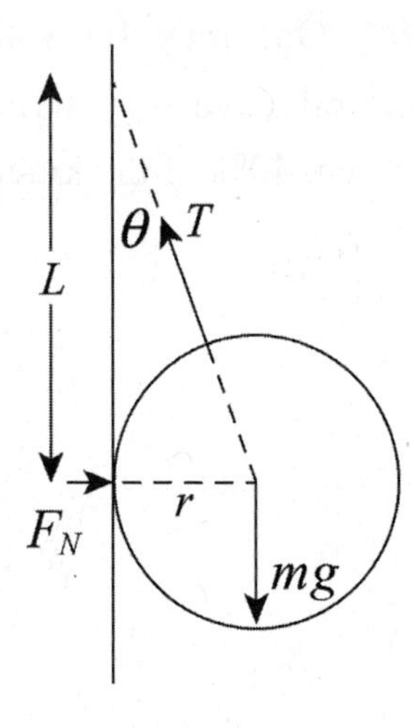

$$\text{vertical component}: \quad T\cos\theta - mg = 0$$
$$\text{horizontal component}: \quad F_N - T\sin\theta = 0.$$

ANALYZE (a) We solve the first equation for the tension: $T = mg/\cos\theta$. We substitute $\cos\theta = L/\sqrt{L^2 + r^2}$ to obtain

$$T = \frac{mg\sqrt{L^2+r^2}}{L} = \frac{(0.85\text{ kg})(9.8\text{ m/s}^2)\sqrt{(0.080\text{ m})^2+(0.042\text{ m})^2}}{0.080\text{ m}} = 9.4\text{ N}.$$

(b) We solve the second equation for the normal force: $F_N = T\sin\theta$. Using $\sin\theta = r/\sqrt{L^2+r^2}$, we obtain

$$F_N = \frac{Tr}{\sqrt{L^2+r^2}} = \frac{mg\sqrt{L^2+r^2}}{L}\frac{r}{\sqrt{L^2+r^2}} = \frac{mgr}{L} = \frac{(0.85\text{ kg})(9.8\text{ m/s}^2)(0.042\text{ m})}{(0.080\text{ m})} = 4.4\text{ N}.$$

LEARN Since the sphere is in static equilibrium, the vector sum of all external forces acting on it must be zero.

12-9

THINK In order for the meter stick to remain in equilibrium, the net force acting on it must be zero. In addition, the net torque about any point must also be zero.

EXPRESS Let the x axis be along the meter stick, with the origin at the zero position on the scale. The forces acting on it are shown to the right. The coins are at $x = x_1 = 0.120$ m, and $m = 10.0$ g is their total mass. The knife edge is at $x = x_2 = 0.455$ m and exerts force $\vec{F}$. The mass of the meter stick is M, and the force of gravity acts at the center of the stick, $x = x_3 = 0.500$ m.

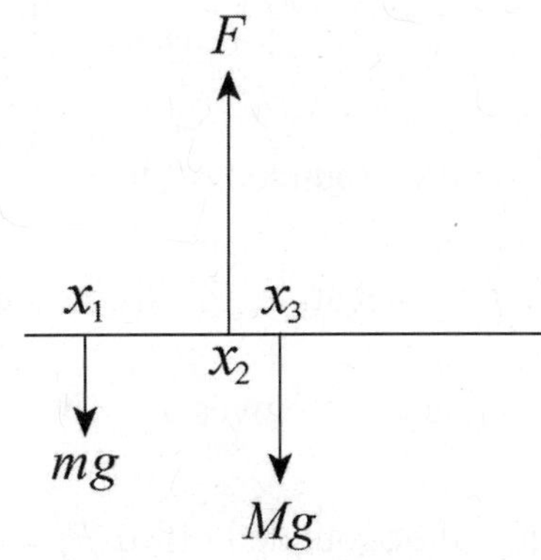

Since the meter stick is in equilibrium, the sum of the torques about x_2 must vanish:

$$Mg(x_3 - x_2) - mg(x_2 - x_1) = 0.$$

ANALYZE Solving the equation above for M, we find the mass of the meter stick to be

$$M = \left(\frac{x_2 - x_1}{x_3 - x_2}\right)m = \left(\frac{0.455\,\text{m} - 0.120\,\text{m}}{0.500\,\text{m} - 0.455\,\text{m}}\right)(10.0\text{g}) = 74.4 \text{ g}.$$

LEARN Since the torque about any point is zero, we could have chosen x_1. In this case, balance of torques requires that

$$F(x_2 - x_1) - Mg(x_3 - x_1) = 0$$

The fact that the net force is zero implies $F = (M + m)g$. Substituting this into the above equation gives the same result as before:

$$M = \left(\frac{x_2 - x_1}{x_3 - x_2}\right)m.$$

12-11

THINK The diving board is in equilibrium, so the net force and net torque must be zero.

EXPRESS We take the force of the left pedestal to be F_1 at $x = 0$, where the x axis is along the diving board. We take the force of the right pedestal to be F_2 and denote its position as $x = d$. Upward direction is taken to be positive and W is the weight of the diver, located at $x = L$. The following two equations result from setting the sum of forces equal to zero (with upwards positive), and the sum of torques (about x_2) equal to zero:

$$F_1 + F_2 - W = 0$$
$$F_1 d + W(L - d) = 0$$

ANALYZE (a) The second equation gives

$$F_1 = -\left(\frac{L-d}{d}\right)W = -\left(\frac{3.0\,\text{m}}{1.5\,\text{m}}\right)(580\,\text{N}) = -1160\,\text{N}$$

which should be rounded off to $F_1 = -1.2\times10^3$ N. Thus, $|F_1| = 1.2\times10^3$ N.

(b) Since F_1 is negative, this force is downward.

(c) The first equation gives $F_2 = W - F_1 = 580\,\text{N} + 1160\,\text{N} = 1740\,\text{N}$.

which should be rounded off to $F_2 = 1.7\times10^3$ N. Thus, $|F_2| = 1.7\times10^3$ N.

(d) The result is positive, indicating that this force is upward.

(e) The force of the diving board on the left pedestal is upward (opposite to the force of the pedestal on the diving board), so this pedestal is being stretched.

(f) The force of the diving board on the right pedestal is downward, so this pedestal is being compressed.

LEARN We may relate F_1 and F_2 via $F_1 = -\left(\frac{L-d}{L}\right)F_2$. The expression makes it clear that the two forces must be of opposite signs, i.e., one acting downward and the other upward.

12-25

THINK At the moment when the wheel leaves the lower floor, the floor no longer exerts a force on it.

EXPRESS As the wheel is raised over the obstacle, the only forces acting are the force F applied horizontally at the axle, the force of gravity mg acting vertically at the center of the wheel, and the force of the step corner, shown as the two components f_h and f_v.

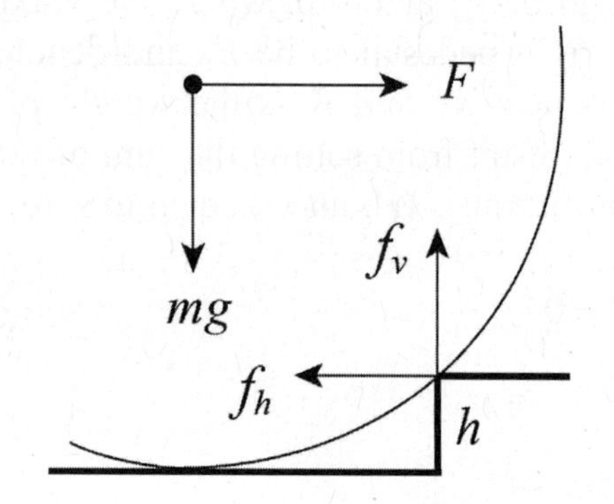

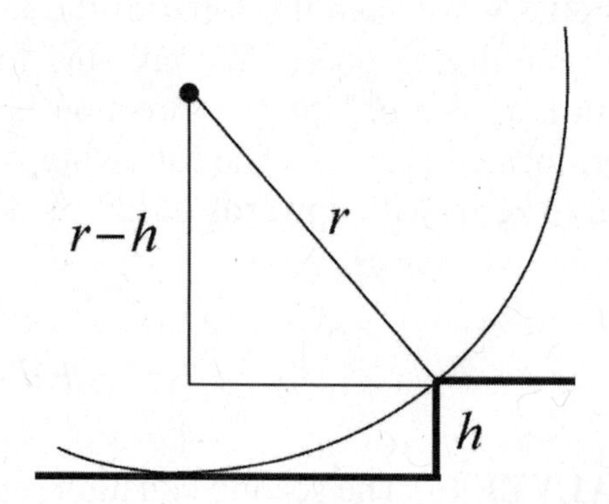

If the minimum force is applied the wheel does not accelerate, so both the total force and the total torque acting on it are zero.

We calculate the torque around the step corner. The second diagram (above right) indicates that the distance from the line of F to the corner is $r - h$, where r is the radius of the wheel and h is the height of the step. The distance from the line of mg to the corner is $\sqrt{r^2 + (r-h)^2} = \sqrt{2rh - h^2}$. Thus,

$$F(r-h) - mg\sqrt{2rh-h^2} = 0.$$

ANALYZE The solution for F is

$$F = \frac{\sqrt{2rh-h^2}}{r-h}mg = \frac{\sqrt{2(6.00\times10^{-2}\,\text{m})(3.00\times10^{-2}\,\text{m})-(3.00\times10^{-2}\,\text{m})^2}}{(6.00\times10^{-2}\,\text{m})-(3.00\times10^{-2}\,\text{m})}(0.800\text{ kg})(9.80\text{ m/s}^2)$$
$$= 13.6\text{ N}.$$

LEARN The applied force here is about 1.73 times the weight of the wheel. If the height is increased, the force that must be applied also goes up. Below we plot F/mg as a function of the ratio h/r. The required force increases rapidly as $h/r \to 1$.

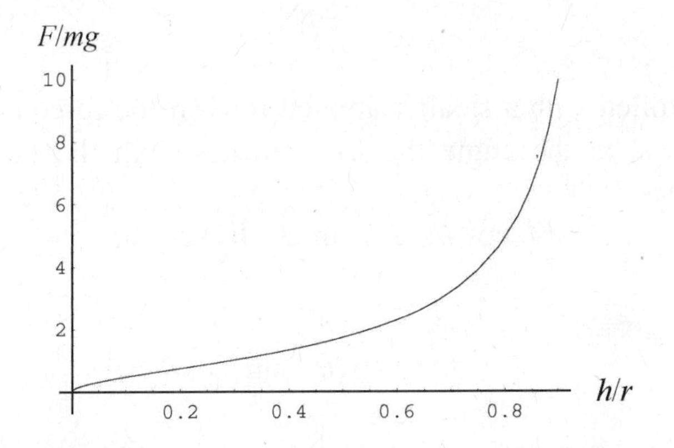

12-35

THINK We examine the box when it is about to tip. Since it will rotate about the lower right edge, this is where the normal force of the floor is exerted.

EXPRESS The free-body diagram is shown below. The normal force is labeled F_N, the force of friction is denoted by f, the applied force by F, and the force of gravity by W. Note that the force of gravity is applied at the center of the box. When the minimum force is applied the box does not accelerate, so the sum of the horizontal force components vanishes: $F - f = 0$, the sum of the vertical force components vanishes: $F_N - W = 0$, and the sum of the torques vanishes:

$$FL - WL/2 = 0.$$

Here L is the length of a side of the box and the origin was chosen to be at the lower right edge.

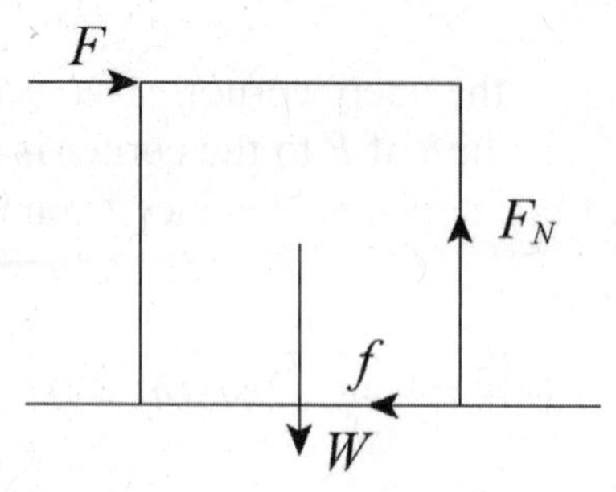

ANALYZE (a) From the torque equation, we find $F = \frac{W}{2} = \frac{890\text{ N}}{2} = 445\text{ N}.$

(b) The coefficient of static friction must be large enough that the box does not slip. The box is on the verge of slipping if $\mu_s = f/F_N$. According to the equations of equilibrium

$$F_N = W = 890\text{ N}$$
$$f = F = 445\text{ N},$$

so

$$\mu_s = \frac{f}{F_N} = \frac{445\text{ N}}{890\text{ N}} = 0.50.$$

(c) The box can be rolled with a smaller applied force if the force points upward as well as to the right. Let θ be the angle the force makes with the horizontal. The torque equation then becomes

$$FL\cos\theta + FL\sin\theta - WL/2 = 0,$$

with the solution

$$F = \frac{W}{2(\cos\theta + \sin\theta)}.$$

We want $\cos\theta + \sin\theta$ to have the largest possible value. This occurs if $\theta = 45°$, a result we can prove by setting the derivative of $\cos\theta + \sin\theta$ equal to zero and solving for θ. The minimum force needed is

$$F = \frac{W}{2(\cos 45° + \sin 45°)} = \frac{890\text{ N}}{2(\cos 45° + \sin 45°)} = 315\text{ N}.$$

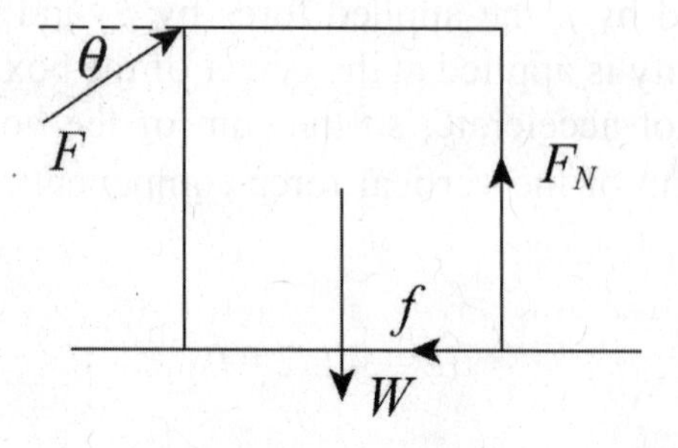

LEARN The applied force as a function of θ is plotted below. From the figure, we readily see that $\theta = 0°$ corresponds to a maximum and $\theta = 45°$ a minimum.

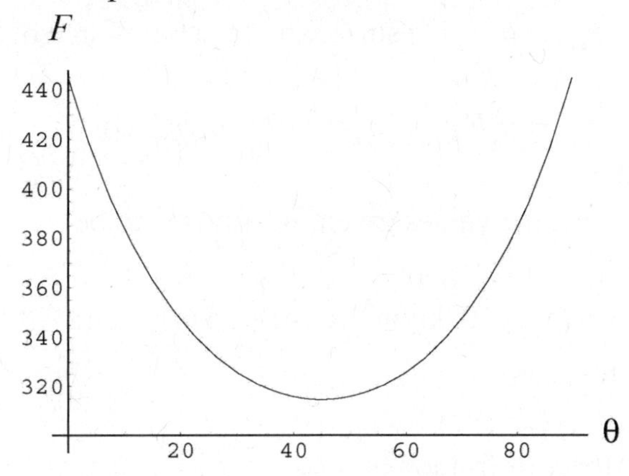

12-43

THINK The weight of the object hung on the end provides the source of shear stress.

EXPRESS The shear stress is given by F/A, where F is the magnitude of the force applied parallel to one face of the aluminum rod and A is the cross–sectional area of the rod. In this case $F = mg$, where m is the mass of the object. The cross-sectional area is $A = \pi r^2$ where r is the radius of the rod.

ANALYZE (a) Substituting the values given, we find the shear stress to be

$$\frac{F}{A} = \frac{mg}{\pi r^2} = \frac{(1200\,\text{kg})(9.8\,\text{m/s}^2)}{\pi(0.024\,\text{m})^2} = 6.5\times10^6\,\text{N/m}^2.$$

(b) The shear modulus G is given by

$$G = \frac{F/A}{\Delta x/L},$$

where L is the protrusion of the rod and Δx is its vertical deflection at its end. Thus,

$$\Delta x = \frac{(F/A)L}{G} = \frac{(6.5\times10^6\,\text{N/m}^2)(0.053\,\text{m})}{3.0\times10^{10}\,\text{N/m}^2} = 1.1\times10^{-5}\,\text{m}.$$

LEARN As expected, the extent of vertical deflection Δx is proportional to F, the weight of the object hung from the end. On the other hand, it is inversely proportional to the shear modulus G.

12-53

THINK The slab can remain in static equilibrium if the combined force of the friction and the bolts is greater than the component of the weight of the slab along the incline.

EXPRESS We denote the mass of the slab as m, its density as ρ, and volume as $V = LTW$. The angle of inclination is $\theta = 26°$. The component of the weight of the slab along the incline is $F_1 = mg\sin\theta = \rho V g \sin\theta$, and the static force of friction is

$$f_s = \mu_s F_N = \mu_s mg\cos\theta = \mu_s \rho V g\cos\theta.$$

ANALYZE (a) Substituting the values given, we find F_1 to be

$$\begin{aligned} F_1 &= \rho V g\sin\theta = (3.2\times10^3\ \text{kg/m}^3)(43\,\text{m})(2.5\,\text{m})(12\,\text{m})(9.8\,\text{m/s}^2)\sin 26° \\ &\approx 1.8\times10^7\ \text{N}. \end{aligned}$$

(b) Similarly, the static force of friction is

$$\begin{aligned} f_s &= \mu_s \rho V g\cos\theta = (0.39)(3.2\times10^3\ \text{kg/m}^3)(43\,\text{m})(2.5\,\text{m})(12\,\text{m})(9.8\,\text{m/s}^2)\cos 26° \\ &\approx 1.4\times10^7\ \text{N}. \end{aligned}$$

(c) The minimum force needed from the bolts to stabilize the slab is

$$F_2 = F_1 - f_s = 1.77\times10^7\,\text{N} - 1.42\times10^7\ \text{N} = 3.5\times10^6\ \text{N}.$$

If the minimum number of bolts needed is n, then $F_2 / nA \le S_G$, where $S_G = 3.6\times10^8\ \text{N/m}^2$ is the shear stress. Solving for n, we find

$$n \ge \frac{3.5\times10^6\ \text{N}}{(3.6\times10^8\ \text{N/m}^2)(6.4\times10^{-4}\ \text{m}^2)} = 15.2$$

Therefore, 16 bolts are needed.

LEARN In general, the number of bolts needed to maintain static equilibrium of the slab is

$$n = \frac{F_1 - f_s}{S_G A}.$$

Thus, no bolt would be necessary if $f_s > F_1$.

12-55

THINK Block A can be in equilibrium if friction is present between the block and the surface in contact.

EXPRESS The free-body diagrams for blocks A, B and the knot (denoted as C) are shown below. The tensions in the three strings are denoted as T_A, T_B and T_C Analyzing forces at C, the conditions for static equilibrium are

$$T_C \cos\theta = T_B$$
$$T_C \sin\theta = T_A$$

which can be combined to give $\tan\theta = T_A / T_B$.

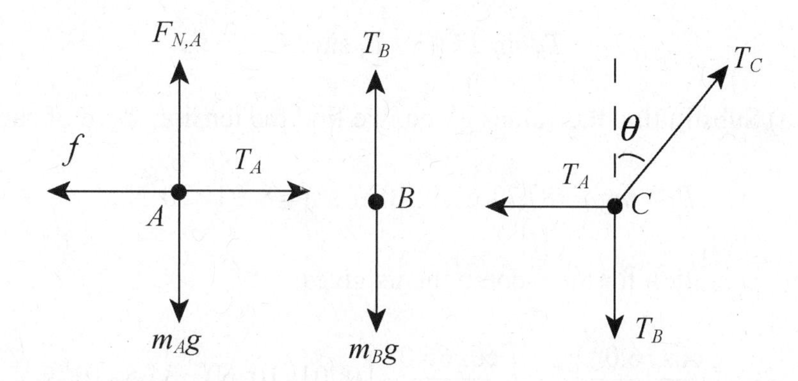

On the other hand, equilibrium condition for block *B* implies $T_B = m_B g$. Similarly, for block A, the conditions are

$$F_{N,A} = m_A g, \quad f = T_A$$

For the static force to be at its maximum value, we have $f = \mu_s F_{N,A} = \mu_s m_A g$. Combining all the equations leads to

$$\tan\theta = \frac{T_A}{T_B} = \frac{\mu_s m_A g}{m_B g} = \frac{\mu_s m_A}{m_B}.$$

ANALYZE Solving for μ_s, we get

$$\mu_s = \left(\frac{m_B}{m_A}\right)\tan\theta = \left(\frac{5.0\text{ kg}}{10\text{ kg}}\right)\tan 30° = 0.29$$

LEARN The greater the mass of block *B*, the greater the static coefficient μ_s would be required for block *A* to be in equilibrium.

12-59
THINK The bucket is in static equilibrium. The forces acting on it are the downward force of gravity and the upward tension force of cable A.

EXPRES Since the bucket is in equilibrium, the tension force of cable A is equal to the weight of the bucket: $T_A = W = mg$. To solve for T_B and T_C, we use the coordinates axes defined in the diagram. Cable A makes an angle of θ_2 = 66.0° with the negative *y* axis, cable B makes an angle of 27.0° with the positive y axis, and cable C is along the x axis. The y components of the forces must sum to zero since the knot is in equilibrium. This means

$$T_B \cos 27.0° - T_A \cos 66.0° = 0.$$

Similarly, the fact that the x components of forces must also sum to zero implies

$$T_C + T_B \sin 27.0° - T_A \sin 66.0° = 0.$$

ANALYZE (a) Substituting the values given, we find the tension force of cable A to be

$$T_A = mg = (817\,\text{kg})(9.80\,\text{m/s}^2) = 8.01\times10^3\,\text{N}.$$

(b) Equilibrium condition for the y-components gives

$$T_B = \left(\frac{\cos 66.0°}{\cos 27.0°}\right) T_A = \left(\frac{\cos 66.0°}{\cos 27.0°}\right)(8.01\times10^3\,\text{N}) = 3.65\times10^3\text{N}.$$

(c) Using the equilibrium condition for the x-components, we have

$$T_C = T_A \sin 66.0° - T_B \sin 27.0° = (8.01\times10^3\,\text{N})\sin 66.0° - (3.65\times10^3\,\text{N})\sin 27.0°$$
$$= 5.66\times10^3\,\text{N}.$$

LEARN One may verify that the tensions obey law of sine:

$$\frac{T_A}{\sin(180° - \theta_1 - \theta_2)} = \frac{T_B}{\sin(90° + \theta_2)} = \frac{T_C}{\sin(90° + \theta_1)}.$$

12-69

THINK Since the rod is in static equilibrium, the net torque about the hinge must be zero.

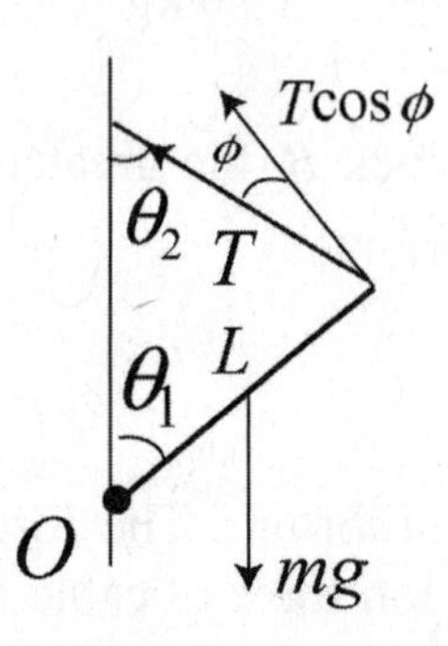

EXPRESS The free-body diagram is shown above (not to scale). The tension in the rope is denoted as T. Since the rod is in rotational equilibrium, the net torque about the hinge, denoted as O, must be zero. This implies

$$-mg\sin\theta_1\tfrac{L}{2} + TL\cos\phi = 0,$$

where $\phi = \theta_1 + \theta_2 - 90°$.

ANALYZE Solving for T gives

$$T = \frac{mg}{2}\frac{\sin\theta_1}{\cos(\theta_1+\theta_2-90°)} = \frac{mg}{2}\frac{\sin\theta_1}{\sin(\theta_1+\theta_2)}.$$

With $\theta_1 = 60°$ and $T = mg/2$, we have $\sin 60° = \sin(60° + \theta_2)$, which yields $\theta_2 = 60°$.

LEARN A plot of T/mg as a function of θ_2 is shown below. The other solution, $\theta_2 = 0°$, is rejected since it corresponds to the limit where the rope becomes infinitely long.

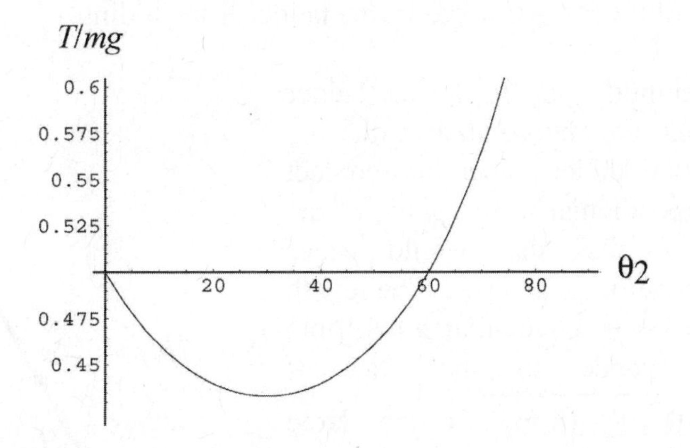

12-71

THINK Upon applying a horizontal force, the cube may tip or slide, depending on the friction between the cube and the floor.

EXPRESS When the cube is about to move, we are still able to apply the equilibrium conditions, but (to obtain the critical condition) we set static friction equal to its maximum value and picture the normal force $\vec{F}_N$ as a concentrated force (upward) at the bottom corner of the cube, directly below the point O where P is being applied. Thus, the line of action of $\vec{F}_N$ passes through point O and exerts no torque about O (of course, a similar observation applied to the pull P). Since $F_N = mg$ in this problem, we have $f_{s\max} = \mu_c mg$ applied a distance h away from O. And the line of action of force of gravity (of magnitude mg), which is best pictured as a concentrated force at the center of the cube, is a distance $L/2$ away from O. Therefore, equilibrium of torques about O produces

$$\mu_c mgh = mg\left(\frac{L}{2}\right) \Rightarrow \mu_c = \frac{L}{2h} = \frac{(8.0\text{ cm})}{2(7.0\text{ cm})} = 0.57$$

for the critical condition we have been considering. We now interpret this in terms of a range of values for μ.

ANALYZE (a) For it to slide but not tip, a value of μ *less* than μ_c is needed, since then — static friction will be exceeded for a smaller value of P, before the pull is strong enough to cause it to tip. Thus, $\mu < \mu_c = L/2h = 0.57$ is required.

(b) And for it to tip but not slide, we need μ *greater* than μ_c is needed, since now — static friction will not be exceeded even for the value of P which makes the cube rotate about its front lower corner. That is, we need to have $\mu > \mu_c = L/2h = 0.57$ in this case.

LEARN Note that the value μ_c depends only on the ratio L/h. The cube will tend to slide when μ is mall (think about the limit of a frictionless floor), and tend to tip over when the friction is sufficiently large.

12-73

THINK The force of the ground prevents the ladder from sliding.

EXPRESS The free-body diagram for the ladder is shown to the right. We choose an axis through O, the top (where the ladder comes into contact with the wall), perpendicular to the plane of the figure and take torques that would cause counterclockwise rotation as positive. The length of the ladder is $L = 10$ m. Given that $h = 8.0$ m, the horizontal distance to the wall is $x = \sqrt{L^2 - h^2} = \sqrt{(10\text{ m})^2 - (8\text{ m})^2} = 6.0\,\text{m}$. Note that the line of action of the applied force $\vec{F}$ intersects the wall at a height of $(8.0\text{ m})/5 = 1.6\,\text{m}$.

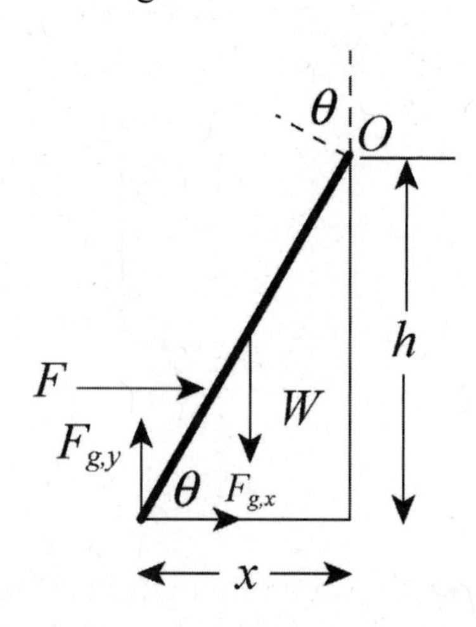

In other words, the *moment arm* for the applied force (in terms of where we have chosen the axis) is

$$r_\perp = (L-d)\sin\theta = (L-d)(h/L) = (8.0\text{ m})(8.0\text{ m}/10.0\text{ m}) = 6.4\,\text{m}.$$

The moment arm for the weight is $x/2 = 3.0\,\text{m}$, half the horizontal distance from the wall to the base of the ladder. Similarly, the moment arms for the x and y components of the force at the ground $(\vec{F}_g)$ are $h = 8.0$ m and $x = 6.0$ m, respectively. Thus, we have

$$\begin{aligned}\sum\tau_z &= Fr_\perp + W(x/2) + F_{g,x}h - F_{g,y}x \\ &= F(6.4\text{ m}) + W(3.0\text{ m}) + F_{g,x}(8.0\text{ m}) - F_{g,y}(6.0\text{ m}) = 0.\end{aligned}$$

In addition, from balancing the vertical forces we find that $W = F_{g,y}$ (keeping in mind that the wall has no friction). Therefore, the above equation can be written as

$$\sum \tau_z = F(6.4\text{ m}) + W(3.0\text{ m}) + F_{g,x}(8.0\text{ m}) - W(6.0\text{ m}) = 0.$$

ANALYZE (a) With F = 50 N and W = 200 N, the above equation yields $F_{g,x} = 35\text{ N}$. Thus, in unit vector notation we obtain

$$\vec{F}_g = (35\text{ N})\hat{\text{i}} + (200\text{ N})\hat{\text{j}}.$$

(b) Similarly, with F = 150 N and W = 200 N, the above equation yields $F_{g,x} = -45\text{ N}$. Therefore, in unit vector notation we obtain

$$\vec{F}_g = (-45\text{ N})\hat{\text{i}} + (200\text{ N})\hat{\text{j}}.$$

(c) Note that the phrase "start to move towards the wall" implies that the friction force is pointed away from the wall (in the $-\hat{\text{i}}$ direction). Now, if $f = -F_{g,x}$ and $F_N = F_{g,y} = 200\text{ N}$ are related by the (maximum) static friction relation ($f = f_{s,\max} = \mu_s F_N$) with μ_s = 0.38, then we find $F_{g,x} = -76\text{ N}$. Returning this to the above equation, we obtain

$$F = \frac{W(x/2) + \mu_s Wh}{r_\perp} = \frac{(200\text{ N})(3.0\text{ m}) + (0.38)(200\text{N})(8.0\text{ m})}{6.4\text{ m}} = 1.9\times10^2\text{ N}.$$

LEARN The force needed to move the ladder toward the wall would decrease with a larger $r_\perp$ or a smaller μ_s.

Chapter 13

13-3

THINK The magnitude of gravitational force between two objects depends on their distance of separation.

EXPRESS The magnitude of the gravitational force of one particle on the other is given by $F = Gm_1m_2/r^2$, where m_1 and m_2 are the masses, r is their separation, and G is the universal gravitational constant.

ANALYZE Solve for r using the values given, we obtain

$$r = \sqrt{\frac{Gm_1m_2}{F}} = \sqrt{\frac{\left(6.67\times10^{-11}\,\text{N}\cdot\text{m}^2/\text{kg}^2\right)\left(5.2\text{kg}\right)\left(2.4\text{kg}\right)}{2.3\times10^{-12}\,\text{N}}} = 19\,\text{m}.$$

LEARN The force of gravitation is inversely proportional to r^2.

13-9

THINK Both the Sun and the Earth exert a gravitational pull on the space probe. The net force can be calculated by using superposition principle.

EXPRESS At the point where the two forces balance, we have $GM_Em/r_1^2 = GM_Sm/r_2^2$, where M_E is the mass of Earth, M_S is the mass of the Sun, m is the mass of the space probe, r_1 is the distance from the center of Earth to the probe, and r_2 is the distance from the center of the Sun to the probe. We substitute $r_2 = d - r_1$, where d is the distance from the center of Earth to the center of the Sun, to find

$$\frac{M_E}{r_1^2} = \frac{M_S}{\left(d - r_1\right)^2}.$$

ANALYZE Using the values for M_E, M_S, and d given in Appendix C, we take the positive square root of both sides to solve for r_1. A little algebra yields

$$r_1 = \frac{d}{1+\sqrt{M_S/M_E}} = \frac{1.50\times10^{11}\text{ m}}{1+\sqrt{(1.99\times10^{30}\text{ kg})/(5.98\times10^{24}\text{ kg})}} = 2.60\times10^{8}\text{ m}.$$

LEARN The fact that $r_1 \ll d$ indicates that the probe is much closer to the Earth than the Sun.

13-19

THINK Earth's gravitational acceleration varies with altitude.

EXPRESS The acceleration due to gravity is given by $a_g = GM/r^2$, where M is the mass of Earth and r is the distance from Earth's center. We substitute $r = R + h$, where R is the radius of Earth and h is the altitude, to obtain

$$a_g = \frac{GM}{r^2} = \frac{GM}{(R_E + h)^2}.$$

ANALYZE Solving for h, we obtain $h = \sqrt{GM / a_g} - R_E$. From Appendix C, $R_E = 6.37 \times 10^6$ m and $M = 5.98 \times 10^{24}$ kg, so

$$h = \sqrt{\frac{(6.67 \times 10^{-11} \text{m}^3/\text{s}^2 \cdot \text{kg})(5.98 \times 10^{24} \text{kg})}{(4.9 \text{m/s}^2)}} - 6.37 \times 10^6 \text{m} = 2.6 \times 10^6 \text{m}.$$

LEARN We may rewrite a_g as

$$a_g = \frac{GM}{r^2} = \frac{GM/R_E^2}{(1 + h/R_E)^2} = \frac{g}{(1 + h/R_E)^2}$$

where $g = 9.83 \text{ m/s}^2$ is the gravitational acceleration on the Surface of the Earth. The plot below depicts how a_g decreases with increasing altitude.

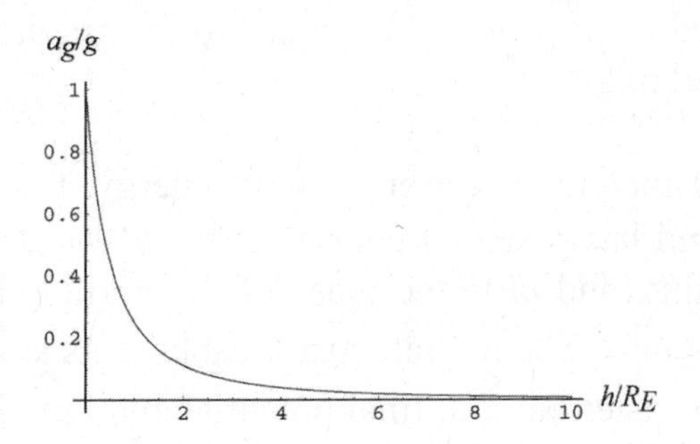

13-31

THINK Given the mass and diameter of Mars, we can calculate its mean density, gravitational acceleration and escape speed.

EXPRESS The density of a uniform sphere is given by $\rho = 3M/4\pi R^3$, where M is its mass and R is its radius. On the other hand, the value of gravitational acceleration a_g at the surface of a planet is given by $a_g = GM/R^2$. for a particle of mass m, its escape speed is given by

$$\frac{1}{2}mv^2 = G\frac{mM}{R} \quad \Rightarrow \quad v = \sqrt{\frac{2GM}{R}}.$$

ANALYZE (a) From the definition of density above, we find the ratio of the density of Mars to the density of Earth to be

$$\frac{\rho_M}{\rho_E} = \frac{M_M}{M_E}\frac{R_E^3}{R_M^3} = 0.11\left(\frac{0.65\times10^4 \text{ km}}{3.45\times10^3 \text{ km}}\right)^3 = 0.74.$$

(b) The value of gravitational acceleration for Mars is

$$a_{gM} = \frac{GM_M}{R_M^2} = \frac{M_M}{R_M^2}\cdot\frac{R_E^2}{M_E}\cdot\frac{GM_E}{R_E^2} = \frac{M_M}{M_E}\frac{R_E^2}{R_M^2}a_{gE} = 0.11\left(\frac{0.65\times10^4 \text{ km}}{3.45\times10^3 \text{ km}}\right)^2 (9.8 \text{ m/s}^2) = 3.8 \text{ m/s}^2.$$

(c) For Mars, the escape speed is

$$v_M = \sqrt{\frac{2GM_M}{R_M}} = \sqrt{\frac{2(6.67\times10^{-11} \text{ m}^3/\text{s}^2\cdot\text{kg})(0.11)(5.98\times10^{24} \text{ kg})}{3.45\times10^6 \text{ m}}} = 5.0\times10^3 \text{ m/s}.$$

LEARN The ratio of the escape speeds on Mars and on Earth is

$$\frac{v_M}{v_E} = \frac{\sqrt{2GM_M/R_M}}{\sqrt{2GM_E/R_E}} = \sqrt{\frac{M_M}{M_E}\cdot\frac{R_E}{R_M}} = \sqrt{(0.11)\cdot\frac{6.5\times10^3 \text{ km}}{3.45\times10^3 \text{ km}}} = 0.455.$$

13-39

THINK The escape speed on the asteroid is related to the gravitational acceleration at the surface of the asteroid and its size.

EXPRESS We use the principle of conservation of energy. Initially the particle is at the surface of the asteroid and has potential energy $U_i = -GMm/R$, where M is the mass of the asteroid, R is its radius, and m is the mass of the particle being fired upward. The initial kinetic energy is $\frac{1}{2}mv^2$. The particle just escapes if its kinetic energy is zero when it is infinitely far from the asteroid. The final potential and kinetic energies are both zero. Conservation of energy yields

$$-GMm/R + \tfrac{1}{2}mv^2 = 0.$$

We replace GM/R with a_gR, where a_g is the acceleration due to gravity at the surface. Then, the energy equation becomes $-a_gR + \frac{1}{2}v^2 = 0$. Solving for v, we have

$$v = \sqrt{2a_gR}.$$

ANALYZE (a) Given that $R = 500$ km and $a_g = 3.0$ m/s^2, we find the escape speed to be

$$v = \sqrt{2a_g R} = \sqrt{2(3.0 \text{ m/s}^2)(500 \times 10^3 \text{ m})} = 1.7 \times 10^3 \text{ m/s}.$$

(b) Initially the particle is at the surface; the potential energy is $U_i = -GMm/R$ and the kinetic energy is $K_i = ½mv^2$. Suppose the particle is a distance h above the surface when it momentarily comes to rest. The final potential energy is $U_f = -GMm/(R + h)$ and the final kinetic energy is $K_f = 0$. Conservation of energy yields

$$-\frac{GMm}{R} + \frac{1}{2}mv^2 = -\frac{GMm}{R+h}.$$

We replace GM with $a_g R^2$ and cancel m in the energy equation to obtain

$$-a_g R + \frac{1}{2}v^2 = -\frac{a_g R^2}{(R+h)}.$$

The solution for h is

$$h = \frac{2a_g R^2}{2a_g R - v^2} - R = \frac{2(3.0 \text{ m/s}^2)(500 \times 10^3 \text{ m})^2}{2(3.0 \text{ m/s}^2)(500 \times 10^3 \text{ m}) - (1000 \text{ m/s})^2} - (500 \times 10^3 \text{ m})$$
$$= 2.5 \times 10^5 \text{ m}.$$

(c) Initially the particle is a distance h above the surface and is at rest. Its potential energy is $U_i = -GMm/(R + h)$ and its initial kinetic energy is $K_i = 0$. Just before it hits the asteroid its potential energy is $U_f = -GMm/R$. Write $\frac{1}{2}mv_f^2$ for the final kinetic energy. Conservation of energy yields

$$-\frac{GMm}{R+h} = -\frac{GMm}{R} + \frac{1}{2}mv^2.$$

We substitute $a_g R^2$ for GM and cancel m, obtaining

$$-\frac{a_g R^2}{R+h} = -a_g R + \frac{1}{2}v^2.$$

The solution for v is

$$v = \sqrt{2a_g R - \frac{2a_g R^2}{R+h}} = \sqrt{2(3.0 \text{ m/s}^2)(500 \times 10^3 \text{ m}) - \frac{2(3.0 \text{ m/s}^2)(500 \times 10^3 \text{ m})^2}{(500 \times 10^3 \text{ m}) + (1000 \times 10^3 \text{ m})}}$$
$$= 1.4 \times 10^3 \text{ m/s}.$$

LEARN The key idea in this problem is to realize that energy is conserved in the process:

$$K_i + U_i = K_f + U_f \quad \Rightarrow \quad \Delta K + \Delta U = 0.$$

The decrease in potential energy is equal to the gain in kinetic energy, and vice versa.

13-41

THINK The two neutron stars are attracted toward each other due to their gravitational interaction.

EXPRESS The momentum of the two-star system is conserved, and since the stars have the same mass, their speeds and kinetic energies are the same. We use the principle of conservation of energy. The initial potential energy is $U_i = -GM^2/r_i$, where M is the mass of either star and r_i is their initial center-to-center separation. The initial kinetic energy is zero since the stars are at rest. The final potential energy is $U_f = -GM^2/r_f$, where the final separation is $r_f = r_i/2$. We write Mv^2 for the final kinetic energy of the system. This is the sum of two terms, each of which is $\frac{1}{2}Mv^2$. Conservation of energy yields

$$-\frac{GM^2}{r_i} = -\frac{2GM^2}{r_i} + Mv^2.$$

ANALYZE (a) The solution for v is

$$v = \sqrt{\frac{GM}{r_i}} = \sqrt{\frac{(6.67 \times 10^{-11}\ \text{m}^3/\text{s}^2 \cdot \text{kg})(10^{30}\ \text{kg})}{10^{10}\ \text{m}}} = 8.2 \times 10^4\ \text{m/s}.$$

(b) Now the final separation of the centers is $r_f = 2R = 2 \times 10^5$ m, where R is the radius of either of the stars. The final potential energy is given by $U_f = -GM^2/r_f$ and the energy equation becomes

$$-GM^2/r_i = -GM^2/r_f + Mv^2.$$

The solution for v is

$$v = \sqrt{GM\left(\frac{1}{r_f} - \frac{1}{r_i}\right)} = \sqrt{(6.67 \times 10^{-11}\text{m}^3/\text{s}^2 \cdot \text{kg})(10^{30}\ \text{kg})\left(\frac{1}{2 \times 10^5\ \text{m}} - \frac{1}{10^{10}\ \text{m}}\right)}$$
$$= 1.8 \times 10^7\ \text{m/s}.$$

LEARN The speed of the stars as a function of their final separation is plotted next. The decrease in gravitational potential energy is accompanied by an increase in kinetic energy, so that the total energy of the two-star system remains conserved.

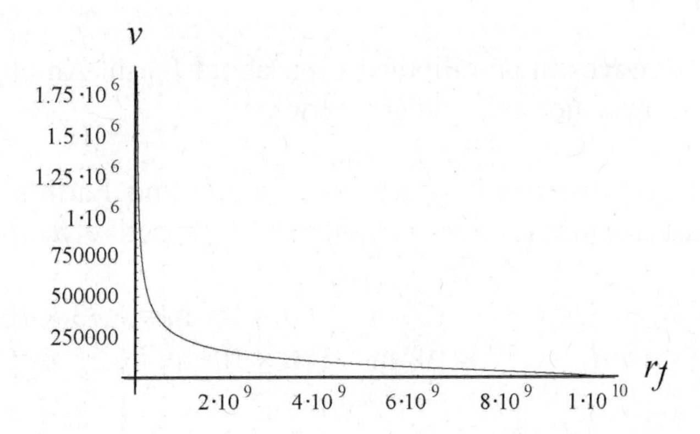

13-47

THINK The centripetal force on the Sun is due to the gravitational attraction between the Sun and the stars at the center of the Galaxy.

EXPRESS Let N be the number of stars in the galaxy, M be the mass of the Sun, and r be the radius of the galaxy. The total mass in the galaxy is $N M$ and the magnitude of the gravitational force acting on the Sun is

$$F_g = \frac{GM(NM)}{R^2} = \frac{GNM^2}{R^2}.$$

The force, pointing toward the galactic center, is the centripetal force on the Sun. Thus,

$$F_c = F_g \Rightarrow \quad \frac{Mv^2}{R} = \frac{GNM^2}{R^2}.$$

The magnitude of the Sun's acceleration is $a = v^2/R$, where v is its speed. If T is the period of the Sun's motion around the galactic center then $v = 2\pi R/T$ and $a = 4\pi^2 R/T^2$. Newton's second law yields

$$GNM^2/R^2 = 4\pi^2 MR/T^2.$$

The solution for N is

$$N = \frac{4\pi^2 R^3}{GT^2 M}.$$

ANALYZE The period is 2.5×10^8 y, which is 7.88×10^{15} s, so

$$N = \frac{4\pi^2 (2.2 \times 10^{20}\text{ m})^3}{(6.67 \times 10^{-11}\text{ m}^3/\text{s}^2 \cdot \text{kg})(7.88 \times 10^{15}\text{ s})^2 (2.0 \times 10^{30}\text{ kg})} = 5.1 \times 10^{10}.$$

LEARN The number of stars in the Milky Way is between 10^{11} to 4×10^{11}. Our simplified model provides a reasonable estimate.

13-51

THINK The satellite moves in an elliptical orbit about Earth. An elliptical orbit can be characterized by its semi-major axis and eccentricity.

EXPRESS The greatest distance between the satellite and Earth's center (the apogee distance) and the least distance (perigee distance) are, respectively,

$$R_a = R_E + d_a = 6.37 \times 10^6 \text{ m} + 360 \times 10^3 \text{ m} = 6.73 \times 10^6 \text{ m}$$
$$R_p = R_E + d_p = 6.37 \times 10^6 \text{ m} + 180 \times 10^3 \text{ m} = 6.55 \times 10^6 \text{ m}.$$

Here $R_E = 6.37 \times 10^6$ m is the radius of Earth.

ANALYZE From Fig. 13-12, we see that the semi-major axis is

$$a = \frac{R_a + R_p}{2} = \frac{6.73 \times 10^6 \text{ m} + 6.55 \times 10^6 \text{ m}}{2} = 6.64 \times 10^6 \text{ m}.$$

(b) The apogee and perigee distances are related to the eccentricity e by $R_a = a(1 + e)$ and $R_p = a(1 - e)$. Add to obtain $R_a + R_p = 2a$ and $a = (R_a + R_p)/2$. Subtract to obtain $R_a - R_p = 2ae$. Thus,

$$e = \frac{R_a - R_p}{2a} = \frac{R_a - R_p}{R_a + R_p} = \frac{6.73 \times 10^6 \text{ m} - 6.55 \times 10^6 \text{ m}}{6.73 \times 10^6 \text{ m} + 6.55 \times 10^6 \text{ m}} = 0.0136.$$

LEARN Since e is very small, the orbit is nearly circular. On the other hand, if e is close to unity, then the orbit would be a long, thin ellipse.

13-63

THINK We apply Kepler's laws to analyze the motion of the asteroid.

EXPRESS We use the law of periods: $T^2 = (4\pi^2/GM)r^3$, where M is the mass of the Sun (1.99×10^{30} kg) and r is the radius of the orbit. On the other hand, the kinetic energy of any asteroid or planet in a circular orbit of radius r is given by $K = GMm/2r$, where m is the mass of the asteroid or planet. We note that it is proportional to m and inversely proportional to r.

ANALYZE (a) The radius of the orbit is twice the radius of Earth's orbit: $r = 2r_{SE} = 2(150 \times 10^9 \text{ m}) = 300 \times 10^9$ m. Thus, the period of the asteroid is

$$T = \sqrt{\frac{4\pi^2 r^3}{GM}} = \sqrt{\frac{4\pi^2 (300 \times 10^9 \text{ m})^3}{(6.67 \times 10^{-11} \text{m}^3/\text{s}^2 \cdot \text{kg})(1.99 \times 10^{30} \text{kg})}} = 8.96 \times 10^7 \text{ s}.$$

Dividing by (365 d/y) (24 h/d) (60 min/h) (60 s/min), we obtain $T = 2.8$ y.

(b) The ratio of the kinetic energy of the asteroid to the kinetic energy of Earth is

$$\frac{K}{K_E} = \frac{GMm/(2r)}{GMM_E/(2r_{SE})} = \frac{m}{M_E} \cdot \frac{r_{SE}}{r} = (2.0\times10^{-4})\left(\frac{1}{2}\right) = 1.0\times10^{-4}.$$

LEARN An alternative way to calculate the ratio of kinetic energies is to use $K = mv^2/2$ and note that $v = 2\pi r/T$. This gives

$$\frac{K}{K_E} = \frac{mv^2/2}{M_E v_E^2/2} = \frac{m}{M_E}\left(\frac{v}{v_E}\right)^2 = \frac{m}{M_E}\left(\frac{r/T}{r_{SE}/T_E}\right)^2 = \frac{m}{M_E}\left(\frac{r}{r_{SE}} \cdot \frac{T_E}{T}\right)^2$$
$$= (2.0\times10^{-4})\left(2 \cdot \frac{1.0\text{ y}}{2.8\text{ y}}\right)^2 = 1.0\times10^{-4}$$

in agreement with what we found in (b).

13-76

THINK We apply Newton's law of gravitation to calculate the force between the meteor and the satellite.

EXPRESS We use $F = Gm_s m_m/r^2$, where m_s is the mass of the satellite, m_m is the mass of the meteor, and r is the distance between their centers. The distance between centers is $r = R + d = 15\text{ m} + 3\text{ m} = 18\text{ m}$. Here R is the radius of the satellite and d is the distance from its surface to the center of the meteor.

ANALYZE The gravitational force between the meteor and the satellite is

$$F = \frac{Gm_s m_s}{r^2} = \frac{(6.67\times10^{-11}\text{N}\cdot\text{m}^2/\text{kg}^2)(20\text{kg})(7.0\text{kg})}{(18\text{m})^2} = 2.9\times10^{-11}\text{N}.$$

LEARN The force of gravitation is inversely proportional to r^2.

13-79

THINK Since the orbit is circular, the net gravitational force on the smaller star is equal to the centripetal force.

EXPRESS The magnitude of the net gravitational force on one of the smaller stars (of mass m) is

$$F = \frac{GMm}{r^2} + \frac{Gmm}{(2r)^2} = \frac{Gm}{r^2}\left(M + \frac{m}{4}\right).$$

This supplies the centripetal force needed for the motion of the star:

$$\frac{Gm}{r^2}\left(M+\frac{m}{4}\right)=m\frac{v^2}{r}$$

where $v = 2\pi r / T$. Combining the two expressions allows us to solve for *T*.

ANALYZE Plugging in for speed *v*, we arrive at an equation for the period *T*:

$$T=\frac{2\pi r^{3/2}}{\sqrt{G(M+m/4)}}.$$

LEARN In the limit where $m \ll M$, we recover the expected result $T=\frac{2\pi r^{3/2}}{\sqrt{GM}}$ for two bodies.

13-81

THINK In a two-star system, the stars rotate about their common center of mass.

EXPRESS The situation is depicted on the right. The gravitational force between the two stars (each having a mass *M*) is

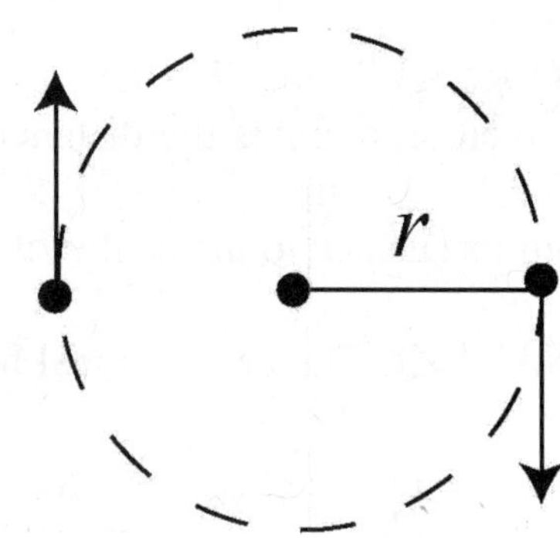

$$F_g=\frac{GM^2}{(2r)^2}=\frac{GM^2}{4r^2}$$

The gravitational force between the stars provides the centripetal force necessary to keep their orbits circular. Thus, writing the centripetal acceleration as $r\omega^2$ where ω is the angular speed, we have

$$F_g=F_c \;\Rightarrow\; \frac{GM^2}{4r^2}=Mr\omega^2.$$

ANALYZE (a) Substituting the values given, we find the common angular speed to be

$$\omega=\frac{1}{2}\sqrt{\frac{GM}{r^3}}=\frac{1}{2}\sqrt{\frac{(6.67\times10^{-11}\ \mathrm{N\cdot m^2/kg^2})(3.0\times10^{30}\ \mathrm{kg})}{(1.0\times10^{11}\ \mathrm{m})^3}}=2.2\times10^{-7}\ \mathrm{rad/s}.$$

(b) To barely escape means to have total energy equal to zero (see discussion prior to Eq. 13-28). If *m* is the mass of the meteoroid, then

$$\frac{1}{2}mv^2-\frac{GmM}{r}-\frac{GmM}{r}=0 \;\Rightarrow\; v=\sqrt{\frac{4GM}{r}}=8.9\times10^4\ \mathrm{m/s}.$$

LEARN Comparing with Eq. 13-28, we see that the escape speed of the two-star system is the same as that of a star with mass $2M$.

12-83

THINK The orbit of the shuttle goes from circular to elliptical after changing its speed by firing the thrusters.

EXPRESS We first use the law of periods: $T^2 = (4\pi^2/GM)r^3$, where M is the mass of the planet and r is the radius of the orbit. After the orbit of the shuttle turns elliptical by firing the thrusters to reduce its speed, the semi-major axis is $a = -GMm/2E$, where $E = K + U$ is the mechanical energy of the shuttle and its new period becomes $T' = \sqrt{4\pi^2 a^3/GM}$.

ANALYZE (a) Using Kepler's law of periods, we find the period to be

$$T = \sqrt{\left(\frac{4\pi^2}{GM}\right)r^3} = \sqrt{\frac{4\pi^2(4.20\times10^7\ \text{m})^3}{(6.67\times10^{-11}\ \text{N}\cdot\text{m}^2/\text{kg}^2)(9.50\times10^{25}\ \text{kg})}} = 2.15\times10^4\ \text{s}.$$

(b) The speed is constant (before she fires the thrusters), so

$$v_0 = \frac{2\pi r}{T} = \frac{2\pi(4.20\times10^7\,\text{m})}{2.15\times10^4\ \text{s}} = 1.23\times10^4\ \text{m/s}.$$

(c) A two percent reduction in the previous value gives

$$v = 0.98v_0 = 0.98(1.23\times10^4\ \text{m/s}) = 1.20\times10^4\ \text{m/s}.$$

(d) The kinetic energy is $K = \frac{1}{2}mv^2 = \frac{1}{2}(3000\,\text{kg})(1.20\times10^4\ \text{m/s})^2 = 2.17\times10^{11}\ \text{J}$.

(e) Immediately after the firing, the potential energy is the same as it was before firing the thruster:

$$U = -\frac{GMm}{r} = -\frac{(6.67\times10^{-11}\ \text{N}\cdot\text{m}^2/\text{kg}^2)(9.50\times10^{25}\ \text{kg})(3000\,\text{kg})}{4.20\times10^7\ \text{m}} = -4.53\times10^{11}\ \text{J}.$$

(f) Adding these two results gives the total mechanical energy:

$$E = K + U = 2.17\times10^{11}\ \text{J} + (-4.53\times10^{11}\ \text{J}) = -2.35\times10^{11}\ \text{J}.$$

(g) Using Eq. 13-42, we find the semi-major axis to be

$$a = -\frac{GMm}{2E} = -\frac{(6.67\times10^{-11}\ \text{N}\cdot\text{m}^2/\text{kg}^2)(9.50\times10^{25}\ \text{kg})(3000\ \text{kg})}{2(-2.35\times10^{11}\ \text{J})} = 4.04\times10^{7}\ \text{m}.$$

(h) Using Kepler's law of periods for elliptical orbits (using a instead of r) we find the new period to be

$$T' = \sqrt{\left(\frac{4\pi^2}{GM}\right)a^3} = \sqrt{\frac{4\pi^2(4.04\times10^7\ \text{m})^3}{(6.67\times10^{-11}\ \text{N}\cdot\text{m}^2/\text{kg}^2)(9.50\times10^{25}\ \text{kg})}} = 2.03\times10^4\ \text{s}.$$

This is smaller than our result for part (a) by $T - T' = 1.22 \times 10^3$ s.

(i) Comparing the results in (a) and (h), we see that elliptical orbit has a smaller period.

LEARN The orbits of the shuttle before and after firing the thruster are shown below. Point P corresponds to the location where the thruster was fired.

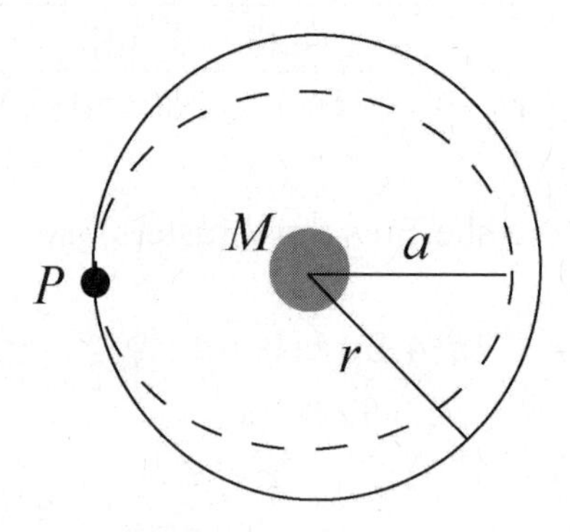

13-89

THINK To compare the kinetic energy, potential energy and the speed of the Earth at aphelion (farthest distance) and perihelion (closest distance), we apply both conservation of energy and conservation of angular momentum.

EXPRESS As Earth orbits about the Sun, its total energy is conserved:

$$\frac{1}{2}mv_a^2 - \frac{GM_SM_E}{R_a} = \frac{1}{2}mv_p^2 - \frac{GM_SM_E}{R_p}.$$

In addition, angular momentum conservation implies $v_aR_a = v_pR_p$.

ANALYZE (a) The total energy is conserved, so there is no difference between its values at aphelion and perihelion.

(b) The difference in potential energy is

$$\Delta U = U_a - U_p = -GM_S M_E \left(\frac{1}{R_a} - \frac{1}{R_p} \right)$$

$$= -(6.67\times10^{-11}\ \text{N}\cdot\text{m}^2/\text{kg}^2)(1.99\times10^{30}\ \text{kg})(5.98\times10^{24}\ \text{kg})\left(\frac{1}{1.52\times10^{11}\ \text{m}} - \frac{1}{1.47\times10^{11}\ \text{m}}\right)$$

$$\approx 1.8\times10^{32}\ \text{J}$$

(c) Since $\Delta K + \Delta U = 0$, $\Delta K = K_a - K_p = -\Delta U \approx -1.8\times10^{32}$ J.

(d) With $v_a R_a = v_p R_p$, the change in kinetic energy may be written as

$$\Delta K = K_a - K_p = \frac{1}{2} M_E \left(v_a^2 - v_p^2\right) = \frac{1}{2} M_E v_a^2 \left(1 - \frac{R_a^2}{R_p^2}\right)$$

from which we find the speed at the aphelion to be

$$v_a = \sqrt{\frac{2(\Delta K)}{M_E(1 - R_a^2/R_p^2)}} = 2.95\times10^4\ \text{m/s}.$$

Thus, the variation in speed is

$$\Delta v = v_a - v_p = \left(1 - \frac{R_a}{R_p}\right) v_a = \left(1 - \frac{1.52\times10^{11}\ \text{m}}{1.47\times10^{11}\ \text{m}}\right)(2.95\times10^4\ \text{m/s})$$
$$= -0.99\times10^3\ \text{m/s} = -0.99\ \text{km/s}$$

The speed at the aphelion is smaller than that at the perihelion by about 1 km/s.

LEARN Since the changes are small, the problem could also be solved by using differentials:

$$dU = \left(\frac{GM_E M_S}{r^2}\right) dr \approx \frac{\left(6.67\times10^{-11}\ \text{N}\cdot\text{m}^2/\text{kg}^2\right)\left(1.99\times10^{30}\ \text{kg}\right)\left(5.98\times10^{24}\ \text{kg}\right)}{\left(1.5\times10^{11}\ \text{m}\right)^2}\left(5\times10^9\ \text{m}\right)$$

This yields $\Delta U \approx 1.8\times10^{32}$ J. Similarly, with $\Delta K \approx dK = M_E v\, dv$, where $v \approx 2\pi R/T$, we have

$$1.8\times10^{32}\ \text{J} \approx \left(5.98\times10^{24}\ \text{kg}\right)\left(\frac{2\pi\left(1.5\times10^{11}\ \text{m}\right)}{3.156\times10^7\ \text{s}}\right)\Delta v$$

which yields a difference of $\Delta v \approx 0.99$ km/s in Earth's speed (relative to the Sun) between aphelion and perihelion.

Chapter 14

14-3

THINK The increase in pressure is equal to the applied force divided by the area.

EXPRESS The change in pressure is given by $\Delta p = F/A = F/\pi r^2$, where r is the radius of the piston.

ANALYZE substituting the values given, we obtain

$$\Delta p = (42 \text{ N})/\pi(0.011 \text{ m})^2 = 1.1 \times 10^5 \text{ Pa}.$$

This is equivalent to 1.1 atm.

LEARN The increase in pressure is proportional to the force applied. In addition, since $\Delta p \sim 1/A$, the smaller the cross-sectional area of the syringe, the greater the pressure increase under the same applied force.

14-5

THINK The pressure difference between two sides of the window results in a net force acting on the window.

EXPRESS The air inside pushes outward with a force given by $p_i A$, where p_i is the pressure inside the room and A is the area of the window. Similarly, the air on the outside pushes inward with a force given by $p_o A$, where p_o is the pressure outside. The magnitude of the net force is $F = (p_i - p_o)A$.

ANALYZE Since 1 atm = 1.013×10^5 Pa, the net force is

$$\begin{aligned} F &= (p_i - p_o)A = (1.0 \text{ atm} - 0.96 \text{ atm})(1.013\times10^5 \text{ Pa/atm})(3.4 \text{ m})(2.1 \text{ m}) \\ &= 2.9 \times 10^4 \text{ N}. \end{aligned}$$

LEARN The net force on the window vanishes when the pressure inside the office is equal to the pressure outside.

14-17

THINK The minimum force that must be applied to open the hatch is equal to the gauge pressure times the area of the hatch.

EXPRESS The pressure p at the depth d of the hatch cover is $p_0 + \rho gd$, where ρ is the density of ocean water and p_0 is atmospheric pressure. Thus, the gauge pressure is $p_{\text{gauge}} = \rho gd$, and the minimum force that must be applied by the crew to open the hatch has magnitude $F = p_{\text{gauge}}A = (\rho gd)A$, where A is the area of the hatch.

Substituting the values given, we find the force to be

$$\begin{aligned}F &= p_{\text{gauge}}A = (\rho gd)A = (1024\text{ kg/m}^3)(9.8\text{ m/s}^2)(100\text{ m})(1.2\text{ m})(0.60\text{ m})\\ &= 7.2\times10^5\text{ N}.\end{aligned}$$

LEARN The downward force of the water on the hatch cover is $(p_0 + \rho gd)A$, and the air in the submarine exerts an upward force of p_0A. The greater the depth of the submarine, the greater the force required to open the hatch.

14-21

THINK Work is done to remove liquid from one vessel to another.

EXPRESS When the levels are the same, the height of the liquid is $h = (h_1 + h_2)/2$, where h_1 and h_2 are the original heights. Suppose h_1 is greater than h_2. The final situation can then be achieved by taking liquid from the first vessel with volume $V = A(h_1 - h)$ and mass $m = \rho V = \rho A(h_1 - h)$, and lowering it a distance $\Delta y = h - h_2$. The work done by the force of gravity is

$$W_g = mg\Delta y = \rho A(h_1 - h)g(h - h_2).$$

ANALYZE We substitute $h = (h_1 + h_2)/2$ to obtain

$$\begin{aligned}W_g &= \frac{1}{4}\rho gA(h_1 - h_2)^2 = \frac{1}{4}(1.30\times10^3\text{kg/m}^3)(9.80\text{ m/s}^2)(4.00\times10^{-4}\text{m}^2)(1.56\text{ m} - 0.854\text{ m})^2\\ &= 0.635\text{ J}\end{aligned}$$

LEARN Since gravitational force is conservative, the work done only depends on the initial and final heights of the vessels, and not on how the liquid is transferred.

14-27

THINK The atmospheric pressure at a given height depends on the density distribution of air.

EXPRESS If the air density were uniform, ρ =const., then the variation of pressure with height may be written as: $p_2 = p_1 - \rho g(y_2 - y_1)$. We take y_1 to be at the surface of Earth, where the pressure is $p_1 = 1.01 \times 10^5$ Pa, and y_2 to be at the top of the atmosphere, where the pressure is $p_2 = 0$. On the other hand, if the density varies with altitude, then

$$p_2 = p_1 - \int_0^h \rho g \, dy .$$

For the case where the density decreases linearly with height, $\rho = \rho_0 (1 - y/h)$, where ρ_0 is the density at Earth's surface and $g = 9.8$ m/s^2 for $0 \le y \le h$, the integral becomes

$$p_2 = p_1 - \int_0^h \rho_0 g \left(1 - \frac{y}{h}\right) dy = p_1 - \frac{1}{2}\rho_0 gh.$$

ANALYZE (a) For uniform density with $\rho = 1.3$ kg/m^3, we find the height of the atmosphere to be

$$y_2 - y_1 = \frac{p_1}{\rho g} = \frac{1.01 \times 10^5 \text{ Pa}}{(1.3 \text{ kg/m}^3)(9.8 \text{ m/s}^2)} = 7.9 \times 10^3 \text{ m} = 7.9 \text{ km}.$$

(b) With density decreasing linearly with height, $p_2 = p_1 - \rho_0 gh/2$. The condition $p_2 = 0$ implies

$$h = \frac{2p_1}{\rho_0 g} = \frac{2(1.01 \times 10^5 \text{ Pa})}{(1.3 \text{ kg/m}^3)(9.8 \text{ m/s}^2)} = 16 \times 10^3 \text{ m} = 16 \text{ km}.$$

LEARN Actually the decrease in air density is approximately exponential, with pressure halved at a height of about 5.6 km.

14-31

THINK The block floats in both water and oil. We apply Archimedes' principle to analyze the problem.

EXPRESS Let V be the volume of the block. Then, the submerged volume in water is $V_s = 2V/3$. Since the block is floating, by Archimedes' principle the weight of the displaced water is equal to the weight of the block, i.e., $\rho_w V_s = \rho_b V$, where ρ_w is the density of water, and ρ_b is the density of the block.

ANALYZE (a) We substitute $V_s = 2V/3$ to obtain the density of the block:

$$\rho_b = 2\rho_w/3 = 2(1000 \text{ kg/m}^3)/3 \approx 6.7 \times 10^2 \text{ kg/m}^3.$$

(b) Now, if ρ_o is the density of the oil, then Archimedes' principle yields $\rho_o V_s' = \rho_b V$. Since the volume submerged in oil is $V_s' = 0.90V$, the density of the oil is

$$\rho_o = \rho_b \left(\frac{V}{V'}\right) = (6.7 \times 10^2 \text{ kg/m}^3)\frac{V}{0.90V} = 7.4 \times 10^2 \text{ kg/m}^3.$$

LEARN Another way to calculate the density of the oil is to note that the mass of the block can be written as

$$m = \rho_b V = \rho_o V_s' = \rho_w V_s .$$

Therefore,

$$\rho_o = \rho_w \left(\frac{V_s}{V_s'}\right) = (1000\,\text{kg/m}^3)\frac{2V/3}{0.90V} = 7.4\times10^2\ \text{kg/m}^3 .$$

That is, by comparing the fraction submerged with that in water (or another liquid with known density), the density of the oil can be deduced.

14-33

THINK The iron anchor is submerged in water, so we apply Archimedes' principle to calculate its volume and weight in air.

EXPRESS The anchor is completely submerged in water of density ρ_w. Its apparent weight is $W_{\text{app}} = W - F_b$, where $W = mg$ is its actual weight and $F_b = \rho_w gV$ is the buoyant force.

ANALYZE (a) Substituting the values given, we find the volume of the anchor to be

$$V = \frac{W - W_{\text{app}}}{\rho_w g} = \frac{F_b}{\rho_w g} = \frac{200\ \text{N}}{\left(1000\ \text{kg/m}^3\right)\left(9.8\ \text{m/s}^2\right)} = 2.04\times10^{-2}\ \text{m}^3 .$$

(b) The mass of the anchor is $m = \rho_{\text{Fe}} g$, where ρ_{Fe} is the density of iron (found in Table 14-1). Therefore, its weight in air is

$$W = mg = \rho_{\text{Fe}} Vg = (7870\ \text{kg/m}^3)(2.04\times10^{-2}\ \text{m}^3)(9.80\ \text{m/s}^2) = 1.57\times10^3\ \text{N} .$$

LEARN In general, the apparent weight of an object of density ρ that is completely submerged in a fluid of density ρ_f can be written as $W_{\text{app}} = (\rho - \rho_f)Vg$.

14-39

THINK The hollow sphere is half submerged in a fluid. We apply Archimedes' principle to calculate its mass and density.

EXPRESS The downward force of gravity mg is balanced by the upward buoyant force of the liquid: $mg = \rho g\, V_s$. Here m is the mass of the sphere, ρ is the density of the liquid, and V_s is the submerged volume. Thus $m = \rho V_s$. The submerged volume is half the total volume of the sphere, so $V_s = \frac{1}{2}(4\pi/3)r_o^3$, where r_o is the outer radius.

ANALYZE (a) Substituting the values given, we find the mass of the sphere to be

$$m = \rho V_s = \rho\left(\frac{1}{2}\cdot\frac{4\pi}{3}r_o^3\right) = \frac{2\pi}{3}\rho r_o^3 = \left(\frac{2\pi}{3}\right)(800\ \text{kg/m}^3)(0.090\ \text{m})^3 = 1.22\ \text{kg}.$$

(b) The density ρ_m of the material, assumed to be uniform, is given by $\rho_m = m/V$, where m is the mass of the sphere and V is its volume. If r_i is the inner radius, the volume is

$$V = \frac{4\pi}{3}(r_o^3 - r_i^3) = \frac{4\pi}{3}\left((0.090\text{ m})^3 - (0.080\text{ m})^3\right) = 9.09\times10^{-4}\text{ m}^3 .$$

The density is

$$\rho_m = \frac{1.22\text{ kg}}{9.09\times10^{-4}\text{ m}^3} = 1.3\times10^3\text{ kg/m}^3 .$$

LEARN Note that $\rho_m > \rho$, i.e., the density of the material is greater that of the fluid. However, the sphere floats (and displaces its own weight of fluid) because it's hollow.

14-51

THINK We use the equation of continuity to solve for the speed of water as it leaves the sprinkler hole.

EXPRESS Let v_1 be the speed of the water in the hose and v_2 be its speed as it leaves one of the holes. The cross-sectional area of the hose is $A_1 = \pi R^2$. If there are N holes and A_2 is the area of a single hole, then the equation of continuity becomes

$$v_1 A_1 = v_2 (N A_2) \Rightarrow v_2 = \frac{A_1}{N A_2} v_1 = \frac{R^2}{N r^2} v_1$$

where R is the radius of the hose and r is the radius of a hole.

ANALYZE Noting that $R/r = D/d$ (the ratio of diameters) we find the speed to be

$$v_2 = \frac{D^2}{N d^2} v_1 = \frac{(1.9\,\text{cm})^2}{24(0.13\,\text{cm})^2}(0.91\text{ m/s}) = 8.1\text{ m/s}.$$

LEARN The equation of continuity implies that the smaller the cross-sectional area of the sprinkler hole, the greater the speed of water as it emerges from the hole.

14-53

THINK The power of the pump is the rate of work done in lifting the water.

EXPRESS Suppose that a mass Δm of water is pumped in time Δt. The pump increases the potential energy of the water by $\Delta U = (\Delta m)gh$, where h is the vertical distance through which it is lifted, and increases its kinetic energy by $\Delta K = \frac{1}{2}(\Delta m)v^2$, where v is its final speed. The work it does is

$$\Delta W = \Delta U + \Delta K = (\Delta m)gh + \frac{1}{2}(\Delta m)v^2$$

and its power is

$$P = \frac{\Delta W}{\Delta t} = \frac{\Delta m}{\Delta t}\left(gh + \frac{1}{2}v^2\right).$$

The rate of mass flow is $\Delta m/\Delta t = \rho_w A v$, where ρ_w is the density of water and A is the area of the hose.

ANALYZE The area of the hose is $A = \pi r^2 = \pi(0.010\text{ m})^2 = 3.14 \times 10^{-4}\text{ m}^2$ and

$$\rho_w A v = (1000\text{ kg/m}^3)\,(3.14 \times 10^{-4}\text{ m}^2)\,(5.00\text{ m/s}) = 1.57\text{ kg/s}.$$

Thus, the power of the pump is

$$P = \rho A v\left(gh + \frac{1}{2}v^2\right) = (1.57\,\text{kg/s})\left(\left(9.8\,\text{m/s}^2\right)(3.0\,\text{m}) + \frac{(5.0\,\text{m/s})^2}{2}\right) = 66\text{ W}.$$

LEARN The work done by the pump is converted into both the potential energy and kinetic energy of the water.

14-57

THINK We use the Bernoulli equation to solve for the flow rate, and the continuity equation to relate cross-sectional area to the vertical distance from the hole.

EXPRESS According to the Bernoulli equation:

$$p_1 + \tfrac{1}{2}\rho v_1^2 + \rho g h_1 = p_2 + \tfrac{1}{2}\rho v_2^2 + \rho g h_2\,,$$

where ρ is the density of water, h_1 is the height of the water in the tank, p_1 is the pressure there, and v_1 is the speed of the water there; h_2 is the altitude of the hole, p_2 is the pressure there, and v_2 is the speed of the water there. The pressure at the top of the tank and at the hole is atmospheric, so $p_1 = p_2$. Since the tank is large we may neglect the water speed at the top; it is much smaller than the speed at the hole. The Bernoulli equation then simplifies to $\rho g h_1 = \tfrac{1}{2}\rho v_2^2 + \rho g h_2$.

ANALYZE (a) With $D = h_1 - h_2 = 0.30\text{ m}$, the speed of water as it emerges from the hole is

$$v_2 = \sqrt{2g\left(h_1 - h_2\right)} = \sqrt{2\left(9.8\,\text{m/s}^2\right)(0.30\,\text{m})} = 2.42\,\text{m/s}.$$

Thus, the flow rate is

$$A_2 v_2 = (6.5 \times 10^{-4}\text{ m}^2)(2.42\text{ m/s}) = 1.6 \times 10^{-3}\text{ m}^3\text{/s}.$$

(b) We use the equation of continuity: $A_2v_2 = A_3v_3$, where $A_3 = \frac{1}{2}A_2$ and v_3 is the water speed where the area of the stream is half its area at the hole (see diagram below).

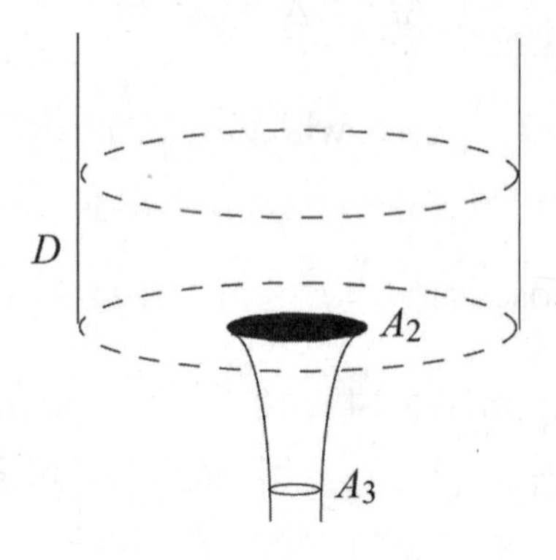

Thus,

$$v_3 = (A_2/A_3)v_2 = 2v_2 = 4.84 \text{ m/s}.$$

The water is in free fall and we wish to know how far it has fallen when its speed is doubled to 4.84 m/s. Since the pressure is the same throughout the fall, $\frac{1}{2}\rho v_2^2 + \rho g h_2 = \frac{1}{2}\rho v_3^2 + \rho g h_3$. Thus,

$$h_2 - h_3 = \frac{v_3^2 - v_2^2}{2g} = \frac{(4.84 \text{ m/s})^2 - (2.42 \text{ m/s})^2}{2(9.8 \text{ m/s}^2)} = 0.90 \text{ m}.$$

LEARN By combing the two expressions obtained from Bernoulli's equation and equation of continuity, the cross-sectional area of the stream may be related to the vertical height fallen as

$$h_2 - h_3 = \frac{v_3^2 - v_2^2}{2g} = \frac{v_2^2}{2g}\left[\left(\frac{A_2}{A_3}\right)^2 - 1\right] = \frac{v_3^2}{2g}\left[1 - \left(\frac{A_3}{A_2}\right)^2\right].$$

14-59

THINK The elevation and cross-sectional area of the pipe are changing, so we apply the Bernoulli equation and continuity equation to analyze the flow of water through the pipe.

EXPRESS To calculate the flow speed at the lower level, we use the equation of continuity: $A_1v_1 = A_2v_2$. Here A_1 is the area of the pipe at the top and v_1 is the speed of the water there; A_2 is the area of the pipe at the bottom and v_2 is the speed of the water there. As for the pressure at the lower level, we use the Bernoulli equation:

$$p_1 + \tfrac{1}{2}\rho v_1^2 + \rho g h_1 = p_2 + \tfrac{1}{2}\rho v_2^2 + \rho g h_2,$$

where ρ is the density of water, h_1 is its initial altitude, and h_2 is its final altitude.

ANALYZE (a) From the continuity equation, we find the speed at the lower level to be

$$v_2 = (A_1/A_2)v_1 = [(4.0\ \text{cm}^2)/(8.0\ \text{cm}^2)]\ (5.0\ \text{m/s}) = 2.5\text{m/s}.$$

(b) Similarly, from the Bernoulli equation, the pressure at the lower level is

$$\begin{aligned} p_2 &= p_1 + \frac{1}{2}\rho\left(v_1^2 - v_2^2\right) + \rho g\left(h_1 - h_2\right) \\ &= 1.5\times10^5\,\text{Pa} + \frac{1}{2}(1000\,\text{kg/m}^3)\left[(5.0\,\text{m/s})^2 - (2.5\,\text{m/s})^2\right] + (1000\,\text{kg/m}^3)(9.8\,\text{m/s}^2)(10\ \text{m}) \\ &= 2.6\times10^5\ \text{Pa}. \end{aligned}$$

LEARN The water at the lower level has a smaller speed ($v_2 < v_1$) but higher pressure ($p_2 > p_1$).

14-65

THINK The design principles of the Venturi meter, a device that measures the flow speed of a fluid in a pipe, involve both the continuity equation and Bernoulli's equation.

EXPRESS The continuity equation yields $AV = av$, and Bernoulli's equation yields $\frac{1}{2}\rho V^2 = \Delta p + \frac{1}{2}\rho v^2$, where $\Delta p = p_2 - p_1$ with p_2 equal to the pressure in the throat and p_1 the pressure in the pipe. The first equation gives $v = (A/a)V$. We use this to substitute for v in the second equation and obtain

$$\tfrac{1}{2}\rho V^2 = \Delta p + \tfrac{1}{2}\rho\left(A/a\right)^2 V^2.$$

The equation can be used to solve for V.

ANALYZE (a) The above equation gives the following expression for V:

$$V = \sqrt{\frac{2\Delta p}{\rho\left(1-(A/a)^2\right)}} = \sqrt{\frac{2a^2\Delta p}{\rho\left(a^2 - A^2\right)}}.$$

(b) We substitute the values given to obtain

$$V = \sqrt{\frac{2a^2\Delta p}{\rho\left(a^2 - A^2\right)}} = \sqrt{\frac{2(32\times10^{-4}\text{m}^2)^2(41\times10^3\,\text{Pa} - 55\times10^3\,\text{Pa})}{(1000\,\text{kg/m}^3)\left((32\times10^{-4}\text{m}^2)^2 - (64\times10^{-4}\text{m}^2)^2\right)}} = 3.06\,\text{m/s}.$$

Consequently, the flow rate is

$$R = AV = (64\times10^{-4}\text{m}^2)(3.06\ \text{m/s}) = 2.0\times10^{-2}\text{m}^3/\text{s}.$$

LEARN The pressure difference Δp between points 1 and 2 is what causes the height difference of the fluid in the two arms of the manometer. Note that $\Delta p = p_2 - p_1 < 0$ (pressure in throat less than that in the pipe), but $a < A$, so the expression inside the square root is positive.

14-81

THINK The U-tube contains two types of liquid in static equilibrium. The pressures at the interface level on both sides of the tube must be the same.

EXPRESS If we examine both sides of the U-tube at the level where the low-density liquid (with $\rho = 0.800$ g/cm^3 = 800 kg/m^3) meets the water (with $\rho_w = 0.998$ g/cm^3 = 998 kg/m^3), then the pressures there on either side of the tube must agree:

$$\rho g h = \rho_w g h_w$$

where h = 8.00 cm = 0.0800 m, and Eq. 14-9 has been used. Thus, the height of the water column (as measured from that level) is h_w = (800/998)(8.00 cm) = 6.41 cm.

ANALYZE The volume of water in that column is

$$V = \pi r^2 h_w = \pi(1.50\text{ cm})^2(6.41\text{ cm}) = 45.3\text{ cm}^3.$$

This is the amount of water that flows out of the right arm.

LEARN As discussed in the Sample Problem – Balancing of pressure in a U-tube, the relationship between the densities of the two liquids can be written as

$$\rho_X = \rho_w \frac{l}{l+d}$$

The liquid in the left arm is higher than the water in the right because the liquid is less dense than water $\rho_X < \rho_w$.

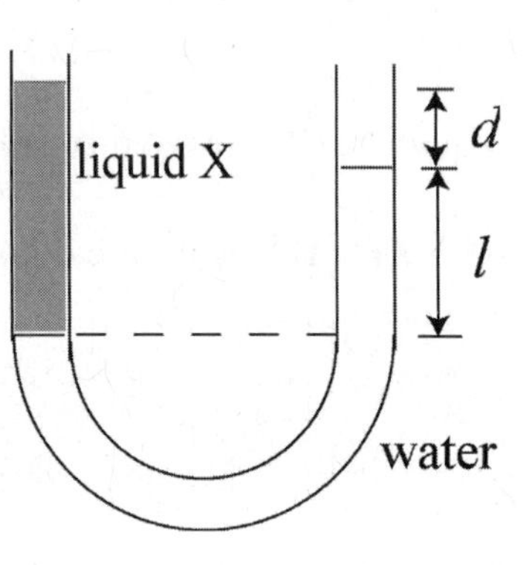

Chapter 15

15-5

THINK The blade of the shaver undergoes simple harmonic motion. We want to find its amplitude, maximum speed and maximum acceleration.

EXPRESS The amplitude x_m is half the range of the displacement D. Once the amplitude is known, the maximum speed v_m is related to the amplitude by $v_m = \omega x_m$, where ω is the angular frequency. Similarly, the maximum acceleration is $a_m = \omega^2 x_m$.

ANALYZE (a) The amplitude is $x_m = D/2 = (2.0 \text{ mm})/2 = 1.0 \text{ mm}$.

(b) Since $\omega = 2\pi f$, where f is the frequency, the maximum speed is

$$v_m = \omega x_m = 2\pi f x_m = 2\pi(120 \text{ Hz})(1.0\times10^{-3} \text{ m}) = 0.75 \text{ m/s}.$$

(c) The maximum acceleration is

$$a_m = \omega^2 x_m = (2\pi f)^2 x_m = (2\pi(120 \text{ Hz}))^2 (1.0\times10^{-3} \text{ m}) = 5.7\times10^2 \text{ m/s}^2.$$

LEARN In SHM, acceleration is proportional to the displacement x_m.

15-7

THINK This problem compares the magnitude of the acceleration of an oscillating diaphragm in a loudspeaker to gravitational acceleration g.

EXPRESS The magnitude of the maximum acceleration is given by $a_m = \omega^2 x_m$, where ω is the angular frequency and x_m is the amplitude.

ANALYZE (a) The angular frequency for which the maximum acceleration has a magnitude g is given by $\omega = \sqrt{g/x_m}$, so the corresponding frequency is

$$f = \frac{\omega}{2\pi} = \frac{1}{2\pi}\sqrt{\frac{g}{x_m}} = \frac{1}{2\pi}\sqrt{\frac{9.8 \text{ m/s}^2}{1.0\times10^{-6} \text{m}}} = 498 \text{ Hz}.$$

(b) For frequencies greater than 498 Hz, the acceleration exceeds g for some part of the motion.

LEARN The acceleration a_m of the diaphragm in a loudspeaker increases with ω^2, or equivalently, with f^2.

15-13

THINK The mass-spring system undergoes simple harmonic motion. Given the amplitude and the period, we can determine the corresponding frequency, angular frequency, spring constant, maximum speed and maximum force.

EXPRESS The angular frequency ω is given by $\omega = 2\pi f = 2\pi/T$, where f is the frequency and T is the period, with $f = 1/T$. The angular frequency is related to the spring constant k and the mass m by $\omega = \sqrt{k/m}$. The maximum speed v_m is related to the amplitude x_m by $v_m = \omega x_m$.

ANALYZE (a) The motion repeats every 0.500 s so the period must be $T = 0.500$ s.

(b) The frequency is the reciprocal of the period: $f = 1/T = 1/(0.500 \text{ s}) = 2.00$ Hz.

(c) The angular frequency is $\omega = 2\pi f = 2\pi(2.00 \text{ Hz}) = 12.6$ rad/s.

(d) We solve for the spring constant k and obtain

$$k = m\omega^2 = (0.500 \text{ kg})(12.6 \text{ rad/s})^2 = 79.0 \text{ N/m}.$$

(e) The amplitude is x_m=35.0 cm = 0.350 m, so the maximum speed is

$$v_m = \omega x_m = (12.6 \text{ rad/s})(0.350 \text{ m}) = 4.40 \text{ m/s}.$$

(f) The maximum force is exerted when the displacement is a maximum. Thus, we have

$$F_m = kx_m = (79.0 \text{ N/m})(0.350 \text{ m}) = 27.6 \text{ N}.$$

LEARN With the maximum acceleration given by $a_m = \omega^2 x_m$, we see that the magnitude of the maximum force can also be written as $F_m = kx_m = m\omega^2 x_m = ma_m$. Maximum acceleration occurs at the endpoints of the path of the block.

15-15

THINK Our system consists of two particles undergoing SHM along a common straight-line segment. Their oscillations are out of phase.

EXPRESS Let

$$x_1 = \frac{A}{2}\cos\left(\frac{2\pi t}{T}\right)$$

be the coordinate as a function of time for particle 1 and

$$x_2 = \frac{A}{2}\cos\left(\frac{2\pi t}{T} + \frac{\pi}{6}\right)$$

be the coordinate as a function of time for particle 2. Here T is the period. Note that since the range of the motion is A, the amplitudes are both $A/2$. The arguments of the cosine functions are in radians. Particle 1 is at one end of its path ($x_1 = A/2$) when $t = 0$. Particle 2 is at $A/2$ when $2\pi t/T + \pi/6 = 0$ or $t = -T/12$. That is, particle 1 lags particle 2 by one-twelfth a period.

ANALYZE (a) The coordinates of the particles 0.50 s later (that is, at $t = 0.50$ s) are

$$x_1 = \frac{A}{2}\cos\left(\frac{2\pi \times 0.50\text{ s}}{1.5\text{ s}}\right) = -0.25A$$

and

$$x_2 = \frac{A}{2}\cos\left(\frac{2\pi \times 0.50\text{ s}}{1.5\text{ s}} + \frac{\pi}{6}\right) = -0.43A.$$

Their separation at that time is $\Delta x = x_1 - x_2 = -0.25A + 0.43A = 0.18A$.

(b) The velocities of the particles are given by

$$v_1 = \frac{dx_1}{dt} = -\frac{\pi A}{T}\sin\left(\frac{2\pi t}{T}\right)$$

and

$$v_2 = \frac{dx_2}{dt} = -\frac{\pi A}{T}\sin\left(\frac{2\pi t}{T} + \frac{\pi}{6}\right).$$

We evaluate these expressions for $t = 0.50$ s and find they are both negative-valued, indicating that the particles are moving in the same direction.

LEARN The plots of x and v as a function of time for particle 1 (solid) and particle 2 (dashed line) are given below.

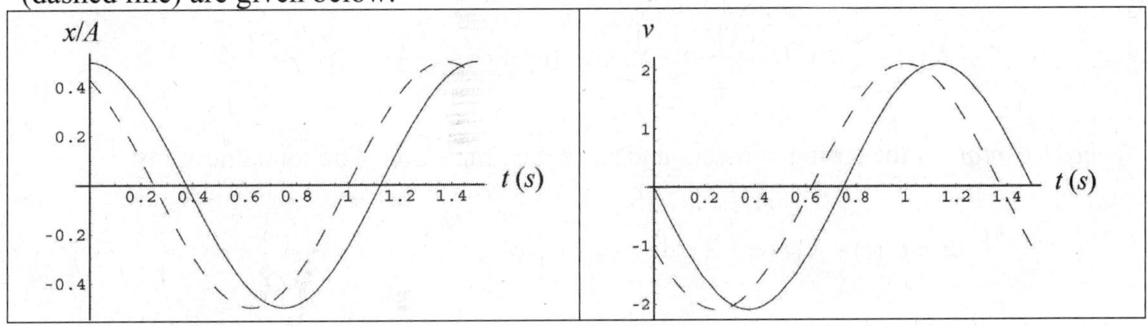

15-23

THINK The maximum force that can be exerted by the surface must be less than the static frictional force or else the block will not follow the surface in its motion.

EXPRESS The static frictional force is given by $f_s = \mu_s F_N$, where μ_s is the coefficient of static friction and F_N is the normal force exerted by the surface on the block. Since the block does not accelerate vertically, we know that $F_N = mg$, where m is the mass of the block. If the block follows the table and moves in simple harmonic motion, the magnitude of the maximum force exerted on it is given by

$$F = ma_m = m\omega^2 x_m = m(2\pi f)^2 x_m,$$

where a_m is the magnitude of the maximum acceleration, ω is the angular frequency, and f is the frequency. The relationship $\omega = 2\pi f$ was used to obtain the last form.

ANALYZE We substitute $F = m(2\pi f)^2 x_m$ and $F_N = mg$ into $F < \mu_s F_N$ to obtain $m(2\pi f)^2 x_m < \mu_s mg$. The largest amplitude for which the block does not slip is

$$x_m = \frac{\mu_s g}{(2\pi f)^2} = \frac{(0.50)(9.8\ \text{m/s}^2)}{(2\pi \times 2.0\ \text{Hz})^2} = 0.031\ \text{m}.$$

LEARN A larger amplitude would require a larger force at the end points of the motion. The block slips if the surface cannot supply a larger force.

15-27

THINK This problem explores the relationship between energies, both kinetic and potential, with amplitude in SHM.

EXPRESS In simple harmonic motion, let the displacement be $x(t) = x_m \cos(\omega t + \phi)$. The corresponding velocity is $v(t) = dx/dt = -\omega x_m \sin(\omega t + \phi)$. Using the expressions for $x(t)$ and $v(t)$, we find the potential and kinetic energies to be

$$U(t) = \frac{1}{2}kx^2(t) = \frac{1}{2}kx_m^2 \cos^2(\omega t + \phi)$$
$$K(t) = \frac{1}{2}mv^2(t) = \frac{1}{2}m\omega^2 x_m^2 \sin^2(\omega t + \phi) = \frac{1}{2}kx_m^2 \sin^2(\omega t + \phi)$$

where $k = m\omega^2$ is the spring constant and x_m is the amplitude. The total energy is

$$E = U(t) + K(t) = \frac{1}{2}kx_m^2\left[\cos^2(\omega t + \phi) + \sin^2(\omega t + \phi)\right] = \frac{1}{2}kx_m^2.$$

ANALYZE (a) The condition $x(t) = x_m / 2$ implies $\cos(\omega t + \phi) = 1/2$, or $\sin(\omega t + \phi) = \sqrt{3}/2$. Thus, the fraction of energy that is kinetic is

$$\frac{K}{E} = \sin^2(\omega t + \phi) = \left(\frac{\sqrt{3}}{2}\right)^2 = \frac{3}{4}.$$

(b) Similarly, we have $\frac{U}{E} = \cos^2(\omega t + \phi) = \left(\frac{1}{2}\right)^2 = \frac{1}{4}$.

(c) Since $E = \frac{1}{2}kx_m^2$ and $U = \frac{1}{2}kx(t)^2$, $U/E = x^2/x_m^2$. Solving $x^2/x_m^2 = 1/2$ for x, we get $x = x_m / \sqrt{2}$.

LEARN The figure to the right depicts the potential energy (solid line) and kinetic energy (dashed line) as a function of time, assuming $x(0) = x_m$. The curves intersect when $K = U = E/2$, or equivalently,

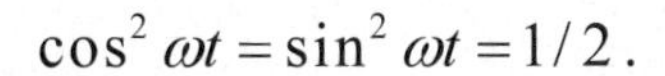

$$\cos^2 \omega t = \sin^2 \omega t = 1/2.$$

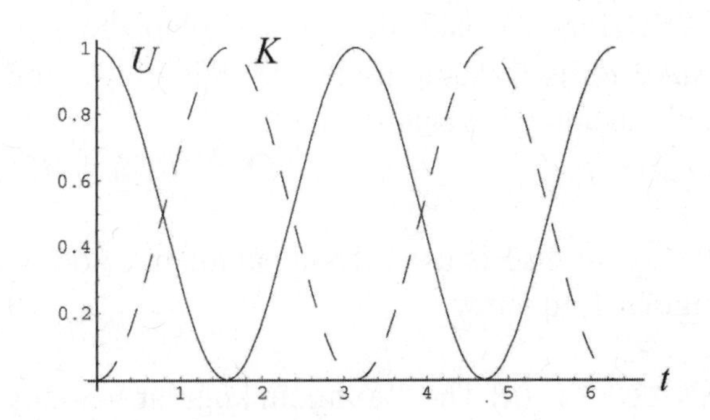

15-29

THINK Knowing the amplitude and the spring constant, we can calculate the mechanical energy of the mass-spring system in simple harmonic motion.

EXPRESS In simple harmonic motion, let the displacement be $x(t) = x_m \cos(\omega t + \phi)$. The corresponding velocity is

$$v(t) = dx/dt = -\omega x_m \sin(\omega t + \phi).$$

Using the expressions for $x(t)$ and $v(t)$, we find the potential and kinetic energies to be

$$U(t) = \frac{1}{2}kx^2(t) = \frac{1}{2}kx_m^2 \cos^2(\omega t + \phi)$$

$$K(t) = \frac{1}{2}mv^2(t) = \frac{1}{2}m\omega^2 x_m^2 \sin^2(\omega t + \phi) = \frac{1}{2}kx_m^2 \sin^2(\omega t + \phi)$$

where $k = m\omega^2$ is the spring constant and x_m is the amplitude. The total energy is

$$E = U(t) + K(t) = \frac{1}{2}kx_m^2\left[\cos^2(\omega t + \phi) + \sin^2(\omega t + \phi)\right] = \frac{1}{2}kx_m^2.$$

ANALYZE With $k = 1.3\ \text{N/cm} = 130\ \text{N/m}$ and $x_m = 2.4\ \text{cm} = 0.024\ \text{m}$, the mechanical energy is

$$E = \frac{1}{2}kx_m^2 = \frac{1}{2}(1.3\times 10^2\ \text{N/m})(0.024\ \text{m})^2 = 3.7\times 10^{-2}\ \text{J}.$$

LEARN An alternative to calculate E is to note that when the block is at the end of its path and is momentarily stopped ($v=0 \Rightarrow K=0$), its displacement is equal to the amplitude and all the energy is potential in nature ($E=U+K=U$). With the spring potential energy taken to be zero when the block is at its equilibrium position, we recover the expression $E = kx_m^2/2$.

15-39

THINK The balance wheel in the watch undergoes angular simple harmonic oscillation. From the amplitude and period, we can calculate the corresponding angular velocity and angular acceleration.

EXPRESS We take the angular displacement of the wheel to be $\theta(t) = \theta_m \cos(2\pi t/T)$, where θ_m is the amplitude and T is the period. We differentiate with respect to time to find the angular velocity:

$$\Omega = d\theta/dt = -(2\pi/T)\theta_m \sin(2\pi t/T).$$

The symbol Ω is used for the angular velocity of the wheel so it is not confused with the angular frequency.

ANALYZE (a) The maximum angular velocity is

$$\Omega_m = \frac{2\pi\theta_m}{T} = \frac{(2\pi)(\pi \text{ rad})}{0.500 \text{ s}} = 39.5 \text{ rad/s}.$$

(b) When $\theta = \pi/2$, then $\theta/\theta_m = 1/2$, $\cos(2\pi t/T) = 1/2$, and

$$\sin(2\pi t/T) = \sqrt{1-\cos^2(2\pi t/T)} = \sqrt{1-(1/2)^2} = \sqrt{3}/2$$

where the trigonometric identity $\cos^2\theta + \sin^2\theta = 1$ is used. Thus,

$$\Omega = -\frac{2\pi}{T}\theta_m \sin\left(\frac{2\pi t}{T}\right) = -\left(\frac{2\pi}{0.500 \text{ s}}\right)(\pi \text{ rad})\left(\frac{\sqrt{3}}{2}\right) = -34.2 \text{ rad/s}.$$

During another portion of the cycle its angular speed is +34.2 rad/s when its angular displacement is $\pi/2$ rad.

(c) The angular acceleration is

$$\alpha = \frac{d^2\theta}{dt^2} = -\left(\frac{2\pi}{T}\right)^2 \theta_m \cos(2\pi t/T) = -\left(\frac{2\pi}{T}\right)^2 \theta.$$

When $\theta = \pi/4$,

$$\alpha = -\left(\frac{2\pi}{0.500\text{ s}}\right)^2\left(\frac{\pi}{4}\right) = -124\text{ rad/s}^2,$$

or $|\alpha| = 124\text{ rad/s}^2$.

LEARN The angular displacement, angular velocity and angular acceleration as a function of time are plotted below.

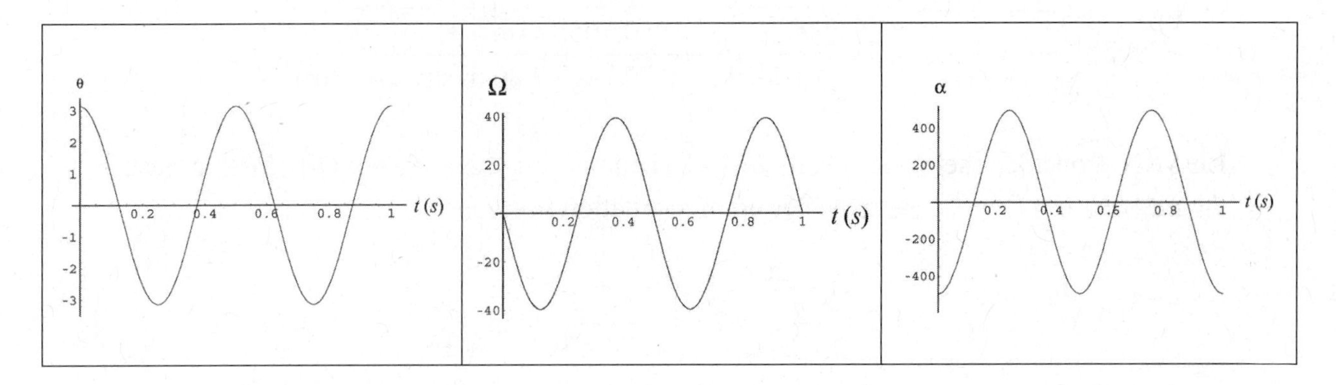

15-41

THINK Our physical pendulum consists of a disk and a rod. To find the period of oscillation, we first calculate the moment of inertia and the distance between the center-of-mass of the disk-rod system to the pivot.

EXPRESS A uniform disk pivoted at its center has a rotational inertia of $\frac{1}{2}Mr^2$, where M is its mass and r is its radius. The disk of this problem rotates about a point that is displaced from its center by $r+L$, where L is the length of the rod, so, according to the parallel-axis theorem, its rotational inertia is $\frac{1}{2}Mr^2 + \frac{1}{2}M(L+r)^2$. The rod is pivoted at one end and has a rotational inertia of $mL^2/3$, where m is its mass.

ANALYZE (a) The total rotational inertia of the disk and rod is

$$\begin{aligned}I &= \frac{1}{2}Mr^2 + M(L+r)^2 + \frac{1}{3}mL^2\\ &= \frac{1}{2}(0.500\text{kg})(0.100\text{m})^2 + (0.500\text{kg})(0.500\text{m}+0.100\text{m})^2 + \frac{1}{3}(0.270\text{kg})(0.500\text{m})^2\\ &= 0.205\text{kg}\cdot\text{m}^2.\end{aligned}$$

(b) We put the origin at the pivot. The center of mass of the disk is

$$\ell_d = L + r = 0.500\text{ m} + 0.100\text{ m} = 0.600\text{ m}$$

away and the center of mass of the rod is $\ell_r = L/2 = (0.500 \text{ m})/2 = 0.250 \text{ m}$ away, on the same line. The distance from the pivot point to the center of mass of the disk-rod system is

$$d = \frac{M\ell_d + m\ell_r}{M+m} = \frac{(0.500 \text{ kg})(0.600 \text{ m}) + (0.270 \text{ kg})(0.250 \text{ m})}{0.500 \text{ kg} + 0.270 \text{ kg}} = 0.477 \text{ m}.$$

(c) The period of oscillation is

$$T = 2\pi\sqrt{\frac{I}{(M+m)gd}} = 2\pi\sqrt{\frac{0.205 \text{ kg}\cdot\text{m}^2}{(0.500 \text{ kg} + 0.270 \text{ kg})(9.80 \text{ m/s}^2)(0.447 \text{ m})}} = 1.50 \text{ s}.$$

LEARN Consider the limit where $M \to 0$ (i.e., uniform disk removed). In this case, $I = mL^2/3$, $d = \ell_r = L/2$ and the period of oscillation becomes

$$T = 2\pi\sqrt{\frac{I}{mgd}} = 2\pi\sqrt{\frac{mL^2/3}{mg(L/2)}} = 2\pi\sqrt{\frac{2L}{3g}}$$

which is the result given in Eq. 15-32.

15-53

THINK By assuming that the torque exerted by the spring on the rod is proportional to the angle of rotation of the rod and that the torque tends to pull the rod toward its equilibrium orientation, we see that the rod will oscillate in simple harmonic motion.

EXPRESS Let $\tau = -C\theta$, where τ is the torque, θ is the angle of rotation, and C is a constant of proportionality, then the angular frequency of oscillation is $\omega = \sqrt{C/I}$ and the period is

$$T = \frac{2\pi}{\omega} = 2\pi\sqrt{\frac{I}{C}},$$

where I is the rotational inertia of the rod. The plan is to find the torque as a function of θ and identify the constant C in terms of given quantities. This immediately gives the period in terms of given quantities. Let ℓ_0 be the distance from the pivot point to the wall. This is also the equilibrium length of the spring. Suppose the rod turns through the angle θ, with the left end moving away from the wall. This end is now $(L/2)\sin\theta$ further from the wall and has moved a distance $(L/2)(1-\cos\theta)$ to the right. The length of the spring is now

$$\ell = \sqrt{(L/2)^2(1-\cos\theta)^2 + [\ell_0 + (L/2)\sin\theta]^2}.$$

If the angle θ is small we may approximate $\cos\theta$ with 1 and $\sin\theta$ with θ in radians. Then the length of the spring is given by $\ell \approx \ell_0 + L\theta/2$ and its elongation is $\Delta x = L\theta/2$. The force it exerts on the rod has magnitude $F = k\Delta x = kL\theta/2$. Since θ is small we may approximate the torque exerted by the spring on the rod by $\tau = -FL/2$, where the pivot point was taken as the origin. Thus, $\tau = -(kL^2/4)\theta$. The constant of proportionality C that relates the torque and angle of rotation is $C = kL^2/4$. The rotational inertia for a rod pivoted at its center is $I = mL^2/12$ (see Table 10-2), where m is its mass.

ANALYZE Substituting the expressions for C and I, we find the period of oscillation to be

$$T = 2\pi\sqrt{\frac{I}{C}} = 2\pi\sqrt{\frac{mL^2/12}{kL^2/4}} = 2\pi\sqrt{\frac{m}{3k}}.$$

With $m = 0.600$ kg and $k = 1850$ N/m, we obtain $T = 0.0653$ s.

LEARN As in the case of a simple linear harmonic oscillator formed by a mass and a spring, the period of the rotating rod is inversely proportional to $\sqrt{k}$. Our result indicates that the rod oscillates very rapidly, with a frequency $f = 1/T = 15.3\,\text{Hz}$, i.e., about 15 times in one second.

15-59

THINK In the presence of a damping force, the amplitude of oscillation of the mass-spring system decreases with time.

EXPRESS As discussed in 15-8, when a damping force is present, we have

$$x(t) = x_m e^{-bt/2m}\cos(\omega' t + \phi)$$

where b is the damping constant and the angular frequency is given by $\omega' = \sqrt{\frac{k}{m} - \frac{b^2}{4m^2}}$.

ANALYZE (a) We want to solve $e^{-bt/2m} = 1/3$ for t. We take the natural logarithm of both sides to obtain $-bt/2m = \ln(1/3)$. Therefore,

$$t = -(2m/b)\ln(1/3) = (2m/b)\ln 3.$$

Thus,

$$t = \frac{2(1.50\text{ kg})}{0.230\text{ kg/s}}\ln 3 = 14.3\text{ s}.$$

(b) The angular frequency is

$$\omega' = \sqrt{\frac{k}{m} - \frac{b^2}{4m^2}} = \sqrt{\frac{8.00\text{ N/m}}{1.50\text{ kg}} - \frac{(0.230\text{ kg/s})^2}{4(1.50\text{ kg})^2}} = 2.31\text{ rad/s}.$$

The period is $T = 2\pi/\omega' = (2\pi)/(2.31\text{ rad/s}) = 2.72$ s and the number of oscillations is

$$t/T = (14.3 \text{ s})/(2.72 \text{ s}) = 5.27.$$

LEARN The displacement $x(t)$ as a function of time is shown below. The amplitude, $x_m e^{-bt/2m}$, decreases exponentially with time.

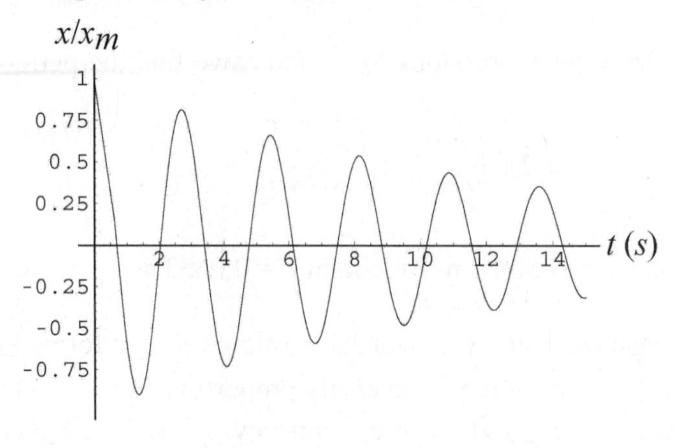

15-69

THINK The piston undergoes simple harmonic motion. Given the amplitude and frequency of oscillation, its maximum speed can be readily calculated.

EXPRESS Let the amplitude be x_m. The maximum speed v_m is related to the amplitude by $v_m = \omega x_m$, where ω is the angular frequency.

ANALYZE We use $v_m = \omega x_m = 2\pi f x_m$, where the frequency is $f = (180 \text{ rev})/(60 \text{ s}) = 3.0$ Hz and the amplitude is half the stroke, or $x_m = 0.38$ m. Thus,

$$v_m = 2\pi(3.0 \text{ Hz})(0.38 \text{ m}) = 7.2 \text{ m/s}.$$

LEARN In a similar manner, the maximum acceleration is

$$a_m = \omega^2 x_m = (2\pi f)^2 x_m = (2\pi(3.0 \text{ Hz}))^2 (0.38 \text{ m}) = 135 \text{ m/s}^2.$$

Acceleration is proportional to the displacement x_m in SHM.

15-73

THINK A mass attached to the end of a vertical spring undergoes simple harmonic motion. Energy is conserved in the process.

EXPRESS The spring stretches until the magnitude of its upward force on the block equals the magnitude of the downward force of gravity: $ky_0 = mg$, where $y_0 = 0.096$ m is the elongation of the spring at equilibrium, k is the spring constant, and $m = 1.3$ kg is the mass of the block. As the block oscillate, its speed is a maximum as it passes the equilibrium point, and zero at the endpoints.

ANALYZE (a) The spring constant is

$$k = mg/y_0 = (1.3\text{ kg})(9.8\text{ m/s}^2)/(0.096\text{ m}) = 1.33\times10^2\text{ N/m}.$$

(b) The period is given by

$$T = \frac{1}{f} = \frac{2\pi}{\omega} = 2\pi\sqrt{\frac{m}{k}} = 2\pi\sqrt{\frac{1.3\text{ kg}}{133\text{ N/m}}} = 0.62\text{ s}.$$

(c) The frequency is $f = 1/T = 1/0.62\text{ s} = 1.6\text{ Hz}$.

(d) The block oscillates in simple harmonic motion about the equilibrium point determined by the forces of the spring and gravity. It is started from rest $\Delta y = 5.0$ cm below the equilibrium point so the amplitude is 5.0 cm.

(e) At the initial position,

$$y_i = y_0 + \Delta y = 9.6\text{ cm} + 5.0\text{ cm} = 14.6\text{ cm} = 0.146\text{ m},$$

the block is not moving but it has potential energy

$$U_i = -mgy_i + \frac{1}{2}ky_i^2 = -(1.3\text{ kg})(9.8\text{ m/s}^2)(0.146\text{ m}) + \frac{1}{2}(133\text{ N/m})(0.146\text{ m})^2 = -0.44\text{ J}.$$

When the block is at the equilibrium point, the elongation of the spring is $y_0 = 9.6$ cm and the potential energy is

$$U_f = -mgy_0 + \frac{1}{2}ky_0^2 = -(1.3\text{ kg})(9.8\text{ m/s}^2)(0.096\text{ m}) + \frac{1}{2}(133\text{ N/m})(0.096\text{ m})^2$$
$$= -0.61\text{ J}.$$

We write the equation for conservation of energy as $U_i = U_f + \frac{1}{2}mv^2$ and solve for v:

$$v = \sqrt{\frac{2(U_i - U_f)}{m}} = \sqrt{\frac{2(-0.44\text{ J} + 0.61\text{ J})}{1.3\text{ kg}}} = 0.51\text{ m/s}.$$

LEARN Both the gravitational force and the spring force are conservative, so the work done by the forces is independent of path. By energy conservation, the kinetic energy of the block is equal to the negative of the change in potential energy of the system:

$$\Delta K = -\Delta U = -(U_f - U_i) = U_i - U_f = -mg(y_i - y_0) + \frac{1}{2}k(y_i^2 - y_0^2)$$
$$= -mg\Delta y + \frac{1}{2}k\left[(y_0 + \Delta y)^2 - y_0^2\right] = -mg\Delta y + \frac{1}{2}k\left[(\Delta y)^2 + 2y_0\Delta y\right]$$
$$= \Delta y(-mg + ky_0) + \frac{1}{2}k(\Delta y)^2$$
$$= \frac{1}{2}k(\Delta y)^2$$

where the relation $ky_0 = mg$ was used.

15-91

THINK This problem explores the oscillation frequency of a pendulum under various accelerating conditions.

EXPRESS In a room, the frequency for small amplitude oscillations is $f = (1/2\pi)\sqrt{g/L}$, where L is the length of the pendulum. Inside an elevator, the forces acting on the pendulum are the tension force $\vec{T}$ of the rod and the force of gravity $m\vec{g}$. Newton's second law yields $\vec{T} + m\vec{g} = m\vec{a}$, where m is the mass and $\vec{a}$ is the acceleration of the pendulum. Let $\vec{a} = \vec{a}_e + \vec{a}'$, where $\vec{a}_e$ is the acceleration of the elevator and $\vec{a}'$ is the acceleration of the pendulum relative to the elevator. Newton's second law can then be written $m(\vec{g} - \vec{a}_e) + \vec{T} = m\vec{a}'$. Relative to the elevator the motion is exactly the same as it would be in an inertial frame where the acceleration due to gravity is $\vec{g}_{\text{eff}} = \vec{g} - \vec{a}_e$.

ANALYZE (a) With $L = 2.0$ m, we find the frequency of the pendulum in a room to be

$$f = \frac{1}{2\pi}\sqrt{\frac{g}{L}} = \frac{1}{2\pi}\sqrt{\frac{9.80\ \text{m/s}^2}{2.0\ \text{m}}} = 0.35\ \text{Hz}.$$

(b) With the elevator accelerating upward, $\vec{g}$ and $\vec{a}_e$ are along the same line but in opposite directions, we can find the frequency for small amplitude oscillations by replacing g with the effective gravitational acceleration $g_{\text{eff}} = g + a_e$ in the expression $f = (1/2\pi)\sqrt{g/L}$. Thus,

$$f = \frac{1}{2\pi}\sqrt{\frac{g + a_e}{L}} = \frac{1}{2\pi}\sqrt{\frac{9.8\ \text{m/s}^2 + 2.0\ \text{m/s}^2}{2.0\ \text{m}}} = 0.39\ \text{Hz}.$$

(c) Now the acceleration due to gravity and the acceleration of the elevator are in the same direction and have the same magnitude. That is, $\vec{g} - \vec{a}_e = 0$. To find the frequency for small amplitude oscillations, replace g with zero in $f = (1/2\pi)\sqrt{g/L}$. The result is zero. The pendulum does not oscillate.

LEARN The frequency of the pendulum increases as g_{eff} increases.

15-101

THINK The block is in simple harmonic motion, so its position relative to the equilibrium position can be written as $x(t) = x_m \cos(\omega t + \phi)$.

EXPRESS The speed of the block is

$$v(t) = dx/dt = -\omega x_m \sin(\omega t + \phi).$$

For a horizontal spring, the relaxed position is the equilibrium position (in a regular simple harmonic motion setting); thus, we infer that the given $v = 5.2$ m/s at $x = 0$ is the maximum value $v_m = \omega x_m$ where

$$\omega = \sqrt{\frac{k}{m}} = \sqrt{\frac{480 \text{ N/m}}{1.2 \text{ kg}}} = 20 \text{ rad/s}.$$

ANALYZE (a) Since $\omega = 2\pi f$, we find $f = 3.2$ Hz.

(b) We have $v_m = 5.2 \text{ m/s} = \omega x_m = (20 \text{ rad/s})x_m$, which leads to $x_m = 0.26$ m.

(c) With meters, seconds and radians understood,

$$x = (0.26 \text{ m})\cos(20t + \phi)$$
$$v = -(5.2 \text{ m/s})\sin(20t + \phi).$$

The requirement that $x = 0$ at $t = 0$ implies (from the first equation above) that either $\phi = +\pi/2$ or $\phi = -\pi/2$. Only one of these choices meets the further requirement that $v > 0$ when $t = 0$; that choice is $\phi = -\pi/2$. Therefore,

$$x = (0.26 \text{ m})\cos\left(20t - \frac{\pi}{2}\right) = (0.26 \text{ m})\sin(20t).$$

LEARN The plots of x and v as a function of time are given below:

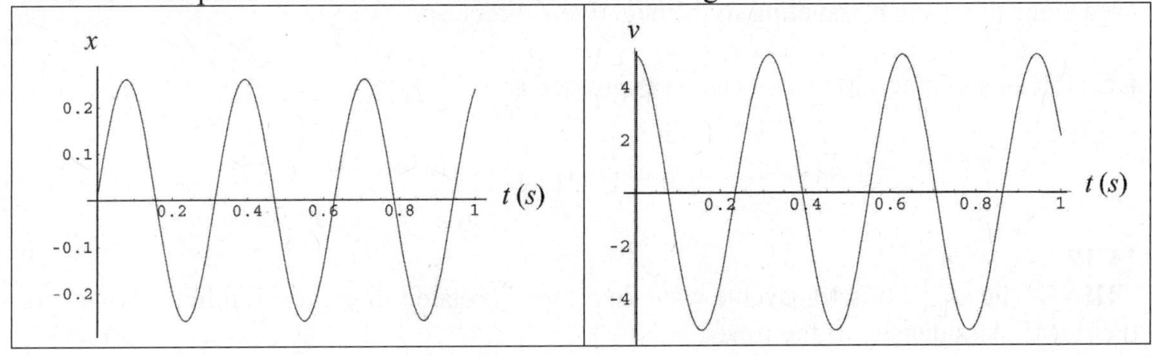

Chapter 16

16-15

THINK Numerous physical properties of a traveling wave can be deduced from its wave function.

EXPRESS We first recall that from Eq. 16-10, a general expression for a sinusoidal wave traveling along the $+x$ direction is

$$y(x,t) = y_m \sin(kx - \omega t + \phi)$$

where y_m is the amplitude, $k = 2\pi/\lambda$ is the angular wave number, $\omega = 2\pi/T$ is the angular frequency and ϕ is the phase constant. The wave speed is given by $v = \sqrt{\tau/\mu}$, where τ is the tension in the string and μ is the linear mass density of the string.

ANALYZE (a) The amplitude of the wave is y_m=0.120 mm.

(b) The wavelength is $\lambda = v/f = \sqrt{\tau/\mu}/f$ and the angular wave number is

$$k = \frac{2\pi}{\lambda} = 2\pi f\sqrt{\frac{\mu}{\tau}} = 2\pi(100\,\text{Hz})\sqrt{\frac{0.50\,\text{kg/m}}{10\,\text{N}}} = 141\,\text{m}^{-1}.$$

(c) The frequency is f = 100 Hz, so the angular frequency is

$$\omega = 2\pi f = 2\pi(100\text{ Hz}) = 628\text{ rad/s}.$$

(d) We may write the string displacement in the form $y = y_m \sin(kx + \omega t)$. The plus sign is used since the wave is traveling in the negative x direction.

LEARN In summary, the wave can be expressed as

$$y = (0.120\,\text{mm})\sin\left[\left(141\,\text{m}^{-1}\right)x + \left(628\,\text{s}^{-1}\right)t\right].$$

16-19

THINK The speed of a transverse wave in a rope is related to the tension in the rope and the linear mass density of the rope.

EXPRESS The wave speed v is given by $v = \sqrt{\tau/\mu}$, where τ is the tension in the rope and μ is the rope's linear mass density, which is defined as the mass per unit length of rope $\mu = m/L$.

ANALYZE With a linear mass density of

$$\mu = m/L = (0.0600 \text{ kg})/(2.00 \text{ m}) = 0.0300 \text{ kg/m},$$

we find the wave speed to be

$$v = \sqrt{\frac{\tau}{\mu}} = \sqrt{\frac{500 \text{ N}}{0.0300 \text{ kg/m}}} = 129 \text{ m/s}.$$

LEARN Since $v \sim 1/\sqrt{\mu}$, the thicker the rope (larger μ), the slower the speed of the rope under the same tension τ.

16-23

THINK Various properties of the sinusoidal wave can be deduced from the plot of its displacement as a function of position.

EXPRESS In analyzing the properties of the wave, we first recall that from Eq. 16-10, a general expression for a sinusoidal wave traveling along the $+x$ direction is

$$y(x,t) = y_m \sin(kx - \omega t + \phi)$$

where y_m is the amplitude, $k = 2\pi/\lambda$ is the angular wave number, $\omega = 2\pi/T$ is the angular frequency and ϕ is the phase constant. The wave speed is given by $v = \sqrt{\tau/\mu}$, where τ is the tension in the string and μ is the linear mass density of the string.

ANALYZE (a) We read the amplitude from the graph. It is about 5.0 cm.

(b) We read the wavelength from the graph. The curve crosses $y = 0$ at about $x = 15$ cm and again with the same slope at about $x = 55$ cm, so

$$\lambda = (55 \text{ cm} - 15 \text{ cm}) = 40 \text{ cm} = 0.40 \text{ m}.$$

(c) The wave speed is

$$v = \sqrt{\frac{\tau}{\mu}} = \sqrt{\frac{3.6 \text{ N}}{25 \times 10^{-3} \text{ kg/m}}} = 12 \text{ m/s}.$$

(d) The frequency is $f = v/\lambda = (12 \text{ m/s})/(0.40 \text{ m}) = 30$ Hz and the period is

$$T = 1/f = 1/(30 \text{ Hz}) = 0.033 \text{ s}.$$

(e) The maximum string speed is

$$u_m = \omega y_m = 2\pi f y_m = 2\pi(30\text{ Hz})(5.0\text{ cm}) = 940\text{ cm/s} = 9.4\text{ m/s}.$$

(f) The angular wave number is $k = 2\pi/\lambda = 2\pi/(0.40\text{ m}) = 16\text{ m}^{-1}$.

(g) The angular frequency is $\omega = 2\pi f = 2\pi(30\text{ Hz}) = 1.9\times10^2\text{ rad/s}$.

(h) According to the graph, the displacement at $x = 0$ and $t = 0$ is 4.0×10^{-2} m. The formula for the displacement gives $y(0, 0) = y_m \sin\phi$. We wish to select ϕ so that

$$(5.0 \times 10^{-2}\text{ m})\sin\phi = (4.0 \times 10^{-2}\text{ m}).$$

The solution is either 0.93 rad or 2.21 rad. In the first case the function has a positive slope at $x = 0$ and matches the graph. In the second case it has negative slope and does not match the graph. We select $\phi = 0.93$ rad.

(i) The string displacement has the form $y(x, t) = y_m \sin(kx + \omega t + \phi)$. A plus sign appears in the argument of the trigonometric function because the wave is moving in the negative x direction.

LEARN Summarizing the results obtained above, the wave function of the traveling wave can be written as

$$y(x,t) = \left(5.0\times10^{-2}\text{m}\right)\sin\left[(16\text{m}^{-1})x + (190\text{s}^{-1})t + 0.93\right].$$

16-31

THINK By superposition principle, the resultant wave is the algebraic sum of the two interfering waves.

EXPRESS The displacement of the string is given by

$$y = y_m \sin(kx - \omega t) + y_m \sin(kx - \omega t + \phi) = 2y_m \cos\left(\tfrac{1}{2}\phi\right)\sin\left(kx - \omega t + \tfrac{1}{2}\phi\right),$$

where we have used

$$\sin\alpha + \sin\beta = 2\sin\frac{1}{2}(\alpha+\beta)\cos\frac{1}{2}(\alpha-\beta).$$

ANALYZE The two waves are out of phase by $\phi = \pi/2$, so the amplitude is

$$A = 2y_m \cos\left(\tfrac{1}{2}\phi\right) = 2y_m \cos(\pi/4) = 1.41y_m .$$

LEARN The interference between two waves can be constructive or destructive, depending on their phase difference.

16-35

THINK We use phasors to add the two waves and calculate the amplitude of the resultant wave.

EXPRESS The phasor diagram is shown below: y_{1m} and y_{2m} represent the original waves and y_m represents the resultant wave. The phasors corresponding to the two constituent waves make an angle of 90° with each other, so the triangle is a right triangle.

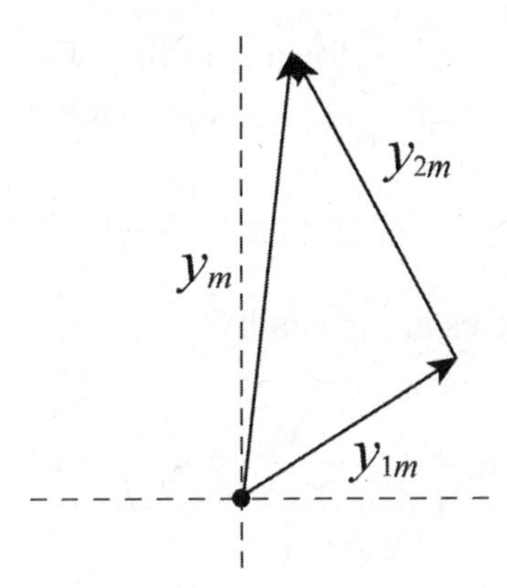

ANALYZE The Pythagorean theorem gives

$$y_m^2 = y_{1m}^2 + y_{2m}^2 = (3.0\,\text{cm})^2 + (4.0\,\text{cm})^2 = (25\,\text{cm})^2 .$$

Thus, the amplitude of the resultant wave is $y_m = 5.0$ cm.

LEARN When adding two waves, it is convenient to represent each wave with a phasor, which is a vector whose magnitude is equal to the amplitude of the wave. The same result, however, could also be obtained as follows: Writing the two waves as $y_1 = 3\sin(kx - \omega t)$ and $y_2 = 4\sin(kx - \omega t + \pi/2) = 4\cos(kx - \omega t)$, we have, after a little algebra,

$$y = y_1 + y_2 = 3\sin(kx - \omega t) + 4\cos(kx - \omega t) = 5\left[\frac{3}{5}\sin(kx - \omega t) + \frac{4}{5}\cos(kx - \omega t)\right]$$

$$= 5\sin(kx - \omega t + \phi)$$

where $\phi = \tan^{-1}(4/3)$. In deducing the phase ϕ, we set $\cos\phi = 3/5$ and $\sin\phi = 4/5$, and use the relation $\cos\phi\sin\theta + \sin\phi\cos\theta = \sin(\theta + \phi)$.

16-41

THINK A string clamped at both ends can be made to oscillate in standing wave patterns.

EXPRESS The wave speed is given by $v = \sqrt{\tau/\mu}$, where τ is the tension in the string and μ is the linear mass density of the string. Since the mass density is the mass per unit length, $\mu = M/L$, where M is the mass of the string and L is its length. The possible wavelengths of a standing wave are given by $\lambda_n = 2L/n$, where L is the length of the string and n is an integer.

ANALYZE (a) The wave speed is

$$v = \sqrt{\frac{\tau L}{M}} = \sqrt{\frac{(96.0\ \text{N})\,(8.40\ \text{m})}{0.120\ \text{kg}}} = 82.0\ \text{m/s}.$$

(b) The longest possible wavelength λ for a standing wave is related to the length of the string by $L = \lambda_1/2$ ($n = 1$), so $\lambda_1 = 2L = 2(8.40\ \text{m}) = 16.8\ \text{m}$.

(c) The corresponding frequency is $f_1 = v/\lambda_1 = (82.0\ \text{m/s})/(16.8\ \text{m}) = 4.88\ \text{Hz}$.

LEARN The resonant frequencies are given by

$$f_n = \frac{v}{\lambda} = \frac{v}{2L/n} = n\frac{v}{2L} = nf_1,$$

where $f_1 = v/\lambda_1 = v/2L$. The oscillation mode with $n = 1$ is called the fundamental mode or the first harmonic.

16-43

THINK A string clamped at both ends can be made to oscillate in standing wave patterns.

EXPRESS Possible wavelengths are given by $\lambda_n = 2L/n$, where L is the length of the wire and n is an integer. The corresponding frequencies are $f_n = v/\lambda_n = nv/2L$, where v is the wave speed. The wave speed is given by $v = \sqrt{\tau/\mu} = \sqrt{\tau L/M}$, where τ is the tension in the wire, μ is the linear mass density of the wire, and M is the mass of the wire. $\mu = M/L$ was used to obtain the last form. Thus,

$$f_n = \frac{n}{2L}\sqrt{\frac{\tau L}{M}} = \frac{n}{2}\sqrt{\frac{\tau}{LM}} = \frac{n}{2}\sqrt{\frac{250\ \text{N}}{(10.0\ \text{m})\,(0.100\ \text{kg})}} = n\ (7.91\ \text{Hz}).$$

ANALYZE (a) The lowest frequency is $f_1 = 7.91\ \text{Hz}$.

(b) The second lowest frequency is $f_2 = 2(7.91\ \text{Hz}) = 15.8\ \text{Hz}$.

(c) The third lowest frequency is $f_3 = 3(7.91\ \text{Hz}) = 23.7\ \text{Hz}$.

LEARN The frequencies are integer multiples of the fundamental frequency f_1. This means that the difference between any successive pair of the harmonic frequencies is equal to the fundamental frequency f_1.

16-45

THINK The difference between any successive pair of the harmonic frequencies is equal to the fundamental frequency.

EXPRESS The resonant wavelengths are given by $\lambda_n = 2L/n$, where L is the length of the string and n is an integer, and the resonant frequencies are

$$f_n = v/\lambda = nv/2L = nf_1,$$

where v is the wave speed. Suppose the lower frequency is associated with the integer n. Then, since there are no resonant frequencies between, the higher frequency is associated with $n + 1$. The frequency difference between successive modes is

$$\Delta f = f_{n+1} - f_n = \frac{v}{2L} = f_1 .$$

ANALYZE (a) The lowest possible resonant frequency is

$$f_1 = \Delta f = f_{n+1} - f_n = 420 \text{ Hz} - 315 \text{ Hz} = 105 \text{ Hz} .$$

(b) The longest possible wavelength is $\lambda_1 = 2L$. If f_1 is the lowest possible frequency then

$$v = \lambda_1 f_1 = (2L)f_1 = 2(0.75 \text{ m})(105 \text{ Hz}) = 158 \text{ m/s}.$$

LEARN Since 315 Hz = 3(105 Hz) and 420 Hz = 4(105 Hz), the two frequencies correspond to $n = 3$ and $n = 4$, respectively.

16-51

THINK In this problem, in order to produce the standing wave pattern, the two waves must have the same amplitude, the same angular frequency, and the same angular wave number, but they travel in opposite directions.

EXPRESS We take the two waves to be

$$y_1 = y_m \sin(kx - \omega t), \quad y_2 = y_m \sin(kx + \omega t).$$

The superposition principle gives

$$y'(x,t) = y_1(x,t) + y_2(x,t) = y_m \sin(kx - \omega t) + y_m \sin(kx + \omega t) = \left[2y_m \sin kx\right] \cos \omega t .$$

ANALYZE (a) The amplitude y_m is half the maximum displacement of the standing wave, or $(0.01 \text{ m})/2 = 5.0 \times 10^{-3}$ m.

(b) Since the standing wave has three loops, the string is three half-wavelengths long: $L = 3\lambda/2$, or $\lambda = 2L/3$. With $L = 3.0$m, $\lambda = 2.0$ m. The angular wave number is

$$k = 2\pi/\lambda = 2\pi/(2.0\text{ m}) = 3.1\text{ m}^{-1}.$$

(c) If v is the wave speed, then the frequency is

$$f = \frac{v}{\lambda} = \frac{3v}{2L} = \frac{3(100\,\text{m/s})}{2(3.0\,\text{m})} = 50\text{ Hz}.$$

The angular frequency is the same as that of the standing wave, or

$$\omega = 2\pi f = 2\pi(50\text{ Hz}) = 314\text{ rad/s}.$$

(d) If one of the waves has the form $y_2(x,t) = y_m \sin(kx + \omega t)$, then the other wave must have the form $y_1(x,t) = y_m \sin(kx - \omega t)$. The sign in front of ω for $y'(x,t)$ is minus.

LEARN Using the results above, the two waves can be written as

$$y_1 = \left(5.0\times10^{-3}\text{ m}\right)\sin\left[\left(3.14\,\text{m}^{-1}\right)x - \left(314\,\text{s}^{-1}\right)t\right]$$

and

$$y_2 = \left(5.0\times10^{-3}\text{ m}\right)\sin\left[\left(3.14\,\text{m}^{-1}\right)x + \left(314\,\text{s}^{-1}\right)t\right].$$

16-69

THINK We use phasors to add the three waves and calculate the amplitude of the resultant wave.

EXPRESS The phasor diagram is shown here: y_1, y_2, and y_3 represent the original waves and y_m represents the resultant wave.

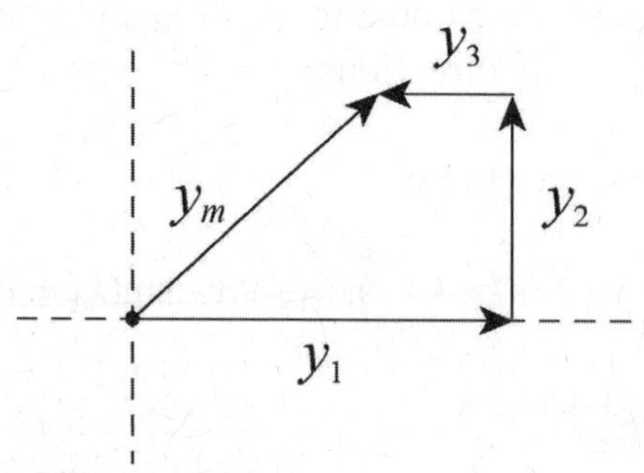

The horizontal component of the resultant is $y_{mh} = y_1 - y_3 = y_1 - y_1/3 = 2y_1/3$. The vertical component is $y_{mv} = y_2 = y_1/2$.

ANALYZE (a) The amplitude of the resultant is

$$y_m = \sqrt{y_{mh}^2 + y_{mv}^2} = \sqrt{\left(\frac{2y_1}{3}\right)^2 + \left(\frac{y_1}{2}\right)^2} = \frac{5}{6}y_1 = 0.83y_1.$$

(b) The phase constant for the resultant is

$$\phi = \tan^{-1}\left(\frac{y_{mv}}{y_{mh}}\right) = \tan^{-1}\left(\frac{y_1/2}{2y_1/3}\right) = \tan^{-1}\left(\frac{3}{4}\right) = 0.644 \text{ rad} = 37°.$$

(c) The resultant wave is

$$y = \frac{5}{6} y_1 \sin(kx - \omega t + 0.644 \text{ rad}).$$

The graph below shows the wave at time $t = 0$. As time goes on it moves to the right with speed $v = \omega/k$.

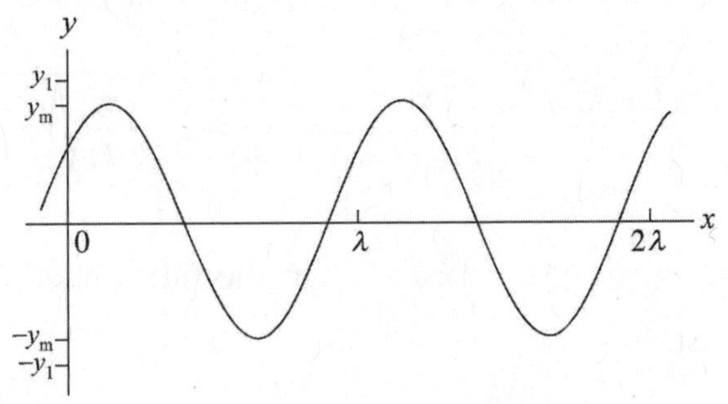

LEARN In adding the three sinusoidal waves, it is convenient to represent each wave with a phasor, which is a vector whose magnitude is equal to the amplitude of the wave. However, adding the three terms explicitly gives, after a little algebra,

$$\begin{aligned}
y_1 + y_2 + y_3 &= y_1 \sin(kx - \omega t) + \frac{1}{2} y_1 \sin(kx - \omega t + \pi/2) + \frac{1}{3} y_1 \sin(kx - \omega t + \pi) \\
&= y_1 \sin(kx - \omega t) + \frac{1}{2} y_1 \cos(kx - \omega t) - \frac{1}{3} y_1 \sin(kx - \omega t) \\
&= \frac{2}{3} y_1 \sin(kx - \omega t) + \frac{1}{2} y_1 \cos(kx - \omega t) \\
&= \frac{5}{6} y_1 \left[\frac{4}{5}\sin(kx - \omega t) + \frac{3}{5}\cos(kx - \omega t)\right] \\
&= \frac{5}{6} y_1 \sin(kx - \omega t + \phi)
\end{aligned}$$

where $\phi = \tan^{-1}(3/4) = 0.644$ rad. In deducing the phase ϕ, we set $\cos\phi = 4/5$ and $\sin\phi = 3/5$, and use the relation $\cos\phi\sin\theta + \sin\phi\cos\theta = \sin(\theta + \phi)$. The result indeed agrees with that obtained in (c).

16-77

THINK The speed of a transverse wave in the stretched rubber band is related to the tension in the band and the linear mass density of the band.

EXPRESS The wave speed v is given by $v = \sqrt{F/\mu}$, where F is the tension in the rubber band and μ is the band's linear mass density, which is defined as the mass per unit length $\mu = m/L$. The fact that the band obeys Hooke's law implies $F = k\Delta\ell$, where k is the spring constant and $\Delta\ell$ is the elongation. Thus, when a force F is applied, the rubber band has a length $L = \ell + \Delta\ell$, where ℓ is the unstretched length, resulting in a linear mass density $\mu = m/(\ell + \Delta\ell)$.

ANALYZE (a) The wave speed is $v = \sqrt{\dfrac{F}{\mu}} = \sqrt{\dfrac{k\Delta\ell}{m/(\ell+\Delta\ell)}} = \sqrt{\dfrac{k\Delta\ell(\ell+\Delta\ell)}{m}}$.

(b) The time required for the pulse to travel the length of the rubber band is

$$t = \frac{2\pi(\ell+\Delta\ell)}{v} = \frac{2\pi(\ell+\Delta\ell)}{\sqrt{k\Delta\ell(\ell+\Delta\ell)/m}} = 2\pi\sqrt{\frac{m}{k}}\sqrt{1+\frac{\ell}{\Delta\ell}}.$$

Thus if $\ell/\Delta\ell \gg 1$, then $t \propto \sqrt{\ell/\Delta\ell} \propto 1/\sqrt{\Delta\ell}$. On the other hand, if $\ell/\Delta\ell \ll 1$, then we have $t \simeq 2\pi\sqrt{m/k} = \text{const}$.

LEARN When $\Delta\ell \ll \ell$, the applied force $F = k\Delta\ell$ is small while $\mu \approx m/\ell = \text{constant}$, leading to a small wave speed. On the other hand, when $\Delta\ell \gg \ell$, $\mu \approx m/\Delta\ell$ and $v = \sqrt{F/\mu} \propto \Delta\ell$, so that $t \simeq 2\pi\sqrt{m/k}$, which is a constant.

16-79

THINK A wire held rigidly at both ends can be made to oscillate in standing wave patterns.

EXPRESS Possible wavelengths are given by $\lambda_n = 2L/n$, where L is the length of the wire and n is an integer. The corresponding frequencies are $f_n = v/\lambda_n = nv/2L$, where v is the wave speed. The wave speed is given by $v = \sqrt{\tau/\mu}$ where τ is the tension in the wire and μ is the linear mass density of the wire.

ANALYZE (a) The wave speed is $v = \sqrt{\dfrac{\tau}{\mu}} = \sqrt{\dfrac{120\,\text{N}}{8.70\times10^{-3}\,\text{kg}/1.50\,\text{m}}} = 144$ m/s.

(b) For the one-loop standing wave we have $\lambda_1 = 2L = 2(1.50\text{ m}) = 3.00$ m.

(c) For the two-loop standing wave $\lambda_2 = L = 1.50$ m.

(d) The frequency for the one-loop wave is $f_1 = v/\lambda_1 = (144\text{ m/s})/(3.00\text{ m}) = 48.0$ Hz.

(e) The frequency for the two-loop wave is $f_2 = v/\lambda_2 = (144\text{ m/s})/(1.50\text{ m}) = 96.0$ Hz.

LEARN The one-loop and two-loop standing wave patterns are plotted below:

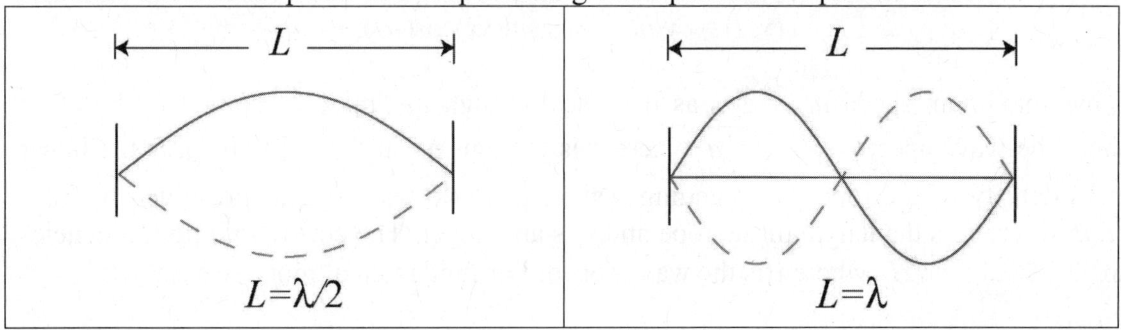

16-83

THINK The speed of a point on the cord is given by $u(x, t) = \partial y/\partial t$, where $y(x, t)$ is displacement.

EXPRESS We take the form of the displacement to be $y(x, t) = y_m \sin(kx - \omega t)$. The speed of a point on the cord is

$$u(x, t) = \partial y/\partial t = -\omega y_m \cos(kx - \omega t),$$

and its maximum value is $u_m = \omega y_m$. The wave speed, on the other hand, is given by $v = \lambda/T = \omega/k$.

(a) The ratio of the maximum particle speed to the wave speed is

$$\frac{u_m}{v} = \frac{\omega y_m}{\omega / k} = k y_m = \frac{2\pi y_m}{\lambda}.$$

(b) The ratio of the speeds depends only on y_m/λ, the ratio of the amplitude to the wavelength.

LEARN Different waves on different cords have the same ratio of speeds if they have the same amplitude and wavelength, regardless of the wave speeds, linear densities of the cords, and the tensions in the cords.

16-91

THINK The rope with both ends fixed and made to oscillate in fundamental mode has wavelength $\lambda = 2L$, where L is the length of the rope.

EXPRESS We first observe that the anti-node at $x = 1.0$ m having zero displacement at $t = 0$ suggests the use of sine instead of cosine for the simple harmonic motion factor. We take the form of the displacement to be

$$y(x, t) = y_m \sin(kx)\sin(\omega t).$$

A point on the rope undergoes simple harmonic motion with a speed

$$u(x, t) = \partial y/\partial t = \omega y_m \sin(kx)\cos(\omega t).$$

It has maximum speed $u_m = \omega y_m$ as it passes through its "middle" point. On the other hand, the wave speed is $v = \sqrt{\tau/\mu}$ where τ is the tension in the rope and μ is the linear mass density of the rope. For standing waves, possible wavelengths are given by $\lambda_n = 2L/n$, where L is the length of the rope and n is an integer. The corresponding frequencies are $f_n = v/\lambda_n = nv/2L$, where v is the wave speed. For fundamental mode, we set $n = 1$.

ANALYZE (a) With $f = 5.0$ Hz, we find the angular frequency to be $\omega = 2\pi f = 10\pi$ rad/s. Thus, if the maximum speed of a point on the rope is $u_m = 5.0$ m/s, then its amplitude is

$$y_m = \frac{u_m}{\omega} = \frac{5.0\text{ m/s}}{10\pi\text{ rad/s}} = 0.16\text{ m}.$$

(b) Since the oscillation is in the *fundamental* mode, we have $\lambda = 2L = 4.0$ m. Therefore, the speed of waves along the rope is $v = f\lambda = 20$ m/s. Then, with $\mu = m/L = 0.60$ kg/m, Eq. 16-26 leads to

$$v = \sqrt{\frac{\tau}{\mu}} \;\Rightarrow\; \tau = \mu v^2 = 240\text{ N} \approx 2.4\times10^2\text{ N}.$$

(c) We note that for the fundamental, $k = 2\pi/\lambda = \pi/L$. Now, *if* the fundamental mode is the only one present (so the amplitude calculated in part (a) is indeed the amplitude of the fundamental wave pattern) then we have

$$y = (0.16\text{ m})\sin\left(\frac{\pi x}{2}\right)\sin(10\pi t) = (0.16\text{ m})\sin[(1.57\text{ m}^{-1})x]\sin[(31.4\text{ rad/s})t]$$

LEARN The period of oscillation is $T = 1/f = 0.20$ s. The snapshots of the patterns at $t = T/4 = 0.05$ s and $t = 3T/4 = 0.15$ s are given below. At $t = T/2$ and T, the displacement is zero everywhere.

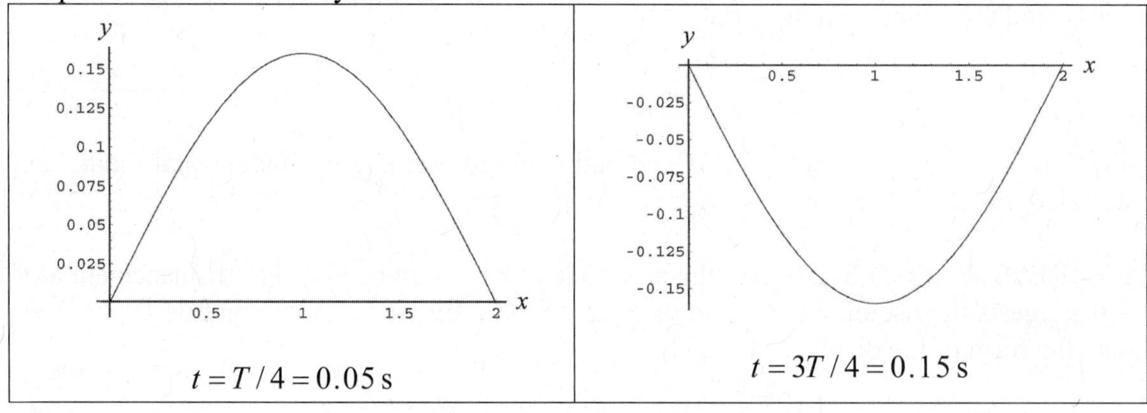

$t = T/4 = 0.05$ s

$t = 3T/4 = 0.15$ s

Chapter 17

17-5

THINK The S and P waves generated by the earthquake travel at different speeds. Knowing the speeds of the waves and the time difference of their arrival to the seismograph allows us to determine the location of the earthquake.

EXPRESS Let d be the distance from the location of the earthquake to the seismograph. If v_s is the speed of the S waves, then the time for these waves to reach the seismograph is $t_s = d/v_s$. Similarly, the time for P waves to reach the seismograph is $t_p = d/v_p$. The time delay is

$$\Delta t = (d/v_s) - (d/v_p) = d(v_p - v_s)/v_s v_p,$$

ANALYZE With $v_s = 4.5\,\text{km/s}$, $v_p = 8.0\,\text{km/s}$ and $\Delta t = 3.0\text{ min} = 180\,\text{s}$, we find the distance to be

$$d = \frac{v_s v_p \Delta t}{(v_p - v_s)} = \frac{(4.5\text{ km/s})(8.0\,\text{km/s})(180\,\text{s})}{8.0\,\text{km/s} - 4.5\,\text{km/s}} = 1.9\times 10^3\,\text{km}.$$

LEARN The distance to the earthquake is proportional to the difference in the arrival times of the P and S waves.

17-7

THINK The time elapsed before hearing the splash is the sum of the time it takes for the stone to hit the water in the well, and the time it takes for the sound wave to travel back to the listener.

EXPRESS Let t_f be the time for the stone to fall to the water and t_s be the time for the sound of the splash to travel from the water to the top of the well. Then, the total time elapsed from dropping the stone to hearing the splash is $t = t_f + t_s$. If d is the depth of the well, then the kinematics of free fall gives

$$d = \frac{1}{2} g t_f^2 \;\Rightarrow\; t_f = \sqrt{2d/g}.$$

The sound travels at a constant speed v_s, so $d = v_s t_s$, or $t_s = d/v_s$. Thus the total time is $t = \sqrt{2d/g} + d/v_s$. This equation is to be solved for d.

ANALYZE Rewriting the above expression as $\sqrt{2d/g} = t - d/v_s$ and squaring both sides, we obtain

$$2d/g = t^2 - 2(t/v_s)d + (1 + v_s^2)d^2.$$

Now multiply by $g v_s^2$ and rearrange to get

$$gd^2 - 2v_s(gt + v_s)d + g v_s^2 t^2 = 0.$$

This is a quadratic equation for *d*. Its solutions are

$$d = \frac{2v_s(gt + v_s) \pm \sqrt{4v_s^2(gt + v_s)^2 - 4g^2v_s^2t^2}}{2g}.$$

The physical solution must yield $d = 0$ for $t = 0$, so we take the solution with the negative sign in front of the square root. Once values are substituted the result $d = 40.7$ m is obtained.

LEARN The relation between the depth of the well and time is plotted below:

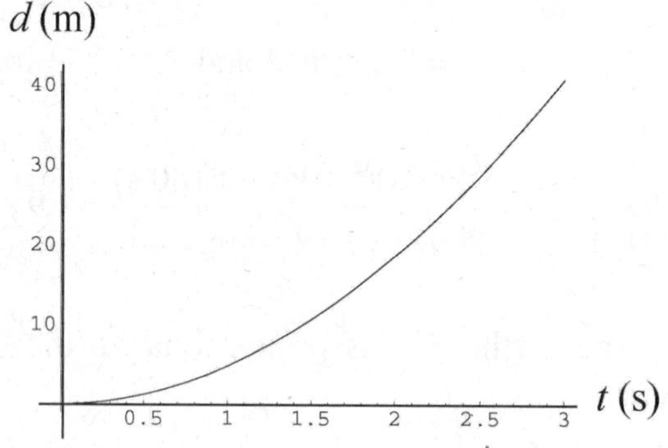

17-11

THINK The speed of sound in a medium is the product of the wavelength and frequency.

EXPRESS The wavelength of the sound wave is given by $\lambda = v/f$, where v is the speed of sound in the medium and f is the frequency,

ANALYZE (a) The speed of sound in air (at 20°C) is $v = 343$ m/s. Thus, we find

$$\lambda = \frac{v}{f} = \frac{343\text{ m/s}}{4.50\times10^6\text{ Hz}} = 7.62\times10^{-5}\text{ m}.$$

(b) The frequency of sound is the same for air and tissue. Now the speed of sound in tissue is $v = 1500$ m/s, the corresponding wavelength is

$$\lambda = \frac{v}{f} = \frac{1500\text{ m/s}}{4.50\times10^6\text{ Hz}} = 3.33\times10^{-4}\text{ m}.$$

LEARN The speed of sound depends on the medium through which it propagates. Table 17-1 provides a list of sound speed in various media.

17-21

THINK The sound waves from the two speakers undergo interference. Whether the interference is constructive or destructive depends on the path length difference, or the phase difference.

EXPRESS From the figure, we see that the distance from the closer speaker to the listener is $L = d_2$, and the distance from the other speaker to the listener is $L' = \sqrt{d_1^2 + d_2^2}$, where d_1 is the distance between the speakers. The phase difference at the location of the listener is $\phi = 2\pi(L' - L)/\lambda$, where λ is the wavelength. For a minimum in intensity at the listener, $\phi = (2n + 1)\pi$, where n is an integer. Thus,

$$\phi = \frac{2\pi(L'-L)}{\lambda_{\text{min}}} = (2n+1)\pi \Rightarrow \lambda_{\text{min}} = \frac{2(L'-L)}{2n+1},$$

and the frequency is

$$f_{\text{min}} = \frac{v}{\lambda_{\text{min}}} = \frac{(2n+1)v}{2\left(\sqrt{d_1^2+d_2^2}-d_2\right)} = \frac{(2n+1)(343\,\text{m/s})}{2\left(\sqrt{(2.00\,\text{m})^2+(3.75\,\text{m})^2}-3.75\,\text{m}\right)} = (2n+1)(343\,\text{Hz}).$$

Now 20,000/343 = 58.3, so $2n + 1$ must range from 0 to 57 for the frequency to be in the audible range (20 Hz to 20 kHz). This means n ranges from 0 to 28.

On the other hand, for a maximum in intensity at the listener, $\phi = 2n\pi$, where n is any positive integer. Thus $\lambda_{\text{max}} = (1/n)\left(\sqrt{d_1^2+d_2^2}-d_2\right)$ and

$$f_{\text{max}} = \frac{v}{\lambda_{\text{max}}} = \frac{nv}{\sqrt{d_1^2+d_2^2}-d_2} = \frac{n(343\,\text{m/s})}{\sqrt{(2.00\,\text{m})^2+(3.75\,\text{m})^2}-3.75\,\text{m}} = n(686\,\text{Hz}).$$

Since 20,000/686 = 29.2, n must be in the range from 1 to 29 for the frequency to be audible.

ANALYZE (a) The lowest frequency that gives minimum signal is (n = 0) $f_{\text{min},1} = 343$ Hz.

(b) The second lowest frequency is (n = 1) $f_{\text{min},2} = [2(1)+1](343\text{ Hz}) = 1029\text{ Hz} = 3f_{\text{min},1}$. Thus, the factor is 3.

(c) The third lowest frequency is (n=2) $f_{\text{min},3} = [2(2)+1](343\text{ Hz}) = 1715\text{ Hz} = 5f_{\text{min},1}$. Thus, the factor is 5.

(d) The lowest frequency that gives maximum signal is (n =1) $f_{\text{max},1} = 686$ Hz.

(e) The second lowest frequency is (n = 2) $f_{\max,2} = 2(686\text{ Hz}) = 1372\text{ Hz} = 2f_{\max,1}$. Thus, the factor is 2.

(f) The third lowest frequency is (n = 3) $f_{\max,3} = 3(686\text{ Hz}) = 2058\text{ Hz} = 3f_{\max,1}$. Thus, the factor is 3.

LEARN We see that the interference of the two sound waves depends on their phase difference $\phi = 2\pi(L' - L)/\lambda$. The interference is fully constructive when ϕ is a multiple of 2π, but fully destructive when ϕ is an odd multiple of π.

17-27

THINK The sound level increases by 10 dB when the intensity increases by a factor of 10.

EXPRESS The sound level β is defined as (see Eq. 17-29):

$$\beta = (10\text{ dB})\log\frac{I}{I_0}$$

where $I_0 = 10^{-12}\text{ W/m}^2$ is the standard reference intensity. In this problem, let I_1 be the original intensity and I_2 be the final intensity. The original sound level is $\beta_1 = (10\text{ dB})\log(I_1/I_0)$ and the final sound level is β_2 = (10 dB) $\log(I_2/I_0)$. With $\beta_2 = \beta_1 + 30\text{ dB}$, we have

$$(10\text{ dB})\log(I_2/I_0) = (10\text{ dB})\log(I_1/I_0) + 30\text{ dB},$$

or

$$(10\text{ dB})\log(I_2/I_0) - (10\text{ dB})\log(I_1/I_0) = 30\text{ dB}.$$

The above equation allows us to solve for the ratio I_2/I_1. On the other hand, combing Eqs. 17-15 and 17-27 leads to the following relation between the intensity I and the pressure amplitude Δp_m: $I = \dfrac{1}{2}\dfrac{(\Delta p_m)^2}{\rho v}$.

ANALYZE (a) Divide by 10 dB and use $\log(I_2/I_0) - \log(I_1/I_0) = \log(I_2/I_1)$ to obtain $\log(I_2/I_1) = 3$. Now use each side as an exponent of 10 and recognize that $10^{\log(I_2/I_1)} = I_2/I_1$. The result is $I_2/I_1 = 10^3$. The intensity is increased by a factor of 1.0×10^3.

(b) The pressure amplitude is proportional to the square root of the intensity so it is increased by a factor of $\sqrt{1000} \approx 32$.

LEARN From the definition of β, we see that doubling sound intensity increases the sound level by $\Delta\beta = (10 \text{ dB})\log 2 = 3.01 \text{ dB}$.

17-29

THINK Power is the time rate of energy transfer, and intensity is the rate of energy flow per unit area perpendicular to the flow.

EXPRESS The rate at which energy flow across every sphere centered at the source is the same, regardless of the sphere radius, and is the same as the power output of the source. If P is the power output and I is the intensity a distance r from the source, then $P = IA = 4\pi r^2 I$, where $A = 4\pi r^2$ is the surface area of a sphere of radius r.

ANALYZE With $r = 2.50$ m and $I = 1.91\times10^{-4}$ W/m^2, we find the power of the source to be

$$P = 4\pi(2.50 \text{ m})^2 (1.91 \times 10^{-4} \text{ W/m}^2) = 1.50 \times 10^{-2} \text{ W}.$$

LEARN Since intensity falls off as $1/r^2$, the further away from the source, the weaker the intensity.

17-39

THINK Violin strings are fixed at both ends. A string clamped at both ends can be made to oscillate in standing wave patterns.

EXPRESS When the string is fixed at both ends and set to vibrate at its fundamental, lowest resonant frequency, exactly one-half of a wavelength fits between the ends (see figure to the right). The wave speed is given by $v = \lambda f = \sqrt{\tau/\mu}$, where τ is the tension in the string and μ is the linear mass density of the string.

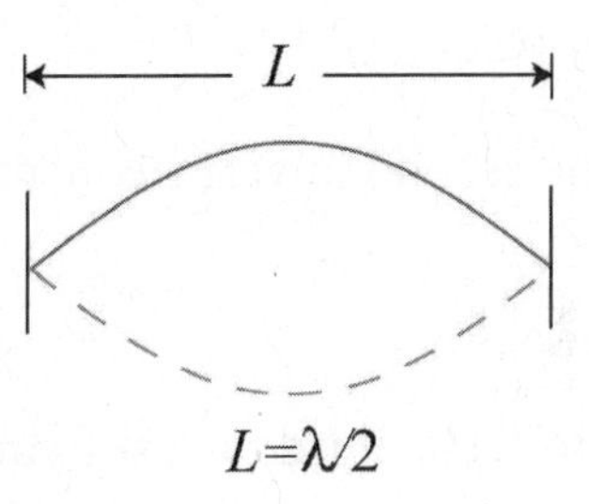

ANALYZE (a) With $\lambda = 2L$, we find the wave speed to be

$$v = f\lambda = 2Lf = 2(0.220 \text{ m})(920 \text{ Hz}) = 405 \text{ m/s}.$$

(b) If M is the mass of the (uniform) string, then $\mu = M/L$. Thus, the string tension is

$$\tau = \mu v^2 = (M/L)v^2 = [(800 \times 10^{-6} \text{ kg})/(0.220 \text{ m})]\,(405 \text{ m/s})^2 = 596 \text{ N}.$$

(c) The wavelength is $\lambda = 2L = 2(0.220 \text{ m}) = 0.440$ m.

(d) If v_a is the speed of sound in air, then the wavelength in air is

$$\lambda_a = v_a/f = (343 \text{ m/s})/(920 \text{ Hz}) = 0.373 \text{ m}.$$

LEARN The frequency of the sound wave in air is the same as the frequency of oscillation of the string. However, the wavelengths of the wave on the string and the sound waves emitted by the string are different because their wave speeds are not the same.

17-43

THINK The pipe is open at both ends so there are displacement antinodes at both ends.

EXPRESS If L is the pipe length and λ is the wavelength then $\lambda = 2L/n$, where n is an integer. That is, an integer number of half-wavelengths fit into the length of the pipe. If v is the speed of sound then the resonant frequencies are given by $f = v/\lambda = nv/2L$. Now L = 0.457 m, so

$$f = \frac{nv}{2L} = \frac{n(344\text{ m/s})}{2(0.457\text{ m})} = (376.4\text{ Hz})n\,.$$

ANALYZE (a) To find the resonant frequencies that lie between 1000 Hz and 2000 Hz, first set f = 1000 Hz and solve for n, then set f = 2000 Hz and again solve for n. The results are 2.66 and 5.32, which imply that n = 3, 4, and 5 are the appropriate values of n. Thus, there are 3 frequencies.

(b) The lowest frequency at which resonance occurs corresponds to n = 3, or

$$f = 3(376.4\text{ Hz}) = 1129\text{ Hz}\,.$$

(c) The second lowest frequency at which resonance occurs corresponds to n = 4, or

$$f = 4(376.4\text{ Hz}) = 1506\text{ Hz}\,.$$

LEARN The third lowest frequency at which resonance occurs corresponds to n = 5, or

$$f = 5(376.4\text{ Hz}) = 1882\text{ Hz}\,.$$

Changing the length of the pipe can affect the number of resonant frequencies.

17-49

THINK Violin strings are fixed at both ends. A string clamped at both ends can be made to oscillate in standing wave patterns.

EXPRESS The resonant wavelengths are given by $\lambda = 2L/n$, where L is the length of the string and n is an integer. The resonant frequencies are given by $f_n = v/\lambda = nv/2L$, where v is the wave speed on the string. Now $v = \sqrt{\tau/\mu}$, where τ is the tension in the string and μ is the linear mass density of the string. Thus $f_n = (n/2L)\sqrt{\tau/\mu}$.

ANALYZE Suppose the lower frequency is associated with n_1 and the higher frequency is associated with $n_2 = n_1 + 1$. There are no resonant frequencies between so you know that the integers associated with the given frequencies differ by 1. Thus, $f_{n_1} = (n_1/2L)\sqrt{\tau/\mu}$ and

$$f_{n_2} = \frac{n_1+1}{2L}\sqrt{\frac{\tau}{\mu}} = \frac{n_1}{2L}\sqrt{\frac{\tau}{\mu}} + \frac{1}{2L}\sqrt{\frac{\tau}{\mu}} = f_{n_1} + \frac{1}{2L}\sqrt{\frac{\tau}{\mu}}.$$

This means $f_{n_2} - f_{n_1} = (1/2L)\sqrt{\tau/\mu}$ and

$$\tau = 4L^2\mu(f_{n_2} - f_{n_1})^2 = 4(0.300\,\text{m})^2(0.650\times10^{-3}\,\text{kg/m})(1320\,\text{Hz} - 880\,\text{Hz})^2$$
$$= 45.3\,\text{N}.$$

LEARN Since the difference between any successive pair of the harmonic frequencies is equal to the fundamental frequency: $\Delta f = f_{n+1} - f_n = \dfrac{v}{2L} = f_1$, we find

$$f_1 = 1320\,\text{Hz} - 880\,\text{Hz} = 440\,\text{Hz}.$$

Since 880 Hz = 2(440 Hz) and 1320 Hz = 3(440 Hz), the two frequencies correspond to $n_1 = 2$ and $n_2 = 3$, respectively.

17-53

THINK Beat arises when two waves detected have slightly different frequencies: $f_{\text{beat}} = f_2 - f_1$.

EXPRESS Each wire is vibrating in its fundamental mode so the wavelength is twice the length of the wire ($\lambda = 2L$) and the frequency is

$$f = v/\lambda = (1/2L)\sqrt{\tau/\mu},$$

where $v = \sqrt{\tau/\mu}$ is the wave speed for the wire, τ is the tension in the wire, and μ is the linear mass density of the wire. Suppose the tension in one wire is τ and the oscillation frequency of that wire is f_1. The tension in the other wire is $\tau + \Delta\tau$ and its frequency is f_2. You want to calculate $\Delta\tau/\tau$ for f_1 = 600 Hz and f_2 = 606 Hz. Now, $f_1 = (1/2L)\sqrt{\tau/\mu}$ and $f_2 = (1/2L)\sqrt{(\tau+\Delta\tau)/\mu}$, so

$$f_2/f_1 = \sqrt{(\tau+\Delta\tau)/\tau} = \sqrt{1+(\Delta\tau/\tau)}.$$

ANALYZE The fractional increase in tension is

$$\Delta\tau/\tau = (f_2/f_1)^2 - 1 = [(606\,\text{Hz})/(600\,\text{Hz})]^2 - 1 = 0.020.$$

LEARN Beat frequency $f_{\text{beat}} = f_2 - f_1$ is zero when $\Delta\tau = 0$. The beat phenomenon is used by musicians to tune musical instruments. The instrument tuned is sounded against a standard frequency until beat disappears.

17-67

THINK The girl and her uncle hear different frequencies because of Doppler effect.

EXPRESS The Doppler shifted frequency is given by

$$f' = f\frac{v \pm v_D}{v \mp v_S},$$

where f is the unshifted frequency, v is the speed of sound, v_D is the speed of the detector (the uncle), and v_S is the speed of the source (the locomotive). All speeds are relative to the air.

ANALYZE (a) The uncle is at rest with respect to the air, so $v_D = 0$. The speed of the source is $v_S = 10$ m/s. Since the locomotive is moving away from the uncle the frequency decreases and we use the plus sign in the denominator. Thus

$$f' = f\frac{v}{v + v_S} = (500.0\,\text{Hz})\left(\frac{343\,\text{m/s}}{343\,\text{m/s} + 10.00\,\text{m/s}}\right) = 485.8\,\text{Hz}.$$

(b) The girl is now the detector. Relative to the air she is moving with speed $v_D = 10.00$ m/s toward the source. This tends to increase the frequency and we use the plus sign in the numerator. The source is moving at $v_S = 10.00$ m/s away from the girl. This tends to decrease the frequency and we use the plus sign in the denominator. Thus, $(v + v_D) = (v + v_S)$ and $f' = f = 500.0$ Hz.

(c) Relative to the air the locomotive is moving at $v_S = 20.00$ m/s away from the uncle. Use the plus sign in the denominator. Relative to the air the uncle is moving at $v_D = 10.00$ m/s toward the locomotive. Use the plus sign in the numerator. Thus

$$f' = f\frac{v + v_D}{v + v_S} = (500.0\,\text{Hz})\left(\frac{343\,\text{m/s} + 10.00\,\text{m/s}}{343\,\text{m/s} + 20.00\,\text{m/s}}\right) = 486.2\,\text{Hz}.$$

(d) Relative to the air the locomotive is moving at $v_S = 20.00$ m/s away from the girl and the girl is moving at $v_D = 20.00$ m/s toward the locomotive. Use the plus signs in both the numerator and the denominator. Thus, $(v + v_D) = (v + v_S)$ and $f' = f = 500.0$ Hz.

LEARN The uncle, standing near the track, hears different frequencies, depending on the direction of the wind. On other hand, since the girl (a detector) is sitting in the train and there's no relative motion between her and the source (locomotive whistle), she hears the same frequency as the source regardless of the wind direction.

17-69

THINK Mach number is the ratio v_S / v, where v_S is the speed of the source and v is the sound speed. A mach number of 1.5 means that the jet plane moves at a supersonic speed.

EXPRESS The half angle θ of the Mach cone is given by sin $\theta = v/v_S$, where v is the speed of sound and v_S is the speed of the plane. To calculate the time it takes for the shock wave to each you after the plane has passed directly overhead, let h be the altitude of the plane and suppose the Mach cone intersects Earth's surface a distance d behind the plane. The situation is shown in the diagram below, with P indicating the plane and O indicating the observer.

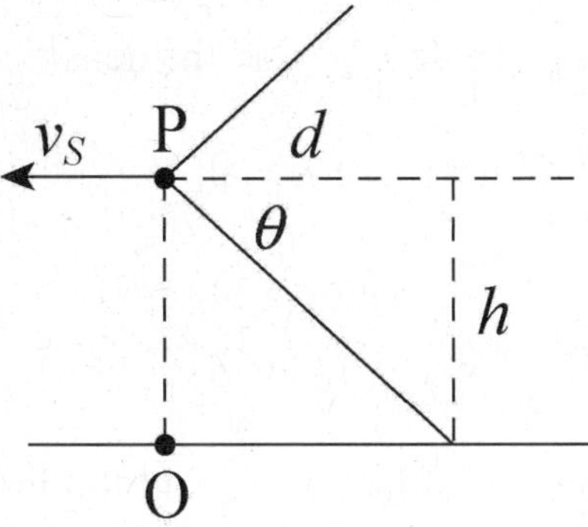

The cone angle is related to h and d by tan $\theta = h/d$, so $d = h/\tan\theta$. The shock wave reaches O in the time the plane takes to fly the distance d.

ANALYZE (a) Since $v_S = 1.5v$, sin $\theta = v/1.5v = 1/1.5$. This means $\theta = 42°$.

(b) The time required for the shock wave to reach you is

$$t = \frac{d}{v} = \frac{h}{v\tan\theta} = \frac{5000\text{ m}}{1.5(331\text{ m/s})\tan 42°} = 11\text{ s}.$$

LEARN The shock wave generated by the supersonic jet produces an explosive sound called sonic boom, in which the air pressure first increases suddenly, and then drops suddenly below normal before returning to normal.

17-81

THINK The pressure amplitude of the sound wave depends on the medium it propagates through.

EXPRESS The intensity of a sound wave is given by $I = \frac{1}{2}\rho v\omega^2 s_m^2$, where ρ is the density of the medium, v is the speed of sound, ω is the angular frequency, and s_m is the displacement amplitude. The displacement and pressure amplitudes are related by $\Delta p_m =$

$\rho v \omega s_m$, so $s_m = \Delta p_m / \rho v \omega$ and $I = (\Delta p_m)^2 / 2\rho v$. For waves of the same frequency the ratio of the intensity for propagation in water to the intensity for propagation in air is

$$\frac{I_w}{I_a} = \left(\frac{\Delta p_{mw}}{\Delta p_{ma}}\right)^2 \frac{\rho_a v_a}{\rho_w v_w},$$

where the subscript a denotes air and the subscript w denotes water.

ANALYZE (a) In case where the intensities are equal, $I_a = I_w$, the ratio of the pressure amplitude is

$$\frac{\Delta p_{mw}}{\Delta p_{ma}} = \sqrt{\frac{\rho_w v_w}{\rho_a v_a}} = \sqrt{\frac{(0.998\times 10^3\ \text{kg/m}^3)(1482\ \text{m/s})}{(1.21\ \text{kg/m}^3)(343\ \text{m/s})}} = 59.7.$$

The speeds of sound are given in Table 17-1 and the densities are given in Table 14-1.

(b) Now, if the pressure amplitudes are equal: $\Delta p_{mw} = \Delta p_{ma}$, then the ratio of the intensities is

$$\frac{I_w}{I_a} = \frac{\rho_a v_a}{\rho_w v_w} = \frac{(1.21\ \text{kg/m}^3)(343\ \text{m/s})}{(0.998\times 10^3\ \text{kg/m}^3)(1482\ \text{m/s})} = 2.81\times 10^{-4}.$$

LEARN The pressure amplitude of sound wave and the intensity depend on the density of the medium and the sound speed in the medium.

17-83

THINK This problem deals with the principle of Doppler ultrasound. The technique can be used to measure blood flow and blood pressure by reflecting high-frequency, ultrasound sound waves off blood cells.

EXPRESS The direction of blood flow can be determined by the Doppler shift in frequency. The reception of the ultrasound by the blood and the subsequent remitting of the signal by the blood back toward the detector is a two-step process which may be compactly written as

$$f + \Delta f = f\left(\frac{v + v_x}{v - v_x}\right)$$

where $v_x = v_{\text{blood}} \cos\theta$. If we write the ratio of frequencies as $R = (f + \Delta f)/f$, then the solution of the above equation for the speed of the blood is

$$v_{\text{blood}} = \frac{(R-1)v}{(R+1)\cos\theta}.$$

ANALYZE (a) The blood is moving towards the right (towards the detector), because the Doppler shift in frequency is an *increase*: $\Delta f > 0$.

(b) With $v = 1540$ m/s, $\theta = 20°$, and

$$R = 1 + (5495\text{ Hz})/(5 \times 10^6\text{ Hz}) = 1.0011,$$

using the expression above, we find the speed of the blood to be

$$v_{\text{blood}} = \frac{(R-1)v}{(R+1)\cos\theta} = 0.90\,\text{m/s}.$$

(c) We interpret the question as asking how Δf (still taken to be positive, since the detector is in the "forward" direction) changes as the detection angle θ changes. Since larger θ means smaller horizontal component of velocity v_x then we expect Δf to decrease towards zero as θ is increased towards 90°.

LEARN The expression for v_{blood} can be inverted to give

$$\Delta f = \left(\frac{2v_{\text{blood}}\cos\theta}{v - v_{\text{blood}}\cos\theta}\right) f.$$

The plot of the frequency shift Δf as a function of θ is given below. Indeed we find Δf to decrease with increasing θ.

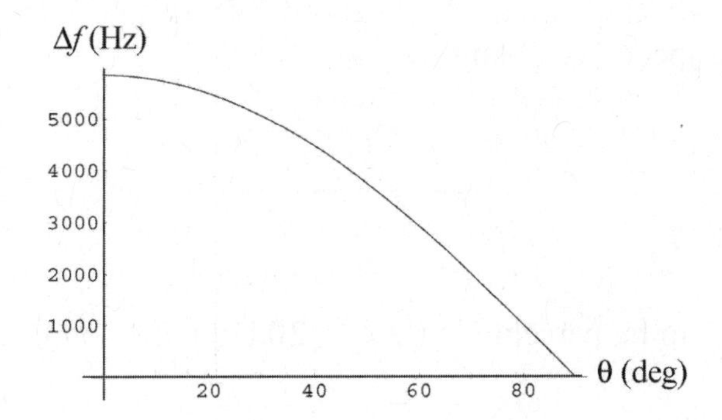

17-87

THINK

The siren is between you and the cliff, moving away from you and towards the cliff. You hear two frequencies, one directly from the siren and the other from the sound reflected off the cliff.

EXPRESS The Doppler shifted frequency is given by

$$f' = f\frac{v \pm v_D}{v \mp v_S},$$

where f is the unshifted frequency, v is the speed of sound, v_D is the speed of the detector, and v_S is the speed of the source. All speeds are relative to the air. Both "detectors" (you and the cliff) are stationary, so $v_D = 0$ in Eq. 17–47. The source is the siren with $v_S = 10\text{ m/s}$. The problem asks us to use $v = 330$ m/s for the speed of sound.

ANALYZE (a) With $f = 1000$ Hz, the frequency f_y you hear becomes

$$f_y = f\left(\frac{v+0}{v+v_S}\right) = (1000\text{ Hz})\left(\frac{330\text{ m/s}}{330\text{ m/s}+10\text{ m/s}}\right) = 970.6\text{ Hz} \approx 9.7\times10^2\text{ Hz}.$$

(b) The frequency heard by an observer at the cliff (and thus the frequency of the sound reflected by the cliff, ultimately reaching your ears at some distance from the cliff) is

$$f_c = f\left(\frac{v+0}{v-v_S}\right) = (1000\text{ Hz})\left(\frac{330\text{ m/s}}{330\text{ m/s}-10\text{ m/s}}\right) = 1031.3\text{ Hz} \approx 1.0\times10^3\text{ Hz}.$$

(c) The beat frequency is $f_{\text{beat}} = f_c - f_y = 60$ beats/s (which, due to specific features of the human ear, is too large to be perceptible).

LEARN The beat frequency in this case can be written as

$$f_{\text{beat}} = f_c - f_y = f\left(\frac{v}{v-v_S}\right) - f\left(\frac{v}{v+v_S}\right) = \frac{2vv_S}{v^2 - v_S^2} f$$

Solving for the source speed, we obtain

$$v_S = \left(\frac{-f + \sqrt{f^2 + f_{\text{beat}}^2}}{f_{\text{beat}}}\right) v$$

For the beat frequency to be perceptible ($f_{\text{beat}} < 20\text{ Hz}$), the source speed would have to be less than 3.3 m/s.

17-93

THINK Acoustic interferometer can be used to demonstrate the interference of sound waves.

EXPRESS When the right side of the instrument is pulled out a distance d the path length for sound waves increases by $2d$. Since the interference pattern changes from a minimum to the next maximum, this distance must be half a wavelength of the sound. So $2d = \lambda/2$, where λ is the wavelength. Thus $\lambda = 4d$.

On the other hand, the intensity is given by $I = \frac{1}{2}\rho v\omega^2 s_m^2$, where ρ is the density of the medium, v is the speed of sound, ω is the angular frequency, and s_m is the displacement amplitude. Thus, s_m is proportional to the square root of the intensity, and we write $\sqrt{I} = Cs_m$, where C is a constant of proportionality. At the minimum, interference is destructive and the displacement amplitude is the difference in the amplitudes of the individual waves: $s_m = s_{SAD} - s_{SBD}$, where the subscripts indicate the paths of the waves. At the maximum, the waves interfere constructively and the displacement amplitude is the sum of the amplitudes of the individual waves: $s_m = s_{SAD} + s_{SBD}$.

ANALYZE (a) The speed of sound is $v = 343$ m/s, so the frequency is

$$f = v/\lambda = v/4d = (343 \text{ m/s})/4(0.0165 \text{ m}) = 5.2 \times 10^3 \text{ Hz}.$$

(b) At intensity minimum, we have $\sqrt{100} = C(s_{SAD} - s_{SBD})$, and $\sqrt{900} = C(s_{SAD} + s_{SBD})$ at the maximum. Adding the equations give

$$s_{SAD} = (\sqrt{100} + \sqrt{900} / 2C = 20 / C,$$

while subtracting them yields

$$s_{SBD} = (\sqrt{900} - \sqrt{100}) / 2C = 10 / C.$$

Thus, the ratio of the amplitudes is $s_{SAD}/s_{SBD} = 2$.

(c) Any energy losses, such as might be caused by frictional forces of the walls on the air in the tubes, result in a decrease in the displacement amplitude. Those losses are greater on path B since it is longer than path A.

LEARN We see that the sound waves propagated along the two paths in the interferometer can interfere constructively or destructively, depending on their path length difference.

17-95

THINK Intensity is power divided by area. For an isotropic source the area is the surface area of a sphere.

EXPRESS If P is the power output and I is the intensity a distance r from the source, then $P = IA = 4\pi r^2 I$, where $A = 4\pi r^2$ is the surface area of a sphere of radius r. On the other hand, the sound level β can be calculated using Eq. 17-29:

$$\beta = (10 \text{ dB}) \log \frac{I}{I_0}$$

where $I_0 = 10^{-12}$ W/m^2 is the standard reference intensity.

ANALYZE (a) With $r = 10$ m and $I = 8.0 \times 10^{-3}$ W/m^2, we have

$$P = 4\pi r^2 I = 4\pi (10)^2 (8.0 \times 10^{-3} \text{ W/m}^2) = 10 \text{ W}.$$

(b) Using the value of P obtained in (a), we find the intensity at $r' = 5.0$ m to be

$$I' = \frac{P}{4\pi r'^2} = \frac{10 \text{ W}}{4\pi (5.0 \text{ m})^2} = 0.032 \text{ W/m}^2.$$

(c) Using Eq. 17–29 with $I = 0.0080 \text{ W/m}^2$, we find the sound level to be

$$\beta = (10 \text{ dB}) \log\left(\frac{8.0 \times 10^{-3} \text{ W/m}^2}{10^{-12} \text{ W/m}^2}\right) = 99 \text{ dB}.$$

LEARN The ratio of the sound intensities at two different locations can be written as

$$\frac{I}{I'} = \frac{P/4\pi r^2}{P/4\pi r'^2} = \left(\frac{r'}{r}\right)^2.$$

Similarly, the difference in sound level is given by $\Delta\beta = \beta - \beta' = (10 \text{ dB}) \log\left(\frac{I}{I'}\right)$.

Chapter 18

18-13

THINK The aluminum sphere expands thermally when being heated, so its volume increases.

EXPRESS Since a volume is the product of three lengths, the change in volume due to a temperature change ΔT is given by $\Delta V = 3\alpha V \Delta T$, where V is the original volume and α is the coefficient of linear expansion (see Eq. 18-11).

ANALYZE With the volume of the sphere given by $V = (4\pi/3)R^3$, where R= 10 cm is the original radius of the sphere and $\alpha = 23\times10^{-6}/\text{C}°$, then

$$\Delta V = 3\alpha\left(\frac{4\pi}{3}R^3\right)\Delta T = (23\times10^{-6}/\text{C}°)(4\pi)(10\,\text{cm})^3(100\,°\text{C}) = 29\,\text{cm}^3.$$

The value for the coefficient of linear expansion is found in Table 18-2.

LEARN The change in volume can be expressed as $\Delta V/V = \beta\Delta T$, where $\beta = 3\alpha$ is the coefficient of volume expansion. For aluminum, we have $\beta = 3\alpha = 69\times10^{-6}/\text{C}°$.

18-17

THINK Since the aluminum cup and the glycerin have different coefficients of thermal expansion, their volumes would change by a different amount under the same ΔT.

EXPRESS If V_c is the original volume of the cup, α_a is the coefficient of linear expansion of aluminum, and ΔT is the temperature increase, then the change in the volume of the cup is $\Delta V_c = 3\alpha_a V_c \Delta T$ (See Eq. 18-11).

On the other hand, if β is the coefficient of volume expansion for glycerin, then the change in the volume of glycerin is $\Delta V_g = \beta V_c \Delta T$. Note that the original volume of glycerin is the same as the original volume of the cup. The volume of glycerin that spills is

$$\Delta V_g - \Delta V_c = (\beta - 3\alpha_a)V_c\Delta T = \left[(5.1\times10^{-4}/\text{C}°) - 3(23\times10^{-6}/\text{C}°)\right](100\,\text{cm}^3)(6.0\,°\text{C})$$
$$= 0.26\,\text{cm}^3.$$

LEARN Glycerin spills over because $\beta > 3\alpha$, which gives $\Delta V_g - \Delta V_c > 0$. Note that since liquids in general have greater coefficients of thermal expansion than solids, heating a cup filled with liquid generally will cause the liquid to spill out.

18-21

THINK The bar expands thermally when heated. Since its two ends are held fixed, the bar buckles upward.

EXPRESS Consider half the bar. Its original length is $\ell_0 = L_0/2$ and its length after the temperature increase is $\ell = \ell_0 + \alpha\ell_0\Delta T$. The old position of the half-bar, its new position, and the distance x that one end is displaced form a right triangle, with a hypotenuse of length ℓ, one side of length ℓ_0, and the other side of length x. The Pythagorean theorem yields

$$x^2 = \ell^2 - \ell_0^2 = \ell_0^2(1+\alpha\Delta T)^2 - \ell_0^2.$$

Since the change in length is small we may approximate $(1 + \alpha\,\Delta T)^2$ by $1 + 2\alpha\,\Delta T$, where the small term $(\alpha\,\Delta T)^2$ was neglected. Then,

$$x^2 = \ell_0^2 + 2\ell_0^2\alpha\,\Delta T - \ell_0^2 = 2\ell_0^2\alpha\,\Delta T$$

and $x \approx \ell_0\sqrt{2\alpha\,\Delta T}$.

ANALYZE Substituting the values given, we obtain

$$x = \ell_0\sqrt{2\alpha\,\Delta T} = \frac{3.77\,\text{m}}{2}\sqrt{2\left(25\times10^{-6}/\text{C}°\right)\left(32°\text{C}\right)} = 7.5\times10^{-2}\,\text{m}.$$

LEARN The length of the bar changes by $\Delta\ell = \alpha\ell_0\Delta T \sim \alpha\Delta T$. However, to the leading order, the vertical distance the bar has risen is proportional to $(\alpha\Delta T)^{1/2}$.

18-23

THINK Electrical energy is supplied and converted into thermal energy to raise the water temperature.

EXPRESS The water has a mass m = 0.100 kg and a specific heat c = 4190 J/kg·K. When raised from an initial temperature T_i = 23°C to its boiling point T_f = 100°C, the heat input is given by $Q = cm(T_f - T_i)$. This must be the power output of the heater P multiplied by the time t: $Q = Pt$.

ANALYZE The time it takes to heat up the water is

$$t = \frac{Q}{P} = \frac{cm(T_f - T_i)}{P} = \frac{(4190\,\text{J/kg}\cdot\text{K})(0.100\,\text{kg})(100°\text{C} - 23°\text{C})}{200\,\text{J/s}} = 160\,\text{s}.$$

LEARN With a fixed power output, the time required is proportional to Q, which is proportional to $\Delta T = T_f - T_i$. In real life, it would take longer because of heat loss.

18-27

THINK Silver is solid at 15.0° C. To melt the sample, we must first raise its temperature to the melting point, and then supply heat of fusion.

EXPRESS The melting point of silver is 1235 K, so the temperature of the silver must first be raised from 15.0° C (= 288 K) to 1235 K. This requires heat

$$Q_1 = cm(T_f - T_i) = (236\,\text{J/kg}\cdot\text{K})(0.130\,\text{kg})(1235°\text{C} - 288°\text{C}) = 2.91\times10^4\ \text{J}.$$

Now the silver at its melting point must be melted. If L_F is the heat of fusion for silver this requires

$$Q_2 = mL_F = (0.130\,\text{kg})(105\times10^3\ \text{J/kg}) = 1.36\times10^4\ \text{J}.$$

ANALYZE The total heat required is

$$Q = Q_1 + Q_2 = 2.91\times10^4\ \text{J} + 1.36\times10^4\ \text{J} = 4.27\times10^4\ \text{J}.$$

LEARN The heating process is associated with the specific heat of silver, while the melting process involves heat of fusion. Both the specific heat and the heat of fusion are chemical properties of the material itself.

18-41

THINK Our system consists of both water and ice cubes. Initially the ice cubes are at $-15°\text{C}$ (below freezing temperatures), so they must first absorb heat until $0°\text{C}$ is reached. The final equilibrium temperature reached is related to the amount of ice melted.

EXPRESS There are three possibilities:

- None of the ice melts and the water-ice system reaches thermal equilibrium at a temperature that is at or below the melting point of ice.

- The system reaches thermal equilibrium at the melting point of ice, with some of the ice melted.

- All of the ice melts and the system reaches thermal equilibrium at a temperature at or above the melting point of ice.

We work in Celsius temperature, which poses no difficulty for the J/kg·K values of specific heat capacity (see Table 18-3) since a change of Kelvin temperature is numerically equal to the corresponding change on the Celsius scale.

First, suppose that no ice melts. The temperature of the water decreases from $T_{Wi} = 25°C$ to some final temperature T_f and the temperature of the ice increases from $T_{Ii} = -15°C$ to T_f. If m_W is the mass of the water and c_W is its specific heat then the water rejects heat

$$|Q| = c_W m_W (T_{Wi} - T_f).$$

If m_I is the mass of the ice and c_I is its specific heat then the ice absorbs heat

$$Q = c_I m_I (T_f - T_{Ii}).$$

Since no energy is lost to the environment, these two heats (in absolute value) must be the same. Consequently,

$$c_W m_W (T_{Wi} - T_f) = c_I m_I (T_f - T_{Ii}).$$

The solution for the equilibrium temperature is

$$\begin{aligned} T_f &= \frac{c_W m_W T_{Wi} + c_I m_I T_{Ii}}{c_W m_W + c_I m_I} \\ &= \frac{(4190\,\text{J/kg}\cdot\text{K})(0.200\,\text{kg})(25°\text{C}) + (2220\,\text{J/kg}\cdot\text{K})(0.100\,\text{kg})(-15°\text{C})}{(4190\,\text{J/kg}\cdot\text{K})(0.200\,\text{kg}) + (2220\,\text{J/kg}\cdot\text{K})(0.100\,\text{kg})} \\ &= 16.6°\text{C}. \end{aligned}$$

This is above the melting point of ice, which invalidates our assumption that no ice has melted. That is, the calculation just completed does not take into account the melting of the ice and is in error. Consequently, we start with a new assumption: that the water and ice reach thermal equilibrium at $T_f = 0°C$, with mass m ($< m_I$) of the ice melted. The magnitude of the heat rejected by the water is

$$|Q| = c_W m_W T_{Wi},$$

and the heat absorbed by the ice is

$$Q = c_I m_I (0 - T_{Ii}) + m L_F,$$

where L_F is the heat of fusion for water. The first term is the energy required to warm all the ice from its initial temperature to 0°C and the second term is the energy required to melt mass m of the ice. The two heats are equal, so

$$c_W m_W T_{Wi} = -c_I m_I T_{Ii} + m L_F.$$

This equation can be solved for the mass m of ice melted.

ANALYZE (a) Solving for m and substituting the values given, we find the amount of ice melted to be

$$m = \frac{c_W m_W T_{Wi} + c_I m_I T_{Ii}}{L_F}$$

$$= \frac{(4190\,\text{J/kg}\cdot\text{K})(0.200\,\text{kg})(25°\text{C}) + (2220\,\text{J/kg}\cdot\text{K})(0.100\,\text{kg})(-15°\text{C})}{333\times 10^3\,\text{J/kg}}$$

$$= 5.3\times 10^{-2}\,\text{kg} = 53\,\text{g}.$$

Since the total mass of ice present initially was 100 g, there *is* enough ice to bring the water temperature down to 0°C. This is then the solution: the ice and water reach thermal equilibrium at a temperature of 0°C with 53 g of ice melted.

(b) Now there is less than 53 g of ice present initially. All the ice melts and the final temperature is above the melting point of ice. The heat rejected by the water is

$$|Q| = c_W m_W (T_{Wi} - T_f)$$

and the heat absorbed by the ice and the water it becomes when it melts is

$$Q = c_I m_I (0 - T_{Ii}) + c_W m_I (T_f - 0) + m_I L_F.$$

The first term is the energy required to raise the temperature of the ice to 0°C, the second term is the energy required to raise the temperature of the melted ice from 0°C to T_f, and the third term is the energy required to melt all the ice. Since the two heats are equal,

$$c_W m_W (T_{Wi} - T_f) = c_I m_I (-T_{Ii}) + c_W m_I T_f + m_I L_F.$$

The solution for T_f is

$$T_f = \frac{c_W m_W T_{Wi} + c_I m_I T_{Ii} - m_I L_F}{c_W (m_W + m_I)}.$$

Inserting the given values, we obtain $T_f = 2.5$°C.

LEARN In order to melt some ice, the energy released by the water must be sufficient to first raise the temperature of the ice to the melting point ($-c_I m_I T_{Ii}$ required, $T_{Ii} < 0$), with the remaining energy contributing to the heat of fusion. If the remaining energy is greater than $m_I L_F$, then all ice will be melted and the final temperature will be above 0°C.

18-45

THINK Over a complete cycle, the internal energy is the same at the beginning and end, so the heat Q absorbed equals the work done: $Q = W$.

EXPRESS Over the portion of the cycle from A to B the pressure p is a linear function of the volume V and we may write $p = a + bV$. The work done over this portion of the cycle is

$$W_{AB} = \int_{V_A}^{V_B} p dV = \int_{V_A}^{V_B} (a + bV) dV = a(V_B - V_A) + \frac{1}{2} b\left(V_B^2 - V_A^2\right).$$

The BC portion of the cycle is at constant pressure and the work done by the gas is

$$W_{BC} = p_B \Delta V_{BC} = p_B (V_C - V_B).$$

The CA portion of the cycle is at constant volume, so no work is done. The total work done by the gas is

$$W = W_{AB} + W_{BC} + W_{CA}.$$

ANALYZE The pressure function can be written as

$$p = \frac{10}{3} \text{ Pa} + \left(\frac{20}{3} \text{ Pa/m}^3\right) V,$$

where the coefficients a and b were chosen so that $p = 10$ Pa when $V = 1.0 \text{ m}^3$ and $p = 30$ Pa when $V = 4.0 \text{ m}^3$. Therefore, the work done going from A to B is

$$\begin{aligned} W_{AB} &= a(V_B - V_A) + \frac{1}{2} b\left(V_B^2 - V_A^2\right) \\ &= \left(\frac{10}{3} \text{ Pa}\right)\left(4.0 \text{ m}^3 - 1.0 \text{ m}^3\right) + \frac{1}{2}\left(\frac{20}{3} \text{ Pa/m}^3\right)\left[(4.0 \text{ m}^3)^2 - (1.0 \text{ m}^3)^2\right] \\ &= 10 \text{ J} + 50 \text{ J} = 60 \text{ J} \end{aligned}$$

Similarly, with $p_B = p_C = 30 \text{ Pa}$, $V_C = 1.0 \text{ m}^3$ and $V_B = 4.0 \text{ m}^3$, we have

$$W_{BC} = p_B (V_C - V_B) = (30 \text{ Pa})(1.0 \text{ m}^3 - 4.0 \text{ m}^3) = -90 \text{ J}.$$

Adding up all contributions, we find the total work done by the gas to be

$$W = W_{AB} + W_{BC} + W_{CA} = 60 \text{ J} - 90 \text{ J} + 0 = -30 \text{ J}.$$

Thus, the total heat absorbed is $Q = W = -30$ J. This means the gas loses 30 J of energy in the form of heat.

LEARN Notice that in calculating the work done by the gas, we always start with Eq. 18-25: $W = \int p dV$. For isobaric process where p = constant, $W = p\Delta V$, and for isochoric process where V = constant, $W = 0$.

18-47

THINK Since the change in internal energy ΔE_{int} only depends on the initial and final states, it is the same for path *iaf* and path *ibf*.

EXPRESS According to the first law of thermodynamics, $\Delta E_{int} = Q - W$, where Q is the heat absorbed and W is the work done by the system. Along *iaf*, we have

$$\Delta E_{int} = Q - W = 50 \text{ cal} - 20 \text{ cal} = 30 \text{ cal}.$$

ANALYZE (a) The work done along path *ibf* is given by

$$W = Q - \Delta E_{int} = 36 \text{ cal} - 30 \text{ cal} = 6.0 \text{ cal}.$$

(b) Since the curved path is traversed from f to i the change in internal energy is $\Delta E_{int} = -30 \text{ cal}$, and

$$Q = \Delta E_{int} + W = -30 \text{ cal} - 13 \text{ cal} = -43 \text{ cal}.$$

(c) Let $\Delta E_{int} = E_{int,\,f} - E_{int,\,i}$. We then have

$$E_{int,\,f} = \Delta E_{int} + E_{int,\,i} = 30 \text{ cal} + 10 \text{ cal} = 40 \text{ cal}.$$

(d) The work W_{bf} for the path *bf* is zero, so

$$Q_{bf} = E_{int,\,f} - E_{int,\,b} = 40 \text{ cal} - 22 \text{ cal} = 18 \text{ cal}.$$

(e) For the path *ibf*, $Q = 36$ cal so $Q_{ib} = Q - Q_{bf} = 36 \text{ cal} - 18 \text{ cal} = 18 \text{ cal}$.

LEARN Work W and heat Q in general are path-dependent quantities, i.e., they depend on how the finial state is reached. However, the combination $\Delta E_{int} = Q - W$ is path independent; it is a *state function.*

18-53

THINK Energy is transferred as heat from the hot reservoir at temperature T_H to the cold reservoir at temperature T_C. The conduction rate is the amount of energy transferred per unit time.

EXPRESS The rate of heat flow is given by

$$P_{cond} = kA\frac{T_H - T_C}{L},$$

where k is the thermal conductivity of copper (401 W/m·K), A is the cross-sectional area (in a plane perpendicular to the flow), L is the distance along the direction of flow between the points where the temperature is T_H and T_C. The thermal conductivity is found in Table 18-6 of the text. Recall that a change in Kelvin temperature is numerically equivalent to a change on the Celsius scale.

ANALYZE Substituting the values given, we find the rate to be

$$P_{\text{cond}} = \frac{(401\,\text{W/m}\cdot\text{K})(90.0\times10^{-4}\,\text{m}^2)(125°\text{C}-10.0°\text{C})}{0.250\,\text{m}} = 1.66\times10^3\,\text{J/s}.$$

LEARN The thermal resistance (R-value) of the copper slab is

$$R = \frac{L}{k} = \frac{0.250\,\text{m}}{401\,\text{W/m}\cdot\text{K}} = 6.23\times10^{-4}\,\text{m}^2\cdot\text{K/W}.$$

The low value of R is an indication that the copper slab is a good conductor.

18-61

THINK As heat continues to leave the water via conduction, more ice is formed and the ice slab gets thicker.

EXPRESS Let h be the thickness of the ice slab and A be its area. Then, the rate of heat flow through the slab is

$$P_{\text{cond}} = \frac{kA(T_H - T_C)}{h},$$

where k is the thermal conductivity of ice, T_H is the temperature of the water (0°C), and T_C is the temperature of the air above the ice (–10°C). The heat leaving the water freezes it, the heat required to freeze mass m of water being $Q = L_F m$, where L_F is the heat of fusion for water. Differentiate with respect to time and recognize that $dQ/dt = P_{\text{cond}}$ to obtain

$$P_{\text{cond}} = L_F\frac{dm}{dt}.$$

Now, the mass of the ice is given by $m = \rho A h$, where ρ is the density of ice and h is the thickness of the ice slab, so $dm/dt = \rho A(dh/dt)$ and

$$P_{\text{cond}} = L_F \rho A\frac{dh}{dt}.$$

We equate the two expressions for P_{cond} and solve for dh/dt:

$$\frac{dh}{dt} = \frac{k(T_H - T_C)}{L_F \rho h}.$$

ANALYZE Since 1 cal = 4.186 J and 1 cm = 1 × 10^{-2} m, the thermal conductivity of ice has the SI value

$$k = (0.0040 \text{ cal/s·cm·K}) (4.186 \text{ J/cal})/(1 \times 10^{-2} \text{ m/cm}) = 1.674 \text{ W/m·K}.$$

The density of ice is $\rho = 0.92 \text{ g/cm}^3 = 0.92 \times 10^3 \text{ kg/m}^3$. Thus, we obtain

$$\frac{dh}{dt} = \frac{(1.674 \text{ W/m}\cdot\text{K})(0°\text{C} + 10°\text{C})}{(333\times10^3 \text{ J/kg})(0.92\times10^3 \text{ kg/m}^3)(0.050 \text{ m})} = 1.1\times10^{-6} \text{ m/s} = 0.40 \text{ cm/h}.$$

LEARN The rate of ice formation is proportional to the conduction rate – the faster the energy leaves the water, the faster the water freezes.

18-77

THINK The heat absorbed by the ice not only raises its temperature but could also change its phase – to water.

EXPRESS Let m_I be the mass of the ice cube and c_I be its specific heat. The energy required to bring the ice cube to the melting temperature (0 C°) is

$$Q_1 = c_I m_I (0 \text{ C}° - T_{Ii}) = (2220 \text{ J/kg}\cdot\text{K})(0.700 \text{ kg})(150 \text{ K}) = 2.331\times10^5 \text{ J}.$$

Since the total amount of energy transferred to the ice is $Q = 6.993\times10^5$ J, and $Q_1 < Q$, some or all the ice will melt. The energy required to melt all the ice is

$$Q_2 = m_I L_F = (0.700 \text{ kg})(3.33\times10^5 \text{ J/kg}) = 2.331\times10^5 \text{ J}.$$

However, since

$$Q_1 + Q_2 = 4.662\times10^5 \text{ J} < Q = 6.993\times10^5 \text{ J},$$

this means that all the ice will melt and the extra energy

$$\Delta Q = Q - (Q_1 + Q_2) = 6.993\times10^5 \text{ J} - 4.662\times10^5 \text{ J} = 2.331\times10^5 \text{ J}$$

would be used to raise the temperature of the water.

ANALYZE The final temperature of the water is given by $\Delta Q = m_I c_{\text{water}} T_f$. Substituting the values given, we have

$$T_f = \frac{\Delta Q}{m_I c_{\text{water}}} = \frac{2.331\times10^5 \text{ J}}{(0.700 \text{ kg})(4186.8 \text{ J/kg}\cdot\text{K})} = 79.5°\text{C}$$

LEARN The key concepts in this problem are outlined in the Sample Problem – "Heat to change temperature and state." An important difference with part (b) of the sample problem is that, in our case, the final state of the H_2O is *all liquid* at $T_f > 0$. As discussed in part (a) of that sample problem, there are three steps to the total process.

18-79

THINK The work done by the expanding gas is given by Eq. 18-24: $W = \int p\, dV$.

EXPRESS Let V_i and V_f be the initial and final volumes, respectively. With $p = aV^2$, the work done by the gas is

$$W = \int_{V_i}^{V_f} p dV = \int_{V_i}^{V_f} aV^2 dV = \frac{1}{3}a\left(V_f^3 - V_i^3\right).$$

ANALYZE With $a = 10\ \text{N/m}^8$, $V_i = 1.0\ \text{m}^3$ and $V_f = 2.0\ \text{m}^3$, we obtain

$$W = \frac{1}{3}a\left(V_f^3 - V_i^3\right) = \frac{1}{3}\left(10\ \text{N/m}^8\right)\left[(2.0\ \text{m}^3)^3 - (1.0\ \text{m}^3)^3\right] = 23\ \text{J}.$$

LEARN In this problem, the initial and final pressures are

$$p_i = aV_i^2 = (10\ \text{N/m}^8)(1.0\ \text{m}^3)^2 = 10\ \text{N/m}^2 = 10\ \text{Pa}$$
$$p_f = aV_f^2 = (10\ \text{N/m}^8)(2.0\ \text{m}^3)^2 = 40\ \text{N/m}^2 = 40\ \text{Pa}$$

In this case, since $p \sim V^2$, the work done would be proportional to V^3 after volume integration.

18-81

THINK The work done is the "area under the curve:" $W = \int p\, dV$.

EXPRESS According to the first law of thermodynamics, $\Delta E_{\text{int}} = Q - W$, where Q is the heat absorbed and W is the work done by the system. For process 1,

$$W_1 = p_i(V_b - V_i) = p_i(5.0V_i - V_i) = 4.0p_iV_i$$

so that

$$\Delta E_{\text{int}} = Q - W_1 = 10p_iV_i - 4.0p_iV_i = 6.0p_iV_i\ .$$

Path 2 involves more work than path 1 (note the triangle in the figure of area $\frac{1}{2}(4V_i)(p_i/2) = p_iV_i$). Thus,

$$W_2 = W_1 + p_iV_i = 5.0p_iV_i\,.$$

Note that $\Delta E_{\text{int}} = 6.0p_iV_i$ is the same for all three paths.

ANALYZE (a) The energy transferred to the gas as heat in process 2 is

$$Q_2 = \Delta E_{\text{int}} + W_2 = 6.0 p_i V_i + 5.0 p_i V_i = 11 p_i V_i.$$

(b) Path 3 starts at *a* and ends at *b* (same as paths 1 and 2), so $\Delta E_{\text{int}} = 6.0 p_i V_i$.

LEARN Work W and heat Q in general are path-dependent quantities, i.e., they depend on how the finial state is reached. However, the combination $\Delta E_{\text{int}} = Q - W$ is path independent; it is a *state function.*

18-83

THINK The Pyrex disk expands as a result of heating, so we expect $\Delta V > 0$.

EXPRESS The initial volume of the disk (thought of as a short cylinder) is $V_0 = \pi r^2 L$ where $L = 0.50$ cm is its thickness and $r = 8.0$ cm is its radius. After heating, the volume becomes

$$V = \pi(r + \Delta r)^2 (L + \Delta L) = \pi r^2 L + \pi r^2 \Delta L + 2\pi r L \Delta r + \ldots$$

where we ignore higher-order terms. Thus, the change in volume of the disk is

$$\Delta V = V - V_0 \approx \pi r^2 \Delta L + 2\pi r L \Delta r$$

ANALYZE With $\Delta L = L\alpha\Delta T$ and $\Delta r = r\alpha\Delta T$, the above expression becomes

$$\Delta V = \pi r^2 L\alpha\Delta T + 2\pi r^2 L\alpha\Delta T = 3\pi r^2 L\alpha\Delta T.$$

Substituting the values given ($\alpha = 3.2 \times 10^{-6}$/C° from Table 18-2), we obtain

$$\begin{aligned}\Delta V &= 3\pi r^2 L\alpha\Delta T = 3\pi(0.080 \text{ m})^2 (0.0050 \text{ m})(3.2\times10^{-6} / °\text{C})(60°\text{C} - 10°\text{C}) \\ &= 4.83\times10^{-8} \text{ m}^3\end{aligned}$$

LEARN All dimensions of the disk expand when heated. So we must take into consideration the change in radius as well as the thickness.

18-85

THINK Since the system remains thermally insulated, the total energy remains unchanged. The energy released by the aluminum lump raises the water temperature.

EXPRESS Let T_f be the final temperature of the aluminum lump-water system. The energy transferred from the aluminum is

$$Q_{Al} = m_{Al} c_{Al} (T_{i,Al} - T_f).$$

Similarly, the energy transferred as heat into water is

$$Q_{\text{water}} = m_{\text{water}} c_{\text{water}} (T_f - T_{i,\text{ water}}).$$

Equating Q_{Al} with Q_{water} allows us to solve for T_f.

ANALYZE With

$$m_{Al} c_{Al} (T_{i,Al} - T_f) = m_{\text{water}} c_{\text{water}} (T_f - T_{i,\text{ water}}),$$

we find the final equilibrium temperature to be

$$\begin{aligned} T_f &= \frac{m_{Al} c_{Al} T_{i,Al} + m_{\text{water}} c_{\text{water}} T_{i,\text{water}}}{m_{Al} c_{Al} + m_{\text{water}} c_{\text{water}}} \\ &= \frac{(2.50\text{ kg})(900\text{ J/kg}\cdot\text{K})(92°\text{C}) + (8.00\text{ kg})(4186.8\text{ J/kg}\cdot\text{K})(5.0°\text{C})}{(2.50\text{ kg})(900\text{ J/kg}\cdot\text{K}) + (8.00\text{ kg})(4186.8\text{ J/kg}\cdot\text{K})} \\ &= 10.5°\text{C}. \end{aligned}$$

LEARN No phase change is involved in this problem, so the thermal energy transferred from the aluminum can only change the water temperature.

Chapter 19

19-3

THINK We treat the oxygen gas in this problem as ideal and apply the ideal-gas law.

EXPRESS In solving the ideal-gas law equation $pV = nRT$ for n, we first convert the temperature to the Kelvin scale: $T_i = (40.0 + 273.15)\,\text{K} = 313.15\,\text{K}$, and the volume to SI units: $V_i = 1000\,\text{cm}^3 = 10^{-3}\,\text{m}^3$.

ANALYZE (a) The number of moles of oxygen present is

$$n = \frac{pV_i}{RT_i} = \frac{(1.01\times10^5\,\text{Pa})(1.000\times10^{-3}\,\text{m}^3)}{(8.31\,\text{J/mol}\cdot\text{K})(313.15\,\text{K})} = 3.88\times10^{-2}\,\text{mol}.$$

(b) Similarly, the ideal gas law $pV = nRT$ leads to

$$T_f = \frac{pV_f}{nR} = \frac{(1.06\times10^5\,\text{Pa})(1.500\times10^{-3}\,\text{m}^3)}{(3.88\times10^{-2}\,\text{mol})(8.31\,\text{J/mol}\cdot\text{K})} = 493\,\text{K}.$$

We note that the final temperature may be expressed in degrees Celsius as 220°C.

LEARN The final temperature can also be calculated by noting that $\frac{p_iV_i}{T_i} = \frac{p_fV_f}{T_f}$, or

$$T_f = \left(\frac{p_f}{p_i}\right)\left(\frac{V_f}{V_i}\right)T_i = \left(\frac{1.06\times10^5\ \text{Pa}}{1.01\times10^5\ \text{Pa}}\right)\left(\frac{1500\ \text{cm}^3}{1000\ \text{cm}^3}\right)(313.15\,\text{K}) = 493\,\text{K}.$$

19-11

THINK The process consists of two steps: isothermal expansion, followed by isobaric (constant-pressure) compression. The total work done by the air is the sum of the works done for the two steps.

EXPRESS Suppose the gas expands from volume V_i to volume V_f during the isothermal portion of the process. The work it does is

$$W_1 = \int_{V_i}^{V_f} p\,dV = nRT \int_{V_i}^{V_f} \frac{dV}{V} = nRT \ln \frac{V_f}{V_i},$$

where the ideal gas law $pV = nRT$ was used to replace p with nRT/V. Now $V_i = nRT/p_i$ and $V_f = nRT/p_f$, so $V_f / V_i = p_i / p_f$. Also replace nRT with p_iV_i to obtain

$$W_1 = p_iV_i \ln \frac{p_i}{p_f}.$$

During the constant-pressure portion of the process the work done by the gas is $W_2 = p_f(V_i - V_f)$. The gas starts in a state with pressure p_f, so this is the pressure throughout this portion of the process. We also note that the volume decreases from V_f to V_i. Now $V_f = p_iV_i/p_f$, so

$$W_2 = p_f\left(V_i - \frac{p_iV_i}{p_f}\right) = (p_f - p_i)V_i.$$

ANALYZE For the first portion, since the initial gauge pressure is 1.03×10^5 Pa,

$$p_i = 1.03 \times 10^5 \text{ Pa} + 1.013 \times 10^5 \text{ Pa} = 2.04 \times 10^5 \text{ Pa}.$$

The final pressure is atmospheric pressure: $p_f = 1.013 \times 10^5$ Pa. Thus,

$$W_1 = \left(2.04 \times 10^5 \text{ Pa}\right)\left(0.14 \text{m}^3\right) \ln\left(\frac{2.04 \times 10^5 \text{ Pa}}{1.013 \times 10^5 \text{ Pa}}\right) = 2.00 \times 10^4 \text{ J}.$$

Similarly, for the second portion, we have

$$W_2 = (p_f - p_i)V_i = (1.013\times10^5 \text{ Pa} - 2.04\times10^5 \text{ Pa})(0.14 \text{ m}^3) = -1.44\times10^4 \text{ J}.$$

The total work done by the gas over the entire process is

$$W = W_1 + W_2 = 2.00 \times 10^4 \text{ J} + (-1.44 \times 10^4 \text{ J}) = 5.60 \times 10^3 \text{ J}.$$

LEARN The work done by the gas is positive when it expands, and negative when it contracts.

19-21

THINK According to kinetic theory, the rms speed is (see Eq. 19-34) $v_{\text{rms}} = \sqrt{3RT/M}$, where T is the temperature and M is the molar mass.

EXPRESS The rms speed is defined as $v_{\text{rms}} = \sqrt{(v^2)_{\text{avg}}}$, where $(v^2)_{\text{avg}} = \int_0^\infty v^2 P(v) dv$, with the Maxwell's speed distribution function given by

$$P(v) = 4\pi \left(\frac{M}{2\pi RT} \right)^{3/2} v^2 e^{-Mv^2/2RT}.$$

According to Table 19-1, the molar mass of molecular hydrogen is 2.02 g/mol = 2.02 × 10^{-3} kg/mol.

ANALYZE At $T = 2.7$ K, we find the rms speed to be

$$v_{\text{rms}} = \sqrt{\frac{3\,(8.31\,\text{J/mol}\cdot\text{K})(2.7\,\text{K})}{2.02 \times 10^{-3}\ \text{kg/mol}}} = 1.8 \times 10^2\ \text{m/s}.$$

LEARN The corresponding average speed and most probable speed are

$$v_{\text{avg}} = \sqrt{\frac{8RT}{\pi M}} = \sqrt{\frac{8(8.31\,\text{J/mol}\cdot\text{K})(2.7\,\text{K})}{\pi(2.02 \times 10^{-3}\ \text{kg/mol})}} = 1.7 \times 10^2\ \text{m/s}$$

and

$$v_p = \sqrt{\frac{2RT}{M}} = \sqrt{\frac{2(8.31\,\text{J/mol}\cdot\text{K})(2.7\,\text{K})}{2.02 \times 10^{-3}\ \text{kg/mol}}} = 1.5 \times 10^2\ \text{m/s},$$

respectively.

19-29

THINK Mean free path is the average distance traveled by a molecule between successive collisions.

EXPRESS According to Eq. 19-25, the mean free path for molecules in a gas is given by

$$\lambda = \frac{1}{\sqrt{2}\pi d^2 N/V},$$

where d is the diameter of a molecule and N is the number of molecules in volume V.

ANALYZE (a) Substituting $d = 2.0 \times 10^{-10}$ m and $N/V = 1 \times 10^6$ molecules/m^3, we obtain

$$\lambda = \frac{1}{\sqrt{2}\pi(2.0 \times 10^{-10}\ \text{m})^2 (1 \times 10^6\ \text{m}^{-3})} = 6 \times 10^{12}\ \text{m}.$$

(b) At this altitude most of the gas particles are in orbit around Earth and do not suffer randomizing collisions. The mean free path has little physical significance.

LEARN Mean free path is inversely proportional to the number density, N/V. The typical value of N/V at room temperature and atmospheric pressure for ideal gas is

$$\frac{N}{V}=\frac{p}{kT}=\frac{1.01\times10^{5}\ \text{Pa}}{(1.38\times10^{-23}\ \text{J/K})(298\ \text{K})}=2.46\times10^{25}\ \text{molecules/m}^3=2.46\times10^{19}\ \text{molecules/cm}^3.$$

This is much higher than that in the outer space.

19-33

THINK We're given the speeds of 10 molecules. The speed distribution is discrete.

EXPRESS The average speed is $\bar{v}=\dfrac{\Sigma v}{N}$, where the sum is over the speeds of the particles and N is the number of particles. Similarly, the rms speed is given by $v_{\text{rms}}=\sqrt{\dfrac{\sum v^2}{N}}$.

ANALYZE (a) From the equation above, we find the average speed to be

$$\bar{v}=\frac{(2.0+3.0+4.0+5.0+6.0+7.0+8.0+9.0+10.0+11.0)\,\text{km/s}}{10}=6.5\,\text{km/s}.$$

(b) With

$$\begin{aligned}\sum v^2 &=[(2.0)^2+(3.0)^2+(4.0)^2+(5.0)^2+(6.0)^2\\ &\quad+(7.0)^2+(8.0)^2+(9.0)^2+(10.0)^2+(11.0)^2]\ \text{km}^2/\text{s}^2=505\ \text{km}^2/\text{s}^2\end{aligned}$$

the rms speed is

$$v_{\text{rms}}=\sqrt{\frac{505\ \text{km}^2/\text{s}^2}{10}}=7.1\ \text{km/s}.$$

LEARN Each speed is weighted equally in calculating the average and the rms values.

19-37

THINK From the distribution function $P(v)$, we can calculate the average and rms speeds.

EXPRESS The distribution function gives the fraction of particles with speeds between v and $v + dv$, so its integral over all speeds is unity: $\int P(v)\,dv = 1$. The average speed is defined as $v_{\text{avg}} = \int_0^\infty vP(v)dv$. Similarly, the rms speed is given by $v_{\text{rms}} = \sqrt{(v^2)_{\text{avg}}}$, where $(v^2)_{\text{avg}} = \int_0^\infty v^2 P(v)dv$.

ANALYZE (a) Evaluate the integral by calculating the area under the curve in Fig. 19-23. The area of the triangular portion is half the product of the base and altitude, or $\frac{1}{2}av_0$. The area of the rectangular portion is the product of the sides, or av_0. Thus,

$$\int P(v)dv = \frac{1}{2}av_0 + av_0 = \frac{3}{2}av_0\,,$$

so $\frac{3}{2}av_0 = 1$ and $av_0 = 2/3 = 0.67$.

(b) For the triangular portion of the distribution $P(v) = av/v_0$, and the contribution of this portion is

$$\frac{a}{v_0}\int_0^{v_0} v^2 dv = \frac{a}{3v_0}v_0^3 = \frac{av_0^2}{3} = \frac{2}{9}v_0,$$

where $2/3v_0$ was substituted for a. $P(v) = a$ in the rectangular portion, and the contribution of this portion is

$$a\int_{v_0}^{2v_0} v\,dv = \frac{a}{2}\left(4v_0^2 - v_0^2\right) = \frac{3a}{2}v_0^2 = v_0.$$

Therefore, we have

$$v_{\text{avg}} = \frac{2}{9}v_0 + v_0 = 1.22v_0 \;\Rightarrow\; \frac{v_{\text{avg}}}{v_0} = 1.22\,.$$

(c) In calculating $v_{\text{avg}}^2 = \int v^2 P(v)\,dv$, we note that the contribution of the triangular section is

$$\frac{a}{v_0}\int_0^{v_0} v^3 dv = \frac{a}{4v_0}v_0^4 = \frac{1}{6}v_0^2.$$

The contribution of the rectangular portion is

$$a\int_{v_0}^{2v_0} v^2 dv = \frac{a}{3}\left(8v_0^3 - v_0^3\right) = \frac{7a}{3}v_0^3 = \frac{14}{9}v_0^2.$$

Thus,

$$v_{\text{rms}} = \sqrt{\frac{1}{6}v_0^2 + \frac{14}{9}v_0^2} = 1.31v_0 \;\Rightarrow\; \frac{v_{\text{rms}}}{v_0} = 1.31\,.$$

(d) The number of particles with speeds between $1.5v_0$ and $2v_0$ is given by $N\int_{1.5v_0}^{2v_0} P(v)dv$. The integral is easy to evaluate since $P(v) = a$ throughout the range of integration. Thus the number of particles with speeds in the given range is

$$Na(2.0v_0 - 1.5v_0) = 0.5N\,av_0 = N/3,$$

where $2/3v_0$ was substituted for a. In other words, the fraction of particles in this range is 1/3 or 0.33.

LEARN From the distribution function shown in Fig. 19-23, it is clear that there are more particles with a speed in the range $v_0 < v < 2v_0$ than $0 < v < v_0$. In fact, straightforward calculation shows that the fraction of particles with speeds between $1.0v_0$ and $2v_0$ is

$$\int_{1.0v_0}^{2v_0} P(v)dv = a(2v_0 - 1.0v_0) = av_0 = \frac{2}{3}.$$

19-49

THINK The molar specific heat at constant volume for a gas is given by Eq. 19-41: $C_V = \Delta E_{\text{int}} / n\Delta T$. Our system consists of three non-interacting gases.

EXPRESS When the temperature changes by ΔT the internal energy of the first gas changes by $n_1 C_{V1} \Delta T$, the internal energy of the second gas changes by $n_2 C_{V2} \Delta T$, and the internal energy of the third gas changes by $n_3 C_{V3} \Delta T$. The change in the internal energy of the composite gas is

$$\Delta E_{\text{int}} = (n_1 C_{V1} + n_2 C_{V2} + n_3 C_{V3})\,\Delta T.$$

This must be $(n_1 + n_2 + n_3)\, C_V\, \Delta T$, where C_V is the molar specific heat of the mixture. Thus,

$$C_V = \frac{n_1 C_{V1} + n_2 C_{V2} + n_3 C_{V3}}{n_1 + n_2 + n_3}.$$

ANALYZE With n_1=2.40 mol, C_{V1}=12.0 J/mol·K for gas 1, n_2=1.50 mol, C_{V2}=12.8 J/mol·K for gas 2, and n_3=3.20 mol, C_{V3}=20.0 J/mol·K for gas 3, we obtain

$$C_V = \frac{(2.40\text{ mol})(12.0\text{ J/mol}\cdot\text{K}) + (1.50\text{ mol})(12.8\text{ J/mol}\cdot\text{K}) + (3.20\text{ mol})(20.0\text{ J/mol}\cdot\text{K})}{2.40\text{ mol} + 1.50\text{ mol} + 3.20\text{ mol}}$$

$$= 15.8\text{ J/mol}\cdot\text{K}$$

for the mixture.

LEARN The molar specific heat of the mixture C_V is the sum of each individual C_{Vi} weighted by the molar fraction.

19-53

THINK The molecules are diatomic, with translational and rotational degrees of freedom. The temperature change is under constant pressure.

EXPRESS Since the process is at constant pressure, energy transferred as heat to the gas is given by $Q = nC_p\,\Delta T$, where n is the number of moles in the gas, C_p is the molar specific heat at constant pressure, and ΔT is the increase in temperature. Similarly, the change in the internal energy is given by $\Delta E_{\text{int}} = nC_V\,\Delta T$, where C_V is the specific heat at constant volume. For a diatomic ideal gas, $C_p = \frac{7}{2}R$ and $C_V = \frac{5}{2}R$ (see Table 19-3).

ANALYZE (a) The heat transferred is

$$Q = nC_p\Delta T = n\left(\frac{7R}{2}\right)\Delta T = \frac{7}{2}(4.00\,\text{mol})(8.31\,\text{J/mol}\cdot\text{K})(60.0\,\text{K}) = 6.98\times10^3\ \text{J}.$$

(b) From the above, we find the change in the internal energy to be

$$\Delta E_{\text{int}} = nC_V\Delta T = n\left(\frac{5R}{2}\right)\Delta T = \frac{5}{2}(4.00\,\text{mol})(8.31\,\text{J/mol.K})(60.0\,\text{K}) = 4.99\times10^3\ \text{J}.$$

(c) According to the first law of thermodynamics, $\Delta E_{\text{int}} = Q - W$, so the work done by the gas is

$$W = Q - \Delta E_{\text{int}} = 6.98\times10^3\ \text{J} - 4.99\times10^3\ \text{J} = 1.99\times10^3\ \text{J}.$$

(d) The change in the total translational kinetic energy is

$$\Delta K = \frac{3}{2}nR\Delta T = \frac{3}{2}(4.00\,\text{mol})(8.31\,\text{J/mol}\cdot\text{K})(60.0\,\text{K}) = 2.99\times10^3\ \text{J}.$$

LEARN The diatomic gas has three translational and two rotational degrees of freedom (making $f = 3+2 = 5$). By equipartition theorem, each degree of freedom accounts for an energy of $RT/2$ per mole. Thus, $C_V = (f/2)R = 5R/2$ and $C_p = C_V + R = 7R/2$.

19-69

THINK The net upward force is the difference between the buoyant force and the weight of the balloon with air inside.

EXPRESS Let ρ_c be the density of the cool air surrounding the balloon and ρ_h be the density of the hot air inside the balloon. The magnitude of the buoyant force on the balloon is $F_b = \rho_c gV$, where V is the volume of the envelope. The force of gravity is $Fg = W + \rho_h gV$, where W is the combined weight of the basket and the envelope. Thus, the net upward force is

$$F_{\text{net}} = F_b - F_g = \rho_c gV - W - \rho_h gV.$$

ANALYZE With F_{net} = 2.67 × 10^3 N, W = 2.45 × 10^3 N, V = 2.18 × 10^3 m^3, $\rho_c g = 11.9 \text{ N/m}^3$, we obtain

$$\rho_h g = \frac{\rho_c gV - W - F_{\text{net}}}{V} = \frac{(11.9 \text{ N/m}^3)(2.18\times10^3 \text{ m}^3) - 2.45\times10^3 \text{ N} - 2.67\times10^3 \text{ N}}{2.18\times10^3 \text{ m}^3} = 9.55 \text{ N/m}^3$$

The ideal gas law gives $p/RT = n/V$. Multiplying both sides by the "molar weight" Mg then leads to

$$\frac{pMg}{RT} = \frac{nMg}{V} = \rho_h g.$$

With $p = 1.01\times10^5$ Pa and M= 0.028 kg/m^3, we find the temperature to be

$$T = \frac{pMg}{R\rho_h g} = \frac{(1.01\times10^5 \text{ Pa})(0.028 \text{ kg/mol})(9.8 \text{ m/s}^2)}{(8.31 \text{ J/mol}\cdot\text{K})(9.55 \text{ N/m}^3)} = 349 \text{ K}.$$

LEARN As can be seen from the results above, increasing the temperature of the gas inside the balloon increases the value of F_{net}, i.e., the lifting capacity.

19-71

THINK An adiabatic process is a process in which the energy transferred as heat is zero.

EXPRESS The change in the internal energy is given by $\Delta E_{\text{int}} = nC_V\,\Delta T$, where C_V is the specific heat at constant volume, n is the number of moles in the gas, and ΔT is the change in temperature. According to the first law of thermodynamics, the work done by the gas is $W = Q - \Delta E_{\text{int}}$. For an adiabatic process, $Q = 0$, and $W = -\Delta E_{\text{int}}$.

ANALYZE (a) The work done by the gas is

$$W = -\Delta E_{\text{int}} = -nC_V\Delta T = -\frac{3}{2}nR\Delta T = -\frac{3}{2}(2.0 \text{ mol})(8.31 \text{ J/mol}\cdot\text{K})(15.0 \text{ K}) = -374 \text{ J}.$$

(b) $Q = 0$ since the process is adiabatic.

(c) The change in internal energy is $\Delta E_{\text{int}} = \frac{3}{2}nR\Delta T = 374 \text{ J}$.

(d) The number of atoms in the gas is $N = nN_A$, where N_A is the Avogadro's number. Thus, the change in the average kinetic energy per atom is

$$\Delta K_1 = \frac{\Delta E_{\text{int}}}{N} = \frac{\Delta E_{\text{int}}}{nN_A} = \frac{374\ \text{J}}{(2.00)(6.02\times10^{23}/\text{mol})} = 3.11\times10^{-22}\ \text{J}.$$

LEARN The work done *on* the system is the negative of the work done *by* the system: $W_{on} = -W = \Delta E_{\text{int}} = +374\ \text{J}$. By work-kinetic energy theorem: $\Delta K = \Delta W_{\text{on}} = \Delta E_{\text{int}}$.

19-73

THINK The collision frequency is related to the mean free path and average speed of the molecules.

EXPRESS According to Eq. 19-25, the mean free path for molecules in a gas is given by

$$\lambda = \frac{1}{\sqrt{2}\pi d^2 N/V},$$

where d is the diameter of a molecule and N is the number of molecules in volume V. Using ideal gas law, the number density can be written as $N/V = p/kT$, where p is the pressure, T is the temperature on the Kelvin scale and k is the Boltzmann constant. The average time between collisions is $\tau = \lambda / v_{\text{avg}}$, where $v_{\text{avg}} = \sqrt{8RT/\pi M}$, where R is the universal gas constant and M is the molar mass. The collision frequency is simply given by $f = 1/\tau$.

ANALYZE With $p = 2.02 \times 10^3$ Pa and $d = 290 \times 10^{-12}$ m, we find the mean free path to be

$$\lambda = \frac{1}{\sqrt{2}\pi d^2 (p/kT)} = \frac{kT}{\sqrt{2}\pi d^2 p} = \frac{(1.38\times10^{-23}\ \text{J/K})(400\ \text{K})}{\sqrt{2}\pi(290\times10^{-12}\ \text{m})^2(1.01\times10^5\ \text{Pa})} = 7.31\times10^{-8}\ \text{m}.$$

Similarly, with $M = 0.032$ kg/mol, we find the average speed to be

$$v_{\text{avg}} = \sqrt{\frac{8RT}{\pi M}} = \sqrt{\frac{8(8.31\ \text{J/mol}\cdot\text{K})(400\ \text{K})}{\pi(32\times10^{-3}\ \text{kg/mol})}} = 514\ \text{m/s}.$$

Thus, the collision frequency is

$$f = \frac{v_{\text{avg}}}{\lambda} = \frac{514\ \text{m/s}}{7.31\times10^{-8}\ \text{m}} = 7.04\times10^9\ \text{collisions/s}$$

LEARN This is very similar to the Sample Problem – "Mean free path, average speed and collision frequency." A general expression for f is

$$f = \frac{\text{speed}}{\text{distance}} = \frac{v_{\text{avg}}}{\lambda} = \frac{pd^2}{k}\sqrt{\frac{16\pi R}{MT}}.$$

19-77

THINK From the distribution function $P(v)$, we can calculate the average and rms speeds of the gas.

EXPRESS The distribution function gives the fraction of particles with speeds between v and $v + dv$, so its integral over all speeds is unity: $\int P(v)\, dv = 1$. The average speed is defined as $v_{\text{avg}} = \int_0^\infty vP(v)dv$. Similarly, the rms speed is given by $v_{\text{rms}} = \sqrt{(v^2)_{\text{avg}}}$, where $(v^2)_{\text{avg}} = \int_0^\infty v^2 P(v)dv$.

ANALYZE (a) By normalizing the distribution function:

$$1 = \int_0^{v_0} P(v)\,dv = \int_0^{v_0} Cv^2\,dv = \frac{C}{3}v_0^3$$

we find the constant C to be $C = 3/v_0^3$.

(b) The average speed is

$$v_{\text{avg}} = \int_0^{v_0} vP(v)\,dv = \int_0^{v_0} v\left(\frac{3v^2}{v_0^3}\right)dv = \frac{3}{v_0^3}\int_0^{v_0} v^3 dv = \frac{3}{4}v_0 .$$

(c) Similarly, the rms speed is the square root of

$$\int_0^{v_0} v^2 P(v)\,dv = \int_0^{v_0} v^2\left(\frac{3v^2}{v_0^3}\right)dv = \frac{3}{v_0^3}\int_0^{v_0} v^4 dv = \frac{3}{5}v_0^2 .$$

Therefore, $v_{\text{rms}} = \sqrt{3/5}v_0 \approx 0.775v_0$.

LEARN The maximum speed of the gas is $v_{\text{max}} = v_0$, as indicated by the distribution function. Using Eq. 19-29, we find the fraction of molecules with speed between v_1 and v_2 to be

$$\text{frac} = \int_{v_1}^{v_2} P(v)\,dv = \int_{v_1}^{v_2}\left(\frac{3v^2}{v_0^3}\right)dv = \frac{3}{v_0^3}\int_{v_1}^{v_2} v^2 dv = \frac{v_2^3 - v_1^3}{v_0^3}.$$

19-79

THINK The compression is isothermal so $\Delta T = 0$. In addition, since the gas is ideal, we can use the ideal gas law: $pV = nRT$.

EXPRESS The work done by the gas during the isothermal compression process from volume V_i to volume V_f is given by

$$W = \int_{V_i}^{V_f} p\,dV = nRT\int_{V_i}^{V_f} \frac{dV}{V} = nRT \ln\left(\frac{V_f}{V_i}\right),$$

where we use the ideal gas law to replace p with nRT/V.

ANALYZE (a) The temperature is T = 10.0°C = 283 K. Then, with n = 3.50 mol, we obtain

$$W = nRT \ln\left(\frac{V_f}{V_0}\right) = (3.50\text{ mol})(8.31\text{ J/mol}\cdot\text{K})(283\text{ K})\ln\left(\frac{3.00\text{ m}^3}{4.00\text{ m}^3}\right) = -2.37\times10^3\text{ J}.$$

(b) The internal energy change ΔE_{int} vanishes (for an ideal gas) when $\Delta T = 0$ so that the First Law of Thermodynamics leads to $Q = W = -2.37$ kJ.

LEARN The work done by the gas is negative since $V_f < V_i$. Also, the negative value in Q implies that the heat transfer is from the sample to its environment.

19-83

THINK For an isothermal expansion, $\Delta T = 0$. However, if the expansion is adiabatic, then $\Delta Q = 0$.

EXPRESS Using ideal gas law: $pV = nRT$, we have $\dfrac{p_f V_f}{p_i V_i} = \dfrac{T_f}{T_i}$. For isothermal process, $T_f = T_i$, which gives $p_f = \dfrac{p_i V_i}{V_f}$. The work done by the gas is

$$W = \int_{V_i}^{V_f} p\,dV = nRT\int_{V_i}^{V_f} \frac{dV}{V} = nRT \ln\left(\frac{V_f}{V_i}\right).$$

Now, for an adiabatic process, $p_i V_i^\gamma = p_f V_f^\gamma$. The final pressures and temperatures are

$$p_f = p_i\left(\frac{V_i}{V_f}\right)^\gamma, \quad T_f = \frac{p_f V_f T_i}{p_i V_i}$$

The work done is $W = Q - \Delta E_{int} = -\Delta E_{int}$.

ANALYZE (a) For the isothermal process, the final pressure is

$$p_f = \frac{p_i V_i}{V_f} = \frac{(32\text{ atm})(1.0\text{ L})}{4.0\text{ L}} = 8.0\text{ atm}.$$

(b) The final temperature of the gas is the same as the initial temperature: $T_f = T_i = 300$ K.

(c) The work done is

$$W = nRT_i \ln\left(\frac{V_f}{V_i}\right) = p_iV_i \ln\left(\frac{V_f}{V_i}\right) = (32\,\text{atm})(1.01\times10^5\,\text{Pa/atm})(1.0\times10^{-3}\,\text{m}^3)\ln\left(\frac{4.0\,\text{L}}{1.0\,\text{L}}\right)$$
$$= 4.4\times10^3\,\text{J}.$$

(d) For the adiabatic process, the final pressure is ($\gamma = 5/3$ for monatomic gas)

$$p_f = p_i\left(\frac{V_i}{V_f}\right)^{\gamma} = (32\,\text{atm})\left(\frac{1.0\,\text{L}}{4.0\,\text{L}}\right)^{5/3} = 3.2\,\text{atm}.$$

(e) The final temperature is

$$T_f = \frac{p_fV_fT_i}{p_iV_i} = \frac{(3.2\,\text{atm})(4.0\,\text{L})(300\,\text{K})}{(32\,\text{atm})(1.0\,\text{L})} = 120\,\text{K}.$$

(f) The work done is

$$W = -\Delta E_{\text{int}} = -\frac{3}{2}nR\Delta T = -\frac{3}{2}(p_fV_f - p_iV_i)$$
$$= -\frac{3}{2}[(3.2\,\text{atm})(4.0\,\text{L}) - (32\,\text{atm})(1.0\,\text{L})](1.01\times10^5\,\text{Pa/atm})(10^{-3}\,\text{m}^3/\text{L})$$
$$= 2.9\times10^3\,\text{J}.$$

(g) If the gas is diatomic, then $\gamma = 1.4$, and the final pressure is

$$p_f = p_i\left(\frac{V_i}{V_f}\right)^{\gamma} = (32\,\text{atm})\left(\frac{1.0\,\text{L}}{4.0\,\text{L}}\right)^{1.4} = 4.6\,\text{atm}.$$

(h) The final temperature is

$$T_f = \frac{p_fV_fT_i}{p_iV_i} = \frac{(4.6\,\text{atm})(4.0\,\text{L})(300\,\text{K})}{(32\,\text{atm})(1.0\,\text{L})} = 170\,\text{K}.$$

(i) The work done is

$$W = Q - \Delta E_{\text{int}} = -\frac{5}{2}nR\Delta T = -\frac{5}{2}(p_fV_f - p_iV_i)$$
$$= -\frac{5}{2}[(4.6\,\text{atm})(4.0\,\text{L}) - (32\,\text{atm})(1.0\,\text{L})](1.01\times10^5\,\text{Pa/atm})(10^{-3}\,\text{m}^3/\text{L})$$
$$= 3.4\times10^3\,\text{J}.$$

LEARN Comparing (c) with (f), we see that more work is done by the gas if the expansion is isothermal rather than adiabatic.

Chapter 20

20-1

THINK If the expansion of the gas is reversible and isothermal, then there's no change in internal energy. However, if the process is reversible and adiabatic, then there would be no change in entropy.

EXPRESS Since the gas is ideal, its pressure p is given in terms of the number of moles n, the volume V, and the temperature T by $p = nRT/V$. If the expansion is isothermal, the work done by the gas is

$$W = \int_{V_1}^{V_2} p\,dV = nRT\int_{V_1}^{V_2}\frac{dV}{V} = nRT\ln\frac{V_2}{V_1},$$

and the corresponding change in entropy is $\Delta S = \int(1/T)\,dQ = Q/T$, where Q is the heat absorbed (see Eq. 20-2).

ANALYZE (a) With $V_2 = 2.00V_1$ and $T = 400$ K, we obtain

$$W = nRT\ln 2.00 = (4.00\text{ mol})(8.31\text{ J/mol}\cdot\text{K})(400\text{ K})\ln 2.00 = 9.22\times10^3\text{ J}.$$

(b) According to the first law of thermodynamics, $\Delta E_{\text{int}} = Q - W$. Now the internal energy of an ideal gas depends only on the temperature and not on the pressure and volume. Since the expansion is isothermal, $\Delta E_{\text{int}} = 0$ and $Q = W$. Thus,

$$\Delta S = \frac{W}{T} = \frac{9.22\times10^3\text{ J}}{400\text{ K}} = 23.1\text{ J/K}.$$

(c) The change in entropy ΔS is zero for all reversible adiabatic processes.

LEARN The general expression for ΔS for reversible processes is given by Eq. 20-4:

$$\Delta S = S_f - S_i = nR\ln\left(\frac{V_f}{V_i}\right) + nC_V\ln\left(\frac{T_f}{T_i}\right).$$

Note that ΔS does not depend on how the gas changes from its initial state i to the final state f.

20-11

THINK The aluminum sample gives off energy as heat to water. Thermal equilibrium is reached when both the aluminum and the water come to a common final temperature T_f.

EXPRESS The energy that leaves the aluminum as heat has magnitude $Q = m_a c_a(T_{ai} - T_f)$, where m_a is the mass of the aluminum, c_a is the specific heat of aluminum, T_{ai} is the initial temperature of the aluminum, and T_f is the final temperature of the aluminum-water system. The energy that enters the water as heat has magnitude $Q = m_w c_w(T_f - T_{wi})$, where m_w is the mass of the water, c_w is the specific heat of water, and T_{wi} is the initial temperature of the water. The two energies are the same in magnitude since no energy is lost. Thus,

$$m_a c_a\left(T_{ai} - T_f\right) = m_w c_w\left(T_f - T_{wi}\right) \Rightarrow T_f = \frac{m_a c_a T_{ai} + m_w c_w T_{wi}}{m_a c_a + m_w c_w}.$$

The change in entropy is $\Delta S = \int dQ/T$.

ANALYZE (a) The specific heat of aluminum is 900 J/kg·K and the specific heat of water is 4190 J/kg·K. Thus,

$$T_f = \frac{(0.200 \text{ kg})(900 \text{ J/kg}\cdot\text{K})(100°\text{C}) + (0.0500 \text{ kg})(4190 \text{ J/kg}\cdot\text{K})(20°\text{C})}{(0.200 \text{ kg})(900 \text{ J/kg}\cdot\text{K}) + (0.0500 \text{ kg})(4190 \text{ J/kg}\cdot\text{K})}$$
$$= 57.0°\text{C} = 330 \text{ K}.$$

(b) Now temperatures must be given in Kelvins: $T_{ai} = 393$ K, $T_{wi} = 293$ K, and $T_f = 330$ K. For the aluminum, $dQ = m_a c_a dT$ and the change in entropy is

$$\Delta S_a = \int \frac{dQ}{T} = m_a c_a \int_{T_{ai}}^{T_f} \frac{dT}{T} = m_a c_a \ln\left(\frac{T_f}{T_{ai}}\right) = (0.200 \text{ kg})(900 \text{ J/kg}\cdot\text{K})\ln\left(\frac{330 \text{ K}}{373 \text{ K}}\right)$$
$$= -22.1 \text{ J/K}.$$

(c) The entropy change for the water is

$$\Delta S_w = \int \frac{dQ}{T} = m_w c_w \int_{T_{wi}}^{T_f} \frac{dT}{T} = m_w c_w \ln\left(\frac{T_f}{T_{wi}}\right) = (0.0500 \text{ kg})(4190 \text{ J/kg.K})\ln\left(\frac{330\,\text{K}}{293\,\text{K}}\right)$$
$$= +24.9\,\text{J/K}.$$

(d) The change in the total entropy of the aluminum-water system is

$$\Delta S = \Delta S_a + \Delta S_w = -22.1 \text{ J/K} + 24.9 \text{ J/K} = +2.8 \text{ J/K}.$$

LEARN The system is closed and the process is irreversible. For aluminum the entropy change is negative ($\Delta S_a < 0$) since $T_f < T_{ai}$. However, for water, entropy increases because $T_f > T_{wi}$. The overall entropy change for the aluminum-water system is positive, in accordance with the second law of thermodynamics.

20-27

THINK The thermal efficiency of the Carnot engine depends on the temperatures of the reservoirs.

EXPRESS The efficiency of the Carnot engine is given by

$$\varepsilon_C = \frac{T_\text{H} - T_\text{L}}{T_\text{H}},$$

where T_H is the temperature of the higher-temperature reservoir, and T_L the temperature of the lower-temperature reservoir, in kelvin scale. The work done by the engine is $|W| = \varepsilon |Q_\text{H}|$.

ANALYZE (a) The efficiency of the engine is

$$\varepsilon_c = \frac{T_\text{H} - T_\text{L}}{T_\text{H}} = \frac{(235-115)\,\text{K}}{(235+273)\,\text{K}} = 0.236 = 23.6\%\ .$$

We note that a temperature difference has the same value on the Kelvin and Celsius scales. Since the temperatures in the equation must be in Kelvins, the temperature in the denominator is converted to the Kelvin scale.

(b) Since the efficiency is given by $\varepsilon = |W|/|Q_\text{H}|$, the work done is given by

$$|W| = \varepsilon |Q_\text{H}| = 0.236(6.30\times10^4\ \text{J}) = 1.49\times10^4\ \text{J}\ .$$

LEARN Expressing the efficiency as $\varepsilon_c = 1 - T_\text{L}/T_\text{H}$, we see that ε_c approaches unity (100% efficiency) in the limit $T_L/T_H \to 0$. This is an impossible dream. An alternative version of the second law of thermodynamics is: *there are no perfect engines*.

20-33

THINK Our engine cycle consists of three steps: isochoric heating (*a* to *b*), adiabatic expansion (*b* to *c*), and isobaric compression (*c* to *a*).

EXPRESS Energy is added as heat during the portion of the process from *a* to *b*. This portion occurs at constant volume (V_b), so $Q_\text{H} = nC_V\,\Delta T$. The gas is a monatomic ideal gas, so $C_V = 3R/2$ and the ideal gas law gives

$$\Delta T = (1/nR)(p_b V_b - p_a V_a) = (1/nR)(p_b - p_a)V_b.$$

Thus, $Q_\text{H} = \frac{3}{2}(p_b - p_a)V_b$. On the other hand, energy leaves the gas as heat during the portion of the process from *c* to *a*. This is a constant pressure process, so

$$Q_L = nC_p\Delta T = nC_p(T_a - T_c) = nC_p\left(\frac{p_aV_a}{nR} - \frac{p_cV_c}{nR}\right) = \frac{C_p}{R}p_a(V_a - V_c)\,.$$

where C_p is the molar specific heat for constant-pressure process.

ANALYZE (a) V_b and p_b are given. We need to find p_a. Now p_a is the same as p_c and points c and b are connected by an adiabatic process. With $p_cV_c^\gamma = p_bV_b^\gamma$ for the adiabat, we have ($\gamma = 5/3$ for monatomic gas)

$$p_a = p_c = \left(\frac{V_b}{V_c}\right)^\gamma p_b = \left(\frac{1}{8.00}\right)^{5/3}(1.013\times10^6\text{ Pa}) = 3.167\times10^4\text{ Pa}.$$

Thus, the energy added as heat is

$$Q_H = \frac{3}{2}(p_b - p_a)V_b = \frac{3}{2}(1.013\times10^6\text{ Pa} - 3.167\times10^4\text{ Pa})(1.00\times10^{-3}\text{ m}^3) = 1.47\times10^3\text{ J}.$$

(b) The energy leaving the gas as heat going from c to a is

$$Q_L = \frac{5}{2}p_a(V_a - V_c) = \frac{5}{2}(3.167\times10^4\text{ Pa})(-7.00)(1.00\times10^{-3}\text{ m}^3) = -5.54\times10^2\text{ J}\,,$$

or $|Q_L| = 5.54\times10^2$ J. The substitutions $V_a - V_c = V_a - 8.00\,V_a = -7.00\,V_a$ and $C_p = \frac{5}{2}R$ were made.

(c) For a complete cycle, the change in the internal energy is zero and

$$W = Q = Q_H - Q_L = 1.47\times10^3\text{ J} - 5.54\times10^2\text{ J} = 9.18\times10^2\text{ J}.$$

(d) The efficiency is

$$\varepsilon = W/Q_H = (9.18\times10^2\text{ J})/(1.47\times10^3\text{ J}) = 0.624 = 62.4\%.$$

LEARN To summarize, the heat engine in this problem intakes energy as heat (from, say, consuming fuel) equal to $|Q_H| = 1.47$ kJ and exhausts energy as heat equal to $|Q_L| =$ 554 J; its efficiency and net work are $\varepsilon = 1 - |Q_L|/|Q_H|$ and $W = |Q_H| - |Q_L|$. The less the exhaust heat $|Q_L|$, the more efficient is the engine.

20-37

THINK The performance of the refrigerator is related to its rate of doing work.

EXPRESS The coefficient of performance for a refrigerator is given by

$$K = \frac{\text{what we want}}{\text{what we pay for}} = \frac{|Q_L|}{|W|},$$

where Q_L is the energy absorbed from the cold reservoir as heat and W is the work done during the refrigeration cycle, a negative value. The first law of thermodynamics yields $Q_H + Q_L - W = 0$ for an integer number of cycles. Here Q_H is the energy ejected to the hot reservoir as heat. Thus, $Q_L = W - Q_H$. Q_H is negative and greater in magnitude than W, so $|Q_L| = |Q_H| - |W|$. Thus,

$$K = \frac{|Q_H| - |W|}{|W|}.$$

The solution for $|W|$ is $|W| = |Q_H|/(K + 1)$.

ANALYZE In one hour, $|Q_H| = 7.54\text{ MJ}$. With $K = 3.8$, the work done is

$$|W| = \frac{7.54\text{ MJ}}{3.8 + 1} = 1.57\text{ MJ}.$$

The rate at which work is done is $P = |W|/\Delta t = (1.57 \times 10^6\text{ J})/(3600\text{ s}) = 440\text{ W}$.

LEARN The greater the value of K, the less the amount of work $|W|$ required to transfer the heat.

20-39

THINK A large (small) value of coefficient of performance K means that less (more) work would be required to transfer the heat

EXPRESS A Carnot refrigerator working between a hot reservoir at temperature T_H and a cold reservoir at temperature T_L has a coefficient of performance K that is given by

$$K = \frac{T_L}{T_H - T_L},$$

where T_H is the temperature of the higher-temperature reservoir, and T_L the temperature of the lower-temperature reservoir, in kelvin scale. Equivalently, the coefficient of performance is the energy Q_L drawn from the cold reservoir as heat divided by the work done: $K = |Q_L|/|W|$.

ANALYZE For the refrigerator of this problem, $T_H = 96°\text{ F} = 309\text{ K}$ and $T_L = 70°\text{ F} = 294\text{ K}$, so

$$K = (294\text{ K})/(309\text{ K} - 294\text{ K}) = 19.6.$$

Thus, with $|W| = 1.0\text{ J}$, the amount of heat removed from the room is

$$|Q_L| = K|W| = (19.6)(1.0\text{ J}) = 20\text{ J}.$$

LEARN The Carnot air conditioner in this problem (with $K = 19.6$) are much more efficient than that of the typical room air conditioners ($K \approx 2.5$).

20-47

THINK The gas molecules inside a box can be distributed in many different ways. The number of microstates associated with each distinct configuration is called the multiplicity.

EXPRESS Given N molecules, if the box is divided into m equal parts, with n_1 molecules in the first, n_2 in the second,…, such that $n_1 + n_2 + \ldots n_m = N$. There are $N!$ arrangements of the N molecules, but $n_1!$ are simply rearrangements of the n_1 molecules in the first part, $n_2!$ are rearrangements of the n_2 molecules in the second,… These rearrangements do not produce a new configuration. Therefore, the multiplicity factor associated with this is

$$W = \frac{N!}{n_1!n_2!n_3!\ldots n_m!}.$$

ANALYZE (a) Suppose there are n_L molecules in the left third of the box, n_C molecules in the center third, and n_R molecules in the right third. Using the argument above, we find the multiplicity to be

$$W = \frac{N!}{n_L!n_C!n_R!}.$$

Note that $n_L + n_C + n_R = N$.

(b) If half the molecules are in the right half of the box and the other half are in the left half of the box, then the multiplicity is

$$W_B = \frac{N!}{(N/2)!(N/2)!}.$$

If one-third of the molecules are in each third of the box, then the multiplicity is

$$W_A = \frac{N!}{(N/3)!(N/3)!(N/3)!}.$$

The ratio is

$$\frac{W_A}{W_B} = \frac{(N/2)!(N/2)!}{(N/3)!(N/3)!(N/3)!}.$$

(c) For $N = 100$,

$$\frac{W_A}{W_B} = \frac{50!50!}{33!33!34!} = 4.2 \times 10^{16}.$$

LEARN The more parts the box is divided into, the greater the number of configurations. This exercise illustrates the statistical view of entropy, which is related to W as $S = k \ln W$.

20-51

THINK Increasing temperature causes a shift of the probability distribution function $P(v)$ toward higher speed.

EXPRESS According to kinetic theory, the rms speed and the most probable speed are (see Eqs. 19-34 and 19035) $v_{\text{rms}} = \sqrt{3RT/M}$, $v_P = \sqrt{2RT/M}$ and where T is the temperature and M is the molar mass. The rms speed is defined as $v_{\text{rms}} = \sqrt{(v^2)_{\text{avg}}}$, where $(v^2)_{\text{avg}} = \int_0^\infty v^2 P(v)dv$, with the Maxwell's speed distribution function given by

$$P(v) = 4\pi\left(\frac{M}{2\pi RT}\right)^{3/2} v^2 e^{-Mv^2/2RT}.$$

Thus, the difference between the two speeds is

$$\Delta v = v_{\text{rms}} - v_P = \sqrt{\frac{3RT}{M}} - \sqrt{\frac{2RT}{M}} = \left(\sqrt{3}-\sqrt{2}\right)\sqrt{\frac{RT}{M}}.$$

ANALYZE (a) With $M = 28$ g/mol $= 0.028$ kg/mol (see Table 19-1), and $T_i = 250$ K, we have

$$\Delta v_i = \left(\sqrt{3}-\sqrt{2}\right)\sqrt{\frac{RT_i}{M}} = \left(\sqrt{3}-\sqrt{2}\right)\sqrt{\frac{(8.31\ \text{J/mol}\cdot\text{K})(250\ \text{K})}{0.028\ \text{kg/mol}}} = 87\ \text{m/s}.$$

(b) Similarly, at $T_f = 500$ K,

$$\Delta v_f = \left(\sqrt{3}-\sqrt{2}\right)\sqrt{\frac{RT_f}{M}} = \left(\sqrt{3}-\sqrt{2}\right)\sqrt{\frac{(8.31\ \text{J/mol}\cdot\text{K})(500\ \text{K})}{0.028\ \text{kg/mol}}} = 122\ \text{m/s} \approx 1.2\times10^2\ \text{m/s}.$$

(c) From Table 19-3 we have $C_V = 5R/2$ (see also Table 19-2). For $n = 1.5$ mol, using Eq. 20-4, we find the change in entropy to be

$$\Delta S = nR\ln\left(\frac{V_f}{V_i}\right) + nC_V\ln\left(\frac{T_f}{T_i}\right) = 0 + (1.5\ \text{mol})(5/2)(8.31\ \text{J/mol}\cdot\text{K})\ln\left(\frac{500\ \text{K}}{250\ \text{K}}\right) = 22\ \text{J/K}.$$

LEARN Notice that the expression for Δv implies $T = \dfrac{M}{R(\sqrt{3}-\sqrt{2})^2}(\Delta v)^2$. Thus, one may also express ΔS as

$$\Delta S = nC_V\ln\left(\frac{T_f}{T_i}\right) = nC_V\ln\left(\frac{(\Delta v_f)^2}{(\Delta v_i)^2}\right) = 2nC_V\ln\left(\frac{\Delta v_f}{\Delta v_i}\right).$$

The entropy of the gas increases as the result of temperature increase.

20-59

THINK The temperature of the ice is first raised to $0°C$, then the ice melts and the temperature of the resulting water is raised to $40°C$. We want to calculate the entropy change in this process.

EXPRESS As the ice warms, the energy it receives as heat when the temperature changes by dT is $dQ = mc_I dT$, where m is the mass of the ice and c_I is the specific heat of ice. If T_i (=−20°C = 253 K) is the initial temperature and T_f (= 273 K) is the final temperature, then the change in its entropy is

$$\Delta S_1 = \int \frac{dQ}{T} = mc_I \int_{T_i}^{T_f} \frac{dT}{T} = mc_I \ln\left(\frac{T_f}{T_i}\right) = (0.60 \text{ kg})(2220 \text{ J/kg}\cdot\text{K})\ln\left(\frac{273 \text{ K}}{253 \text{ K}}\right) = 101 \text{ J/K}.$$

Melting is an isothermal process. The energy leaving the ice as heat is mL_F, where L_F is the heat of fusion for ice. Thus,

$$\Delta S_2 = \frac{Q}{T} = \frac{mL_F}{T} = \frac{(0.60 \text{ kg})(333\times10^3 \text{ J/kg})}{273 \text{ K}} = 732 \text{ J/K}.$$

For the warming of the water from the melted ice, the change in entropy is

$$\Delta S_3 = mc_w \ln\left(\frac{T_f}{T_i}\right) = (0.600 \text{ kg})(4190 \text{ J/kg}\cdot\text{K})\ln\left(\frac{313 \text{ K}}{273 \text{ K}}\right) = 344 \text{ J/K},$$

where c_w = 4190 J/kg · K is the specific heat of water.

ANALYZE The total change in entropy for the ice and the water it becomes is

$$\Delta S = \Delta S_1 + \Delta S_2 + \Delta S_3 = 101 \text{ J/K} + 732 \text{ J/K} + 344 \text{ J/K} = 1.18\times10^3 \text{ J/K}.$$

LEARN From the above, we readily see that the biggest increase in entropy comes from ΔS_2, which accounts for the melting process.

20-73

THINK The performance of the Carnot refrigerator is related to its rate of doing work.

EXPRESS The coefficient of performance for a refrigerator is defined as

$$K = \frac{\text{what we want}}{\text{what we pay for}} = \frac{|Q_L|}{|W|},$$

where Q_L is the energy absorbed from the cold reservoir (interior of refrigerator) as heat and W is the work done during the refrigeration cycle, a negative value. The first law of thermodynamics yields $Q_H + Q_L - W = 0$ for an integer number of cycles. Here Q_H is the energy ejected as heat to the hot reservoir (the room). Thus, $Q_L = W - Q_H$. Q_H is negative and greater in magnitude than W, so $|Q_L| = |Q_H| - |W|$. Thus,

$$K = \frac{|Q_H| - |W|}{|W|}.$$

The solution for $|Q_H| = |W|(1+K) = |Q_L|(1+K)/K$.

ANALYZE (a) From the expression above, the energy per cycle transferred as heat to the room is

$$|Q_H| = |Q_L|\frac{1+K}{K} = (35.0\text{ kJ})\frac{1+4.60}{4.60} = 42.6\text{ kJ}.$$

(b) Similarly, the work done per cycle is

$$|W| = \frac{|Q_L|}{K} = \frac{35.0\text{ kJ}}{4.60} = 7.61\text{ kJ}.$$

LEARN A Carnot refrigerator is a Carnot engine operating in reverse. Its coefficient of performance can also be written as

$$K = \frac{T_L}{T_H - T_L}$$

The value of K is higher when the temperatures of the two reservoirs are close to each other.

20-75

THINK The gas molecules inside a box can be distributed in many different ways. The number of microstates associated with each distinct configuration is called the multiplicity.

EXPRESS In general, if there are N molecules and if the box is divided into two halves, with n_L molecules in the left half and n_R in the right half, such that $n_L + n_R = N$. There are $N!$ arrangements of the N molecules, but $n_L!$ are simply rearrangements of the n_L molecules in the left half, and $n_R!$ are rearrangements of the n_R molecules in the right half. These rearrangements do not produce a new configuration. Therefore, the multiplicity factor associated with this is

$$W = \frac{N!}{n_L!\,n_R!}.$$

The entropy is given by $S = k \ln W$.

ANALYZE (a) The least multiplicity configuration is when all the particles are in the same half of the box. In this case, for system A with with $N = 3$, we have

$$W = \frac{3!}{3!0!} = 1.$$

(b) Similarly for box *B,* with $N = 5$, $W = 5!/(5!0!) = 1$ in the "least" case.

(c) The most likely configuration in the 3 particle case is to have 2 on one side and 1 on the other. Thus,

$$W = \frac{3!}{2!1!} = 3.$$

(d) The most likely configuration in the 5 particle case is to have 3 on one side and 2 on the other. Thus,

$$W = \frac{5!}{3!2!} = 10.$$

(e) We use Eq. 20-21 with our result in part (c) to obtain

$$S = k\ln W = \left(1.38\times10^{-23}\right)\ln 3 = 1.5\times10^{-23} \text{ J/K}.$$

(f) Similarly for the 5 particle case (using the result from part (d)), we find

$$S = k \ln 10 = 3.2 \times 10^{-23} \text{ J/K}.$$

LEARN The least multiplicity is $W = 1$; this happens when $n_L = N$ or $n_L = 0$. On the other hand, the greatest multiplicity occurs when $n_L = (N-1)/2$ or $n_L = (N+1)/2$.

Chapter 21

21-1

THINK After the transfer, the charges on the two spheres are $Q-q$ and q.

EXPRESS The magnitude of the electrostatic force between two charges q_1 and q_2 separated by a distance r is given by the Coulomb's law (see Eq. 21-1):

$$F = k\frac{q_1 q_2}{r^2},$$

where $k = 1/4\pi\varepsilon_0 = 8.99\times10^9\,\text{N}\cdot\text{m}^2/\text{C}^2$. In our case, $q_1 = Q-q$ and $q_2 = q$, so the magnitude of the force of either of the charges on the other is

$$F = \frac{1}{4\pi\varepsilon_0}\frac{q(Q-q)}{r^2}.$$

We want the value of q that maximizes the function $f(q) = q(Q-q)$.

ANALYZE Setting the derivative df/dq equal to zero leads to $Q - 2q = 0$, or $q = Q/2$. Thus, $q/Q = 0.500$.

LEARN The force between the two spheres is a maximum when charges are distributed evenly between them.

21-3

THINK The magnitude of the electrostatic force between two charges q_1 and q_2 separated by a distance r is given by Coulomb's law.

EXPRESS Equation 21-1 gives Coulomb's law, $F = k\frac{|q_1||q_2|}{r^2}$, which can be used to solve for the distance:

$$r = \sqrt{\frac{k|q_1||q_2|}{F}}.$$

ANALYZE With $F = 5.70\text{ N}$, $q_1 = 2.60\times10^{-6}$ C and $q_2 = -47.0\times10^{-6}$ C, the distance between the two charges is

$$r=\sqrt{\frac{k|q_1||q_2|}{F}}=\sqrt{\frac{(8.99\times10^9\,\text{N}\cdot\text{m}^2/\text{C}^2)(26.0\times10^{-6}\,\text{C})(47.0\times10^{-6}\,\text{C})}{5.70\text{N}}}=1.39\text{ m}.$$

LEARN The electrostatic force between two charges falls as $1/r^2$. The same inverse-square nature is also seen in the gravitational force between two masses.

21-9

THINK Since opposite charges attract, the initial charge configurations must be of opposite signs. Similarly, since like charges repel, the final charge configurations must carry the same sign.

EXPRESS We assume that the spheres are far apart. Then the charge distribution on each of them is spherically symmetric and Coulomb's law can be used. Let q_1 and q_2 be the original charges. We choose the coordinate system so the force on q_2 is positive if it is repelled by q_1. Then the force on q_2 is

$$F_a=-\frac{1}{4\pi\varepsilon_0}\frac{q_1q_2}{r^2}=-k\frac{q_1q_2}{r^2}$$

where $k=1/4\pi\varepsilon_0=8.99\times10^9\,\text{N}\cdot\text{m}^2/\text{C}^2$ and $r = 0.500$ m. The negative sign indicates that the spheres attract each other. After the wire is connected, the spheres, being identical, acquire the same charge. Since charge is conserved, the total charge is the same as it was originally. This means the charge on each sphere is $(q_1 + q_2)/2$. The force is now repulsive and is given by

$$F_b=\frac{1}{4\pi\varepsilon_0}\frac{\left(\frac{q_1+q_2}{2}\right)\left(\frac{q_1+q_2}{2}\right)}{r^2}=k\frac{(q_1+q_2)^2}{4r^2}.$$

We solve the two force equations simultaneously for q_1 and q_2.

ANALYZE The first equation gives the product

$$q_1q_2=-\frac{r^2F_a}{k}=-\frac{(0.500\,\text{m})^2(0.108\,\text{N})}{8.99\times10^9\,\text{N}\cdot\text{m}^2/\text{C}^2}=-3.00\times10^{-12}\,\text{C}^2,$$

and the second gives the sum

$$q_1+q_2=2r\sqrt{\frac{F_b}{k}}=2(0.500\,\text{m})\sqrt{\frac{0.0360\,\text{N}}{8.99\times10^9\,\text{N}\cdot\text{m}^2/\text{C}^2}}=2.00\times10^{-6}\,\text{C}$$

where we have taken the positive root (which amounts to assuming $q_1 + q_2 \geq 0$). Thus, the product result provides the relation

$$q_2 = \frac{-\left(3.00\times10^{-12}\,\text{C}^2\right)}{q_1}$$

which we substitute into the sum result, producing

$$q_1 - \frac{3.00\times10^{-12}\,\text{C}^2}{q_1} = 2.00\times10^{-6}\,\text{C}.$$

Multiplying by q_1 and rearranging, we obtain a quadratic equation

$$q_1^2 - \left(2.00\times10^{-6}\,\text{C}\right)q_1 - 3.00\times10^{-12}\,\text{C}^2 = 0.$$

The solutions are

$$q_1 = \frac{2.00\times10^{-6}\,\text{C} \pm \sqrt{\left(-2.00\times10^{-6}\,\text{C}\right)^2 - 4\left(-3.00\times10^{-12}\,\text{C}^2\right)}}{2}.$$

If the positive sign is used, $q_1 = 3.00 \times 10^{-6}$ C, and if the negative sign is used, $q_1 = -1.00\times10^{-6}\,\text{C}$.

(a) Using $q_2 = (-3.00 \times 10^{-12})/q_1$ with $q_1 = 3.00 \times 10^{-6}$ C, we get $q_2 = -1.00\times10^{-6}\,\text{C}$.

(b) If we instead work with the $q_1 = -1.00 \times 10^{-6}$ C root, then we find $q_2 = 3.00\times10^{-6}\,\text{C}$.

LEARN Note that since the spheres are identical, the solutions are essentially the same: one sphere originally had charge -1.00×10^{-6} C and the other had charge $+3.00 \times 10^{-6}$ C. What happens if we had not made the assumption, above, that $q_1 + q_2 \geq 0$? If the signs of the charges were reversed (so $q_1 + q_2 < 0$), then the forces remain the same, so a charge of $+1.00 \times 10^{-6}$ C on one sphere and a charge of -3.00×10^{-6} C on the other also satisfies the conditions of the problem.

21-19

THINK Our system consists of two charges in a straight line. We'd like to place a third charge so that all three charges are in equilibrium.

EXPRESS If the system of three charges is to be in equilibrium, the force on each charge must be zero. The third charge q_3 must lie between the other two or else the forces acting on it due to the other charges would be in the same direction and q_3 could not be in equilibrium. Suppose q_3 is at a distance x from q, and $L - x$ from $4.00q$. The force acting on it is then given by

$$F_3 = \frac{1}{4\pi\varepsilon_0}\left(\frac{qq_3}{x^2} - \frac{4qq_3}{(L-x)^2}\right)$$

where the positive direction is rightward. We require $F_3 = 0$ and solve for x.

ANALYZE (a) Canceling common factors yields $1/x^2 = 4/(L-x)^2$ and taking the square root yields $1/x = 2/(L-x)$. The solution is $x = L/3$. With $L = 9.00$ cm, we have $x = 3.00$ cm.

(b) Similarly, the y coordinate of q_3 is $y = 0$.

(c) The force on q is

$$F_q = \frac{-1}{4\pi\varepsilon_0}\left(\frac{qq_3}{x^2} + \frac{4.00q^2}{L^2}\right).$$

The signs are chosen so that a negative force value would cause q to move leftward. We require $F_q = 0$ and solve for q_3:

$$q_3 = -\frac{4qx^2}{L^2} = -\frac{4}{9}q \;\Rightarrow\; \frac{q_3}{q} = -\frac{4}{9} = -0.444$$

where $x = L/3$ is used.

LEARN We may also verify that the force on $4.00q$ also vanishes:

$$F_{4q} = \frac{1}{4\pi\varepsilon_0}\left(\frac{4q^2}{L^2} + \frac{4qq_0}{(L-x)^2}\right) = \frac{1}{4\pi\varepsilon_0}\left(\frac{4q^2}{L^2} + \frac{4(-4/9)q^2}{(4/9)L^2}\right) = \frac{1}{4\pi\varepsilon_0}\left(\frac{4q^2}{L^2} - \frac{4q^2}{L^2}\right) = 0.$$

21-27

THINK The magnitude of the electrostatic force between two charges q_1 and q_2 separated by a distance r is given by Coulomb's law.

EXPRESS Let the charge of the ions be q. With $q_1 = q_2 = +q$, the magnitude of the force between the (positive) ions is given by

$$F = \frac{(q)(q)}{4\pi\varepsilon_0 r^2} = k\frac{q^2}{r^2},$$

where $k = 1/4\pi\varepsilon_0 = 8.99\times10^9\,\mathrm{N\cdot m^2/C^2}$.

ANALYZE (a) We solve for the charge:

$$q = r\sqrt{\frac{F}{k}} = (5.0\times10^{-10}\,\mathrm{m})\sqrt{\frac{3.7\times10^{-9}\,\mathrm{N}}{8.99\times10^9\,\mathrm{N\cdot m^2/C^2}}} = 3.2\times10^{-19}\,\mathrm{C}.$$

(b) Let n be the number of electrons missing from each ion. Then, $ne = q$, or

$$n = \frac{q}{e} = \frac{3.2\times10^{-9}\ \text{C}}{1.6\times10^{-19}\ \text{C}} = 2.$$

LEARN Electric charge is quantized. This means that any charge can be written as $q = ne$, where n is an integer (positive or negative), and $e = 1.6\times10^{-19}$ C is the elementary charge.

21-35

THINK Our system consists of 8 Cs^+ ions at the corners of a cube and a Cl^- ion at the cube's center. To calculate the electrostatic force on the Cl^- ion, we apply the superposition principle and make use of the symmetry property of the configuration.

EXPRESS In (a) where all 8 Cs^+ ions are present, every cesium ion at a corner of the cube exerts a force of the same magnitude on the chlorine ion at the cube center. Each force is attractive and is directed toward the cesium ion that exerts it, along the body diagonal of the cube. We can pair every cesium ion with another, diametrically positioned at the opposite corner of the cube.

In (b) where one Cs^+ ion is missing at the corner, rather than remove a cesium ion, we superpose charge $-e$ at the position of one cesium ion. This neutralizes the ion, and as far as the electrical force on the chlorine ion is concerned, it is equivalent to removing the ion. The forces of the eight cesium ions at the cube corners sum to zero, so the only force on the chlorine ion is the force of the added charge.

ANALYZE (a) Since the two Cs^+ ions in such a pair exert forces that have the same magnitude but are oppositely directed, the two forces sum to zero and, since every cesium ion can be paired in this way, the total force on the chlorine ion is zero.

(b) The length of a body diagonal of a cube is $\sqrt{3}a$, where a is the length of a cube edge. Thus, the distance from the center of the cube to a corner is $d = \left(\sqrt{3}/2\right)a$. The force has magnitude

$$F = k\frac{e^2}{d^2} = \frac{ke^2}{(3/4)a^2} = \frac{\left(8.99\times10^9\ \text{N}\cdot\text{m}^2/\text{C}^2\right)\left(1.60\times10^{-19}\text{C}\right)^2}{(3/4)\left(0.40\times10^{-9}\ \text{m}\right)^2} = 1.9\times10^{-9}\text{N}.$$

Since both the added charge and the chlorine ion are negative, the force is one of repulsion. The chlorine ion is pushed away from the site of the missing cesium ion.

LEARN When solving electrostatic problems involving a discrete number of charges, symmetry argument can often be applied to simplify the problem.

21-37

THINK Charges are conserved in nuclear reactions.

EXPRESS We note that none of the reactions given include a beta decay (see Chapter 42), so the number of protons (*Z*), the number of neutrons (*N*), and the number of electrons are each conserved. The mass number (total number of nucleons) is defined as $A = N + Z$. Atomic numbers (number of protons) and molar masses can be found in Appendix F of the text.

ANALYZE (a) ^{1}H has 1 proton, 1 electron, and 0 neutrons and ^{9}Be has 4 protons, 4 electrons, and 9 – 4 = 5 neutrons, so X has 1 + 4 = 5 protons, 1 + 4 = 5 electrons, and 0 + 5 – 1 = 4 neutrons. One of the neutrons is freed in the reaction. X must be boron with a molar mass of 5 + 4 = 9 g/mol: ^{9}B.

(b) ^{12}C has 6 protons, 6 electrons, and 12 – 6 = 6 neutrons and ^{1}H has 1 proton, 1 electron, and 0 neutrons, so X has 6 + 1 = 7 protons, 6 + 1 = 7 electrons, and 6 + 0 = 6 neutrons. It must be nitrogen with a molar mass of 7 + 6 = 13 g/mol: ^{13}N.

(c) ^{15}N has 7 protons, 7 electrons, and 15 – 7 = 8 neutrons; ^{1}H has 1 proton, 1 electron, and 0 neutrons; and ^{4}He has 2 protons, 2 electrons, and 4 – 2 = 2 neutrons; so X has 7 + 1 – 2 = 6 protons, 6 electrons, and 8 + 0 – 2 = 6 neutrons. It must be carbon with a molar mass of 6 + 6 = 12: ^{12}C.

LEARN A general expression for the reaction can be expressed as

$$^{A_1}_{Z_1}X1_{N_1} + {}^{A_2}_{Z_2}X2_{N_2} \rightarrow {}^{A_3}_{Z_3}X3_{N_3} + {}^{A_4}_{Z_4}X4_{N_4}$$

where $A_i = Z_i + N_i$. Since the number of protons (*Z*), the number of neutrons (*N*), and the number of nucleons (*A*) are each conserved, we have $A_1 + A_2 = A_3 + A_4$, $Z_1 + Z_2 = Z_3 + Z_4$ and $N_1 + N_2 = N_3 + N_4$.

21-39

THINK We have two discrete charges in the *xy*-plane. The electrostatic force on particle 2 due to particle 1 has both *x* and *y* components.

EXPRESS Using Coulomb's law, the magnitude of the force of particle 1 on particle 2 is $F_{21} = k\dfrac{q_1 q_2}{r^2}$, where $r = \sqrt{d_1^2 + d_2^2}$ and $k = 1/4\pi\varepsilon_0 = 8.99\times10^9\,\text{N}\cdot\text{m}^2/\text{C}^2$. Since both q_1 and q_2 are positively charged, particle 2 is repelled by particle 1, so the direction of $\vec{F}_{21}$ is away from particle 1 and toward 2. In unit-vector notation, $\vec{F}_{21} = F_{21}\hat{\text{r}}$, where

$$\hat{r} = \frac{\vec{r}}{r} = \frac{d_2\hat{\text{i}} - d_1\hat{\text{j}}}{\sqrt{d_1^2 + d_2^2}} .$$

The x component of $\vec{F}_{21}$ is $F_{21,x} = F_{21}d_2 / \sqrt{d_1^2 + d_2^2}$.

ANALYZE Combining the expressions above, we obtain

$$\begin{aligned} F_{21,x} &= k\frac{q_1q_2d_2}{r^3} = k\frac{q_1q_2d_2}{(d_1^2 + d_2^2)^{3/2}} \\ &= \frac{(8.99\times10^9\,\text{N}\cdot\text{m}^2/\text{C}^2)(4\cdot1.60\times10^{-19}\text{ C})(6\cdot1.60\times10^{-19}\text{ C})(6.00\times10^{-3}\text{ m})}{\left[(2.00\times10^{-3}\text{ m})^2 + (6.00\times10^{-3}\text{ m})^2\right]^{3/2}} \\ &= 1.31\times10^{-22}\text{ N} \end{aligned}$$

LEARN In a similar manner, we find the y component of $\vec{F}_{21}$ to be

$$\begin{aligned} F_{21,y} &= -k\frac{q_1q_2d_1}{r^3} = -k\frac{q_1q_2d_1}{(d_1^2 + d_2^2)^{3/2}} \\ &= -\frac{(8.99\times10^9\,\text{N}\cdot\text{m}^2/\text{C}^2)(4\cdot1.60\times10^{-19}\text{ C})(6\cdot1.60\times10^{-19}\text{ C})(2.00\times10^{-3}\text{ m})}{\left[(2.00\times10^{-3}\text{ m})^2 + (6.00\times10^{-3}\text{ m})^2\right]^{3/2}} \\ &= -0.437\times10^{-22}\text{ N}. \end{aligned}$$

Thus, $\vec{F}_{21} = (1.31\times10^{-22}\text{ N})\hat{\text{i}} - (0.437\times10^{-22}\text{ N})\hat{\text{j}}$.

21-44

THINK The problem compares the electrostatic force between two protons and the gravitational force by Earth on a proton.

EXPRESS The magnitude of the gravitational force on a proton near the surface of the Earth is $F_g = mg$, where $m = 1.67\times10^{-27}$ kg is the mass of the proton. On the other hand, the electrostatic force between two protons separated by a distance r is $F_e = kq^2 / r$. When the two forces are equal, we have $kq^2/r^2 = mg$.

ANALYZE Solving for r, we obtain

$$r = q\sqrt{\frac{k}{mg}} = \left(1.60\times10^{-19}\,\text{C}\right)\sqrt{\frac{8.99\times10^9\text{ N}\cdot\text{m}^2/\text{C}^2}{\left(1.67\times10^{-27}\,\text{kg}\right)\left(9.8\text{ m/s}^2\right)}} = 0.119\text{ m}.$$

LEARN The electrostatic force at this distance is $F_e = F_g = 1.64\times10^{-26}$ N.

21-62

THINK We have four discrete charges in the *xy*-plane. We use superposition principle to calculate the net electrostatic force on particle 4 due to the other three particles.

EXPRESS Using Coulomb's law, the magnitude of the force on particle 4 by particle i is $F_{4i} = k\dfrac{q_4 q_i}{r_{4i}^2}$. For example, the magnitude of $\vec{F}_{41}$ is

$$F_{41} = k\frac{|q_4||q_1|}{r_{41}^2} = \frac{(8.99\times10^9\,\text{N}\cdot\text{m}^2/\text{C}^2)(3.20\times10^{-19}\ \text{C})(3.20\times10^{-19}\ \text{C})}{(0.0300\ \text{m})^2} = 1.02\times10^{-24}\ \text{N}.$$

Since the force is attractive, $\hat{r}_{41} = -\cos\theta_1\hat{\text{i}} - \sin\theta_1\hat{\text{j}} = -\cos35°\hat{\text{i}} - \sin35°\hat{\text{j}} = -0.82\hat{\text{i}} - 0.57\hat{\text{j}}$. In unit-vector notation, we have

$$\vec{F}_{41} = F_{41}\hat{r}_{41} = (1.02\times10^{-24}\ \text{N})(-0.82\hat{\text{i}} - 0.57\hat{\text{j}}) = -(8.36\times10^{-25}\ \text{N})\hat{\text{i}} - (5.85\times10^{-24}\ \text{N})\hat{\text{j}}.$$

Similarly,

$$\vec{F}_{42} = -k\frac{|q_4||q_2|}{r_{42}^2}\hat{\text{j}} = -\frac{(8.99\times10^9\,\text{N}\cdot\text{m}^2/\text{C}^2)(3.20\times10^{-19}\ \text{C})(3.20\times10^{-19}\ \text{C})}{(0.0200\ \text{m})^2}\hat{\text{j}}$$
$$= -(2.30\times10^{-24}\ \text{N})\hat{\text{j}}$$

and

$$\vec{F}_{43} = -k\frac{|q_4||q_3|}{r_{43}^2}\hat{\text{i}} = -\frac{(8.99\times10^9\,\text{N}\cdot\text{m}^2/\text{C}^2)(6.40\times10^{-19}\ \text{C})(3.20\times10^{-19}\ \text{C})}{(0.0200\ \text{m})^2}\hat{\text{i}}$$
$$= -(4.60\times10^{-24}\ \text{N})\hat{\text{j}}.$$

ANALYZE (a) The net force on particle 4 is

$$\vec{F}_{4,\text{net}} = \vec{F}_{41} + \vec{F}_{42} + \vec{F}_{43} = -(5.44\times10^{-24}\ \text{N})\hat{\text{i}} - (2.89\times10^{-24}\ \text{N})\hat{\text{j}}.$$

The magnitude of the force is

$$F_{4,\text{net}} = \sqrt{(-5.44\times10^{-24}\ \text{N})^2 + (-2.89\times10^{-24}\ \text{N})^2} = 6.16\times10^{-24}\ \text{N}.$$

(b) The direction of the net force is at an angle of

$$\varphi = \tan^{-1}\left(\frac{F_{4y,\text{net}}}{F_{4x,\text{net}}}\right) = \tan^{-1}\left(\frac{-2.89\times10^{-24}\ \text{N}}{-5.44\times10^{-24}\ \text{N}}\right) = 208°,$$

measured counterclockwise from the +x axis.

LEARN A nonzero net force indicates that particle 4 will be accelerated in the direction of the force.

21-67

THINK Our system consists of two charges along a straight line. We'd like to place a third charge so that the net force on it due to charges 1 and 2 vanishes.

EXPRESS The net force on particle 3 is the vector sum of the forces due to particles 1 and 2: $\vec{F}_{3,\text{net}} = \vec{F}_{31} + \vec{F}_{32}$. In order that $\vec{F}_{3,\text{net}} = 0$, particle 3 must be on the x axis and be attracted by one and repelled by another. As the result, it cannot be between particles 1 and 2, but instead either to the left of particle 1 or to the right of particle 2. Let q_3 be placed a distance x to the right of $q_1 = -5.00q$. Then its attraction to q_1 particle will be exactly balanced by its repulsion from $q_2 = +2.00q$:

$$F_{3x,\text{net}} = k\left[\frac{q_1q_3}{x^2} + \frac{q_2q_3}{(x-L)^2}\right] = kq_3q\left[\frac{-5}{x^2} + \frac{2}{(x-L)^2}\right] = 0\,.$$

ANALYZE (a) Cross-multiplying and taking the square root, we obtain

$$\frac{x}{x-L} = \sqrt{\frac{5}{2}}$$

which can be rearranged to produce

$$x = \frac{L}{1-\sqrt{2/5}} \approx 2.72\,L\,.$$

(b) The y coordinate of particle 3 is $y = 0$.

LEARN We can use the result obtained above for consistency check. We find the force on particle 3 due to particle 1 to be

$$F_{31} = k\frac{q_1q_3}{x^2} = k\frac{(-5.00q)(q_3)}{(2.72L)^2} = -0.675\frac{kqq_3}{L^2}\,.$$

Similarly, the force on particle 3 due to particle 2 is

$$F_{32} = k\frac{q_2q_3}{x^2} = k\frac{(+2.00q)(q_3)}{(2.72L-L)^2} = +0.675\frac{kqq_3}{L^2}\,.$$

Indeed, the sum of the two forces is zero.

Chapter 22

22-3
THINK Since the nucleus is treated as a sphere with uniform surface charge distribution, the electric field at the surface is exactly the same as it would be if the charge were all at the center.

EXPRESS The nucleus has a radius R = 6.64 fm and a total charge $q = Ze$, where $Z = 94$ for Pu. Thus, the magnitude of the electric field at the nucleus surface is

$$E = \frac{q}{4\pi\varepsilon_0 R^2} = \frac{Ze}{4\pi\varepsilon_0 R^2}.$$

ANALYZE (a) Substituting the values given, we find the field to be

$$E = \frac{Ze}{4\pi\varepsilon_0 R^2} = \frac{(8.99\times10^9\ \text{N}\cdot\text{m}^2/\text{C}^2)(94)(1.60\times10^{-19}\,\text{C})}{(6.64\times10^{-15}\,\text{m})^2} = 3.07\times10^{21}\ \text{N/C}.$$

(b) The field is normal to the surface. In addition, since the charge is positive, it points outward from the surface.

LEARN The direction of electric field lines is radially outward for a positive charge, and radially inward for a negative charge. The field lines of our nucleus are shown on the right.

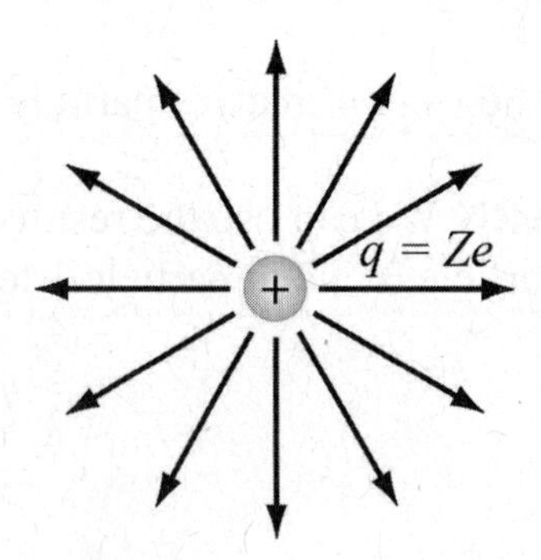

22-5
THINK The magnitude of the electric field produced by a point charge q is given by $E = |q|/4\pi\varepsilon_0 r^2$, where r is the distance from the charge to the point where the field has magnitude E.

EXPRESS From $E = |q|/4\pi\varepsilon_0 r^2$, the magnitude of the charge is $|q| = 4\pi\varepsilon_0 r^2 E$.

ANALYZE With $E = 2.0$ N/C at $r = 50$ cm $= 0.50$ m, we obtain

$$|q| = 4\pi\varepsilon_0 r^2 E = \frac{(0.50\,\text{m})^2(2.0\,\text{N/C})}{8.99\times10^9\,\text{N}\cdot\text{m}^2/\text{C}^2} = 5.6\times10^{-11}\,\text{C}.$$

LEARN To determine the sign of the charge, we would need to know the direction of the field. The field lines extend away from a positive charge and toward a negative charge.

22-7

THINK Our system consists of four point charges that are placed at the corner of a square. The total electric field at a point is the vector sum of the electric fields of individual charges.

EXPRESS Applying the superposition principle, the net electric field at the center of the square is

$$\vec{E} = \sum_{i=1}^{4} \vec{E}_i = \sum_{i=1}^{4} \frac{1}{4\pi\varepsilon_0}\frac{q_i}{r_i^2}\hat{\mathrm{r}}_i .$$

With $q_1 = +10$ nC, $q_2 = -20$ nC, $q_3 = +20$ nC, and $q_4 = -10$ nC, the x component of the electric field at the center of the square is given by, taking the signs of the charges into consideration,

$$\begin{aligned} E_x &= \frac{1}{4\pi\varepsilon_0}\left[\frac{|q_1|}{(a/\sqrt{2})^2} + \frac{|q_2|}{(a/\sqrt{2})^2} - \frac{|q_3|}{(a/\sqrt{2})^2} - \frac{|q_4|}{(a/\sqrt{2})^2}\right]\cos 45° \\ &= \frac{1}{4\pi\varepsilon_0}\frac{1}{a^2/2}\left(|q_1| + |q_2| - |q_3| - |q_4|\right)\frac{1}{\sqrt{2}}. \end{aligned}$$

Similarly, the y component of the electric field is

$$\begin{aligned} E_y &= \frac{1}{4\pi\varepsilon_0}\left[-\frac{|q_1|}{(a/\sqrt{2})^2} + \frac{|q_2|}{(a/\sqrt{2})^2} + \frac{|q_3|}{(a/\sqrt{2})^2} - \frac{|q_4|}{(a/\sqrt{2})^2}\right]\cos 45° \\ &= \frac{1}{4\pi\varepsilon_0}\frac{1}{a^2/2}\left(-|q_1| + |q_2| + |q_3| - |q_4|\right)\frac{1}{\sqrt{2}}. \end{aligned}$$

The magnitude of the net electric field is $E = \sqrt{E_x^2 + E_y^2}$.

ANALYZE Substituting the values given, we obtain

$$E_x = \frac{1}{4\pi\varepsilon_0}\frac{\sqrt{2}}{a^2}\left(|q_1| + |q_2| - |q_3| - |q_4|\right) = \frac{1}{4\pi\varepsilon_0}\frac{\sqrt{2}}{a^2}\left(10\text{ nC} + 20\text{ nC} - 20\text{ nC} - 10\text{ nC}\right) = 0$$

and

$$E_y = \frac{1}{4\pi\varepsilon_0}\frac{\sqrt{2}}{a^2}\left(-|q_1|+|q_2|+|q_3|-|q_4|\right) = \frac{1}{4\pi\varepsilon_0}\frac{\sqrt{2}}{a^2}\left(-10\text{ nC}+20\text{ nC}+20\text{ nC}-10\text{ nC}\right)$$

$$= \frac{\left(8.99\times10^9\text{ N}\cdot\text{m}^2/\text{C}^2\right)(2.0\times10^{-8}\text{ C})\sqrt{2}}{(0.050\text{ m})^2} = 1.02\times10^5\text{ N/C}.$$

Thus, the electric field at the center of the square is $\vec{E} = E_y\hat{\text{j}} = (1.02\times10^5\text{ N/C})\hat{\text{j}}$.

LEARN The net electric field at the center of the square is depicted in the figure below (not to scale). The field, pointing to the $+y$ direction, is the vector sum of the electric fields of individual charges.

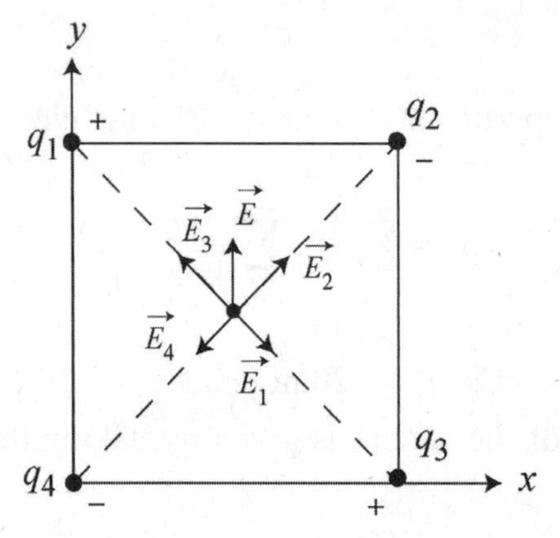

22-11

THINK Our system consists of two point charges of opposite signs fixed to the x axis. Since the net electric field at a point is the vector sum of the electric fields of individual charges, there exists a location where the net field is zero.

EXPRESS At points between the charges, the individual electric fields are in the same direction and do not cancel. Since charge $q_2 = -4.00\,q_1$ located at $x_2 = 70$ cm has a greater magnitude than $q_1 = 2.1\times10^{-8}$ C located at $x_1 = 20$ cm, a point of zero field must be closer to q_1 than to q_2. It must be to the left of q_1.

Let x be the coordinate of P, the point where the field vanishes. Then, the total electric field at P is given by

$$E = \frac{1}{4\pi\varepsilon_0}\left(\frac{|q_2|}{(x-x_2)^2} - \frac{|q_1|}{(x-x_1)^2}\right).$$

ANALYZE If the field is to vanish, then

$$\frac{|q_2|}{(x-x_2)^2} = \frac{|q_1|}{(x-x_1)^2} \Rightarrow \frac{|q_2|}{|q_1|} = \frac{(x-x_2)^2}{(x-x_1)^2}.$$

Taking the square root of both sides, noting that $|q_2|/|q_1| = 4$, we obtain

$$\frac{x-70\text{ cm}}{x-20\text{ cm}} = \pm 2.0\,.$$

Choosing –2.0 for consistency, the value of x is found to be $x = -30$ cm.

LEARN The results are depicted in the figure below. At P, the field $\vec{E}_1$ due to q_1 points to the left, while the field $\vec{E}_2$ due to q_2 points to the right. Since $|\vec{E}_1| = |\vec{E}_2|$, the net field at P is zero.

$\vec{E}_1$ P $\vec{E}_2$ q_1 $q_2 = -4q_1$ x

$x_P = -30$ cm $x_1 = 20$ cm $x_2 = 70$ cm

22-21
THINK The electric quadrupole is composed of two dipoles, each with a dipole moment of magnitude $p = qd$. The dipole moments point in the opposite directions and produce fields in the opposite directions at points on the quadrupole axis.

EXPRESS Consider the point P on the axis, a distance z to the right of the quadrupole center and take a rightward pointing field to be positive. Then the field produced by the right dipole of the pair is given by $qd/2\pi\varepsilon_0(z - d/2)^3$ while the field produced by the left dipole is $-qd/2\pi\varepsilon_0(z + d/2)^3$.

ANALYZE Use the binomial expansions

$$(z - d/2)^{-3} \approx z^{-3} - 3z^{-4}(-d/2)$$

$$(z + d/2)^{-3} \approx z^{-3} - 3z^{-4}(d/2)$$

we obtain

$$E = \frac{qd}{2\pi\varepsilon_0(z-d/2)^3} - \frac{qd}{2\pi\varepsilon_0(z+d/2)^3} \approx \frac{qd}{2\pi\varepsilon_0}\left[\frac{1}{z^3} + \frac{3d}{2z^4} - \frac{1}{z^3} + \frac{3d}{2z^4}\right] = \frac{6qd^2}{4\pi\varepsilon_0 z^4}.$$

Since the quadrupole moment is $Q = 2qd^2$, we have

$$E = \frac{3Q}{4\pi\varepsilon_0 z^4}.$$

LEARN For a quadrupole moment Q, the electric field varies with z as $E \sim Q/z^4$. For a point charge q, the dependence is $E \sim q/z^2$, and for a dipole p, we have $E \sim p/z^3$.

22-31

THINK Our system is a nonconducting rod with uniform charge density. Since the rod is an extended object and not a point charge, the calculation of electric field requires an integration.

EXPRESS The linear charge density λ is the charge per unit length of rod. Since the total charge $-q$ is uniformly distributed on the rod of length L, we have $\lambda = -q/L$. To calculate the electric at the point P shown in Fig. 22-49, we position the x-axis along the rod with the origin at the left end of the rod, as shown in the diagram below.

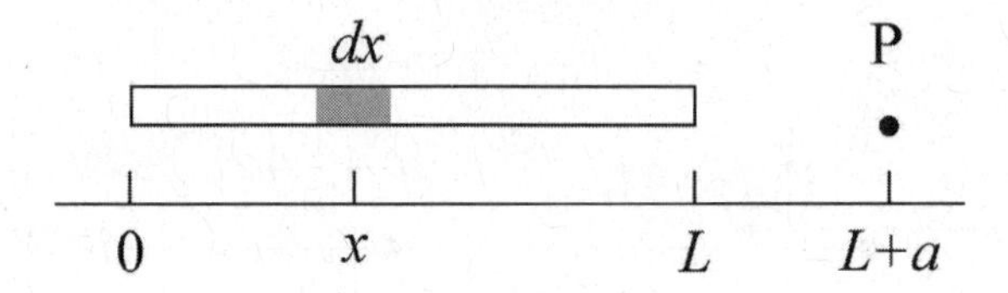

Let dx be an infinitesimal length of rod at x. The charge in this segment is $dq = \lambda\, dx$. The charge dq may be considered to be a point charge. The electric field it produces at point P has only an x component and this component is given by

$$dE_x = \frac{1}{4\pi\varepsilon_0}\frac{\lambda\, dx}{(L+a-x)^2}.$$

The total electric field produced at P by the whole rod is the integral

$$E_x = \frac{\lambda}{4\pi\varepsilon_0}\int_0^L \frac{dx}{(L+a-x)^2} = \frac{\lambda}{4\pi\varepsilon_0}\frac{1}{L+a-x}\Bigg|_0^L = \frac{\lambda}{4\pi\varepsilon_0}\left(\frac{1}{a}-\frac{1}{L+a}\right)$$
$$= \frac{\lambda}{4\pi\varepsilon_0}\frac{L}{a(L+a)} = -\frac{1}{4\pi\varepsilon_0}\frac{q}{a(L+a)},$$

upon substituting $-q = \lambda L$.

ANALYZE (a) With q = 4.23 × 10^{-15} C, L = 0.0815 m, and a = 0.120 m, the linear charge density of the rod is

$$\lambda = \frac{-q}{L} = \frac{-4.23\times10^{-15}\ \text{C}}{0.0815\ \text{m}} = -5.19\times10^{-14}\ \text{C/m}.$$

(b) Similarly, we obtain

$$E_x = -\frac{1}{4\pi\varepsilon_0}\frac{q}{a(L+a)} = -\frac{(8.99\times10^9\ \text{N}\cdot\text{m}^2/\text{C}^2)(4.23\times10^{-15}\ \text{C})}{(0.120\ \text{m})(0.0815\ \text{m}+0.120\ \text{m})} = -1.57\times10^{-3}\ \text{N/C},$$

or $|E_x| = 1.57\times10^{-3}$ N/C.

(c) The negative sign in E_x indicates that the field points in the $-x$ direction, or $-180°$ counterclockwise from the $+x$ axis.

(d) If a is much larger than L, the quantity $L + a$ in the denominator can be approximated by a, and the expression for the electric field becomes

$$E_x = -\frac{q}{4\pi\varepsilon_0 a^2}.$$

Since $a = 50\text{ m} \gg L = 0.0815\text{ m}$, the above approximation applies and we have $E_x = -1.52\times10^{-8}\text{ N/C}$, or $|E_x| = 1.52\times10^{-8}\text{ N/C}$.

(e) For a particle of charge $-q = -4.23\times10^{-15}$ C, the electric field at a distance a = 50 m away has a magnitude $|E_x| = 1.52\times10^{-8}\text{ N/C}$.

LEARN At a distance much greater than the length of the rod ($a \gg L$), the rod can be effectively regarded as a point charge $-q$, and the electric field can be approximated as $E_x \approx \frac{-q}{4\pi\varepsilon_0 a^2}$.

22-35
THINK Our system is a uniformly charged disk of radius R. We compare the field strengths at different points on its axis of symmetry.

EXPRESS At a point on the axis of a uniformly charged disk a distance z above the center of the disk, the magnitude of the electric field is given by Eq. 22-26:

$$E = \frac{\sigma}{2\varepsilon_0}\left[1 - \frac{z}{\sqrt{z^2+R^2}}\right]$$

where R is the radius of the disk and σ is the surface charge density on the disk. The magnitude of the field at the center of the disk (z = 0) is $E_c = \sigma/2\varepsilon_0$. We want to solve for the value of z such that $E/E_c = 1/2$. This means

$$1 - \frac{z}{\sqrt{z^2+R^2}} = \frac{1}{2} \Rightarrow \frac{z}{\sqrt{z^2+R^2}} = \frac{1}{2}.$$

ANALYZE Squaring both sides, then multiplying them by $z^2 + R^2$, we obtain $z^2 = (z^2/4) + (R^2/4)$. Thus, $z^2 = R^2/3$, or $z = R/\sqrt{3}$. With R = 0.600 m, we have z = 0.346 m.

LEARN The ratio of the electric field strengths, $E/E_c = 1-(z/R)/\sqrt{(z/R)^2+1}$, as a function of z/R, is plotted below. From the plot, we readily see that at $z/R = (0.346\text{ m})/(0.600\text{ m}) = 0.577$, the ratio indeed is 1/2.

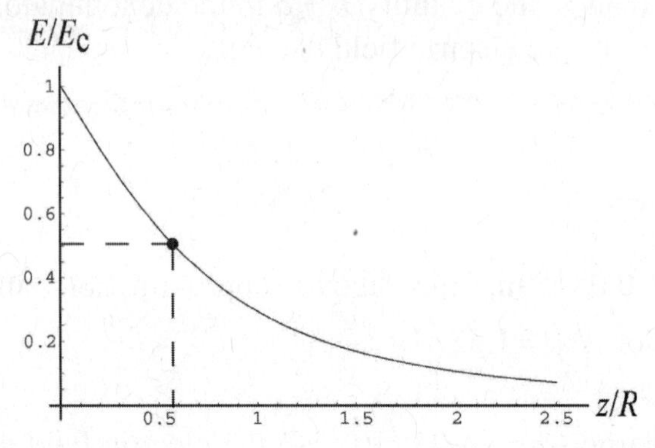

22-41

THINK In this problem we compare the strengths between the electrostatic force and the gravitational force.

EXPRESS The magnitude of the electrostatic force on a point charge of magnitude q is given by $F = qE$, where E is the magnitude of the electric field at the location of the particle. On the other hand, the force of gravity on a particle of mass m is $F_g = mg$.

ANALYZE (a) With $q = -2.0\times10^{-9}$ C and $F = 3.0\times10^{-6}$ N, the magnitude of the electric field strength is

$$E = \frac{F}{q} = \frac{3.0\times10^{-6}\text{ N}}{2.0\times10^{-9}\text{ C}} = 1.5\times10^{3}\text{ N/C}.$$

In vector notation, $\vec{F} = q\vec{E}$. Since the force points downward and the charge is negative, the field $\vec{E}$ must points upward (in the opposite direction of $\vec{F}$).

(b) The magnitude of the electrostatic force on a proton is

$$F_{el} = eE = \left(1.60\times10^{-19}\text{ C}\right)\left(1.5\times10^{3}\text{ N/C}\right) = 2.4\times10^{-16}\text{N}.$$

(c) A proton is positively charged, so the force is in the same direction as the field, upward.

(d) The magnitude of the gravitational force on the proton is

$$F_g = mg = \left(1.67\times10^{-27}\text{ kg}\right)\left(9.8\text{ m/s}^2\right) = 1.6\times10^{-26}\text{N}.$$

The force is downward.

(e) The ratio of the forces is

$$\frac{F_{el}}{F_g} = \frac{2.4\times10^{-16}\,\text{N}}{1.64\times10^{-26}\,\text{N}} = 1.5\times10^{10}.$$

LEARN The force of gravity on the proton is much smaller than the electrostatic force on the proton due to the field of strength $E = 1.5\times10^3$ N/C. For the two forces to have equal strength, the electric field would have to be very small:

$$E = \frac{mg}{q} = \frac{(1.67\times10^{-27}\text{ kg})(9.8\text{ m/s}^2)}{1.6\times10^{-19}\text{ C}} = 1.02\times10^{-7}\text{ N/C}.$$

22-43

THINK The acceleration of the electron is given by Newton's second law: $F = ma$, where F is the electrostatic force.

EXPRESS The magnitude of the force acting on the electron is $F = eE$, where E is the magnitude of the electric field at its location. Using Newton's second law, the acceleration of the electron is

$$a = \frac{F}{m} = \frac{eE}{m}.$$

ANALYZE With $e = 1.6\times10^{-19}$ C, $E = 2.00\times10^4$ N/C, and $m = 9.11\times10^{-31}$ kg, we find the acceleration to be

$$a = \frac{eE}{m} = \frac{\left(1.60\times10^{-19}\text{ C}\right)\left(2.00\times10^4\text{ N/C}\right)}{9.11\times10^{-31}\text{ kg}} = 3.51\times10^{15}\text{ m/s}^2.$$

LEARN In vector notation, $\vec{a} = \vec{F}/m = -e\vec{E}/m$, so $\vec{a}$ is in the opposite direction of $\vec{E}$. The magnitude of electron's acceleration is proportional to the field strength E: the greater the value of E, the greater the acceleration.

22-47

THINK The acceleration of the proton is given by Newton's second law: $F = ma$, where F is the electrostatic force.

EXPRESS The magnitude of the force acting on the proton is $F = eE$, where E is the magnitude of the electric field. According to Newton's second law, the acceleration of the proton is $a = F/m = eE/m$, where m is the mass of the proton. Thus,

$$a = \frac{F}{m} = \frac{eE}{m}.$$

We assume that the proton starts from rest ($v_0 = 0$) and apply the kinematic equation $v^2 = v_0^2 + 2ax$ (or else $x = \frac{1}{2}at^2$ and $v = at$). Thus, the speed of the proton after having traveling a distance x is $v = \sqrt{2ax}$.

ANALYZE (a) With $e = 1.6\times10^{-19}$ C, $E = 2.00\times10^4$ N/C, and $m = 1.67\times10^{-27}$ kg, we find the acceleration to be

$$a = \frac{eE}{m} = \frac{(1.60\times10^{-19}\text{ C})(2.00\times10^4\text{ N/C})}{1.67\times10^{-27}\text{ kg}} = 1.92\times10^{12}\text{ m/s}^2.$$

(b) With $x = 1.00\text{ cm} = 1.0\times10^{-2}\text{ m}$, the speed of the proton is

$$v = \sqrt{2ax} = \sqrt{2(1.92\times10^{12}\text{ m/s}^2)(0.0100\text{ m})} = 1.96\times10^5\text{ m/s}.$$

LEARN The time it takes for the proton to attain the final speed is

$$t = \frac{v}{a} = \frac{1.96\times10^5\text{ m/s}}{1.92\times10^{12}\text{ m/s}^2} = 1.02\times10^{-7}\text{ s}.$$

The distance the proton travels can be written as

$$x = \frac{1}{2}at^2 = \frac{1}{2}\left(\frac{eE}{m}\right)t^2.$$

22-57

THINK The potential energy of the electric dipole placed in an electric field depends on its orientation relative to the electric field.

EXPRESS The magnitude of the electric dipole moment is $p = qd$, where q is the magnitude of the charge, and d is the separation between the two charges. When placed in an electric field, the potential energy of the dipole is given by Eq. 22-38:

$$U(\theta) = -\vec{p}\cdot\vec{E} = -pE\cos\theta.$$

Therefore, if the initial angle between $\vec{p}$ and $\vec{E}$ is θ_0 and the final angle is θ, then the change in potential energy would be

$$\Delta U = U(\theta) - U_0(\theta) = -pE(\cos\theta - \cos\theta_0).$$

ANALYZE (a) With $q = 1.50\times10^{-9}$ C and $d = 6.20\times10^{-6}$ m, we find the magnitude of the dipole moment to be

$$p = qd = \left(1.50\times10^{-9}\ \text{C}\right)\left(6.20\times10^{-6}\ \text{m}\right) = 9.30\times10^{-15}\ \text{C}\cdot\text{m}.$$

(b) The initial and the final angles are $\theta_0 = 0$ (parallel) and $\theta = 180°$ (anti-parallel), so we find ΔU to be

$$\Delta U = U(180°) - U(0) = 2pE = 2\left(9.30\times10^{-15}\text{C}\cdot\text{m}\right)\left(1100\ \text{N/C}\right) = 2.05\times10^{-11}\ \text{J}.$$

LEARN The potential energy is a maximum ($U_{\text{max}} = +pE$) when the dipole is oriented antiparallel to $\vec{E}$, and is a minimum ($U_{\text{min}} = -pE$) when it is parallel to $\vec{E}$.

22-73

THINK We have a positive charge in the *xy* plane. From the electric fields it produces at two different locations, we can determine the position and the magnitude of the charge.

EXPRESS Let the charge be placed at (x_0, y_0). In Cartesian coordinates, the electric field at a point (x, y) can be written as

$$\vec{E} = E_x\hat{\text{i}} + E_y\hat{\text{j}} = \frac{q}{4\pi\varepsilon_0}\frac{(x-x_0)\hat{\text{i}} + (y-y_0)\hat{\text{j}}}{\left[(x-x_0)^2 + (y-y_0)^2\right]^{3/2}}.$$

The ratio of the field components is

$$\frac{E_y}{E_x} = \frac{y-y_0}{x-x_0}.$$

ANALYZE (a) The fact that the second measurement at the location (2.0 cm, 0) gives $\vec{E} = (100\ \text{N/C})\hat{\text{i}}$ indicates that $y_0 = 0$, that is, the charge must be somewhere on the *x* axis. Thus, the above expression can be simplified to

$$\vec{E} = \frac{q}{4\pi\varepsilon_0}\frac{(x-x_0)\hat{\text{i}} + y\hat{\text{j}}}{\left[(x-x_0)^2 + y^2\right]^{3/2}}.$$

On the other hand, the field at (3.0 cm, 3.0 cm) is

$$\vec{E} = (7.2\ \text{N/C})(4.0\hat{\text{i}} + 3.0\hat{\text{j}}),$$

which gives $E_y / E_x = 3/4$. Thus, we have

$$\frac{3}{4} = \frac{3.0\text{ cm}}{3.0\text{ cm} - x_0}$$

which implies that $x_0 = -1.0$ cm.

(b) As shown above, the y coordinate is $y_0 = 0$.

(c) To calculate the magnitude of the charge, we note that the field magnitude measured at (2.0 cm, 0) (which is $r = 0.030$ m from the charge) is

$$\left|\vec{E}\right| = \frac{1}{4\pi\varepsilon_0}\frac{q}{r^2} = 100\text{ N/C}.$$

Therefore,

$$q = 4\pi\varepsilon_0\left|\vec{E}\right|r^2 = \frac{(100\text{ N/C})(0.030\text{ m})^2}{8.99\times10^9\text{ N}\cdot\text{m}^2/\text{C}^2} = 1.0\times10^{-11}\text{C}.$$

LEARN Alternatively, we may calculate q by noting that at (3.0 cm, 3.00 cm)

$$E_x = 28.8\text{ N/C} = \frac{q}{4\pi\varepsilon_0}\frac{0.040\text{ m}}{\left[(0.040\text{ m})^2 + (0.030\text{ m})^2\right]^{3/2}} = \frac{q}{4\pi\varepsilon_0}\left(320/\text{m}^2\right).$$

This gives

$$q = \frac{28.8\text{ N/C}}{(8.99\times10^9\text{ N}\cdot\text{m}^2/\text{C}^2)(320/\text{m}^2)} = 1.0\times10^{-11}\text{ C},$$

in agreement with that calculated above.

22-83

THINK The potential energy of the electric dipole placed in an electric field depends on its orientation relative to the electric field. The field causes a torque that tends to align the dipole with the field.

EXPRESS When placed in an electric field $\vec{E}$, the potential energy of the dipole $\vec{p}$ is given by Eq. 22-38:

$$U(\theta) = -\vec{p}\cdot\vec{E} = -pE\cos\theta.$$

The torque caused by the electric field is (see Eq. 22-34) $\vec{\tau} = \vec{p}\times\vec{E}$.

ANALYZE (a) From Eq. 22-38 (and the facts that $\hat{\text{i}}\cdot\hat{\text{i}} = 1$ and $\hat{\text{j}}\cdot\hat{\text{i}} = 0$), the potential energy is

$$\begin{aligned} U &= -\vec{p}\cdot\vec{E} = -\left[\left(3.00\hat{\text{i}} + 4.00\hat{\text{j}}\right)\left(1.24\times10^{-30}\text{C}\cdot\text{m}\right)\right]\cdot\left[(4000\text{ N/C})\hat{\text{i}}\right] \\ &= -1.49\times10^{-26}\text{ J}. \end{aligned}$$

(b) From Eq. 22-34 (and the facts that $\hat{\mathrm{i}} \times \hat{\mathrm{i}} = 0$ and $\hat{\mathrm{j}} \times \hat{\mathrm{i}} = -\hat{\mathrm{k}}$), the torque is

$$\begin{aligned}\vec{\tau} = \vec{p} \times \vec{E} &= \left[\left(3.00\hat{\mathrm{i}} + 4.00\hat{\mathrm{j}}\right)\left(1.24\times10^{-30}\,\mathrm{C\cdot m}\right)\right]\times\left[\left(4000\,\mathrm{N/C}\right)\hat{\mathrm{i}}\right]\\ &= \left(-1.98\times10^{-26}\,\mathrm{N\cdot m}\right)\hat{\mathrm{k}}.\end{aligned}$$

(c) The work done is

$$\begin{aligned}W = \Delta U = \Delta\left(-\vec{p}\cdot\vec{E}\right) &= \left(\vec{p}_i - \vec{p}_f\right)\cdot\vec{E}\\ &= \left[\left(3.00\hat{\mathrm{i}} + 4.00\hat{\mathrm{j}}\right) - \left(-4.00\hat{\mathrm{i}} + 3.00\hat{\mathrm{j}}\right)\right]\left(1.24\times10^{-30}\,\mathrm{C\cdot m}\right)\right]\cdot\left[\left(4000\,\mathrm{N/C}\right)\hat{\mathrm{i}}\right]\\ &= 3.47\times10^{-26}\ \mathrm{J}.\end{aligned}$$

LEARN The work done by the agent is equal to the change in the potential energy of the dipole.

Chapter 23

23-1
THINK This exercise deals with electric flux through a square surface.

EXPRESS The vector area $\vec{A}$ and the electric field $\vec{E}$ are shown on the diagram below.

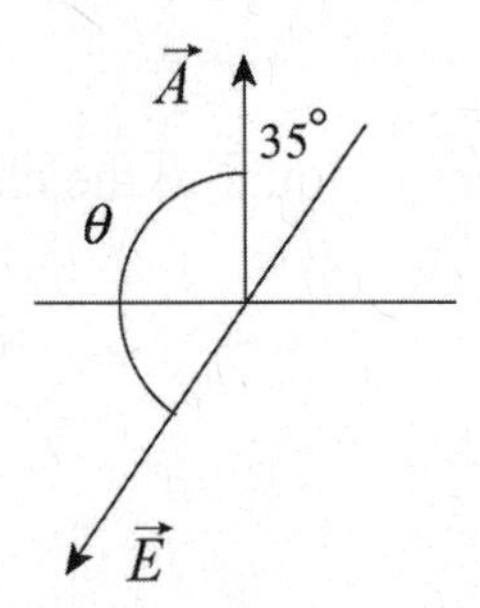

The electric flux through the surface is given by $\Phi = \vec{E} \cdot \vec{A} = EA\cos\theta$.

EXPRESS The angle θ between $\vec{A}$ and $\vec{E}$ is 180° – 35° = 145°, so the electric flux through the area is

$$\Phi = EA\cos\theta = (1800\ \text{N/C})(3.2\times10^{-3}\ \text{m})^2 \cos 145° = -1.5\times10^{-2}\ \text{N}\cdot\text{m}^2/\text{C}.$$

LEARN The flux is a maximum when $\vec{A}$ and $\vec{E}$ points in the same direction ($\theta = 0$), and is zero when the two vectors are perpendicular to each other ($\theta = 90$).

23-13
THINK A cube has six surfaces. The total flux through the cube is the sum of fluxes through each individual surface. We use Gauss' law to find the net charge inside the cube.

EXPRESS Let A be the area of one face of the cube, E_u be the magnitude of the electric field at the upper face, and E_l be the magnitude of the field at the lower face. Since the field is downward, the flux through the upper face is negative and the flux through the lower face is positive. The flux through the other faces is zero (because their area vectors are parallel to the field), so the total flux through the cube surface is

$$\Phi = A(E_\ell - E_u).$$

The net charge inside the cube is given by Gauss' law: $q = \varepsilon_0 \Phi$.

ANALYZE Substituting the values given, we find the net charge to be

$$\begin{aligned} q &= \varepsilon_0 \Phi = \varepsilon_0 A(E_\ell - E_u) = (8.85\times10^{-12}\ \text{C}^2/\text{N}\cdot\text{m}^2)(100\ \text{m})^2(100\ \text{N/C} - 60.0\ \text{N/C}) \\ &= 3.54\times10^{-6}\ \text{C} = 3.54\ \mu\text{C}. \end{aligned}$$

LEARN Since $\Phi > 0$, we conclude that the cube encloses a net positive charge.

23-17

THINK The system has spherical symmetry, so our Gaussian surface is a sphere of radius R with a surface area $A = 4\pi R^2$.

EXPRESS The charge on the surface of the sphere is the product of the surface charge density σ and the surface area of the sphere: $q = \sigma A = \sigma(4\pi R^2)$. We calculate the total electric flux leaving the surface of the sphere using Gauss' law: $q = \varepsilon_0 \Phi$.

ANALYZE (a) With $R = (1.20\ \text{m})/2 = 0.60\ \text{m}$ and $\sigma = 8.1\times10^{-6}\ \text{C/m}^2$, the charge on the surface is

$$q = 4\pi R^2\sigma = 4\pi(0.60\ \text{m})^2(8.1\times10^{-6}\ \text{C/m}^2) = 3.7\times10^{-5}\ \text{C}.$$

(b) We choose a Gaussian surface in the form of a sphere, concentric with the conducting sphere and with a slightly larger radius. By Gauss's law, the flux is

$$\Phi = \frac{q}{\varepsilon_0} = \frac{3.66\times10^{-5}\ \text{C}}{8.85\times10^{-12}\ \text{C}^2/\text{N}\cdot\text{m}^2} = 4.1\times10^6\ \text{N}\cdot\text{m}^2/\text{C}.$$

LEARN Since there is no charge inside the conducting sphere, the total electric flux through the surface of the sphere only depends on the charge residing on the surface of the sphere.

23-25

THINK Our system is an infinitely long line of charge. Since the system possesses cylindrical symmetry, we may apply Gauss' law and take the Gaussian surface to be in the form of a closed cylinder.

EXPRESS We imagine a cylindrical Gaussian surface A of radius r and length h concentric with the metal tube. Then by symmetry,

$$\oint_A \vec{E}\cdot d\vec{A} = 2\pi rhE = \frac{q}{\varepsilon_0},$$

where q is the amount of charge enclosed by the Gaussian cylinder. Thus, the magnitude of the electric field produced by a uniformly charged infinite line is

$$E = \frac{q/h}{2\pi\varepsilon_0 r} = \frac{\lambda}{2\pi\varepsilon_0 r}$$

where λ is the linear charge density and r is the distance from the line to the point where the field is measured.

ANALYZE Substituting the values given, we have

$$\lambda = 2\pi\varepsilon_0 Er = 2\pi\left(8.85\times10^{-12}\ \text{C}^2/\text{N}\cdot\text{m}^2\right)\left(4.5\times10^4\ \text{N/C}\right)\left(2.0\ \text{m}\right) = 5.0\times10^{-6}\ \text{C/m}.$$

LEARN Since $\lambda > 0$, the direction of $\vec{E}$ is radially outward from the line of charge. Note that the field varies with r as $E \sim 1/r$, in contrast to the $1/r^2$ dependence due to a point charge.

23-29

THINK The charge densities of both the conducting cylinder and the shell are uniform, and we neglect fringing effect. Symmetry can be used to show that the electric field is radial, both between the cylinder and the shell and outside the shell. It is zero, of course, inside the cylinder and inside the shell.

EXPRESS We take the Gaussian surface to be a cylinder of length L, coaxial with the given cylinders and of radius r. The flux through this surface is $\Phi = 2\pi rLE$, where E is the magnitude of the field at the Gaussian surface. We may ignore any flux through the ends. Gauss' law yields $q_{\text{enc}} = \varepsilon_0\Phi = 2\pi r\varepsilon_0 LE$, where q_{enc} is the charge enclosed by the Gaussian surface.

ANALYZE (a) In this case, we take the radius of our Gaussian cylinder to be

$$r = 2.00R_2 = 20.0R_1 = (20.0)(1.3\times10^{-3}\ \text{m}) = 2.6\times10^{-2}\ \text{m}.$$

The charge enclosed is $q_{\text{enc}} = Q_1+Q_2 = -Q_1 = -3.40\times10^{-12}$ C. Consequently, Gauss' law yields

$$E = \frac{q_{\text{enc}}}{2\pi\varepsilon_0 Lr} = \frac{-3.40\times10^{-12}\ \text{C}}{2\pi(8.85\times10^{-12}\ \text{C}^2/\text{N}\cdot\text{m}^2)(11.0\ \text{m})(2.60\times10^{-2}\text{m})} = -0.214\ \text{N/C},$$

or $|E| = 0.214$ N/C.

(b) The negative sign in E indicates that the field points inward.

(c) Next, for r = 5.00 R_1, the charge enclosed by the Gaussian surface is $q_{enc} = Q_1 = 3.40\times10^{-12}$ C. Consequently, Gauss' law yields $2\pi r\varepsilon_0 LE = q_{enc}$, or

$$E = \frac{q_{enc}}{2\pi\varepsilon_0 Lr} = \frac{3.40\times10^{-12}\ \text{C}}{2\pi(8.85\times10^{-12}\ \text{C}^2/\text{N}\cdot\text{m}^2)(11.0\ \text{m})(5.00\times1.30\times10^{-3}\text{m})} = 0.855\ \text{N/C}.$$

(d) The positive sign indicates that the field points outward.

(e) We consider a cylindrical Gaussian surface whose radius places it within the shell itself. The electric field is zero at all points on the surface since any field within a conducting material would lead to current flow (and thus to a situation other than the electrostatic ones being considered here), so the total electric flux through the Gaussian surface is zero and the net charge within it is zero (by Gauss' law). Since the central rod has charge Q_1, the inner surface of the shell must have charge $Q_{in} = -Q_1 = -3.40\times10^{-12}$ C.

(f) Since the shell is known to have total charge $Q_2 = -2.00Q_1$, it must have charge $Q_{out} = Q_2 - Q_{in} = -Q_1 = -3.40\times10^{-12}$ C on its outer surface.

LEARN Cylindrical symmetry of the system allows us to apply Gauss' law to the problem. Since electric field is zero inside the conducting shell, by Gauss' law, any net charge must be distributed on the surfaces of the shells.

23-37

THINK To calculate the electric field at a point very close to the center of a large, uniformly charged conducting plate, we replace the finite plate with an infinite plate having the same charge density. Planar symmetry then allows us to apply Gauss' law to calculate the electric field.

EXPRESS Using Gauss' law, we find the magnitude of the field to be $E = \sigma/\varepsilon_0$, where σ is the area charge density for the surface just under the point. The charge is distributed uniformly over both sides of the original plate, with half being on the side near the field point. Thus, $\sigma = q/2A$.

ANALYZE (a) With $q = 6.0\times10^{-6}$ C and $A = (0.080\ \text{m})^2$, we obtain

$$\sigma = \frac{q}{2A} = \frac{6.0\times10^{-6}\ \text{C}}{2(0.080\ \text{m})^2} = 4.69\times10^{-4}\ \text{C/m}^2.$$

The magnitude of the field is

$$E = \frac{\sigma}{\varepsilon_0} = \frac{4.69\times10^{-4}\ \text{C/m}^2}{8.85\times10^{-12}\ \text{C}^2/\text{N}\cdot\text{m}^2} = 5.3\times10^{7}\ \text{N/C}.$$

The field is normal to the plate and since the charge on the plate is positive, it points away from the plate.

(b) At a point far away from the plate, the electric field is nearly that of a point particle with charge equal to the total charge on the plate. The magnitude of the field is $E = q/4\pi\varepsilon_0 r^2 = kq/r^2$, where r is the distance from the plate. Thus,

$$E = \frac{(8.99\times10^9\ \text{N}\cdot\text{m}^2/\text{C}^2)(6.0\times10^{-6}\ \text{C})}{(30\ \text{m})^2} = 60\ \text{N/C}.$$

LEARN In summary, the electric field is nearly uniform ($E = \sigma/\varepsilon_0$) close to the plate, but resembles that of a point charge far away from the plate.

23-39

THINK Since the nonconducting charged ball is in equilibrium with the nonconducting charged sheet (see Fig. 23-45), both the vertical and horizontal components of the net force on the ball must be zero.

EXPRESS The forces acting on the ball are shown in the diagram below.

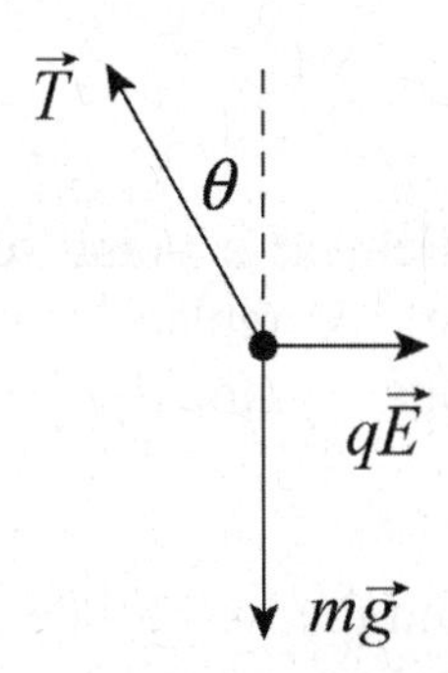

The gravitational force has magnitude *mg*, where *m* is the mass of the ball; the electrical force has magnitude *qE*, where *q* is the charge on the ball and *E* is the magnitude of the electric field at the position of the ball; and the tension in the thread is denoted by *T*. The electric field produced by the plate is normal to the plate and points to the right. Since the ball is positively charged, the electric force on it also points to the right. The tension in the thread makes the angle θ (= 30°) with the vertical. Since the ball is in equilibrium the net force on it vanishes. The sum of the horizontal components yields

$$qE - T\sin\theta = 0$$

and the sum of the vertical components yields

$$T\cos\theta - mg = 0\,.$$

We solve for the electric field E and deduce σ, the charge density of the sheet, from $E = \sigma/2\varepsilon_0$ (see Eq. 23-13).

ANALYZE The expression $T = qE/\sin\theta$, from the first equation, is substituted into the second to obtain $qE = mg\tan\theta$. The electric field produced by a large uniform sheet of charge is given by $E = \sigma/2\varepsilon_0$, so

$$\frac{q\sigma}{2\varepsilon_0} = mg\tan\theta$$

and we have

$$\sigma = \frac{2\varepsilon_0 mg\tan\theta}{q} = \frac{2(8.85\times10^{-12}\ \text{C}^2/\text{N}\cdot\text{m}^2)(1.0\times10^{-6}\ \text{kg})(9.8\ \text{m/s}^2)\tan 30°}{2.0\times10^{-8}\ \text{C}}$$
$$= 5.0\times10^{-9}\ \text{C/m}^2.$$

LEARN Since both the sheet and the ball are positively charged, the force between them is repulsive. This is balanced by the horizontal component of the tension in the thread. The angle the thread makes with the vertical direction increases with the charge density of the sheet.

23-47

THINK The unknown charge is distributed uniformly over the surface of the conducting solid sphere.

EXPRESS The electric field produced by the unknown charge at points outside the sphere is like the field of a point particle with charge equal to the net charge on the sphere. That is, the magnitude of the field is given by $E = |q|/4\pi\varepsilon_0 r^2$, where $|q|$ is the magnitude of the charge on the sphere and r is the distance from the center of the sphere to the point where the field is measured.

ANALYZE Thus, we have

$$|q| = 4\pi\varepsilon_0 r^2 E = \frac{(0.15\ \text{m})^2(3.0\times10^3\ \text{N/C})}{8.99\times10^9\ \text{N}\cdot\text{m}^2/\text{C}^2} = 7.5\times10^{-9}\ \text{C}.$$

The field points inward, toward the sphere center, so the charge is negative, i.e., $q = -7.5\times10^{-9}\text{C}$.

LEARN The electric field strength as a function of r is shown to the right. Inside the metal sphere, $E = 0$; outside the sphere, $E = k|q|/r^2$, where $k = 1/4\pi\varepsilon_0$.

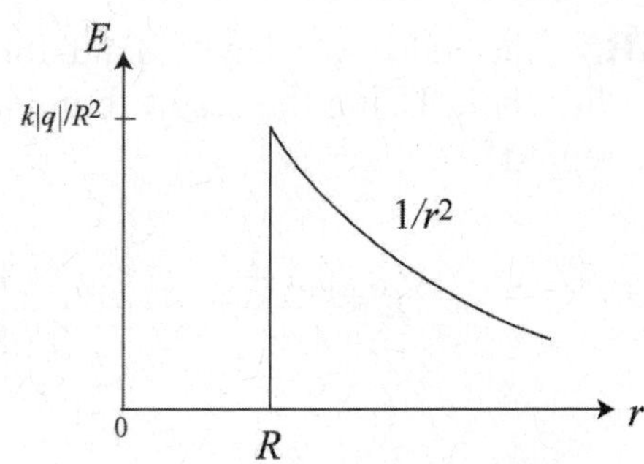

23-51

THINK Since our system possesses spherical symmetry, to calculate the electric field strength, we may apply Gauss' law and take the Gaussian surface to be in the form of a sphere of radius r.

EXPRESS To find an expression for the electric field inside the shell in terms of A and the distance from the center of the shell, choose A so the field does not depend on the distance. We use a Gaussian surface in the form of a sphere with radius r_g, concentric with the spherical shell and within it ($a < r_g < b$). Gauss' law will be used to find the magnitude of the electric field a distance r_g from the shell center. The charge that is both in the shell and within the Gaussian sphere is given by the integral $q_s = \int \rho\, dV$ over the portion of the shell within the Gaussian surface. Since the charge distribution has spherical symmetry, we may take dV to be the volume of a spherical shell with radius r and infinitesimal thickness dr: $dV = 4\pi r^2\, dr$. Thus,

$$q_s = 4\pi \int_a^{r_g} \rho r^2 dr = 4\pi \int_a^{r_g} \frac{A}{r} r^2 dr = 4\pi\, A \int_a^{r_g} r\, dr = 2\pi\, A \left(r_g^2 - a^2\right).$$

The total charge inside the Gaussian surface is

$$q_{\text{enc}} = q + q_s = q + 2\pi A(r_g^2 - a^2).$$

The electric field is radial, so the flux through the Gaussian surface is $\Phi = 4\pi r_g^2 E$, where E is the magnitude of the field. Gauss' law yields

$$\Phi = q_{\text{enc}} / \varepsilon_0 \;\Rightarrow\; 4\pi\varepsilon_0 E r_g^2 = q + 2\pi A(r_g^2 - a^2).$$

We solve for E:

$$E = \frac{1}{4\pi\varepsilon_0}\left[\frac{q}{r_g^2} + 2\pi\, A - \frac{2\pi A a^2}{r_g^2}\right].$$

ANALYZE For the field to be uniform, the first and last terms in the brackets must cancel. They do if $q - 2\pi A a^2 = 0$ or $A = q/2\pi a^2$. With $a = 2.00 \times 10^{-2}$ m and $q = 45.0 \times 10^{-15}$ C, we have $A = 1.79\times 10^{-11}\,\text{C/m}^2$.

LEARN The value we have found for A ensures the uniformity of the field strength inside the shell. Using the result found above, we can readily show that the electric field in the region $a \le r \le b$ is

$$E = \frac{2\pi A}{4\pi\varepsilon_0} = \frac{A}{2\varepsilon_0} = \frac{1.79\times 10^{-11}\,\text{C/m}^2}{2(8.85\times 10^{-12}\ \text{C}^2/\text{N}\cdot\text{m}^2)} = 1.01\ \text{N/C}.$$

23-61

THINK Our system consists of two concentric metal shells. We apply the superposition principle and Gauss' law to calculate the electric field everywhere.

EXPRESS At all points where there is an electric field, it is radially outward. For each part of the problem, use a Gaussian surface in the form of a sphere that is concentric with the metal shells of charge and passes through the point where the electric field is to be found. The field is uniform on the surface, so

$$\Phi = \oint \vec{E} \cdot d\vec{A} = 4\pi r^2 E = \frac{q_{\text{enc}}}{\varepsilon_0},$$

where r is the radius of the Gaussian surface.

ANALYZE (a) For $r < a$, the charge enclosed is $q_{\text{enc}} = 0$, so $E = 0$ in the region inside the shell.

(b) For $a < r < b$, the charged enclosed by the Gaussian surface is $q_{\text{enc}} = q_a$, so the field strength is $E = q_a/4\pi\varepsilon_0 r^2$.

(c) For $r > b$, the charged enclosed by the Gaussian surface is $q_{\text{enc}} = q_a + q_b$, so the field strength is $E = (q_a + q_b)/4\pi\varepsilon_0 r^2$.

(d) Since $E = 0$ for $r < a$ the charge on the inner surface of the inner shell is always zero. The charge on the outer surface of the inner shell is therefore q_a. Since $E = 0$ inside the metallic outer shell the net charge enclosed in a Gaussian surface that lies in between the inner and outer surfaces of the outer shell is zero. Thus the inner surface of the outer shell must carry a charge $-q_a$, leaving the charge on the outer surface of the outer shell to be $q_b + q_a$.

LEARN In the case of a single shell of radius R and charge q, the field strength is $E = 0$ for $r < R$, and $E = q/4\pi\varepsilon_0 r^2$ for $r > R$ (see Eqs. 23-15 and 23-16).

23-67

THINK The electric field at P is due to the charge on the surface of the metallic conductor and the point charge Q.

EXPRESS The initial field (evaluated "just outside the outer surface" which means it is evaluated at $R_2 = 0.20$ m, the outer radius of the conductor) is related to the charge q on the hollow conductor by Eq. 23-15: $E_{\text{initial}} = q/4\pi\varepsilon_0 R_2^2$. After the point charge Q is placed at the geometric center of the hollow conductor, the final field at that point is a combination of the initial and that due to Q (determined by Eq. 22-3):

$$E_{\text{final}} = E_{\text{initial}} + \frac{Q}{4\pi\varepsilon_0 R_2^2}.$$

ANALYZE (a) The charge on the spherical shell is

$$q = 4\pi\varepsilon_0 R_2^2 E_{\text{initial}} = \frac{(0.20\text{ m})^2(450\text{ N/C})}{8.99\times10^9\text{ N}\cdot\text{m}^2/\text{C}^2} = 2.0\times10^{-9}\text{ C}.$$

(b) Similarly, using the equation above, we find the point charge to be

$$Q = 4\pi\varepsilon_0 R_2^2\left(E_{\text{final}} - E_{\text{initial}}\right) = \frac{(0.20\text{ m})^2(180\text{ N/C} - 450\text{ N/C})}{8.99\times10^9\text{ N}\cdot\text{m}^2/\text{C}^2}$$
$$= -1.2\times10^{-9}\text{ C}.$$

(c) In order to cancel the field (due to Q) within the conducting material, there must be an amount of charge equal to $-Q$ distributed uniformly on the inner surface (of radius R_1). Thus, the answer is $+1.2 \times 10^{-9}$ C.

(d) Since the total excess charge on the conductor is q and is located on the surfaces, then the outer surface charge must equal the total minus the inner surface charge. Thus, the answer is $2.0 \times 10^{-9}\text{ C} - 1.2 \times 10^{-9}\text{ C} = +0.80 \times 10^{-9}$ C.

LEARN The key idea here is to realize that the electric field inside the conducting shell ($R_1 < r < R_2$) must be zero, so the charge must be distributed in such a way that the charge enclosed by a Gaussian sphere of radius r ($R_1 < r < R_2$) is zero.

23-77

THINK The total charge on the conducting shell is equal to the sum of the charges on the shell's inner surface and the outer surface.

EXPRESS Let q_{in} be the charge on the inner surface and q_{out} the charge on the outer surface. The net charge on the shell is $Q = q_{\text{in}} + q_{\text{out}}$.

ANALYZE (a) In order to have net charge $Q = -10\ \mu$C when the charge on the outer surface is $q_{\text{out}} = -14\ \mu$C, then there must be

$$q_{\text{in}} = Q - q_{\text{out}} = -10\ \mu\text{C} - (-14\ \mu\text{C}) = +4\ \mu\text{C}$$

on the inner surface (since charges reside on the surfaces of a conductor in electrostatic situations).

(b) Let q be the charge of the particle. In order to cancel the electric field inside the conducting material, the contribution from the $q_{\text{in}} = +4\ \mu\text{C}$ on the inner surface must be canceled by that of the charged particle in the hollow, that is, $q_{\text{enc}} = q + q_{\text{in}} = 0$. Thus, the particle's charge is $q = -q_{\text{in}} = -4\ \mu\text{C}$.

LEARN The key idea here is to realize that the electric field inside the conducting shell must be zero. Thus, in the presence of a point charge in the hollow, the charge on the shell must be redistributed between its inner and outer surfaces in such a way that the net charge enclosed by a Gaussian sphere of radius r ($R_1 < r < R_2$, where R_1 is the inner radius and R_2 is the outer radius) remains zero.

Chapter 24

24-1

THINK Ampere is the SI unit for current. An ampere is one coulomb per second.

EXPRESS To calculate the total charge through the circuit, we note that $1\ \text{A} = 1\ \text{C/s}$ and $1\ \text{h} = 3600\ \text{s}$.

ANALYZE (a) Thus,

$$84\ \text{A}\cdot\text{h} = \left(84\frac{\text{C}\cdot\text{h}}{\text{s}}\right)\left(3600\frac{\text{s}}{\text{h}}\right) = 3.0\times10^{5}\ \text{C}.$$

(b) The change in potential energy is $\Delta U = q\,\Delta V = (3.0\times10^{5}\ \text{C})(12\ \text{V}) = 3.6\times10^{6}\ \text{J}$.

LEARN Potential difference is the change of potential energy per unit charge. Unlike electric field, potential difference is a scalar quantity.

24-5

THINK The electric field produced by an infinite sheet of charge is normal to the sheet and is uniform.

EXPRESS The magnitude of the electric field produced by the infinite sheet of charge is $E = \sigma/2\varepsilon_0$, where σ is the surface charge density. Place the origin of a coordinate system at the sheet and take the x axis to be parallel to the field and positive in the direction of the field. Then the electric potential is

$$V = V_s - \int_0^x E\,dx = V_s - Ex,$$

where V_s is the potential at the sheet. The equipotential surfaces are surfaces of constant x; that is, they are planes that are parallel to the plane of charge. If two surfaces are separated by Δx then their potentials differ in magnitude by

$$\Delta V = E\Delta x = (\sigma/2\varepsilon_0)\Delta x.$$

ANALYZE Thus, for $\sigma = 0.10\times10^{-6}\ \text{C/m}^2$ and $\Delta V = 50\ \text{V}$, we have

$$\Delta x = \frac{2\varepsilon_0 \Delta V}{\sigma} = \frac{2\left(8.85\times10^{-12}\ \text{C}^2/\text{N}\cdot\text{m}^2\right)(50\ \text{V})}{0.10\times10^{-6}\ \text{C/m}^2} = 8.8\times10^{-3}\ \text{m}.$$

LEARN Equipotential surfaces are always perpendicular to the electric field lines. Figure 24-3(a) depicts the electric field lines and equipotential surfaces for a uniform electric field.

24-15

THINK The electric potential for a spherically symmetric charge distribution falls off as $1/r$, where r is the radial distance from the center of the charge distribution.

EXPRESS The electric potential V at the surface of a drop of charge q and radius R is given by $V = q/4\pi\varepsilon_0 R$.

ANALYZE (a) With $V = 500$ V and $q = 30\times10^{-12}$ C, we find the radius to be

$$R = \frac{q}{4\pi\varepsilon_0 V} = \frac{\left(8.99\times10^9\ \text{N}\cdot\text{m}^2/\text{C}^2\right)\left(30\times10^{-12}\ \text{C}\right)}{500\ \text{V}} = 5.4\times10^{-4}\ \text{m}.$$

(b) After the two drops combine to form one big drop, the total volume is twice the volume of an original drop, so the radius R' of the combined drop is given by $(R')^3 = 2R^3$ and $R' = 2^{1/3}R$. The charge is twice the charge of the original drop: $q' = 2q$. Thus,

$$V' = \frac{1}{4\pi\varepsilon_0}\frac{q'}{R'} = \frac{1}{4\pi\varepsilon_0}\frac{2q}{2^{1/3}R} = 2^{2/3}V = 2^{2/3}(500\ \text{V}) \approx 790\ \text{V}.$$

LEARN A positively charged configuration produces a positive electric potential, and a negatively charged configuration produces a negative electric potential. Adding more charge increases the electric potential.

24-31

THINK Since the disk is uniformly charged, when the full disk is present each quadrant contributes equally to the electric potential at P.

EXPRESS Electrical potential is a scalar quantity. The potential at P due to a single quadrant is one-fourth the potential due to the entire disk. We first find an expression for the potential at P due to the entire disk. To do so, consider a ring of charge with radius r and (infinitesimal) width dr. Its area is $2\pi r\, dr$ and it contains charge $dq = 2\pi\sigma r\, dr$. All the charge in it is at a distance $\sqrt{r^2 + D^2}$ from P, so the potential it produces at P is

$$dV = \frac{1}{4\pi\varepsilon_0}\frac{2\pi\sigma r dr}{\sqrt{r^2 + D^2}} = \frac{\sigma r dr}{2\varepsilon_0\sqrt{r^2 + D^2}}.$$

ANALYZE Integrating over r, the total potential at P is

$$V=\frac{\sigma}{2\varepsilon_0}\int_0^R\frac{rdr}{\sqrt{r^2+D^2}}=\frac{\sigma}{2\varepsilon_0}\sqrt{r^2+D^2}\Big|_0^R=\frac{\sigma}{2\varepsilon_0}\left[\sqrt{R^2+D^2}-D\right].$$

Therefore, the potential V_{sq} at P due to a single quadrant is

$$V_{sq}=\frac{V}{4}=\frac{\sigma}{8\varepsilon_0}\left[\sqrt{R^2+D^2}-D\right]=\frac{(7.73\times10^{-15}\,\text{C/m}^2)}{8(8.85\times10^{-12}\,\text{C}^2/\text{N}\cdot\text{m}^2)}\left[\sqrt{(0.640\text{ m})^2+(0.259\text{ m})^2}-0.259\text{ m}\right]$$
$$=4.71\times10^{-5}\text{ V}.$$

LEARN Consider the limit $D \gg R$. The potential becomes

$$V_{sq}=\frac{\sigma}{8\varepsilon_0}\left[\sqrt{R^2+D^2}-D\right]\approx\frac{\sigma}{8\varepsilon_0}\left[D\left(1+\frac{1}{2}\frac{R^2}{D^2}+\cdots\right)-D\right]=\frac{\sigma}{8\varepsilon_0}\frac{R^2}{2D}=\frac{\pi R^2\sigma/4}{4\pi\varepsilon_0 D}=\frac{q_{sq}}{4\pi\varepsilon_0 D}$$

where $q_{sq}=\pi R^2\sigma/4$ is the charge on the quadrant. In this limit, we see that the potential resembles that due to a point charge q_{sq}.

24-37

THINK The component of the electric field $\vec{E}$ in a given direction is the negative of the rate at which potential changes with distance in that direction.

EXPRESS With $V=2.00xyz^2$, we apply Eq. 24-41 to calculate the x, y, and z components of the electric field:

$$E_x=-\frac{\partial V}{\partial x}=-2.00yz^2$$
$$E_y=-\frac{\partial V}{\partial y}=-2.00xz^2$$
$$E_z=-\frac{\partial V}{\partial z}=-4.00xyz$$

which, at (x, y, z) = (3.00 m, –2.00 m, 4.00 m), gives

$$(E_x, E_y, E_z) = (64.0\text{ V/m}, -96.0\text{ V/m}, 96.0\text{ V/m}).$$

ANALYZE The magnitude of the field is therefore

$$|\vec{E}|=\sqrt{E_x^2+E_y^2+E_z^2}=\sqrt{(64.0\text{ V/m})^2+(-96.0\text{ V/m})^2+(96.0\text{ V/m})^2}=150\text{ V/m}=150\text{ N/C}.$$

LEARN If the electric potential increases along some direction, say x, with $\partial V/\partial x > 0$, then there is a corresponding nonvanishing component of $\vec{E}$ in the opposite direction $(-E_x \neq 0)$.

24-43

THINK The work required to set up the arrangement is equal to the potential energy of the system.

EXPRESS We choose the zero of electric potential to be at infinity. The initial electric potential energy U_i of the system before the particles are brought together is therefore zero. After the system is set up the final potential energy is

$$U_f = \frac{q^2}{4\pi\varepsilon_0}\left(-\frac{1}{a}-\frac{1}{a}+\frac{1}{\sqrt{2}a}-\frac{1}{a}-\frac{1}{a}+\frac{1}{\sqrt{2}a}\right)=\frac{2q^2}{4\pi\varepsilon_0 a}\left(\frac{1}{\sqrt{2}}-2\right).$$

Thus the amount of work required to set up the system is given by

$$W = \Delta U = U_f - U_i = U_f = \frac{2q^2}{4\pi\varepsilon_0 a}\left(\frac{1}{\sqrt{2}}-2\right)=\frac{2(8.99\times10^9\ \text{N}\cdot\text{m}^2/\text{C}^2)(2.30\times10^{-12}\ \text{C})^2}{0.640\ \text{m}}\left(\frac{1}{\sqrt{2}}-2\right)$$

$$= -1.92\times10^{-13}\ \text{J}.$$

LEARN The work done in assembling the system is negative. This means that an external agent would have to supply $W_{\text{ext}} = +1.92\times10^{-13}$ J in order to take apart the arrangement completely.

24-57

THINK Mechanical energy is conserved in the process.

EXPRESS The electric potential at (0, y) due to the two charges Q held fixed at $(\pm x, 0)$ is

$$V = \frac{2Q}{4\pi\varepsilon_0\sqrt{x^2+y^2}}.$$

Thus, the potential energy of the particle of charge q at (0, y) is

$$U = qV = \frac{2Qq}{4\pi\varepsilon_0\sqrt{x^2+y^2}}.$$

Conservation of mechanical energy ($K_i + U_i = K_f + U_f$) gives

$$K_f = K_i + U_i - U_f = K_i + \frac{2Qq}{4\pi\varepsilon_0}\left(\frac{1}{\sqrt{x^2+y_i^2}} - \frac{1}{\sqrt{x^2+y_f^2}}\right),$$

where y_i and y_f are the initial and final coordinates of the moving charge along the y axis.

ANALYZE (a) With $q = -15\times10^{-6}$ C, $Q = 50\times10^{-6}$ C, $x = \pm3$ m, $y_i = 4$ m, and $y_f = 0$, we obtain

$$K_f = 1.2\text{ J} + \frac{2(50\times10^{-6}\text{ C})(-15\times10^{-6}\text{ C})}{4\pi(8.85\times10^{-12}\text{ C}^2/\text{N}\cdot\text{m}^2)}\left(\frac{1}{\sqrt{(3.0\text{ m})^2+(4.0\text{ m})^2}} - \frac{1}{\sqrt{(3.0\text{ m})^2}}\right)$$
$$= 3.0\text{ J}.$$

(b) We set $K_f = 0$ and solve for y_f (choosing the negative root, as indicated in the problem statement):

$$K_i + U_i = U_f \Rightarrow 1.2\text{ J} + \frac{2Qq}{4\pi\varepsilon_0\sqrt{x^2+y_i^2}} = \frac{2Qq}{4\pi\varepsilon_0\sqrt{x^2+y_f^2}}.$$

Substituting the values given, we have $U_i = -2.7$ J, and $y_f = -8.5$ m.

LEARN The dependence of the final kinetic energy of the particle on y is plotted below.

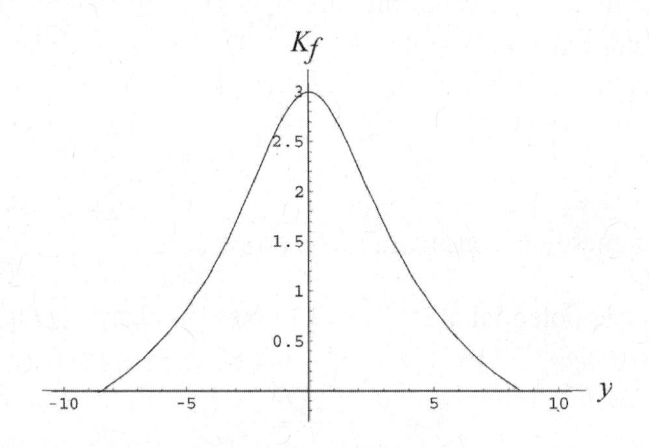

From the plot, we see that $K_f = 3.0$ J at $y = 0$, and $K_f = 0$ at $y = -8.5$ m. The particle oscillates between the two end-points $y_f = \pm8.5$ m.

24-63

THINK The electric potential is the sum of the contributions of the individual spheres.

EXPRESS Let q_1 be the charge on one, q_2 be the charge on the other, and d be their separation. The point halfway between them is the same distance $d/2$ (= 1.0 m) from the center of each sphere.

For parts (b) and (c), we note that the distance from the center of one sphere to the surface of the other is $d - R$, where R is the radius of either sphere. The potential of either one of the spheres is due to the charge on that sphere as well as the charge on the other sphere.

ANALYZE (a) The potential at the halfway point is

$$V = \frac{q_1 + q_2}{4\pi\varepsilon_0 d/2} = \frac{(8.99\times10^9\ \text{N}\cdot\text{m}^2/\text{C}^2)(1.0\times10^{-8}\text{C} - 3.0\times10^{-8}\text{C})}{1.0\ \text{m}} = -1.8\times10^2\,\text{V}.$$

(b) The potential at the surface of sphere 1 is

$$V_1 = \frac{1}{4\pi\varepsilon_0}\left[\frac{q_1}{R} + \frac{q_2}{d-R}\right] = (8.99\times10^9\text{N}\cdot\text{m}^2/\text{C}^2)\left[\frac{1.0\times10^{-8}\text{C}}{0.030\,\text{m}} - \frac{3.0\times10^{-8}\text{C}}{2.0\,\text{m} - 0.030\,\text{m}}\right] = 2.9\times10^3\,\text{V}.$$

(c) Similarly, the potential at the surface of sphere 2 is

$$V_2 = \frac{1}{4\pi\varepsilon_0}\left[\frac{q_1}{d-R} + \frac{q_2}{R}\right] = (8.99\times10^9\text{N}\cdot\text{m}^2/\text{C}^2)\left[\frac{1.0\times10^{-8}\text{C}}{2.0\,\text{m} - 0.030\,\text{m}} - \frac{3.0\times10^{-8}\text{C}}{0.030\,\text{m}}\right] = -8.9\times10^3\,\text{V}.$$

LEARN In the limit where $d \to \infty$, the spheres are isolated from each other and the electric potentials at the surface of each individual sphere become

$$V_{10} = \frac{q_1}{4\pi\varepsilon_0 R} = \frac{(8.99\times10^9\ \text{N}\cdot\text{m}^2/\text{C}^2)(1.0\times10^{-8}\text{C})}{0.030\,\text{m}} = 3.0\times10^3\,\text{V},$$

and

$$V_{20} = \frac{q_2}{4\pi\varepsilon_0 R} = \frac{(8.99\times10^9\ \text{N}\cdot\text{m}^2/\text{C}^2)(-3.0\times10^{-8}\text{C})}{0.030\,\text{m}} = -8.99\times10^3\,\text{V}.$$

24-65

THINK If the electric potential is zero at infinity, then the potential at the surface of the sphere is given by $V = Q/4\pi\varepsilon_0 R$, where Q is the charge on the sphere and R is its radius.

EXPRESS From $V = Q/4\pi\varepsilon_0 R$, we find the charge to be $Q = 4\pi\varepsilon_0 RV$.

ANALYZE With $R = 0.15$ m and $V = 1500$ V, we have

$$Q = 4\pi\varepsilon_0 RV = \frac{(0.15\text{ m})(1500\text{ V})}{8.99\times10^9\text{ N}\cdot\text{m}^2/\text{C}^2} = 2.5\times10^{-8}\text{C}.$$

LEARN A plot of the electric potential as a function of r is shown below ($k = 1/4\pi\varepsilon_0$).

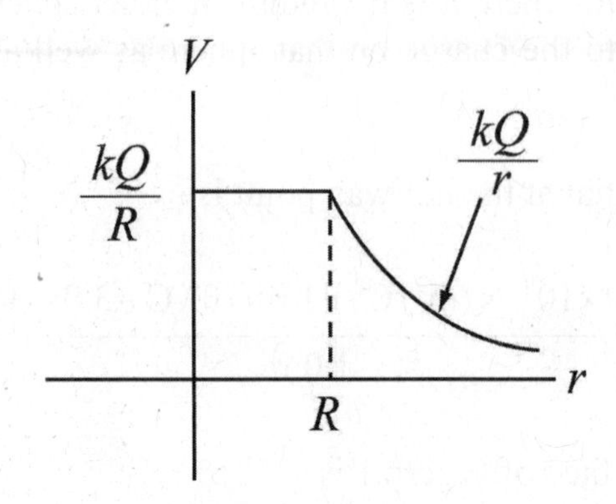

Note that the potential is constant inside the conducting sphere.

24-69

THINK To calculate the potential, we first apply Gauss' law to calculate the electric field of the charged cylinder of radius R. The Gaussian surface is a cylindrical surface that is concentric with the cylinder.

EXPRESS We imagine a cylindrical Gaussian surface A of radius r and length h concentric with the cylinder. Then, by Gauss' law,

$$\oint_A \vec{E}\cdot d\vec{A} = 2\pi rhE = \frac{q_{\text{enc}}}{\varepsilon_0},$$

where q_{enc} is the amount of charge enclosed by the Gaussian cylinder. Inside the charged cylinder ($r < R$), $q_{\text{enc}} = 0$, so the electric field is zero. On the other hand, outside the cylinder ($r > R$), $q_{\text{enc}} = \lambda h$ so the magnitude of the electric field is

$$E = \frac{q/h}{2\pi\varepsilon_0 r} = \frac{\lambda}{2\pi\varepsilon_0 r}$$

where λ is the linear charge density and r is the distance from the line to the point where the field is measured. The potential difference between two points 1 and 2 is

$$V(r_2) - V(r_1) = -\int_{r_1}^{r_2} E(r)\,dr.$$

ANALYZE (a) The radius of the cylinder (0.020 m, the same as R_B) is denoted R, and the field magnitude there (160 N/C) is denoted E_B. From the equation above, we see that the electric field beyond the surface of the cylinder is inversely proportional with r:

$$E = E_B \frac{R_B}{r}, \quad r \geq R_B .$$

Thus, if $r = R_C = 0.050$ m, we obtain

$$E_C = E_B \frac{R_B}{R_C} = (160 \text{ N/C})\left(\frac{0.020 \text{ m}}{0.050 \text{ m}}\right) = 64 \text{ N/C}.$$

(b) The potential difference between V_B and V_C is

$$V_B - V_C = -\int_{R_C}^{R_B} \frac{E_B R_B}{r} dr = E_B R_B \ln\left(\frac{R_C}{R_B}\right) = (160 \text{ N/C})(0.020 \text{ m}) \ln\left(\frac{0.050 \text{ m}}{0.020 \text{ m}}\right) = 2.9 \text{ V}.$$

(c) The electric field throughout the conducting volume is zero, which implies that the potential there is constant and equal to the value it has on the surface of the charged cylinder: $V_A - V_B = 0$.

LEARN The electric potential at a distance $r > R_B$ can be written as

$$V(r) = V_B - E_B R_B \ln\left(\frac{r}{R_B}\right).$$

We see that $V(r)$ decreases logarithmically with r.

24-71

THINK The component of the electric field $\vec{E}$ in any direction is the negative of the rate at which potential changes with distance in that direction.

EXPRESS According to Eq. 24-30, the electric potential of a dipole at a point a distance r away is

$$V = \frac{1}{4\pi\varepsilon_0} \frac{p\cos\theta}{r^2}$$

where p is the magnitude of the dipole moment $\vec{p}$ and θ is the angle between $\vec{p}$ and the position vector of the point. The potential at infinity is taken to be zero.

ANALYZE On the dipole axis $\theta = 0$ or π, so $|\cos\theta| = 1$. Therefore, magnitude of the electric field is

$$|E(r)| = \left|-\frac{\partial V}{\partial r}\right| = \frac{p}{4\pi\varepsilon_0}\left|\frac{d}{dr}\left(\frac{1}{r^2}\right)\right| = \frac{p}{2\pi\varepsilon_0 r^3}.$$

LEARN Take the z axis to be the dipole axis. For $r = z > 0\,(\theta = 0)$, $E = p/2\pi\varepsilon_0 z^3$. On the other hand, for $r = -z < 0\ (\theta = \pi)$, $E = -p/2\pi\varepsilon_0 z^3$.

24-87

THINK The work done is equal to the change in potential energy.

EXPRESS The initial potential energy of the system is

$$U_i = \frac{2q^2}{4\pi\varepsilon_0 L} + U_0$$

where q is the charge on each particle, L is the length of the triangle side, and U_0 is the potential energy associated with the interaction of the two fixed charges. After moving to the midpoint of the line joining the two fixed charges, the final energy of the configuration is

$$U_f = \frac{2q^2}{4\pi\varepsilon_0 (L/2)} + U_0 .$$

Thus, the work done by the external agent is

$$W = \Delta U = U_f - U_i = \frac{2q^2}{4\pi\varepsilon_0}\left(\frac{2}{L} - \frac{1}{L}\right) = \frac{2q^2}{4\pi\varepsilon_0 L}.$$

ANALYZE Substituting the values given, we have

$$W = \frac{2q^2}{4\pi\varepsilon_0 L} = \frac{2(8.99\times10^9\ \text{N}\cdot\text{m}^2/\text{C}^2)(0.12\ \text{C})^2}{1.7\ \text{m}} = 1.5\times10^8\ \text{J}.$$

At a rate of $P = 0.83 \times 10^3$ joules per second, it would take $W/P = 1.8 \times 10^5$ seconds or about 2.1 days to do this amount of work.

LEARN Since all three particles are positively charged, positive work is required by the external agent in order to bring them closer.

24-93

THINK To calculate the potential at point B due to the charged ring, we note that all points on the ring are at the same distance from B.

EXPRESS Let point B be at $(0, 0, z)$. The electric potential at B is given by

$$V = \frac{q}{4\pi\varepsilon_0\sqrt{z^2+R^2}}$$

where q is the charge on the ring. The potential at infinity is taken to be zero.

ANALYZE With $q = 16 \times 10^{-6}$ C, $z = 0.040$ m, and $R = 0.0300$ m, we find the potential difference between points A (located at the origin) and B to be

$$\begin{aligned} V_B - V_A &= \frac{q}{4\pi\varepsilon_0}\left(\frac{1}{\sqrt{z^2+R^2}} - \frac{1}{R}\right) \\ &= (8.99\times10^9\ \text{N}\cdot\text{m}^2/\text{C}^2)(16.0\times10^{-6}\ \text{C})\left(\frac{1}{\sqrt{(0.030\ \text{m})^2+(0.040\ \text{m})^2}} - \frac{1}{0.030\ \text{m}}\right) \\ &= -1.92\times10^6\ \text{V}. \end{aligned}$$

LEARN In the limit $z \gg R$, the potential approaches its "point-charge" limit:

$$V \approx \frac{q}{4\pi\varepsilon_0 z}.$$

24-95

THINK To calculate the electric potential, we first apply Gauss' law to calculate the electric field of the spherical shell. The Gaussian surface is a sphere that is concentric with the shell.

EXPRESS At all points where there is an electric field, it is radially outward. For each part of the problem, use a Gaussian surface in the form of a sphere that is concentric with the sphere of charge and passes through the point where the electric field is to be found. The field is uniform on the surface, so the flux through the surface is given by $\Phi = \oint \vec{E}\cdot d\vec{A} = 4\pi r^2 E = q_{\text{enc}}/\varepsilon_0$, where r is the radius of the Gaussian surface and q_{enc} is the charge enclosed. (i) In the region $r < r_1$, the enclosed charge is $q_{\text{enc}} = 0$ and therefore, $E = 0$. (ii) In the region $r_1 < r < r_2$, the volume of the shell is $(4\pi/3)(r_2^3 - r_1^3)$, so the charge density is

$$\rho = \frac{3Q}{4\pi(r_2^3 - r_1^3)},$$

where Q is the total charge on the spherical shell. Thus, the charge enclosed by the Gaussian surface is

$$q_{\text{enc}} = \left(\frac{4\pi}{3}\right)\left(r^3 - r_1^3\right)\rho = Q\left(\frac{r^3 - r_1^3}{r_2^3 - r_1^3}\right).$$

Gauss' law yields

$$4\pi\varepsilon_0 r^2 E = Q\left(\frac{r^3 - r_1^3}{r_2^3 - r_1^3}\right) \Rightarrow E = \frac{Q}{4\pi\varepsilon_0} \frac{r^3 - r_1^3}{r^2\left(r_2^3 - r_1^3\right)}.$$

(iii) In the region $r > r_2$, the charge enclosed is $q_{\text{enc}} = Q$, and the electric field is like that of a point charge:

$$E = \frac{1}{4\pi\varepsilon_0}\frac{Q}{r^2}.$$

ANALYZE (a) For $r > r_2$ the field is like that of a point charge, and so is the potential:

$$V = \frac{1}{4\pi\varepsilon_0}\frac{Q}{r},$$

where the potential was taken to be zero at infinity.

(b) In the region $r_1 < r < r_2$, we have

$$E = \frac{Q}{4\pi\varepsilon_0}\frac{r^3 - r_1^3}{r^2\left(r_2^3 - r_1^3\right)}.$$

If V_s is the electric potential at the outer surface of the shell ($r = r_2$) then the potential a distance r from the center is given by

$$V = V_s - \int_{r_2}^{r} E\,dr = V_s - \frac{Q}{4\pi\varepsilon_0}\frac{1}{r_2^3 - r_1^3}\int_{r_2}^{r}\left(r - \frac{r_1^3}{r^2}\right)dr$$

$$= V_s - \frac{Q}{4\pi\varepsilon_0}\frac{1}{r_2^3 - r_1^3}\left(\frac{r^2}{2} - \frac{r_2^2}{2} + \frac{r_1^3}{r} - \frac{r_1^3}{r_2}\right).$$

The potential at the outer surface is found by placing $r = r_2$ in the expression found in part (a). It is $V_s = Q/4\pi\varepsilon_0 r_2$. We make this substitution and collect terms to find

$$V = \frac{Q}{4\pi\varepsilon_0}\frac{1}{r_2^3 - r_1^3}\left(\frac{3r_2^2}{2} - \frac{r^2}{2} - \frac{r_1^3}{r}\right).$$

Since $\rho = 3Q/4\pi\left(r_2^3 - r_1^3\right)$ this can also be written as

$$V(r)=\frac{\rho}{3\varepsilon_0}\left(\frac{3r_2^2}{2}-\frac{r^2}{2}-\frac{r_1^3}{r}\right).$$

(c) For $r<r_1$, the electric field vanishes in the cavity, so the potential is everywhere the same inside and has the same value as at a point on the inside surface of the shell. We put $r = r_1$ in the result of part (b). After collecting terms the result is

$$V=\frac{Q}{4\pi\varepsilon_0}\frac{3(r_2^2-r_1^2)}{2(r_2^3-r_1^3)},$$

or in terms of the charge density $V=\frac{\rho}{2\varepsilon_0}(r_2^2-r_1^2)$.

(d) Using the expression for $V(r)$ found in (b), we have

$$V(r_1)=\frac{\rho}{3\varepsilon_0}\left(\frac{3r_2^2}{2}-\frac{r_1^2}{2}-\frac{r_1^3}{r_1}\right)=\frac{\rho}{3\varepsilon_0}\left(\frac{3r_2^2}{2}-\frac{3r_1^2}{2}\right)=\frac{\rho}{2\varepsilon_0}(r_2^2-r_1^2)$$

and

$$V(r_2)=\frac{\rho}{3\varepsilon_0}\left(\frac{3r_2^2}{2}-\frac{r_2^2}{2}-\frac{r_1^3}{r_2}\right)=\frac{\rho}{3\varepsilon_0}\left(r_2^2-\frac{r_1^3}{r_2}\right)=\frac{\rho}{3\varepsilon_0 r_2}(r_2^3-r_1^3)=\frac{3Q/4\pi}{3\varepsilon_0 r_2}=\frac{Q}{4\pi\varepsilon_0 r_2}.$$

So the solutions agree at $r = r_1$ and at $r = r_2$.

LEARN Electric potential must be continuous at the boundaries at $r = r_1$ and $r = r_2$. In the region where the electric field is zero, no work is required to move the charge around. Thus, there's no change in potential energy and the electric potential is constant.

23-97

THINK The increase in electric potential at the surface of the copper sphere is proportional to the increase in electric charge.

EXPRESS The electric potential at the surface of a sphere of radius R is given by $V=q/4\pi\varepsilon_0 R$, where q is the charge on the sphere. Thus, $q=4\pi\varepsilon_0 RV$. The number of electrons entering the copper sphere is $N=q/e$, but this must be equal to $(\lambda/2)t$, where λ is the decay rate of the nickel.

ANALYZE (a) With $R = 0.010$ m, when $V = 1000$ V, the net charge on the sphere is

$$q = 4\pi\varepsilon_0 RV = \frac{(0.010\text{ m})(1000\text{ V})}{8.99\times10^9\text{ N}\cdot\text{m}^2/\text{C}^2} = 1.11\times10^{-9}\text{ C}.$$

Dividing q by e yields $N = (1.11\times10^{-9}\text{ C})/(1.6\times10^{-19}\text{ C}) = 6.95\times10^9$ electrons that entered the copper sphere. So the time required is

$$t = \frac{N}{\lambda/2} = \frac{6.95\times10^9}{(3.7\times10^8\text{ /s})/2} = 38\text{ s}.$$

(b) The energy deposited by each electron that enters the sphere is $E_0 = 100\text{ keV} = 1.6\times10^{-14}$ J. Using the given heat capacity, we note that a temperature increase of $\Delta T = 5.0$ K = 5.0 ºC required

$$E = C\Delta T = (14\text{ J/K})(5.0\text{ K}) = 70\text{ J}$$

of energy. Dividing this by E_0 gives the number of electrons needed to enter the sphere (in order to achieve that temperature change):

$$N' = \frac{E}{E_0} = \frac{70\text{ J}}{1.6\times10^{-14}\text{ J}} = 4.375\times10^{15}$$

Thus, the time needed is

$$t' = \frac{N'}{\lambda/2} = \frac{4.375\times10^{15}}{(3.7\times10^8\text{ /s})/2} = 2.36\times10^7\text{ s}$$

or roughly 270 days.

LEARN As more electrons get into copper, more energy is deposited, and the copper sample gets hotter.

Chapter 25

25-3

THINK The capacitance of a parallel-plate capacitor is given by $C = \varepsilon_0 A/d$, where A is the area of each plate and d is the plate separation.

EXPRESS Since the plates are circular, the plate area is $A = \pi R^2$, where R is the radius of a plate. The charge on the positive plate is given by $q = CV$, where V is the potential difference across the plates.

ANALYZE (a) Substituting the values given, the capacitance is

$$C = \frac{\varepsilon_0 \pi R^2}{d} = \frac{(8.85\times10^{-12}\ \text{F/m})\pi(8.2\times10^{-2}\ \text{m})^2}{1.3\times10^{-3}\ \text{m}} = 1.44\times10^{-10}\ \text{F} = 144\ \text{pF}.$$

(b) Similarly, the charge on the plate when $V = 120$ V is

$$q = (1.44 \times 10^{-10}\ \text{F})(120\ \text{V}) = 1.73 \times 10^{-8}\ \text{C} = 17.3\ \text{nC}.$$

LEARN Capacitance depends only on geometric factors, namely, the plate area and plate separation.

25-13

THINK Charge remains conserved when a fully charged capacitor is connected to an uncharged capacitor.

EXPRESS The charge initially on the charged capacitor is given by $q = C_1V_0$, where $C_1 = 100$ pF is the capacitance and $V_0 = 50$ V is the initial potential difference. After the battery is disconnected and the second capacitor wired in parallel to the first, the charge on the first capacitor is $q_1 = C_1V$, where $V = 35$ V is the new potential difference. Since charge is conserved in the process, the charge on the second capacitor is $q_2 = q - q_1$, where C_2 is the capacitance of the second capacitor.

ANALYZE Substituting C_1V_0 for q and C_1V for q_1, we obtain $q_2 = C_1(V_0 - V)$. The potential difference across the second capacitor is also V, so the capacitance of the second capacitor is

$$C_2 = \frac{q_2}{V} = \frac{V_0 - V}{V} C_1 = \frac{50\,\text{V} - 35\,\text{V}}{35\,\text{V}} (100\,\text{pF}) = 42.86\ \text{pF} \approx 43\,\text{pF}.$$

LEARN Capacitors in parallel have the same potential difference. To verify charge conservation explicitly, we note that the initial charge on the first capacitor is $q = C_1 V_0 = (100\ \text{pF})(50\ \text{V}) = 5000\ \text{pC}$. After the connection, the charges on each capacitor are

$$q_1 = C_1 V = (100\ \text{pF})(35\ \text{V}) = 3500\ \text{pC}$$
$$q_2 = C_2 V = (42.86\ \text{pF})(35\ \text{V}) = 1500\ \text{pC}.$$

Indeed, $q = q_1 + q_2$.

25-21

THINK After the switches are closed, the potential differences across the capacitors are the same and they are connected in parallel.

EXPRESS The potential difference from a to b is given by $V_{ab} = Q/C_{eq}$, where Q is the net charge on the combination and C_{eq} is the equivalent capacitance.

ANALYZE (a) The equivalent capacitance is $C_{eq} = C_1 + C_2 = 4.0 \times 10^{-6}$ F. The total charge on the combination is the net charge on either pair of connected plates. The initial charge on capacitor 1 is

$$q_1 = C_1 V = (1.0\times10^{-6}\ \text{F})(100\,\text{V}) = 1.0\times10^{-4}\,\text{C}$$

and the initial charge on capacitor 2 is

$$q_2 = C_2 V = (3.0\times10^{-6}\,\text{F})(100\,\text{V}) = 3.0\times10^{-4}\,\text{C}.$$

With opposite polarities, the net charge on the combination is

$$Q = 3.0 \times 10^{-4}\ \text{C} - 1.0 \times 10^{-4}\ \text{C} = 2.0 \times 10^{-4}\ \text{C}.$$

The potential difference is

$$V_{ab} = \frac{Q}{C_{eq}} = \frac{2.0\times10^{-4}\,\text{C}}{4.0\times10^{-6}\,\text{F}} = 50\,\text{V}.$$

(b) The charge on capacitor 1 is now $q_1' = C_1 V_{ab} = (1.0 \times 10^{-6}\ \text{F})(50\ \text{V}) = 5.0 \times 10^{-5}\ \text{C}$.

(c) The charge on capacitor 2 is now $q_2' = C_2 V_{ab} = (3.0 \times 10^{-6}\ \text{F})(50\ \text{V}) = 1.5 \times 10^{-4}\ \text{C}$.

LEARN The potential difference $V_{ab} = 50$ V is half of the original V (= 100 V), so the final charges on the capacitors are also halved.

25-31

THINK The total electrical energy is the sum of the energies stored in the individual capacitors.

EXPRESS The energy stored in a charged capacitor is

$$U = \frac{q^2}{2C} = \frac{1}{2}CV^2.$$

Since we have two capacitors that are connected in parallel, the potential difference V across the capacitors is the same and the total energy is

$$U_{\text{tot}} = U_1 + U_2 = \frac{1}{2}(C_1 + C_2)V^2.$$

ANALYZE Substituting the values given, we have

$$U = \frac{1}{2}(C_1 + C_2)V^2 = \frac{1}{2}\left(2.0\times10^{-6}\,\text{F} + 4.0\times10^{-6}\,\text{F}\right)(300\,\text{V})^2 = 0.27\,\text{J}.$$

LEARN The energy stored in a capacitor is equal to the amount of work required to charge the capacitor.

25-37

THINK The potential difference between the plates of a parallel-plate capacitor depends on their distance of separation.

EPXRESS Let q be the charge on the positive plate. Since the capacitance of a parallel-plate capacitor is given by $C_i = \varepsilon_0 A/d_i$, the charge is $q_i = C_iV_i = \varepsilon_0 AV_i/d_i$. After the plates are pulled apart, their separation is d_f and the final potential difference is V_f. Thus, the final charge is $q_f = \varepsilon_0 AV_f/2d_f$. Since charge remains unchanged, $q_i = q_f$, we have

$$V_f = \frac{q_f}{C_f} = \frac{d_f}{\varepsilon_0 A}q_f = \frac{d_f}{\varepsilon_0 A}\frac{\varepsilon_0 A}{d_i}V_i = \frac{d_f}{d_i}V_i.$$

ANALYZE (a) With $d_i = 3.00\times10^{-3}\,\text{m}$, $V_i = 6.00\ \text{V}$ and $d_f = 8.00\times10^{-3}\,\text{m}$, the final potential difference is $V_f = 16.0\ \text{V}$.

(b) The initial energy stored in the capacitor is

$$U_i = \frac{1}{2}CV_i^2 = \frac{\varepsilon_0 AV_i^2}{2d_i} = \frac{(8.85\times10^{-12}\ \text{C}^2/\text{N}\cdot\text{m}^2)(8.50\times10^{-4}\ \text{m}^2)(6.00\ \text{V})^2}{2(3.00\times10^{-3}\ \text{m})}$$
$$= 4.51\times10^{-11}\ \text{J}.$$

(c) The final energy stored is

$$U_f = \frac{1}{2}C_f V_f^2 = \frac{1}{2}\frac{\varepsilon_0 A}{d_f}V_f^2 = \frac{1}{2}\frac{\varepsilon_0 A}{d_f}\left(\frac{d_f}{d_i}V_i\right)^2 = \frac{d_f}{d_i}\left(\frac{\varepsilon_0 AV_i^2}{d_i}\right) = \frac{d_f}{d_i}U_i.$$

With $d_f/d_i = 8.00/3.00$, we have $U_f = 1.20\times10^{-10}$ J.

(d) The work done to pull the plates apart is the difference in the energy:

$$W = U_f - U_i = 7.52\times10^{-11}\ \text{J}.$$

LEARN In a parallel-plate capacitor, the energy density (energy per unit volume) is given by $u = \varepsilon_0 E^2/2$, where E is constant at all points between the plates. Thus, increasing the plate separation increases the volume (= Ad), and hence the total energy of the system.

25-41

THINK Our system, a coaxial cable, is a cylindrical capacitor filled with polystyrene, a dielectric.

EXPRESS The capacitance of a cylindrical capacitor can be written as

$$C = \kappa C_0 = \frac{2\pi\kappa\varepsilon_0 L}{\ln(b/a)},$$

where C_0 is the capacitance without the dielectric, κ is the dielectric constant, L is the length, a is the inner radius, and b is the outer radius.

ANALYZE With $\kappa = 2.6$ for polystyrene, the capacitance per unit length of the cable is

$$\frac{C}{L} = \frac{2\pi\kappa\varepsilon_0}{\ln(b/a)} = \frac{2\pi(2.6)(8.85\times10^{-12}\ \text{F/m})}{\ln[(0.60\ \text{mm})/(0.10\ \text{mm})]} = 8.1\times10^{-11}\,\text{F/m} = 81\ \text{pF/m}.$$

LEARN When the space between the plates of a capacitor is completely filled with a dielectric material, the capacitor increases by a factor κ, the dielectric constant characteristic of the material.

25-47

THINK Dielectric strength is the maximum value of the electric field a dielectric material can tolerate without breakdown.

EXPRESS The capacitance is given by $C = \kappa C_0 = \kappa\varepsilon_0 A/d$, where C_0 is the capacitance without the dielectric, κ is the dielectric constant, A is the plate area, and d is the plate separation. The electric field between the plates is given by $E = V/d$, where V is the potential difference between the plates. Thus, $d = V/E$ and $C = \kappa\varepsilon_0 AE/V$. Therefore, we find the plate area to be

$$A = \frac{CV}{\kappa\varepsilon_0 E}.$$

ANALYZE For the area to be a minimum, the electric field must be the greatest it can be without breakdown occurring. That is,

$$A = \frac{(7.0\times10^{-8}\,\text{F})(4.0\times10^{3}\,\text{V})}{2.8(8.85\times10^{-12}\,\text{F/m})(18\times10^{6}\ \text{V/m})} = 0.63\ \text{m}^2.$$

LEARN If the area is smaller than the minimum value found above, then electric breakdown occurs and the dielectric is no longer insulating and will start to conduct.

25-51

THINK We have a parallel-plate capacitor, so the capacitance is given by $C = \kappa C_0 = \kappa\varepsilon_0 A/d$, where C_0 is the capacitance without the dielectric, κ is the dielectric constant, A is the plate area, and d is the plate separation.

EXPRESS The electric field in the region between the plates is given by $E = V/d$, where V is the potential difference between the plates and d is the plate separation. Since the separation can be written as $d = \kappa\varepsilon_0 A/C$, we have $E = VC/\kappa\varepsilon_0 A$. The free charge on the plates is $q_f = CV$.

ANALYZE (a) Substituting the values given, we find the magnitude of the field strength to be

$$E = \frac{VC}{\kappa\varepsilon_0 A} = \frac{(50\ \text{V})(100\times10^{-12}\ \text{F})}{5.4(8.85\times10^{-12}\ \text{F/m})(100\times10^{-4}\ \text{m}^2)} = 1.0\times10^{4}\ \text{V/m}.$$

(b) Similarly, we have $q_f = CV = (100 \times 10^{-12}\ \text{F})(50\ \text{V}) = 5.0 \times 10^{-9}\ \text{C}$.

(c) The electric field is produced by both the free and induced charge. Since the field of a large uniform layer of charge is $q/2\varepsilon_0 A$, the field between the plates is

$$E=\frac{q_f}{2\varepsilon_0 A}+\frac{q_f}{2\varepsilon_0 A}-\frac{q_i}{2\varepsilon_0 A}-\frac{q_i}{2\varepsilon_0 A},$$

where the first term is due to the positive free charge on one plate, the second is due to the negative free charge on the other plate, the third is due to the positive induced charge on one dielectric surface, and the fourth is due to the negative induced charge on the other dielectric surface. Note that the field due to the induced charge is opposite the field due to the free charge, so they tend to cancel. The induced charge is therefore

$$\begin{aligned} q_i &= q_f-\varepsilon_0 AE = 5.0\times10^{-9}\,\text{C}-\left(8.85\times10^{-12}\ \text{F/m}\right)\left(100\times10^{-4}\ \text{m}^2\right)\left(1.0\times10^{4}\ \text{V/m}\right) \\ &= 4.1\times10^{-9}\,\text{C} = 4.1\,\text{nC}. \end{aligned}$$

LEARN An alternative way to calculate the induced charge is to apply Eq. 25-35:

$$q_i = q_f\left(1-\frac{1}{\kappa}\right)=(5.0\,\text{nC})\left(1-\frac{1}{5.4}\right)=4.1\,\text{nC}.$$

Note that there's no induced charge ($q_i=0$) in the absence of dielectric ($\kappa=1$).

25-57

THINK The problem consists of a system of capacitors. The pair C_3 and C_4 are in parallel.

EXPRESS Since C_3 and C_4 are in parallel, we replace them with an equivalent capacitance $C_{34}=C_3+C_4=30\ \mu\text{F}$. Now, C_1, C_2, and C_{34} are in series, and all are numerically 30 μF, we observe that each has one-third the battery voltage across it. Hence, 3.0 V is across C_4.

ANALYZE The charge on capacitor 4 is $q_4 = C_4V_4 = (15\ \mu\text{F})(3.0\ \text{V}) = 45\ \mu\text{C}$.

LEARN Alternatively, one may show that the equivalent capacitance of the arrangement is given by

$$\frac{1}{C_{1234}}=\frac{1}{C_1}+\frac{1}{C_2}+\frac{1}{C_{34}}=\frac{1}{30\ \mu\text{F}}+\frac{1}{30\ \mu\text{F}}+\frac{1}{30\ \mu\text{F}}=\frac{1}{10\ \mu\text{F}}$$

or $C_{1234}=10\ \mu\text{F}$. Thus, the charge across C_1, C_2, and C_{34} are

$$q_1=q_2=q_{34}=q_{1234}=C_{1234}V=(10\ \mu\text{F})(9.0\ \text{V})=90\ \text{nC}.$$

Now, since C_3 and C_4 are in parallel, and $C_3=C_4$, the charge on C_4 (as well as on C_3) is $q_3=q_4=q_{34}/2=(90\ \mu\text{F})/2=45\ \mu\text{F}$.

25-65

THINK We may think of the arrangement as two capacitors connected in series.

EXPRESS Let the capacitances be C_1 and C_2, with the former filled with the $\kappa_1 = 3.00$ material and the latter with the κ_2 = 4.00 material. Upon using Eq. 25-9, Eq. 25-27, and reducing C_1 and C_2 to an equivalent capacitance, we have

$$\frac{1}{C_{eq}} = \frac{1}{C_1} + \frac{1}{C_2} = \frac{1}{\kappa_1 \varepsilon_0 A/d} + \frac{1}{\kappa_2 \varepsilon_0 A/d} = \left(\frac{\kappa_1 + \kappa_2}{\kappa_1 \kappa_2}\right)\frac{d}{\varepsilon_0 A}$$

or $C_{eq} = \left(\frac{\kappa_1 \kappa_2}{\kappa_1 + \kappa_2}\right)\frac{\varepsilon_0 A}{d}$. The charge stored on the capacitor is $q = C_{eq}V$.

ANALYZE Substituting the values given, we find

$$C_{eq} = \left(\frac{\kappa_1 \kappa_2}{\kappa_1 + \kappa_2}\right)\frac{\varepsilon_0 A}{d} = 1.52 \times 10^{-10} \text{ F},$$

Therefore, $q = C_{eq}V = 1.06 \times 10^{-9}$ C.

LEARN In the limit where $\kappa_1 = \kappa_2 = \kappa$, our expression for C_{eq} becomes $C_{eq} = \frac{\kappa \varepsilon_0 A}{2d}$, where $2d$ is the plate separation.

25-77

THINK We have two parallel-plate capacitors that are connected in parallel. They both have the same plate separation and same potential difference across their plates.

EXPRESS The magnitude of the electric field in the region between the plates is given by $E = V/d$, where V is the potential difference between the plates and d is the plate separation. The surface charge density on the plate is $\sigma = q/A$.

ANALYZE (a) With d = 0.00300 m and V = 600 V, we have

$$E_A = \frac{V}{d} = \frac{600 \text{ V}}{3.00 \times 10^{-3} \text{ m}} = 2.00 \times 10^5 \text{ V/m}.$$

(b) Since d = 0.00300 m and V = 600 V in capacitor B as well, $E_B = 2.00 \times 10^5$ V/m.

(c) For the air-filled capacitor, Eq. 25-4 leads to

$$\sigma_A = \frac{q_A}{A} = \frac{C_A V}{A} = \frac{(\varepsilon_0 A/d)V}{A} = \frac{\varepsilon_0 V}{d} = \varepsilon_0 E_A = (8.85\times10^{-12}\ \mathrm{C^2/N\cdot m^2})(2.00\times10^5\ \mathrm{V/m})$$
$$= 1.77\times10^{-6}\ \mathrm{C/m^2}.$$

(d) For the dielectric-filled capacitor, we use Eq. 25-29:

$$\sigma_B = \kappa\varepsilon_0 E_B = (2.60)(8.85\times10^{-12}\ \mathrm{C^2/N\cdot m^2})(2.00\times10^5\ \mathrm{V/m}) = 4.60\times10^{-6}\ \mathrm{C/m^2}.$$

(e) Although the discussion in Section 25-6 of the textbook is in terms of the charge being held fixed (while a dielectric is inserted), it is readily adapted to this situation (where comparison is made of two capacitors that have the same *voltage* and are identical except for the fact that one has a dielectric). The fact that capacitor B has a relatively large charge but only produces the field that A produces (with its smaller charge) is in line with the point being made (in the text) with Eq. 25-34 and in the material that follows. Adapting Eq. 25-35 to this problem, we see that the difference in charge densities between parts (c) and (d) is due, in part, to the (negative) layer of charge at the top surface of the dielectric; consequently,

$$\sigma_{\text{ind}} = \sigma_A - \sigma_B = (1.77\times10^{-6}\ \mathrm{C/m^2}) - (4.60\times10^{-6}\ \mathrm{C/m^2}) = -2.83\times10^{-6}\ \mathrm{C/m^2}.$$

LEARN We note that the electric field in capacitor B is produced by both the charge on the plates ($\sigma_B A$) and the induced charges ($\sigma_{\text{ind}} A$), while the field in capacitor A is produced by the charge on the plates alone ($\sigma_A A$). Since $E_A = E_B$, we have $\sigma_A = \sigma_B + \sigma_{\text{ind}}$, or $\sigma_{\text{ind}} = \sigma_A - \sigma_B$.

Chapter 26

26-5

THINK The magnitude of the current density is given by $J = nqv_d$, where n is the number of particles per unit volume, q is the charge on each particle, and v_d is the drift speed of the particles.

EXPRESS In vector form, we have (see Eq. 26-7) $\vec{J} = nq\vec{v}_d$. Current density $\vec{J}$ is related to the current i by (see Eq. 26-4): $i = \int \vec{J} \cdot d\vec{A}$.

ANALYZE (a) The particle concentration is n = 2.0 × 10^8/cm^3 = 2.0 × 10^{14} m^{-3}, the charge is

$$q = 2e = 2(1.60 \times 10^{-19}\ \text{C}) = 3.20 \times 10^{-19}\ \text{C},$$

and the drift speed is 1.0 × 10^5 m/s. Thus, we find the current density to be

$$J = \left(2\times10^{14}\ /\ \text{m}\right)\left(3.2\times10^{-19}\ \text{C}\right)\left(1.0\times10^{5}\ \text{m}/\text{s}\right) = 6.4\ \text{A}/\text{m}^2.$$

(b) Since the particles are positively charged the current density is in the same direction as their motion, to the north.

(c) The current cannot be calculated unless the cross-sectional area of the beam is known. Then $i = JA$ can be used.

LEARN That the current density is in the direction of the motion of the *positive* charge carriers means that it is in the opposite direction of the motion of the negatively charged electrons.

26-15

THINK The resistance of the coil is given by $R = \rho L/A$, where L is the length of the wire, ρ is the resistivity of copper, and A is the cross-sectional area of the wire.

EXPRESS Since each turn of wire has length $2\pi r$, where r is the radius of the coil, then

$$L = (250)2\pi r = (250)(2\pi)(0.12\ \text{m}) = 188.5\ \text{m}.$$

If r_w is the radius of the wire itself, then its cross-sectional area is

$$A = \pi r_w^2 = \pi(0.65 \times 10^{-3}\ \text{m})^2 = 1.33 \times 10^{-6}\ \text{m}^2.$$

According to Table 26-1, the resistivity of copper is $\rho = 1.69 \times 10^{-8}\ \Omega \cdot \text{m}$.

ANALYZE Thus, the resistance of the copper coil is

$$R = \frac{\rho L}{A} = \frac{(1.69 \times 10^{-8}\ \Omega \cdot \text{m})(188.5\ \text{m})}{1.33 \times 10^{-6}\ \text{m}^2} = 2.4\ \Omega.$$

LEARN Resistance R is the property of an object (depending on quantities such as L and A), while resistivity is a property of the material.

26-19

THINK The resistance of the wire is given by $R = \rho L / A$, where ρ is the resistivity of the material, L is the length of the wire, and A is its cross-sectional area.

EXPRESS In this case, the cross-sectional area is

$$A = \pi r^2 = \pi\left(0.50 \times 10^{-3}\ \text{m}\right)^2 = 7.85 \times 10^{-7}\ \text{m}^2.$$

ANALYZE Thus, the resistivity of the wire is

$$\rho = \frac{RA}{L} = \frac{\left(50 \times 10^{-3}\ \Omega\right)\left(7.85 \times 10^{-7}\ \text{m}^2\right)}{2.0\ \text{m}} = 2.0 \times 10^{-8}\ \Omega \cdot \text{m}.$$

LEARN Resistance R is the property of an object (depending on quantities such as L and A), while resistivity is a property of the material itself. The equation $R = \rho L / A$ implies that the larger the cross-sectional area A, the smaller the resistance R.

26-25

THINK The resistance of an object depends on its length and the cross-sectional area.

EXPRESS Since the mass and density of the material do not change, the volume remains the same. If L_0 is the original length, L is the new length, A_0 is the original cross-sectional area, and A is the new cross-sectional area, then $L_0 A_0 = LA$ and

$$A = L_0 A_0 / L = L_0 A_0 / 3L_0 = A_0/3.$$

ANALYZE The new resistance is

$$R = \frac{\rho L}{A} = \frac{\rho 3L_0}{A_0/3} = 9\frac{\rho L_0}{A_0} = 9R_0,$$

where R_0 is the original resistance. Thus, $R = 9(6.0\ \Omega) = 54\ \Omega$.

LEARN In general, the resistances of two objects made of the same material but different cross-sectional areas and lengths may be related by

$$R_2 = R_1\left(\frac{A_1}{A_2}\right)\left(\frac{L_2}{L_1}\right).$$

26-27

THINK In this problem we compare the resistances of two conductors that are made of the same materials.

EXPRESS The resistance of conductor A is given by

$$R_A = \frac{\rho L}{\pi r_A^2},$$

where r_A is the radius of the conductor. If r_o is the outside diameter of conductor B and r_i is its inside diameter, then its cross-sectional area is $\pi(r_o^2 - r_i^2)$, and its resistance is

$$R_B = \frac{\rho L}{\pi\left(r_o^2 - r_i^2\right)}.$$

ANALYZE The ratio of the resistances is

$$\frac{R_A}{R_B} = \frac{r_o^2 - r_i^2}{r_A^2} = \frac{(1.0\,\text{mm})^2 - (0.50\,\text{mm})^2}{(0.50\,\text{mm})^2} = 3.0.$$

LEARN The resistance R of an object depends on how the electric potential is applied to the object. Also, R depends on the ratio L/A, according to $R = \rho L/A$.

26-41

THINK In an electrical circuit, the electrical energy is dissipated through the resistor as heat.

EXPRESS Electrical energy is converted to heat at a rate given by $P = V^2/R$, where V is the potential difference across the heater and R is the resistance of the heater.

ANALYZE With $V = 120$ V and $R = 14\ \Omega$, we have

$$P = \frac{(120\ \text{V})^2}{14\ \Omega} = 1.0\times 10^3\ \text{W} = 1.0\ \text{kW}.$$

(b) The cost is given by $(1.0\,\text{kW})(5.0\,\text{h})(5.0\,\text{cents/kW}\cdot\text{h}) = \text{US\$}0.25$.

LEARN The energy transferred is lost because the process is irreversible. The thermal energy causes the temperature of the resistor to rise.

26-45

THINK Let P be the power dissipated, i be the current in the heater, and V be the potential difference across the heater. The three quantities are related by $P = iV$.

EXPRESS The current is given by $i = P/V$. Using Ohm's law $V = iR$, the resistance of the heater can be written as

$$R = \frac{V}{i} = \frac{V}{P/V} = \frac{V^2}{P}.$$

ANALYZE (a) Substituting the values given, we have

$$i = \frac{P}{V} = \frac{1250\ \text{W}}{115\ \text{V}} = 10.9\ \text{A}.$$

(b) Similarly, the resistance is

$$R = \frac{V^2}{P} = \frac{(115\ \text{V})^2}{1250\ \text{W}} = 10.6\ \Omega.$$

(c) The thermal energy E generated by the heater in time $t = 1.0\ \text{h} = 3600\ \text{s}$ is

$$E = Pt = (1250\,\text{W})(3600\,\text{s}) = 4.50\times 10^6\ \text{J}.$$

LEARN Current in the heater produces a transfer of mechanical energy to thermal energy, with a rate of the transfer equal to $P = iV = V^2/R$.

26-51

THINK Our system is made up of two wires that are joined together. To calculate the electrical potential difference between two points, we first calculate their resistances.

EXPRESS The potential difference between points 1 and 2 is $\Delta V_{12} = iR_C$, where R_C is the resistance of wire C. Similarly, the potential difference between points 2 and 3 is

$\Delta V_{23} = iR_D$, where R_D is the resistance of wire D. The corresponding rates of energy dissipation are $P_{12} = i^2 R_C$ and $P_{23} = i^2 R_D$, respectively.

ANALYZE (a) Using Eq. 26-16, we find the resistance of wire C to be

$$R_C = \rho_C \frac{L_C}{\pi r_C^2} = (2.0\times10^{-6}\,\Omega\cdot\text{m})\frac{1.0\text{ m}}{\pi(0.00050\text{ m})^2} = 2.55\,\Omega.$$

Thus, $\Delta V_{12} = iR_C = (2.0\text{ A})(2.55\,\Omega) = 5.1\text{V}$.

(b) Similarly, the resistance for wire D is

$$R_D = \rho_D \frac{L_D}{\pi r_D^2} = (1.0\times10^{-6}\,\Omega\cdot\text{m})\frac{1.0\text{ m}}{\pi(0.00025\text{ m})^2} = 5.09\,\Omega$$

and the potential difference is $\Delta V_{23} = iR_D = (2.0\text{ A})(5.09\,\Omega) = 10.2\text{V} \approx 10\text{V}$.

(c) The power dissipated between points 1 and 2 is $P_{12} = i^2 R_C = 10\text{W}$.

(d) Similarly, the power dissipated between points 2 and 3 is $P_{23} = i^2 R_D = 20\text{W}$.

LEARN The results may be summarized in terms of the following ratios:

$$\frac{P_{23}}{P_{12}} = \frac{\Delta V_{23}}{\Delta V_{12}} = \frac{R_D}{R_C} = \frac{\rho_D}{\rho_C}\cdot\frac{L_D}{L_C}\cdot\left(\frac{r_C}{r_D}\right)^2 = \frac{1}{2}\cdot 1\cdot(2)^2 = 2.$$

26-55

THINK Since the resistivity of Nichrome varies with temperature, the power dissipated through the Nichrome wire will also depend on temperature.

EXPRESS Let R_H be the resistance at the higher temperature (800°C) and let R_L be the resistance at the lower temperature (200°C). Since the potential difference is the same for the two temperatures, the power dissipated at the lower temperature is $P_L = V^2/R_L$, and the power dissipated at the higher temperature is $P_H = V^2/R_H$, so $P_L = (R_H/R_L)P_H$. Now,

$$R_H = \frac{\rho_H L}{A} = \frac{\rho_0 L}{A}[1+\alpha(T_H - T_0)]$$

$$R_L = \frac{\rho_L L}{A} = \frac{\rho_0 L}{A}[1+\alpha(T_L - T_0)]$$

so that

$$R_L = R_H + \alpha R_H \Delta T,$$

where ΔT is the temperature difference: $T_L - T_H = -600$ C° = –600 K.

ANALYZE Thus, the dissipation rate at 200°C is

$$P_L = \frac{R_H}{R_H + \alpha R_H \Delta T} P_H = \frac{P_H}{1+\alpha\Delta T} = \frac{500\text{ W}}{1+(4.0\times10^{-4}\text{ / K})(-600\text{ K})} = 660\text{ W}.$$

LEARN Since the power dissipated is inversely proportional to R, at lower temperature where $R_L < R_H$, we expect a higher rate of energy dissipation: $P_L > P_H$.

26-61

THINK The amount of charge that strikes the surface in time Δt is given by $\Delta q = i\,\Delta t$, where i is the current.

EXPRESS Since each alpha particle carries charge $q = +2e$, the number of particles that strike the surface is

$$N = \frac{\Delta q}{2e} = \frac{i\Delta t}{2e}.$$

For part (b), let N' be the number of particles in a length L of the beam. They will all pass through the beam cross section at one end in time $t = L/v$, where v is the particle speed. The current is the charge that moves through the cross section per unit time. That is,

$$i = \frac{2eN'}{t} = \frac{2eN'v}{L}.$$

Thus $N' = iL/2ev$.

ANALYZE (a) Substituting the values given, we have

$$N = \frac{\Delta q}{2e} = \frac{i\Delta t}{2e} = \frac{(0.25\times10^{-6}\text{ A})(3.0\text{ s})}{2(1.6\times10^{-19}\text{ C})} = 2.34\times10^{12}.$$

(b) To find the particle speed, we note the kinetic energy of a particle is

$$K = 20\text{ MeV} = (20\times10^{6}\text{ eV})(1.60\times10^{-19}\text{ J / eV}) = 3.2\times10^{-12}\text{ J}.$$

Since $K = \frac{1}{2}mv^2$, the speed is $v = \sqrt{2K/m}$. The mass of an alpha particle is (very nearly) 4 times the mass of a proton, or

$$m = 4(1.67\times10^{-27}\text{ kg}) = 6.68\times10^{-27}\text{ kg},$$

so

$$v = \sqrt{\frac{2(3.2\times10^{-12}\ \text{J})}{6.68\times10^{-27}\ \text{kg}}} = 3.1\times10^{7}\ \text{m/s}.$$

Therefore, the number of particles in a length L = 20 cm of the beam is

$$N' = \frac{iL}{2ev} = \frac{(0.25\times10^{-6})(20\times10^{-2}\ \text{m})}{2(1.60\times10^{-19}\ \text{C})(3.1\times10^{7}\ \text{m/s})} = 5.0\times10^{3}.$$

(c) We use conservation of energy, where the initial kinetic energy is zero and the final kinetic energy is 20 MeV = 3.2×10^{-12} J. We note too, that the initial potential energy is

$$U_i = qV = 2eV,$$

and the final potential energy is zero. Here V is the electric potential through which the particles are accelerated. Consequently, $K_f = U_i = 2eV$, which gives

$$V = \frac{K_f}{2e} = \frac{3.2\times10^{-12}\ \text{J}}{2(1.60\times10^{-19}\ \text{C})} = 1.0\times10^{7}\ \text{V}.$$

LEARN By the work-kinetic energy theorem, the work done on 2.34×10^{12} such alpha particles is

$$W = (2.34\times10^{12})(20\ \text{MeV}) = (2.34\times10^{12})(3.2\times10^{-12}\ \text{J}) = 7.5\ \text{J}.$$

The same result can also be obtained from

$$W = q\Delta V = (i\Delta t)\Delta V = (0.25\times10^{-6}\ \text{A})(3.0\ \text{s})(1.0\times10^{7}\ \text{V}) = 7.5\ \text{J}.$$

26-71

THINK The resistance of copper increases with temperature.

EXPRESS According to Eq. 26-17, the resistance of copper at temperature T can be written as

$$R = \frac{\rho L}{A} = \frac{\rho_0 L}{A}[1+\alpha(T-T_0)]$$

where $T_0 = 20\ °\text{C}$ is the reference temperature. Thus, the resistance is $R_0 = \rho_0 L/A$ at $T_0 = 20\ °\text{C}$. The temperature at which $R = 2R_0$ (or equivalently, $\rho = 2\rho_0$) can be found by solving

$$2 = \frac{R}{R_0} = 1 + \alpha(T - T_0) \;\Rightarrow\; \alpha(T - T_0) = 1.$$

ANALYZE (a) From the above equation, we find the temperature to be

$$T = T_0 + \frac{1}{\alpha} = 20°\,\text{C} + \frac{1}{4.3\times 10^{-3}\,/\,\text{K}} \approx 250°\,\text{C}.$$

(b) Since a change in Celsius is equivalent to a change on the Kelvin temperature scale, the value of α used in this calculation is not inconsistent with the other units involved.

LEARN It is worth noting that our result agrees well with Fig. 26-10.

Chapter 27

27-1

THINK The circuit consists of two batteries and two resistors. We apply Kirchhoff's loop rule to solve for the current.

EXPRESS Let i be the current in the circuit and take it to be positive if it is to the left in R_1. Kirchhoff's loop rule gives

$$\varepsilon_1 - iR_2 - iR_1 - \varepsilon_2 = 0.$$

For parts (b) and (c), we note that if i is the current in a resistor R, then the power dissipated by that resistor is given by $P = i^2 R$.

ANALYZE (a) We solve for i:

$$i = \frac{\varepsilon_1 - \varepsilon_2}{R_1 + R_2} = \frac{12\ \text{V} - 6.0\ \text{V}}{4.0\Omega + 8.0\Omega} = 0.50\ \text{A}.$$

A positive value is obtained, so the current is counterclockwise around the circuit.

(b) For R_1, the dissipation rate is $P_1 = i^2 R_1 = (0.50\ \text{A})^2(4.0\ \Omega) = 1.0\ \text{W}$.

(c) For R_2, the rate is $P_2 = i^2 R_2 = (0.50\ \text{A})^2\ (8.0\ \Omega) = 2.0\ \text{W}$.

If i is the current in a battery with emf ε, then the battery supplies energy at the rate $P = i\varepsilon$ provided the current and emf are in the same direction. On the other hand, the battery absorbs energy at the rate $P = i\varepsilon$ if the current and emf are in opposite directions.

(d) For ε_1, $P_1 = i\varepsilon_1 = (0.50\ \text{A})(12\ \text{V}) = 6.0\ \text{W}$.

(e) For ε_2, $P_2 = i\varepsilon_2 = (0.50\ \text{A})(6.0\ \text{V}) = 3.0\ \text{W}$.

(f) In battery 1 the current is in the same direction as the emf. Therefore, this battery supplies energy to the circuit; the battery is discharging.

(g) The current in battery 2 is opposite the direction of the emf, so this battery absorbs energy from the circuit. It is charging.

LEARN Multiplying the equation obtained from Kirchhoff's loop rule by *idt* leads to the "energy-method" equation discussed in the book:

$$i\varepsilon_1 dt - i^2 R_1 dt - i^2 R_2 dt - i\varepsilon_2 dt = 0.$$

The first term represents the rate of work done by battery 1, the second and third terms the thermal energies that appear in resistors R_1 and R_2, and the last term the work done *on* battery 2.

27-11

THINK As shown in Fig. 27-29, the circuit contains an emf device *X*. How it is connected to the rest of the circuit can be deduced from the power dissipated and the potential drop across it.

EXPRESS The power absorbed by a circuit element is given by $P = i\Delta V$, where i is the current and ΔV is the potential difference across the element. The end-to-end potential difference is given by

$$V_A - V_B = +iR + \varepsilon,$$

where ε is the emf of device *X* and is taken to be positive if it is to the left in the diagram.

ANALYZE (a) The potential difference between *A* and *B* is

$$\Delta V = \frac{P}{i} = \frac{50 \text{ W}}{1.0 \text{ A}} = 50 \text{ V}.$$

Since the energy of the charge decreases, point *A* is at a higher potential than point *B*; that is, $V_A - V_B = 50$ V.

(b) From the equation above, we find the emf of device *X* to be

$$\varepsilon = V_A - V_B - iR = 50 \text{ V} - (1.0 \text{ A})(2.0 \ \Omega) = 48 \text{ V}.$$

(c) A positive value was obtained for ε, so it is toward the left. The negative terminal is at *B*.

LEARN Writing the potential difference as $V_A - iR - \varepsilon = V_B$, we see that our result is consistent with the resistance and emf rules. Namely, starting at point *A*, the change in potential is $-iR$ for a move through a resistance *R* in the direction of the current, and the change in potential is $-\varepsilon$ for a move through an emf device in the opposite direction of the emf arrow (which points from negative to positive terminals).

27-17

THINK A zero terminal-to-terminal potential difference implies that the emf of the battery is equal to the voltage drop across its internal resistance, that is, $\varepsilon = ir$.

EXPRESS To be as general as possible, we refer to the individual emf's as ε_1 and ε_2 and wait until the latter steps to equate them ($\varepsilon_1 = \varepsilon_2 = \varepsilon$). The batteries are placed in series in such a way that their voltages add; that is, they do not "oppose" each other. The total resistance in the circuit is therefore $R_{total} = R + r_1 + r_2$ (where the problem tells us $r_1 > r_2$), and the "net emf" in the circuit is $\varepsilon_1 + \varepsilon_2$. Since battery 1 has the higher internal resistance, it is the one capable of having a zero terminal voltage, as the computation in part (a) shows.

ANALYZE (a) The current in the circuit is

$$i = \frac{\varepsilon_1 + \varepsilon_2}{r_1 + r_2 + R},$$

and the requirement of zero terminal voltage leads to $\varepsilon_1 = ir_1$, or

$$R = \frac{\varepsilon_2 r_1 - \varepsilon_1 r_2}{\varepsilon_1} = \frac{(12.0\text{ V})(0.016\,\Omega) - (12.0\text{ V})(0.012\,\Omega)}{12.0\text{ V}} = 0.0040\ \Omega.$$

Note that $R = r_1 - r_2$ when we set $\varepsilon_1 = \varepsilon_2$.

(b) As mentioned above, this occurs in battery 1.

LEARN If we assume the potential difference across battery 2 to be zero and repeat the calculation above, we would find $R = r_2 - r_1 < 0$, which is physically impossible. Thus, only the potential difference across the battery with the larger internal resistance can be made zero with suitable choice of R.

27-25

THINK The resistance of a copper wire varies with its cross-sectional area, or its diameter.

EXPRESS Let r be the resistance of each of the narrow wires. Since they are in parallel the equivalent resistance R_{eq} of the composite is given by

$$\frac{1}{R_{eq}} = \frac{9}{r},$$

or $R_{eq} = r/9$. Now each thin wire has a resistance $r = 4\rho\ell / \pi d^2$, where ρ is the resistivity of copper, and $A = \pi d^2/4$ is the cross-sectional area of a single thin wire. On the other hand, the resistance of the thick wire of diameter D is $R = 4\rho\ell/\pi D^2$, where the cross-sectional area is $\pi D^2/4$.

ANALYZE If the single thick wire is to have the same resistance as the composite of 9 thin wires, $R = R_{eq}$, then

$$\frac{4\rho\ell}{\pi D^2} = \frac{4\rho\ell}{9\pi d^2}.$$

Solving for D, we obtain $D = 3d$.

LEARN The equivalent resistance R_{eq} is smaller than r by a factor of 9. Since $r \sim 1/A \sim 1/d^2$, increasing the diameter of the wire threefold will also reduce the resistance by a factor of 9.

27-31

THINK This problem involves a multi-loop circuit. We first simplify the circuit by finding the equivalent resistance. We then apply Kirchhoff's loop rule to calculate the current in the loop, and the potentials at various points in the circuit.

EXPRESS We first reduce the parallel pair of identical 2.0-Ω resistors (on the right side) to $R' = 1.0\ \Omega$, and we reduce the series pair of identical 2.0-Ω resistors (on the upper left side) to $R'' = 4.0\ \Omega$. With R denoting the 2.0-Ω resistor at the bottom (between V_2 and V_1), we now have three resistors in series which are equivalent to

$$R_{eq} = R + R' + R'' = 7.0\ \Omega$$

across which the voltage is $\varepsilon_2 - \varepsilon_1 = 7.0$ V (by the loop rule, this is 12 V – 5.0 V), implying that the current is

$$i = \frac{\varepsilon_2 - \varepsilon_1}{R_{eq}} = \frac{7.0\ \text{V}}{7.0\ \Omega} = 1.0\ \text{A}.$$

The direction of i is upward in the right-hand emf device. Knowing i allows us to solve for V_1 and V_2.

ANALYZE (a) The voltage across R' is (1.0 A)(1.0 Ω) = 1.0 V, which means that (examining the right side of the circuit) the voltage difference between *ground* and V_1 is 12 V – 1.0 V = 11 V. Noting the orientation of the battery, we conclude that $V_1 = -11$ V.

(b) The voltage across R'' is (1.0 A)(4.0 Ω) = 4.0 V, which means that (examining the left side of the circuit) the voltage difference between *ground* and V_2 is 5.0 V + 4.0 V = 9.0 V. Noting the orientation of the battery, we conclude $V_2 = -9.0$ V.

LEARN The potential difference between points 1 and 2 is

$$V_2 - V_1 = -9.0\text{ V} - (-11.0\text{ V}) = 2.0\text{ V},$$

which is equal to $iR = (1.0\text{ A})(2.0\ \Omega) = 2.0\text{ V}$.

27-47

THINK The copper wire and the aluminum sheath are connected in parallel, so the potential difference is the same for them.

EXPRESS Since the potential difference is the product of the current and the resistance, $i_C R_C = i_A R_A$, where i_C is the current in the copper, i_A is the current in the aluminum, R_C is the resistance of the copper, and R_A is the resistance of the aluminum. The resistance of either component is given by $R = \rho L/A$, where ρ is the resistivity, L is the length, and A is the cross-sectional area. The resistance of the copper wire is $R_C = \rho_C L/\pi a^2$, and the resistance of the aluminum sheath is $R_A = \rho_A L/\pi(b^2 - a^2)$. We substitute these expressions into $i_C R_C = i_A R_A$, and cancel the common factors L and π to obtain

$$\frac{i_C \rho_C}{a^2} = \frac{i_A \rho_A}{b^2 - a^2}.$$

We solve this equation simultaneously with $i = i_C + i_A$, where i is the total current. We find

$$i_C = \frac{r_C^2 \rho_C i}{(r_A^2 - r_C^2)\rho_C + r_C^2 \rho_A}$$

and

$$i_A = \frac{(r_A^2 - r_C^2)\rho_C i}{(r_A^2 - r_C^2)\rho_C + r_C^2 \rho_A}.$$

ANALYZE (a) The denominators are the same and each has the value

$$\begin{aligned}(b^2 - a^2)\rho_C + a^2\rho_A &= \left[(0.380\times10^{-3}\text{ m})^2 - (0.250\times10^{-3}\text{ m})^2\right](1.69\times10^{-8}\ \Omega\cdot\text{m}) \\ &\quad + (0.250\times10^{-3}\text{ m})^2(2.75\times10^{-8}\ \Omega\cdot\text{m}) \\ &= 3.10\times10^{-15}\ \Omega\cdot\text{m}^3.\end{aligned}$$

Thus,

$$i_C = \frac{(0.250\times10^{-3}\text{ m})^2(2.75\times10^{-8}\ \Omega\cdot\text{m})(2.00\text{ A})}{3.10\times10^{-15}\ \Omega\cdot\text{m}^3} = 1.11\text{ A}.$$

(b) Similarly,

$$i_A = \frac{\left[(0.380\times10^{-3}\,\text{m})^2-(0.250\times10^{-3}\,\text{m})^2\right](1.69\times10^{-8}\,\Omega\cdot\text{m})(2.00\,\text{A})}{3.10\times10^{-15}\,\Omega\cdot\text{m}^3}=0.893\,\text{A}.$$

(c) Consider the copper wire. If V is the potential difference, then the current is given by $V = i_C R_C = i_C \rho_C L/\pi a^2$, so the length of the composite wire is

$$L=\frac{\pi a^2 V}{i_C\rho_C}=\frac{(\pi)(0.250\times10^{-3}\,\text{m})^2(12.0\text{ V})}{(1.11\text{ A})(1.69\times10^{-8}\,\Omega\cdot\text{m})}=126\text{ m}.$$

LEARN The potential difference can also be written as $V = i_A R_A = i_A \rho_A L/\pi(b^2 - a^2)$. Thus,

$$L=\frac{\pi(b^2-a^2)V}{i_A\rho_A}=\frac{\pi\left[(0.380\times10^{-3}\text{ m})^2-(0.250\times10^{-3}\text{ m})^2\right](12.0\text{ V})}{(0.893\text{ A})(2.75\times10^{-8}\,\Omega\cdot\text{m})}=126\text{ m},$$

in agreement with the result found in (c).

27-59

THINK We have an RC circuit that is being charged. When fully charged, the charge on the capacitor is equal to $C\varepsilon$.

EXPRESS During charging, the charge on the positive plate of the capacitor is given by

$$q=C\varepsilon\left(1-e^{-t/\tau}\right),$$

where C is the capacitance, ε is applied emf, and $\tau = RC$ is the capacitive time constant. The equilibrium charge is $q_{eq} = C\varepsilon$, so we require $q = 0.99q_{eq} = 0.99C\varepsilon$.

ANALYZE The time required to reach 99% of its final charge is given by

$$0.99=1-e^{-t/\tau}.$$

Thus, $e^{-t/\tau} = 0.01$. Taking the natural logarithm of both sides, we obtain $t/\tau = -\ln 0.01 = 4.61$ or $t = 4.61\,\tau$.

LEARN The corresponding current in a charging capacitor is given by

$$i=\frac{dq}{dt}=\frac{\varepsilon}{R}e^{-t/\tau}.$$

The current has an initial value ε/R but decays exponentially to zero as the capacitor becomes fully charged. The plots of $q(t)$ and $i(t)$ are shown in Fig. 27-16 of the text.

27-63

THINK We have a multi-loop circuit with a capacitor that's being charged. Since at $t = 0$ the capacitor is completely uncharged, the current in the capacitor branch is as it would be if the capacitor were replaced by a wire.

EXPRESS Let i_1 be the current in R_1 and take it to be positive if it is to the right. Let i_2 be the current in R_2 and take it to be positive if it is downward. Let i_3 be the current in R_3 and take it to be positive if it is downward. The junction rule produces $i_1 = i_2 + i_3$, the loop rule applied to the left-hand loop produces

$$\varepsilon - i_1R_1 - i_2R_2 = 0,$$

and the loop rule applied to the right-hand loop produces

$$i_2R_2 - i_3R_3 = 0.$$

Since the resistances are all the same we can simplify the mathematics by replacing R_1, R_2, and R_3 with R.

ANALYZE (a) Solving the three simultaneous equations, we find

$$i_1 = \frac{2\varepsilon}{3R} = \frac{2(1.2\times10^3\ \text{V})}{3(0.73\times10^6\ \Omega)} = 1.1\times10^{-3}\ \text{A},$$

(b) $i_2 = \dfrac{\varepsilon}{3R} = \dfrac{1.2\times10^3\ \text{V}}{3(0.73\times10^6\ \Omega)} = 5.5\times10^{-4}\ \text{A}$, and

(c) $i_3 = i_2 = 5.5\times10^{-4}\ \text{A}$.

At $t = \infty$ the capacitor is fully charged and the current in the capacitor branch is 0. Thus, $i_1 = i_2$, and the loop rule yields

$$\varepsilon - i_1R_1 - i_1R_2 = 0.$$

(d) The solution is

$$i_1 = \frac{\varepsilon}{2R} = \frac{1.2\times10^3\ \text{V}}{2(0.73\times10^6\ \Omega)} = 8.2\times10^{-4}\ \text{A}$$

(e) and $i_2 = i_1 = 8.2\times10^{-4}\ \text{A}$.

(f) As stated before, the current in the capacitor branch is $i_3 = 0$.

We take the upper plate of the capacitor to be positive. This is consistent with current flowing into that plate. The junction equation is $i_1 = i_2 + i_3$, and the loop equations are

$$\varepsilon - i_1 R - i_2 R = 0$$

$$-\frac{q}{C} - i_3 R + i_2 R = 0.$$

We use the first equation to substitute for i_1 in the second and obtain $\varepsilon - 2i_2R - i_3R = 0$. Thus $i_2 = (\varepsilon - i_3R)/2R$. We substitute this expression into the third equation above to obtain

$$-(q/C) - (i_3R) + (\varepsilon/2) - (i_3R/2) = 0.$$

Now we replace i_3 with dq/dt to obtain

$$\frac{3R}{2}\frac{dq}{dt} + \frac{q}{C} = \frac{\varepsilon}{2}.$$

This is just like the equation for an RC series circuit, except that the time constant is $\tau = 3RC/2$ and the impressed potential difference is $\varepsilon/2$. The solution is

$$q = \frac{C\varepsilon}{2}\left(1 - e^{-2t/3RC}\right).$$

The current in the capacitor branch is

$$i_3(t) = \frac{dq}{dt} = \frac{\varepsilon}{3R} e^{-2t/3RC}.$$

The current in the center branch is

$$i_2(t) = \frac{\varepsilon}{2R} - \frac{i_3}{2} = \frac{\varepsilon}{2R} - \frac{\varepsilon}{6R} e^{-2t/3RC} = \frac{\varepsilon}{6R}\left(3 - e^{-2t/3RC}\right)$$

and the potential difference across R_2 is

$$V_2(t) = i_2 R = \frac{\varepsilon}{6}\left(3 - e^{-2t/3RC}\right).$$

(g) For $t = 0$, $e^{-2t/3RC} = 1$ and $V_2 = \varepsilon/3 = \left(1.2\times10^3\ \text{V}\right)/3 = 4.0\times10^2\ \text{V}$.

(h) For $t = \infty$, $e^{-2t/3RC} \to 0$ and $V_2 = \varepsilon/2 = \left(1.2\times20^3\ \text{V}\right)/2 = 6.0\times10^2\ \text{V}$.

(i) A plot of V_2 as a function of time is shown in the following graph.

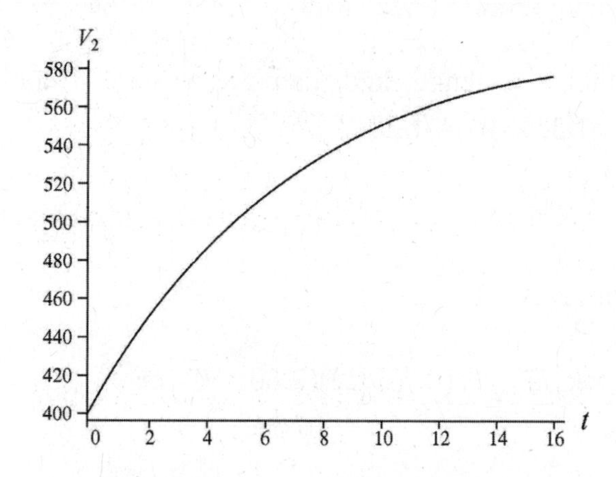

LEARN A capacitor that is being charged initially behaves like an ordinary connecting wire relative to the charging current. However, a long time later after it's fully charged, it acts like a broken wire.

27-73

THINK Since the wires are connected in series, the current is the same in both wires.

EXPRESS Let i be the current in the wires and V be the applied potential difference. Using Kirchhoff's loop rule, we have $V - iR_A - iR_B = 0$. Thus, the current is $i = V/(R_A + R_B)$, and the corresponding current density is

$$J = \frac{i}{A} = \frac{V}{R_A + R_B}.$$

ANALYZE (a) For wire A, the magnitude of the current density vector is

$$J_A = \frac{i}{A} = \frac{V}{(R_A + R_B)A} = \frac{4V}{(R_1 + R_2)\pi D^2} = \frac{4(60.0\text{V})}{\pi(0.127\Omega + 0.729\Omega)(2.60\times10^{-3}\text{m})^2}$$

$$= 1.32\times10^7\ \text{A/m}^2.$$

(b) The potential difference across wire A is

$$V_A = iR_A = V R_A/(R_A + R_B) = (60.0\text{ V})(0.127\ \Omega)/(0.127\ \Omega + 0.729\ \Omega) = 8.90\text{ V}.$$

(c) The resistivity of wire A is

$$\rho_A = \frac{R_A A}{L_A} = \frac{\pi R_A D^2}{4L_A} = \frac{\pi(0.127\,\Omega)(2.60\times10^{-3}\text{ m})^2}{4(40.0\,\text{m})} = 1.69\times10^{-8}\ \Omega\cdot\text{m}.$$

So wire A is made of copper.

(d) Since wire B has the same length and diameter as wire A, and the currents are the same, we have $J_B = J_A = 1.32\times10^7\ \text{A/m}^2$.

(e) The potential difference across wire B is $V_B = V - V_A = 60.0\ \text{V} - 8.9\ \text{V} = 51.1\ \text{V}$.

(f) The resistivity of wire B is

$$\rho_B = \frac{R_B A}{L_B} = \frac{\pi R_B D^2}{4L_B} = \frac{\pi(0.729\ \Omega)(2.60\times10^{-3}\ \text{m})^2}{4(40.0\,\text{m})} = 9.68\times10^{-8}\ \Omega\cdot\text{m},$$

so wire B is made of iron.

LEARN Resistance R is the property of an object (depending on quantities such as L and A), while resistivity is a property of the material itself. Knowing the value of ρ allows us to deduce what material the wire is made of.

27-77

THINK The silicon resistor and the iron resistor are connected in series. Both resistors are temperature-dependent, but we want the combination to be independent of temperature.

EXPRESS We denote silicon with subscript s and iron with i. Let T_0 = 20°. The resistances of the two resistors can be written as

$$R_s(T) = R_s(T_0)\left[1+\alpha_s(T-T_0)\right], \qquad R_i(T) = R_i(T_0)\left[1+\alpha_i(T-T_0)\right].$$

The resistors are in series connection so

$$\begin{aligned} R(T) &= R_s(T) + R_i(T) = R_s(T_0)\left[1+\alpha_s(T-T_0)\right] + R_i(T_0)\left[1+\alpha_i(T-T_0)\right] \\ &= R_s(T_0) + R_i(T_0) + \left[R_s(T_0)\alpha_s + R_i(T_0)\alpha_i\right](T-T_0). \end{aligned}$$

Now, if $R(T)$ is to be temperature-independent, we must require that $R_s(T_0)\alpha_s + R_i(T_0)\alpha_i = 0$. Also note that $R_s(T_0) + R_i(T_0) = R = 1000\ \Omega$.

ANALYZE (a) We solve for $R_s(T_0)$ and $R_i(T_0)$ to obtain

$$R_s(T_0) = \frac{R\alpha_i}{\alpha_i - \alpha_s} = \frac{(1000\Omega)\left(6.5\times10^{-3}\,/\text{K}\right)}{(6.5\times10^{-3}\,/\text{K}) - (-70\times10^{-3}\,/\text{K})} = 85.0\Omega.$$

(b) Similarly, $R_i(T_0) = 1000\ \Omega - 85.0\ \Omega = 915\ \Omega$.

LEARN The temperature independence of the combined resistor was possible because α_i and α_s, the temperature coefficients of resistivity of the two materials have opposite signs, so their temperature dependences can cancel.

27-79

THINK As the capacitor in an *RC* circuit is being charged, some energy supplied by the emf device also goes to the resistor as thermal energy.

EXPRESS The charge q on the capacitor as a function of time is $q(t) = (\varepsilon C)(1 - e^{-t/RC})$, so the charging current is $i(t) = dq/dt = (\varepsilon/R)e^{-t/RC}$. The rate at which the emf device supplies energy is $P_\varepsilon = i\varepsilon dt$.

ANALYZE (a) The energy supplied by the emf is then

$$U = \int_0^\infty P_\varepsilon dt = \int_0^\infty \varepsilon i\, dt = \frac{\varepsilon^2}{R}\int_0^\infty e^{-t/RC} dt = C\varepsilon^2 = 2U_C$$

where $U_C = \frac{1}{2}C\varepsilon^2$ is the energy stored in the capacitor.

(b) By directly integrating i^2R we obtain

$$U_R = \int_0^\infty i^2 R dt = \frac{\varepsilon^2}{R}\int_0^\infty e^{-2t/RC} dt = \frac{1}{2}C\varepsilon^2.$$

LEARN Half of the energy supplied by the emf device is stored in the capacitor as electrical energy, while the other half is dissipated in the resistor as thermal energy.

27-83

THINK The time constant in an *RC* circuit is $\tau = RC$, where R is the resistance and C is the capacitance. A greater value of τ means a longer discharging time.

EXPRESS The potential difference across the capacitor varies as a function of time t as $V(t) = V_0 e^{-t/\tau}$, where $\tau = RC$. Thus, $R = \dfrac{t}{C\ln(V_0/V)}$.

ANALYZE (a) Then, for the smaller time interval $t_{\min} = 10.0\ \mu s$

$$R_{\min} = \frac{10.0\ \mu s}{(0.220\ \mu F)\ln(5.00/0.800)} = 24.8\ \Omega.$$

(b) Similarly, for the larger time interval $t_{max} = 6.00$ ms,

$$R_{max} = \frac{6.00\times10^{-3}\ \text{s}}{(0.220\ \mu\text{F})\ln(5.00\ \text{V}/0.800\ \text{V})} = 1.49\times10^{4}\ \Omega.$$

LEARN The two extrema of the resistances are related by

$$\frac{R_{max}}{R_{min}} = \frac{t_{max}}{t_{min}}.$$

The larger the value of R for a given capacitance, the longer the discharging time.

27-85

THINK One of the three parts could be defective: the battery, the motor, or the cable.

EXPRESS All three circuit elements are connected in series, so the current is the same in all of them. The battery is discharging, so the potential drop across the terminals is $V_{battery} = \varepsilon - ir$, where ε is the emf and r is the internal resistance. On the other hand, the resistances in the cable and the motor are $R_{cable} = V_{cable}/i$ and $R_{motor} = V_{motor}/i$, respectively.

ANALYZE The internal resistance of the battery is

$$r = \frac{\varepsilon - V_{battery}}{i} = \frac{12\ \text{V} - 11.4\ \text{V}}{50\ \text{A}} = 0.012\ \Omega$$

which is less than 0.020 Ω. So the battery is OK. For the motor, we have

$$R_{motor} = \frac{V_{motor}}{i} = \frac{11.4\ \text{V} - 3.0\ \text{V}}{50\ \text{A}} = 0.17\ \Omega$$

which is less than 0.20 Ω. So the motor is OK. Now, the resistance of the cable is

$$R_{cable} = \frac{V_{cable}}{i} = \frac{3.0\ \text{V}}{50\ \text{A}} = 0.060\ \Omega$$

which is greater than 0.040 Ω. So the cable is defective.

LEARN In this exercise, we see that a defective component has a resistance outside its the range of acceptance.

27-97

THINK To calculate the current in the resistor R, we first find the equivalent resistance of the N batteries.

EXPRESS When all the batteries are connected in parallel, the emf is ε and the equivalent resistance is $R_{\text{parallel}} = R + r/N$, so the current is

$$i_{\text{parallel}} = \frac{\varepsilon}{R_{\text{parallel}}} = \frac{\varepsilon}{R + r/N} = \frac{N\varepsilon}{NR + r}.$$

Similarly, when all the batteries are connected in series, the total emf is $N\varepsilon$ and the equivalent resistance is $R_{\text{series}} = R + Nr$. Therefore,

$$i_{\text{series}} = \frac{N\varepsilon}{R_{\text{series}}} = \frac{N\varepsilon}{R + Nr}.$$

ANALYZE Comparing the two expressions, we see that the two currents i_{parallel} and i_{series} are equal if $R = r$, with

$$i_{\text{parallel}} = i_{\text{series}} = \frac{N\varepsilon}{(N+1)r}.$$

LEARN In general, the current difference is

$$i_{\text{parallel}} - i_{\text{series}} = \frac{N\varepsilon}{NR + r} - \frac{N\varepsilon}{R + Nr} = \frac{N\varepsilon(N-1)(r-R)}{(NR+r)(R+Nr)}.$$

If $R > r$, then $i_{\text{parallel}} < i_{\text{series}}$.

27-98

THINK The rate of energy supplied by the battery is $i\varepsilon$. So we first calculate the current in the circuit.

EXPRESS With R_2 and R_3 in parallel, and the combination in series with R_1, the equivalent resistance for the circuit is

$$R_{\text{eq}} = R_1 + \frac{R_2 R_3}{R_2 + R_3} = \frac{R_1 R_2 + R_1 R_3 + R_2 R_3}{R_2 + R_3}$$

and the current is

$$i = \frac{\varepsilon}{R_{\text{eq}}} = \frac{(R_2 + R_3)\varepsilon}{R_1 R_2 + R_1 R_3 + R_2 R_3}.$$

The rate at which the battery supplies energy is

$$P = i\varepsilon = \frac{(R_2 + R_3)\varepsilon^2}{R_1R_2 + R_1R_3 + R_2R_3}.$$

To find the value of R_3 that maximizes P, we differentiate P with respect to R_3.

ANALYZE (a) With a little algebra, we find

$$\frac{dP}{dR_3} = -\frac{R_2^2\varepsilon^2}{(R_1R_2 + R_1R_3 + R_2R_3)^2}.$$

The derivative is negative for all positive value of R_3. Thus, we see that P is maximized when $R_3 = 0$.

(b) With the value of R_3 set to zero, we obtain $P = \frac{\varepsilon^2}{R_1} = \frac{(12.0\text{ V})^2}{10.0\ \Omega} = 14.4\text{ W}.$

LEARN Mathematically speaking, the function P is a monotonically decreasing function of R_3 (as well as R_2 and R_1), so P is a maximum at $R_3 = 0$.

27-99

THINK A capacitor that is being charged initially behaves like an ordinary connecting wire relative to the charging current.

EXPRESS The capacitor is *initially* uncharged. So immediately after the switch is closed, by the Kirchhoff's loop rule, there is zero voltage (at $t = 0$) across the $R_2 = 10\text{ k}\Omega$ resistor, and that $\varepsilon = 30$ V is across the $R_1 = 20\text{ k}\Omega$ resistor.

ANALYZE (a) By Ohm's law, the initial current in R_1 is

$$i_{10} = \varepsilon / R_1 = (30\text{ V})/(20\text{ k}\Omega) = 1.5 \times 10^{-3}\text{ A}.$$

(b) Similarly, the initial current in R_2 is $i_{20} = 0$.

(c) As $t \to \infty$ the current to the capacitor reduces to zero and the $R_1 = 20\text{ k}\Omega$ and $R_2 = 10\text{ k}\Omega$ resistors behave more like a series pair (having the same current), equivalent to $R_{eq} = R_1 + R_2 = 30\text{ k}\Omega$. The current through them, then, at long times, is

$$i = \varepsilon / R_{eq} = (30\text{ V})/(30\text{ k}\Omega) = 1.0 \times 10^{-3}\text{ A}.$$

LEARN A long time later after a capacitor is being fully charged, it acts like a broken wire.

Chapter 28

28-1

THINK The magnetic force on a charged particle is given by $\vec{F}_B = q\vec{v}\times\vec{B}$, where $\vec{v}$ is the velocity of the charged particle and $\vec{B}$ is the magnetic field.

EXPRESS The magnitude of the magnetic force on the proton (of charge $+e$) is $F_B = evB\sin\phi$, where ϕ is the angle between $\vec{v}$ and $\vec{B}$.

ANALYZE (a) The speed of the proton is

$$v = \frac{F_B}{eB\sin\phi} = \frac{6.50\times10^{-17}\,\text{N}}{(1.60\times10^{-19}\,\text{C})(2.60\times10^{-3}\,\text{T})\sin 23.0°} = 4.00\times10^{5}\ \text{m/s}.$$

(b) The kinetic energy of the proton is

$$K = \frac{1}{2}mv^2 = \frac{1}{2}(1.67\times10^{-27}\,\text{kg})(4.00\times10^{5}\ \text{m/s})^2 = 1.34\times10^{-16}\ \text{J},$$

which is equivalent to $K = (1.34\times10^{-16}\ \text{J})/(1.60\times10^{-19}\ \text{J/eV}) = 835\ \text{eV}$.

LEARN from the definition of $\vec{B}$ given by the expression $\vec{F}_B = q\vec{v}\times\vec{B}$, we see that the magnetic force $\vec{F}_B$ is always perpendicular to $\vec{v}$ and $\vec{B}$.

28-21

THINK The electron is in circular motion because the magnetic force acting on it points toward the center of the circle.

EXPRESS The kinetic energy of the electron is given by $K = \frac{1}{2}m_e v^2$, where m_e is the mass of electron and v is its speed. The magnitude of the magnetic force on the electron is $F_B = evB$ which is equal to the centripetal force:

$$evB = \frac{m_e v^2}{r}.$$

ANALYZE (a) From $K = \frac{1}{2} m_e v^2$ we get

$$v = \sqrt{\frac{2K}{m_e}} = \sqrt{\frac{2(1.20\times10^3\,\text{eV})(1.60\times10^{-19}\,\text{eV/J})}{9.11\times10^{-31}\,\text{kg}}} = 2.05\times10^7\,\text{m/s}.$$

(b) Since $evB = m_e v^2/r$, we find the magnitude of the magnetic field to be

$$B = \frac{m_e v}{er} = \frac{(9.11\times10^{-31}\,\text{kg})(2.05\times10^7\,\text{m/s})}{(1.60\times10^{-19}\,\text{C})(25.0\times10^{-2}\,\text{m})} = 4.67\times10^{-4}\,\text{T}.$$

(c) The "orbital" frequency is

$$f = \frac{v}{2\pi r} = \frac{2.07\times10^7\,\text{m/s}}{2\pi(25.0\times10^{-2}\,\text{m})} = 1.31\times10^7\,\text{Hz}.$$

(d) The period is simply equal to the reciprocal of frequency:

$$T = 1/f = (1.31\times10^7\,\text{Hz})^{-1} = 7.63\times10^{-8}\,\text{s}.$$

LEARN The period of the electron's circular motion can be written as

$$T = \frac{2\pi r}{v} = \frac{2\pi}{v}\frac{mv}{|e|B} = \frac{2\pi m}{|e|B}.$$

The period is inversely proportional to B.

28-33

THINK The path of the positron is helical because its velocity $\vec{v}$ has components parallel and perpendicular to the magnetic field $\vec{B}$.

EXPRESS If v is the speed of the positron then $v \sin\phi$ is the component of its velocity in the plane that is perpendicular to the magnetic field. Here $\phi = 89°$ is the angle between the velocity and the field. Newton's second law yields $eBv \sin\phi = m_e(v \sin\phi)^2/r$, where r is the radius of the orbit. Thus $r = (m_e v/eB)\sin\phi$. The period is given by

$$T = \frac{2\pi r}{v\sin\phi} = \frac{2\pi m_e}{eB}.$$

The equation for r is substituted to obtain the second expression for T. For part (b), the pitch is the distance traveled along the line of the magnetic field in a time interval of one period. Thus $p = vT\cos\phi$.

ANALYZE (a) Substituting the values given, we find the period to be

$$T = \frac{2\pi m_e}{eB} = \frac{2\pi\left(9.11\times10^{-31}\,\text{kg}\right)}{\left(1.60\times10^{-19}\,\text{C}\right)\left(0.100\text{T}\right)} = 3.58\times10^{-10}\,\text{s}.$$

(b) We use the kinetic energy, $K = \frac{1}{2}m_e v^2$, to find the speed:

$$v = \sqrt{\frac{2K}{m_e}} = \sqrt{\frac{2\left(2.00\times10^3\,\text{eV}\right)\left(1.60\times10^{-19}\,\text{J/eV}\right)}{9.11\times10^{-31}\,\text{kg}}} = 2.65\times10^7\,\text{m/s}.$$

Thus, the pitch is

$$p = \left(2.65\times10^7\,\text{m/s}\right)\left(3.58\times10^{-10}\,\text{s}\right)\cos 89° = 1.66\times10^{-4}\,\text{m}.$$

(c) The orbit radius is

$$R = \frac{m_e v\sin\phi}{eB} = \frac{\left(9.11\times10^{-31}\,\text{kg}\right)\left(2.65\times10^7\,\text{m/s}\right)\sin 89°}{\left(1.60\times10^{-19}\,\text{C}\right)\left(0.100\,\text{T}\right)} = 1.51\times10^{-3}\,\text{m}.$$

LEARN The parallel component of the velocity, $v_\parallel = v\cos\phi$, is what determines the pitch of the helix. On the other hand, the perpendicular component, $v_\perp = v\sin\phi$, determines the radius of the helix.

28-39

THINK The magnetic force on a wire that carries a current i is given by $\vec{F}_B = i\vec{L}\times\vec{B}$, where $\vec{L}$ is the length vector of the wire and $\vec{B}$ is the magnetic field.

EXPRESS The magnitude of the magnetic force on the wire is given by $F_B = iLB\sin\phi$, where ϕ is the angle between the current and the field.

ANALYZE (a) With $\phi = 70°$, we have

$$F_B = \left(5000\,\text{A}\right)\left(100\,\text{m}\right)\left(60.0\times10^{-6}\,\text{T}\right)\sin 70° = 28.2\,\text{N}.$$

(b) We apply the right-hand rule to the vector product $\vec{F}_B = i\vec{L} \times \vec{B}$ to show that the force is to the west.

LEARN From the expression $\vec{F}_B = i\vec{L} \times \vec{B}$, we see that the magnetic force acting on a current-carrying wire is a maximum when $\vec{L}$ is perpendicular to $\vec{B}$ ($\phi = 90°$), and is zero when $\vec{L}$ is parallel to $\vec{B}$ ($\phi = 0°$).

28-49

THINK Magnetic forces on the loop produce a torque that rotates it about the hinge line. Our applied field has two components: $B_x > 0$ and $B_z > 0$.

EXPRESS Considering each straight segment of the rectangular coil, we note that Eq. 28-26 produces a nonzero force only for the component of $\vec{B}$ which is perpendicular to that segment; we also note that the equation is effectively multiplied by $N = 20$ due to the fact that this is a 20-turn coil. Since we wish to compute the torque about the hinge line, we can ignore the force acting on the straight segment of the coil that lies along the y axis (forces acting at the axis of rotation produce no torque about that axis). The top and bottom straight segments experience forces due to Eq. 28-26 (caused by the B_z component), but these forces are (by the right-hand rule) in the $\pm y$ directions and are thus unable to produce a torque about the y axis. Consequently, the torque derives completely from the force exerted on the straight segment located at $x = 0.050$ m, which has length $L = 0.10$ m and is shown in Fig. 28-45 carrying current in the $-y$ direction.

Now, the B_z component will produce a force on this straight segment which points in the $-x$ direction (back toward the hinge) and thus will exert no torque about the hinge. However, the B_x component (which is equal to $B\cos\theta$ where $B = 0.50$ T and $\theta = 30°$) produces a force equal to $F = NiLB_x$ which points (by the right-hand rule) in the $+z$ direction.

ANALYZE Since the action of the force F is perpendicular to the plane of the coil, and is located a distance x away from the hinge, then the torque has magnitude

$$\tau = (NiLB_x)(x) = NiLxB\cos\theta = (20)(0.10\text{ A})(0.10\text{ m})(0.050\text{ m})(0.50\text{ T})\cos 30°$$
$$= 0.0043\text{ N}\cdot\text{m}.$$

Since $\vec{\tau} = \vec{r} \times \vec{F}$, the direction of the torque is $-y$. In unit-vector notation, the torque is $\vec{\tau} = (-4.3\times10^{-3}\text{ N}\cdot\text{m})\hat{\text{j}}$

LEARN An alternative way to do this problem is through the use of Eq. 28-37: $\vec{\tau} = \vec{\mu} \times \vec{B}$. The magnetic moment vector is

$$\vec{\mu} = -(NiA)\,\hat{k} = -(20)(0.10\,\text{A})(0.0050\,\text{m}^2)\,\hat{k} = -(0.01\,\text{A}\cdot\text{m}^2)\hat{k}.$$

The torque on the loop is

$$\begin{aligned}\vec{\tau} &= \vec{\mu}\times\vec{B} = (-\mu\,\hat{k})\times(B\cos\theta\,\hat{i} + B\sin\theta\,\hat{k}) = -(\mu B\cos\theta)\hat{j}\\ &= -(0.01\,\text{A}\cdot\text{m}^2)(0.50\,\text{T})\cos 30°\hat{j}\\ &= (-4.3\times10^{-3}\ \text{N}\cdot\text{m})\hat{j}.\end{aligned}$$

28-55

THINK Our system consists of two concentric current-carrying loops. The net magnetic dipole moment is the vector sum of the individual contributions.

EXPRESS The magnitude of the magnetic dipole moment is given by $\mu = NiA$, where N is the number of turns, i is the current in each turn, and A is the area of a loop. Each of the loops is a circle, so the area is $A = \pi r^2$, where r is the radius of the loop.

ANALYZE (a) Since the currents are in the same direction, the magnitude of the magnetic moment vector is

$$\mu = \sum_n i_n A_n = \pi r_1^2 i_1 + \pi r_2^2 i_2 = \pi(7.00\text{A})\left[(0.200\text{m})^2 + (0.300\text{m})^2\right] = 2.86\text{A}\cdot\text{m}^2.$$

(b) Now, the two currents flow in the opposite directions, so the magnitude of the magnetic moment vector is

$$\mu = \pi r_2^2 i_2 - \pi r_1^2 i_1 = \pi(7.00\text{A})\left[(0.300\text{m})^2 - (0.200\text{m})^2\right] = 1.10\text{A}\cdot\text{m}^2.$$

LEARN In both cases, the directions of the dipole moments are into the page. The direction of $\vec{\mu}$ is that of the normal vector $\vec{n}$ to the plane of the coil, in accordance with the right-hand rule shown in Fig. 28-19.

28-57

THINK Magnetic forces on a current-carrying loop produce a torque that tends to align the magnetic dipole moment with the magnetic field.

EXPRESS The magnitude of the magnetic dipole moment is given by $\mu = NiA$, where N is the number of turns, i is the current in each turn, and A is the area of a loop. In this case the loops are circular, so $A = \pi r^2$, where r is the radius of a turn.

ANALYZE (a) Thus, the current is

$$i = \frac{\mu}{N\pi r^2} = \frac{2.30\,\text{A}\cdot\text{m}^2}{(160)(\pi)(0.0190\,\text{m})^2} = 12.7\,\text{A}.$$

(b) The maximum torque occurs when the dipole moment is perpendicular to the field (or the plane of the loop is parallel to the field). It is given by

$$\tau_{\max} = \mu B = (2.30\ \text{A}\cdot\text{m}^2)(35.0\times10^{-3}\ \text{T}) = 8.05\times10^{-2}\ \text{N}\cdot\text{m}.$$

LEARN The torque on the coil can be written as $\vec{\tau} = \vec{\mu}\times\vec{B}$, with $\tau = |\vec{\tau}| = \mu B\sin\theta$, where θ is the angle between $\vec{\mu}$ and $\vec{B}$. Thus, τ is a maximum when $\theta = 90°$, and zero when $\theta = 0°$.

28-61

THINK Magnetic forces on a current-carrying coil produce a torque that tends to align the magnetic dipole moment with the field. The magnetic energy of the dipole depends on its orientation relative to the field.

EXPRESS The magnetic potential energy of the dipole is given by $U = -\vec{\mu}\cdot\vec{B}$, where $\vec{\mu}$ is the magnetic dipole moment of the coil and $\vec{B}$ is the magnetic field. The magnitude of $\vec{\mu}$ is $\mu = NiA$, where i is the current in the coil, N is the number of turns, A is the area of the coil. On the other hand, the torque on the coil is given by the vector product $\vec{\tau} = \vec{\mu}\times\vec{B}$.

ANALYZE (a) By using the right-hand rule, we see that $\vec{\mu}$ is in the $-y$ direction. Thus, we have

$$\vec{\mu} = (NiA)(-\hat{\text{j}}) = -(3)(2.00\ \text{A})(4.00\times10^{-3}\ \text{m}^2)\hat{\text{j}} = -(0.0240\ \text{A}\cdot\text{m}^2)\hat{\text{j}}.$$

The corresponding magnetic energy is

$$U = -\vec{\mu}\cdot\vec{B} = -\mu_y B_y = -(-0.0240\ \text{A}\cdot\text{m}^2)(-3.00\times10^{-3}\ \text{T}) = -7.20\times10^{-5}\ \text{J}.$$

(b) Using the fact that $\hat{\text{j}}\cdot\hat{\text{i}} = 0$, $\hat{\text{j}}\times\hat{\text{j}} = 0$, and $\hat{\text{j}}\times\hat{\text{k}} = \hat{\text{i}}$, the torque on the coil is

$$\begin{aligned}\vec{\tau} &= \vec{\mu}\times\vec{B} = \mu_y B_z\hat{\text{i}} - \mu_y B_x\hat{\text{k}}\\ &= (-0.0240\ \text{A}\cdot\text{m}^2)(-4.00\times10^{-3}\ \text{T})\hat{\text{i}} - (-0.0240\ \text{A}\cdot\text{m}^2)(2.00\times10^{-3}\ \text{T})\hat{\text{k}}\\ &= (9.60\times10^{-5}\ \text{N}\cdot\text{m})\hat{\text{i}} + (4.80\times10^{-5}\ \text{N}\cdot\text{m})\hat{\text{k}}.\end{aligned}$$

LEARN The magnetic energy is highest when $\vec{\mu}$ is in the opposite direction of $\vec{B}$, and lowest when $\vec{\mu}$ lines up with $\vec{B}$.

28-65

THINK The torque on a current-carrying coil is a maximum when its dipole moment is perpendicular to the magnetic field.

EXPRESS The magnitude of the torque on the coil is given by $\tau = |\vec{\tau}| = \mu B \sin\theta$, where θ is the angle between $\vec{\mu}$ and $\vec{B}$. The magnitude of $\vec{\mu}$ is $\mu = NiA$, where i is the current in the coil, N is the number of turns, A is the area of the coil. Thus, if N closed loops are formed from the wire of length L, the circumference of each loop is L/N, the radius of each loop is $R = L/2\pi N$, and the area of each loop is

$$A = \pi R^2 = \pi\left(L/2\pi N\right)^2 = L^2/4\pi N^2.$$

ANALYZE (a) For maximum torque, we orient the plane of the loops parallel to the magnetic field, so the dipole moment is perpendicular (i.e., at a 90° angle) to the field.

(b) The magnitude of the torque is then

$$\tau = NiAB = \left(Ni\right)\left(\frac{L^2}{4\pi N^2}\right)B = \frac{iL^2B}{4\pi N}.$$

To maximize the torque, we take the number of turns N to have the smallest possible value, 1. Then $\tau = iL^2B/4\pi$.

(c) The magnitude of the maximum torque is

$$\tau = \frac{iL^2B}{4\pi} = \frac{(4.51\times10^{-3}\text{ A})(0.250\text{ m})^2(5.71\times10^{-3}\text{T})}{4\pi} = 1.28\times10^{-7}\text{ N}\cdot\text{m}.$$

LEARN The torque tends to align $\vec{\mu}$ with $\vec{B}$. The magnitude of the torque is a maximum when the angle between $\vec{\mu}$ and $\vec{B}$ is $\theta = 90°$, and is zero when $\theta = 0°$.

28-73

THINK The electron moving in the Earth's magnetic field is being accelerated by the magnetic force acting on it.

EXPRESS Since the electron is moving in a line that is parallel to the horizontal component of the Earth's magnetic field, the magnetic force on the electron is due to the vertical component of the field only. The magnitude of the force acting on the electron is given by $F = evB$, where B represents the downward component of Earth's field. With $F = m_e a$, the acceleration of the electron is $a = evB/m_e$.

ANALYZE (a) The electron speed can be found from its kinetic energy $K = \frac{1}{2}m_e v^2$:

$$v = \sqrt{\frac{2K}{m_e}} = \sqrt{\frac{2\left(12.0\times10^3\,\text{eV}\right)\left(1.60\times10^{-19}\,\text{J/eV}\right)}{9.11\times10^{-31}\,\text{kg}}} = 6.49\times10^7\,\text{m/s}.$$

Therefore,

$$a = \frac{evB}{m_e} = \frac{(1.60\times10^{-19}\,\text{C})(6.49\times10^7\,\text{m/s})(55.0\times10^{-6}\,\text{T})}{9.11\times10^{-31}\,\text{kg}} = 6.27\times10^{14}\,\text{m/s}^2 \approx 6.3\times10^{14}\,\text{m/s}^2.$$

(b) We ignore any vertical deflection of the beam that might arise due to the horizontal component of Earth's field. Then, the path of the electron is a circular arc. The radius of the path is given by $a = v^2/R$, or

$$R = \frac{v^2}{a} = \frac{(6.49\times10^7\,\text{m/s})^2}{6.27\times10^{14}\,\text{m/s}^2} = 6.72\ \text{m}.$$

The dashed curve shown represents the path. Let the deflection be h after the electron has traveled a distance d along the x axis. With $d = R\sin\theta$, we have

$$h = R(1-\cos\theta) = R\left(1-\sqrt{1-\sin^2\theta}\right)$$
$$= R\left(1-\sqrt{1-(d/R)^2}\right).$$

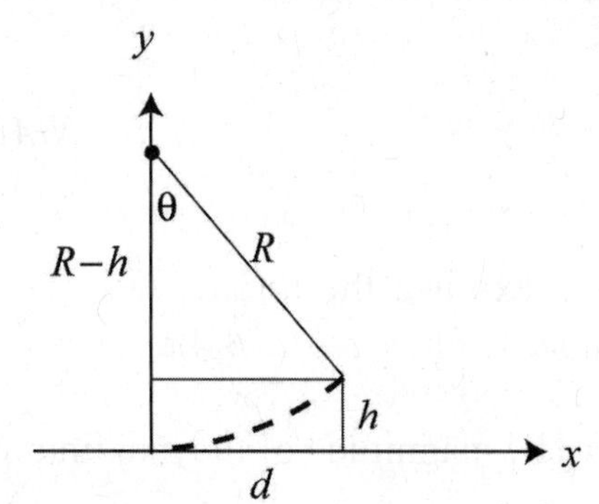

Substituting $R = 6.72$ m and $d = 0.20$ m into the expression, we obtain $h = 0.0030$ m.

LEARN The deflection is so small that many of the technicalities of circular geometry may be ignored, and a calculation along the lines of projectile motion analysis (see Chapter 4) provides an adequate approximation:

$$d = vt \;\Rightarrow\; t = \frac{d}{v} = \frac{0.200\ \text{m}}{6.49\times10^7\ \text{m/s}} = 3.08\times10^{-9}\ \text{s}.$$

Then, with our y axis oriented eastward,

$$h = \frac{1}{2}at^2 = \frac{1}{2}\left(6.27\times10^{14}\right)\left(3.08\times10^{-9}\right)^2 = 0.00298\text{m} \approx 0.0030\ \text{m}.$$

28-77

THINK Since both electric and magnetic fields are present, the net force on the electron is the vector sum of the electric force and the magnetic force.

EXPRESS The force on the electron is given by $\vec{F} = -e(\vec{E} + \vec{v} \times \vec{B})$, where $\vec{E}$ is the electric field, $\vec{B}$ is the magnetic field, and $\vec{v}$ is the velocity of the electron. The fact that the fields are uniform with the feature that the charge moves in a straight line, implies that the speed is constant. Thus, the net force must vanish.

ANALYZE The condition $\vec{F} = 0$ implies that

$$E = vB = 500\,\text{V/m}.$$

Its direction (so that $\vec{F} = 0$) is downward, or $-\hat{\text{j}}$, in the "page" coordinates. In unit-vector notation, $\vec{E} = (-500\ \text{V/m})\hat{\text{j}}$

LEARN Electron moves in a straight line only when the condition $E = vB$ is met. In many experiments, a velocity selector can be set up so that only electrons with a speed given by $v = E/B$ can pass through.

28-79

THINK We have charged particles that are accelerated through an electric potential difference, and then moved through a region of uniform magnetic field. Energy is conserved in the process.

EXPRESS The kinetic energy of a particle is given by $K = qV$, where q is the particle's charge and V is the potential difference. With $K = mv^2/2$, the speed of the particle is

$$v = \sqrt{\frac{2K}{m}} = \sqrt{\frac{2qV}{m}}.$$

In the region with uniform magnetic field, the magnetic force on a particle of charge q is qvB, which according to Newton's second law, is equal to mv^2/r, where r is the radius of the orbit. Thus, we have

$$r = \frac{mv}{qB} = \frac{m}{qB}\sqrt{\frac{2K}{m}} = \frac{\sqrt{2mK}}{qB}.$$

ANALYZE (a) Since $K = qV$ we have $K_p = \frac{1}{2}K_\alpha$ (as $q_\alpha = 2K_p$), or $K_p / K_\alpha = 0.50$.

(b) Similarly, $q_\alpha = 2K_d$, $K_d / K_\alpha = 0.50$.

(c) Since $r \propto \sqrt{mK}/q$, we have

$$r_d = \sqrt{\frac{m_d K_d}{m_p K_p}}\frac{q_p}{q_d} r_p = \sqrt{\frac{(2.00\ \text{u})K_p}{(1.00\ \text{u})K_p}} r_p = 10\sqrt{2}\ \text{cm} = 14\ \text{cm}.$$

(d) Similarly, for the alpha particle, we have

$$r_\alpha = \sqrt{\frac{m_\alpha K_\alpha}{m_p K_p}}\frac{q_p}{q_\alpha} r_p = \sqrt{\frac{(4.00\text{u})K_\alpha}{(1.00\text{u})\,(K_\alpha/2)}}\frac{e}{2e} r_p = 10\sqrt{2}\ \text{cm} = 14\ \text{cm}.$$

LEARN The radius of the particle's path, given by $r = \sqrt{2mK}/qB$, depends on its mass. kinetic energy, and charge, in addition to the field strength.

28-83

THINK The force on the charged particle is given by $\vec{F} = q\vec{v} \times \vec{B}$, where q is the charge, $\vec{B}$ is the magnetic field, and $\vec{v}$ is the velocity of the electron.

EXPRESS We write $\vec{B} = B\hat{\text{i}}$ and take the velocity of the particle to be $\vec{v} = v_x\hat{\text{i}} + v_y\hat{\text{j}}$. Thus,

$$\vec{F} = q\vec{v} \times \vec{B} = q(v_x\hat{\text{i}} + v_y\hat{\text{j}}) \times (B\hat{\text{i}}) = -qv_yB\hat{\text{k}}.$$

For the force to point along $+\hat{\text{k}}$, we must have $q < 0$.

ANALYZE The charge of the particle is

$$q = -\frac{F}{v_yB} = -\frac{0.48\ \text{N}}{(4.0\times10^3\ \text{m/s})(\sin 37°)(0.0050\ \text{T})} = -4.0\times10^{-2}\ \text{C}.$$

LEARN The component of the velocity, v_x, being parallel to the magnetic field, does not contribute to the magnetic force $\vec{F}$; only v_y, the component of $\vec{v}$ that is perpendicular to $\vec{B}$, contributes to $\vec{F}$.

Chapter 29

29-3

THINK The magnetic field produced by a current-carrying wire can be calculated using the Biot-Savart law.

EXPRESS The magnitude of the magnetic field at a distance r from a long straight wire carrying current i is, using the Biot-Savart law, $B = \mu_0 i/2\pi r$.

ANALYZE (a) The field due to the wire, at a point 8.0 cm from the wire, must be 39 μT and must be directed due south. Therefore,

$$i = \frac{2\pi r B}{\mu_0} = \frac{2\pi(0.080\,\text{m})(39\times10^{-6}\,\text{T})}{4\pi\times10^{-7}\,\text{T}\cdot\text{m/A}} = 16\,\text{A}.$$

(b) The current must be from west to east to produce a field that is directed southward at points below it.

LEARN The direction of the current is given by the right-hand rule: grasp the element in your right hand with your thumb pointing in the direction of the current. The direction of the field due to the current-carrying element corresponds to the direction your fingers naturally curl.

29-9

THINK The net magnetic field at a point half way between the two long straight wires is the vector sum of the magnetic fields due to the currents in the two wires.

EXPRESS Since the magnitude of the magnetic field at a distance r from a long straight wire carrying current i is given by $B = \mu_0 i/2\pi r$, at a point half way between the two sires, the magnetic field is $\vec{B} = \vec{B}_1 + \vec{B}_2$, where $B_1 = B_2 = \mu_0 i/2\pi r$ (assuming the two wires to be $2r$ apart). The directions of $\vec{B}_1$ and $\vec{B}_2$ are determined by the right-hand rule.

ANALYZE (a) The currents must be opposite or anti-parallel, so that the resulting fields are in the same direction in the region between the wires. If the currents are parallel, then the two fields are in opposite directions in the region between the wires. Since the currents are the same, the total field is zero along the line that runs halfway between the wires.

(b) The total field at the midpoint has magnitude $B = \mu_0 i/\pi r$ and

$$i = \frac{\pi r B}{\mu_0} = \frac{\pi(0.040\,\text{m})(300\times10^{-6}\,\text{T})}{4\pi\times10^{-7}\,\text{T}\cdot\text{m/A}} = 30\,\text{A}.$$

LEARN For two parallel wires carrying currents in the opposite directions, a point that is a distance d from one wire and $2r - d$ from the other, the magnitude of the magnetic field is

$$B = B_1 + B_2 = \frac{\mu_0 i}{2\pi d} + \frac{\mu_0 i}{2\pi(2r-d)} = \frac{\mu_0 i}{2\pi}\left(\frac{1}{d} + \frac{1}{2r-d}\right).$$

29-17

THINK We apply the Biot-Savart law to calculate the magnetic field at point P_2. An integral is required since the length of the wire is finite.

EXPRESS We take the x axis to be along the wire with the origin at the right endpoint. The current is in the $+x$ direction. All segments of the wire produce magnetic fields at P_2 that are out of the page. According to the Biot-Savart law, the magnitude of the field any (infinitesimal) segment produces at P_2 is given by

$$dB = \frac{\mu_0 i}{4\pi}\frac{\sin\theta}{r^2}dx$$

where θ (the angle between the segment and a line drawn from the segment to P_2) and r (the length of that line) are functions of x. Replacing r with $\sqrt{x^2+R^2}$ and $\sin\theta$ with $R/r = R/\sqrt{x^2+R^2}$, we integrate from $x = -L$ to $x = 0$.

ANALYZE The total field is

$$B = \frac{\mu_0 iR}{4\pi}\int_{-L}^{0}\frac{dx}{(x^2+R^2)^{3/2}} = \frac{\mu_0 iR}{4\pi}\frac{1}{R^2}\frac{x}{(x^2+R^2)^{1/2}}\Bigg|_{-L}^{0} = \frac{\mu_0 i}{4\pi R}\frac{L}{\sqrt{L^2+R^2}}$$
$$= \frac{(4\pi\times10^{-7}\,\text{T}\cdot\text{m/A})(0.693\,\text{A})}{4\pi(0.251\,\text{m})}\frac{0.136\,\text{m}}{\sqrt{(0.136\,\text{m})^2+(0.251\,\text{m})^2}} = 1.32\times10^{-7}\,\text{T}.$$

LEARN In calculating B at P_2, we could have chosen the origin to be at the left endpoint. This only changes the integration limit, but the result remains the same:

$$B = \frac{\mu_0 i R}{4\pi}\int_0^L \frac{dx}{\left(x^2+R^2\right)^{3/2}} = \frac{\mu_0 i R}{4\pi}\frac{1}{R^2}\frac{x}{\left(x^2+R^2\right)^{1/2}}\Bigg|_0^L = \frac{\mu_0 i}{4\pi R}\frac{L}{\sqrt{L^2+R^2}}.$$

29-25

THINK The magnetic field at the center of the circle is the vector sum of the fields of the two straight wires and the arc.

EXPRESS Each of the semi-infinite straight wires contributes $B_{\text{straight}} = \mu_0 i/4\pi R$ (Eq. 29-7) to the field at the center of the circle (both contributions pointing "out of the page"). The current in the arc contributes a term given by Eq. 29-9:

$$B_{\text{arc}} = \frac{\mu_0 i \phi}{4\pi R}$$

pointing into the page.

ANALYZE The total magnetic field is

$$B = 2B_{\text{straight}} - B_{\text{arc}} = 2\left(\frac{\mu_0 i}{4\pi R}\right) - \frac{\mu_0 i \phi}{4\pi R} = \frac{\mu_0 i}{4\pi R}(2-\phi).$$

Therefore, $\phi = 2.00$ rad would produce zero total field at the center of the circle.

LEARN The total contribution of the two semi-infinite wires is the same as that of an infinite wire. Note that the angle ϕ is in radians rather than degrees.

29-29

THINK Our system consists of four long straight wires whose cross section form a square of length a. The magnetic field at the center of the square is the vector sum of the fields of the four wires.

EXPRESS Each wire produces a field with magnitude given by $B = \mu_0 i/2\pi r$, where r is the distance from the corner of the square to the center. According to the Pythagorean theorem, the diagonal of the square has length $\sqrt{2}a$, so $r = a/\sqrt{2}$ and $B = \mu_0 i/\sqrt{2}\pi a$. The fields due to the wires at the upper left (wire 1) and lower right (wire 3) corners both point toward the upper right corner of the square. The fields due to the wires at the upper right (wire 2) and lower left (wire 4) corners both point toward the upper left corner.

ANALYZE The horizontal components of the fields cancel and the vertical components sum to

$$B_{\text{net}} = 4\frac{\mu_0 i}{\sqrt{2}\pi a}\cos 45° = \frac{2\mu_0 i}{\pi a} = \frac{2\left(4\pi\times10^{-7}\ \text{T}\cdot\text{m/A}\right)(20\ \text{A})}{\pi(0.20\ \text{m})} = 8.0\times10^{-5}\ \text{T}.$$

In the calculation, cos 45° was replaced with $1/\sqrt{2}$. The total field points upward, or in the $+y$ direction. Thus, $\vec{B}_{\text{net}} = (8.0\times10^{-5}\ \text{T})\hat{\text{j}}$.

LEARN In the figure below, we show the contributions from the individual wires. The directions of the fields are deduced using the right-hand rule.

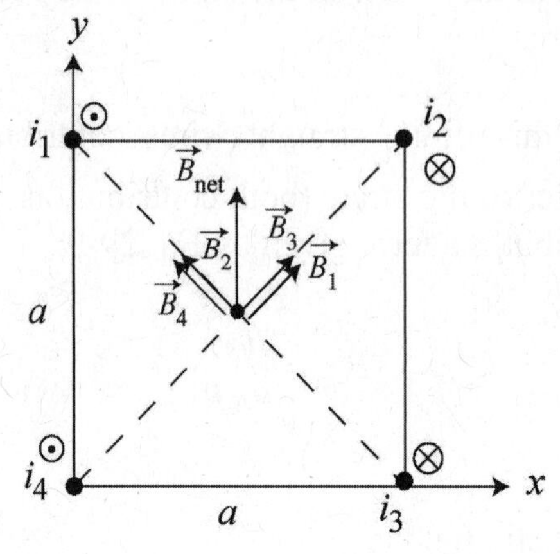

29-33

THINK The magnetic field at point P produced by the current-carrying ribbon (shown in Fig. 29-60) can be calculated using the Biot-Savart law.

EXPRESS Consider a section of the ribbon of thickness dx located a distance x away from point P. The current it carries is $di = i\,dx/w$, and its contribution to B_P is

$$dB_P = \frac{\mu_0 di}{2\pi x} = \frac{\mu_0 i dx}{2\pi x w}.$$

ANALYZE Integrating over the length of the ribbon, we obtain

$$\begin{aligned} B_P &= \int dB_P = \frac{\mu_0 i}{2\pi w}\int_d^{d+w}\frac{dx}{x} = \frac{\mu_0 i}{2\pi w}\ln\left(1+\frac{w}{d}\right) \\ &= \frac{(4\pi\times10^{-7}\ \text{T}\cdot\text{m/A})(4.61\times10^{-6}\ \text{A})}{2\pi(0.0491\ \text{m})}\ln\left(1+\frac{0.0491}{0.0216}\right) \\ &= 2.23\times10^{-11}\ \text{T}. \end{aligned}$$

and $\vec{B}_P$ points upward. In unit-vector notation, $\vec{B}_P = (2.23\times10^{-11}\ \text{T})\hat{\text{j}}$.

LEARN In the limit where $d \gg w$, using

$$\ln(1+x) = x - x^2/2 + \cdots,$$

the magnetic field becomes

$$B_P = \frac{\mu_0 i}{2\pi w} \ln\left(1 + \frac{w}{d}\right) \approx \frac{\mu_0 i}{2\pi w} \cdot \frac{w}{d} = \frac{\mu_0 i}{2\pi d}$$

which is the same as that due to a thin wire.

29-35

THINK The magnitude of the force of wire 1 on wire 2 is given by $F_{21} = \mu_0 i_1 i_2 L / 2\pi r$, where i_1 is the current in wire 1, i_2 is the current in wire 2, and r is the distance between the wires.

EXPRESS The distance between the wires is $r = \sqrt{d_1^2 + d_2^2}$. The x component of the force is $F_{21,x} = F_{21}\cos\phi$, where $\cos\phi = d_2 / \sqrt{d_1^2 + d_2^2}$.

ANALYZE Substituting the values given, the x component of the force per unit length is

$$\frac{F_{21,x}}{L} = \frac{\mu_0 i_1 i_2 d_2}{2\pi(d_1^2 + d_2^2)} = \frac{(4\pi\times10^{-7}\,\text{T}\cdot\text{m/A})(4.00\times10^{-3}\,\text{A})(6.80\times10^{-3}\,\text{A})(0.050\,\text{m})}{2\pi[(0.0240\,\text{m})^2 + (0.050\,\text{m})^2]}$$
$$= 8.84\times10^{-11}\,\text{N/m}.$$

LEARN Since the two currents flow in the opposite directions, the force between the wires is repulsive. Thus, the direction of $\vec{F}_{21}$ is along the line that joins the wire and is away from wire 1.

29-45

THINK The value of the line integral $\oint \vec{B}\cdot d\vec{s}$ is proportional to the net current enclosed.

EXPRESS By Ampere's law, we have $\oint \vec{B}\cdot d\vec{s} = \mu_0 i_{\text{enc}}$, where i_{enc} is the current enclosed by the closed path.

ANALYZE (a) Two of the currents are out of the page and one is into the page, so the net current enclosed by the path, or "Amperian loop" is 2.0 A, out of the page. Since the path is traversed in the clockwise sense, a current into the page is positive and a current out of the page is negative, as indicated by the right-hand rule associated with Ampere's law. Thus,

$$\oint \vec{B}\cdot d\vec{s} = -\mu_0 i = -(4\pi\times10^{-7}\,\text{T}\cdot\text{m/A})(2.0\,\text{A}) = -2.5\times10^{-6}\,\text{T}\cdot\text{m}.$$

(b) The net current enclosed by the path is zero (two currents are out of the page and two are into the page), so $\oint \vec{B}\cdot d\vec{s} = \mu_0 i_{enc} = 0$.

LEARN The value of $\oint \vec{B}\cdot d\vec{s}$ depends only on the current enclosed, and not the shape of the Amperian loop.

29-55

THINK The net field at a point inside the solenoid is the vector sum of the fields of the solenoid and that of the long straight wire along the central axis of the solenoid.

EXPRESS The magnetic field at a point P is given by $\vec{B} = \vec{B}_s + \vec{B}_w$, where $\vec{B}_s$ and $\vec{B}_w$ are the fields due to the solenoid and the wire, respectively. The direction of $\vec{B}_s$ is along the axis of the solenoid, and the direction of $\vec{B}_w$ is perpendicular to it, so the two fields are perpendicular to each other, $\vec{B}_s \perp \vec{B}_w$. For the net field $\vec{B}$ to be at 45° with the axis, we must have $B_s = B_w$.

ANALYZE (a) Thus,

$$B_s = B_w \Rightarrow \quad \mu_0 i_s n = \frac{\mu_0 i_w}{2\pi d},$$

which gives the separation d to point P on the axis:

$$d = \frac{i_w}{2\pi i_s n} = \frac{6.00\,\text{A}}{2\pi(20.0\times10^{-3}\,\text{A})(10\,\text{turns/cm})} = 4.77\,\text{cm}.$$

(b) The magnetic field strength is

$$B = \sqrt{2}B_s = \sqrt{2}\left(4\pi\times10^{-7}\,\text{T}\cdot\text{m/A}\right)\left(20.0\times10^{-3}\,\text{A}\right)\left(10\text{ turns}/0.0100\text{ m}\right) = 3.55\times10^{-5}\,\text{T}.$$

LEARN In general, the angle $\vec{B}$ makes with the solenoid axis is give by

$$\phi = \tan^{-1}\left(\frac{B_w}{B_s}\right) = \tan^{-1}\left(\frac{\mu_0 i_w/2\pi d}{\mu_0 i_s n}\right) = \tan^{-1}\left(\frac{i_w}{2\pi d\, n i_s}\right).$$

29-57

THINK The magnitude of the magnetic dipole moment is given by $\mu = NiA$, where N is the number of turns, i is the current, and A is the area.

EXPRESS The cross-sectional area is a circle, so $A = \pi R^2$, where R is the radius. The magnetic field on the axis of a magnetic dipole, a distance z away, is given by Eq. 29-27:

$$B = \frac{\mu_0}{2\pi}\frac{\mu}{z^3}.$$

ANALYZE (a) Substituting the values given, we find the magnitude of the dipole moment to be

$$\mu = Ni\pi R^2 = (300)(4.0\,\text{A})\pi(0.025\,\text{m})^2 = 2.4\,\text{A}\cdot\text{m}^2.$$

(b) Solving for z, we obtain

$$z = \left(\frac{\mu_0}{2\pi}\frac{\mu}{B}\right)^{1/3} = \left(\frac{(4\pi\times10^{-7}\,\text{T}\cdot\text{m/A})(2.36\,\text{A}\cdot\text{m}^2)}{2\pi(5.0\times10^{-6}\,\text{T})}\right)^{1/3} = 46\,\text{cm}.$$

LEARN Note the similarity between $B = \frac{\mu_0}{2\pi}\frac{\mu}{z^3}$, the magnetic field of a magnetic dipole μ and $E = \frac{1}{2\pi\varepsilon_0}\frac{p}{z^3}$, the electric field of an electric dipole p (see Eq. 22-9).

29-59

THINK The magnitude of the magnetic dipole moment is given by $\mu = NiA$, where N is the number of turns, i is the current, and A is the area.

EXPRESS The cross-sectional area is a circle, so $A = \pi R^2$, where R is the radius.

ANALYZE With $N = 200$, $i = 0.30$ A, and $R = 0.050$ m, the magnitude of the dipole moment is

$$\mu = (200)(0.30\,\text{A})\pi(0.050\,\text{m})^2 = 0.47\,\text{A}\cdot\text{m}^2.$$

LEARN The direction of $\vec{\mu}$ is that of the normal vector $\vec{n}$ to the plane of the coil, in accordance with the right-hand rule shown in Fig. 28-19.

29-75

THINK In this problem, we apply the Biot-Savart law to calculate the magnetic field due to a current-carrying segment at various locations.

EXPRESS The Biot-Savart law can be written as

$$\vec{B}(x,y,z) = \frac{\mu_0}{4\pi}\frac{i\Delta\vec{s}\times\hat{\text{r}}}{r^2} = \frac{\mu_0}{4\pi}\frac{i\Delta\vec{s}\times\vec{r}}{r^3}.$$

With $\Delta\vec{s} = \Delta s\hat{j}$ and $\vec{r} = x\hat{i} + y\hat{j} + z\hat{k}$, their cross product is

$$\Delta\vec{s} \times \vec{r} = (\Delta s\hat{j}) \times (x\hat{i} + y\hat{j} + z\hat{k}) = \Delta s(z\hat{i} - x\hat{k})$$

where we have used $\hat{j}\times\hat{i} = -\hat{k}$, $\hat{j}\times\hat{j} = 0$, and $\hat{j}\times\hat{k} = \hat{i}$. Thus, the Biot-Savart equation becomes

$$\vec{B}(x, y, z) = \frac{\mu_0 i\,\Delta s\left(z\hat{i} - x\hat{k}\right)}{4\pi\left(x^2 + y^2 + z^2\right)^{3/2}}.$$

ANALYZE (a) The field on the z axis (at $x = 0$, $y = 0$, and $z = 5.0$ m) is

$$\vec{B}(0, 0, 5.0\text{ m}) = \frac{\left(4\pi\times10^{-7}\text{T}\cdot\text{m/A}\right)\left(2.0\text{A}\right)\left(3.0\times10^{-2}\text{m}\right)\left(5.0\text{ m}\right)\hat{i}}{4\pi\left(0^2 + 0^2 + (5.0\text{ m})^2\right)^{3/2}} = (2.4\times10^{-10}\text{ T})\hat{i}.$$

(b) Similarly, $\vec{B}$ (0, 6.0 m, 0) = 0, since $x = z = 0$.

(c) The field in the xy plane, at (x, y, z) = (7 m, 7 m, 0), is

$$\vec{B}(7.0\text{ m}, 7.0\text{ m}, 0) = \frac{(4\pi\times10^{-7}\text{ T}\cdot\text{m/A})(2.0\text{ A})(3.0\times10^{-2}\text{ m})(-7.0\text{ m})\hat{k}}{4\pi\left((7.0\text{ m})^2 + (7.0\text{ m})^2 + 0^2\right)^{3/2}} = (-4.3\times10^{-11}\text{ T})\hat{k}.$$

(d) The field in the xy plane, at (x, y, z) = (–3, –4, 0), is

$$\vec{B}(-3.0\text{m}, -4.0\text{m}, 0) = \frac{(4\pi\times10^{-7}\text{ T}\cdot\text{m/A})(2.0\text{ A})(3.0\times10^{-2}\text{ m})(3.0\text{ m})\hat{k}}{4\pi\left((-3.0\text{m})^2 + (-4.0\text{m})^2 + 0^2\right)^{3/2}} = (1.4\times10^{-10}\text{ T})\hat{k}.$$

LEARN Along the x and z axes, the expressions for $\vec{B}$ simplify to

$$\vec{B}(x,0,0) = -\frac{\mu_0}{4\pi}\frac{i\,\Delta s}{x^2}\hat{k}, \quad \vec{B}(0,0,z) = \frac{\mu_0}{4\pi}\frac{i\,\Delta s}{z^2}\hat{i}.$$

The magnetic field at any point on the y axis vanishes because the current flows in the $+y$ direction, so $d\vec{s}\times\hat{r} = 0$.

29-81

THINK The objective of this problem is to calculate the magnetic field due to an infinite current sheet by applying Ampere's law.

EXPRESS The "current per unit x-length" may be viewed as current density multiplied by the thickness Δy of the sheet; thus, $\lambda = J\Delta y$. Ampere's law may be (and often is) expressed in terms of the current density vector as follows:

$$\oint \vec{B} \cdot d\vec{s} = \mu_0 \int \vec{J} \cdot d\vec{A}$$

where the area integral is over the region enclosed by the path relevant to the line integral (and $\vec{J}$ is in the $+z$ direction, out of the paper). With J uniform throughout the sheet, then it is clear that the right-hand side of this version of Ampere's law should reduce, in this problem, to

$$\mu_0 JA = \mu_0 J \Delta y \Delta x = \mu_0 \lambda \Delta x.$$

ANALYZE (a) Figure 29-83 certainly has the horizontal components of $\vec{B}$ drawn correctly at points P and P', so the question becomes: is it possible for $\vec{B}$ to have vertical components in the figure?

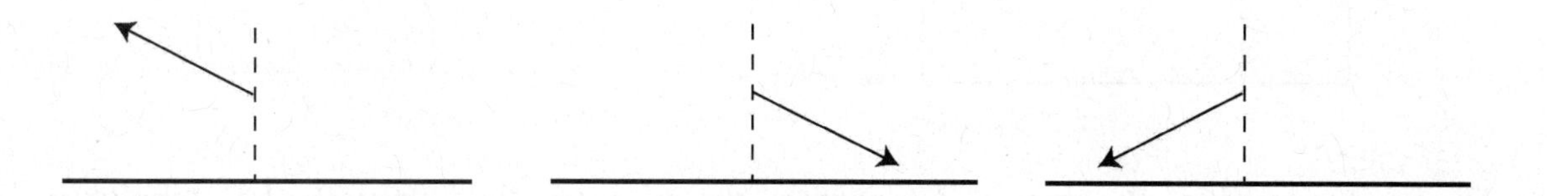

Our focus is on point P. Suppose the magnetic field is not parallel to the sheet, as shown in the upper left diagram. If we reverse the direction of the current, then the direction of the field will also be reversed (as shown in the upper middle diagram). Now, if we rotate the sheet by 180° about a line that is perpendicular to the sheet, the field will rotate and point in the direction shown in the diagram on the upper right. The current distribution now is exactly the same as the original; however, comparing the upper left and upper right diagrams, we see that the fields are not the same, unless the original field is parallel to the sheet and only has a horizontal component. That is, the field at P must be purely horizontal, as drawn in Fig. 29-83.

(b) The path used in evaluating $\oint \vec{B} \cdot d\vec{s}$ is rectangular, of horizontal length Δx (the horizontal sides passing through points P and P', respectively) and vertical size $\delta y > \Delta y$. The vertical sides have no contribution to the integral since $\vec{B}$ is purely horizontal (so the scalar dot product produces zero for those sides), and the horizontal sides contribute two equal terms, as shown next. Ampere's law yields

$$2B\Delta x = \mu_0 \lambda \Delta x \Rightarrow B = \frac{1}{2}\mu_0 \lambda.$$

LEARN In order to apply Ampere's law, the system must possess certain symmetry. In the case of an infinite current sheet, the symmetry is planar.

29-83

THINK The magnetic field at P is the vector sum of the fields of the individual wire segments.

EXPRESS The two small wire segments, each of length $a/4$, shown in Fig. 29-85 nearest to point P, are labeled 1 and 8 in the figure (below left). Let $-\hat{k}$ be a unit vector pointing into the page.

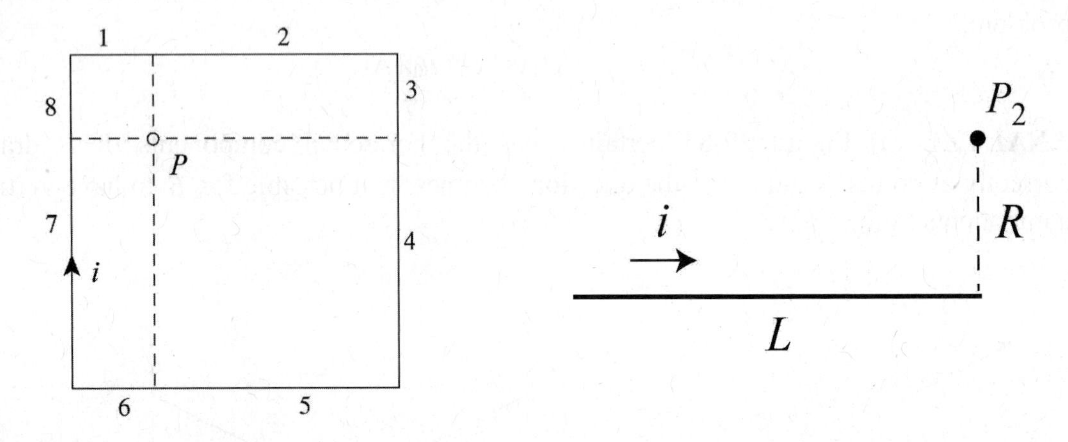

We use the result of Problem 29-17: namely, the magnetic field strength at P_2 (shown in Fig. 29-43 and upper right) is

$$B_{P_2} = \frac{\mu_0 i}{4\pi R}\frac{L}{\sqrt{L^2 + R^2}}.$$

Therefore, the magnetic fields due to the 8 segments are

$$B_{P1} = B_{P8} = \frac{\sqrt{2}\mu_0 i}{8\pi(a/4)} = \frac{\sqrt{2}\mu_0 i}{2\pi a},$$

$$B_{P4} = B_{P5} = \frac{\sqrt{2}\mu_0 i}{8\pi(3a/4)} = \frac{\sqrt{2}\mu_0 i}{6\pi a},$$

$$B_{P2} = B_{P7} = \frac{\mu_0 i}{4\pi(a/4)} \cdot \frac{3a/4}{\left[(3a/4)^2 + (a/4)^2\right]^{1/2}} = \frac{3\mu_0 i}{\sqrt{10}\pi a},$$

and

$$B_{P3} = B_{P6} = \frac{\mu_0 i}{4\pi(3a/4)} \cdot \frac{a/4}{\left[(a/4)^2 + (3a/4)^2\right]^{1/2}} = \frac{\mu_0 i}{3\sqrt{10}\pi a}.$$

ANALYZE Adding up all the contributions, the total magnetic field at P is

$$\vec{B}_P = \sum_{n=1}^{8} B_{Pn}(-\hat{k}) = 2\frac{\mu_0 i}{\pi a}\left(\frac{\sqrt{2}}{2}+\frac{\sqrt{2}}{6}+\frac{3}{\sqrt{10}}+\frac{1}{3\sqrt{10}}\right)(-\hat{k})$$
$$= \frac{2\left(4\pi\times10^{-7}\,\text{T}\cdot\text{m/A}\right)(10\text{A})}{\pi\left(8.0\times10^{-2}\text{m}\right)}\left(\frac{\sqrt{2}}{2}+\frac{\sqrt{2}}{6}+\frac{3}{\sqrt{10}}+\frac{1}{3\sqrt{10}}\right)(-\hat{k})$$
$$= \left(2.0\times10^{-4}\text{T}\right)(-\hat{k}).$$

LEARN If point P is located at the center of the square, then each segment would contribute

$$B_{P1} = B_{P2} = \cdots B_{P8} = \frac{\sqrt{2}\mu_0 i}{4\pi a},$$

making the total field

$$B_{\text{center}} = 8B_{P1} = \frac{8\sqrt{2}\mu_0 i}{4\pi a}.$$

29-85

THINK The hollow conductor has cylindrical symmetry, so Ampere's law can be applied to calculate the magnetic field due to the current distribution.

EXPRESS Ampere's law states that $\oint \vec{B}\cdot d\vec{s} = \mu_0 i_{\text{enc}}$, where i_{enc} is the current enclosed by the closed path, or Amperian loop. We choose the Amperian loop to be a circle of radius r and concentric with the cylindrical shell. Since the current is uniformly distributed throughout the cross section of the shell, the enclosed current is

$$i_{\text{enc}} = i\frac{\pi(r^2-b^2)}{\pi(a^2-b^2)} = i\left(\frac{r^2-b^2}{a^2-b^2}\right).$$

ANALYZE (a) Thus, in the region $b < r < a$, we have

$$\oint \vec{B}\cdot d\vec{s} = 2\pi r B = \mu_0 i_{\text{enc}} = \mu_0 i\left(\frac{r^2-b^2}{a^2-b^2}\right)$$

which gives $B = \dfrac{\mu_0 i}{2\pi\left(a^2-b^2\right)}\left(\dfrac{r^2-b^2}{r}\right).$

(b) At $r = a$, the magnetic field strength is

$$\frac{\mu_0 i}{2\pi\left(a^2-b^2\right)}\left(\frac{a^2-b^2}{a}\right) = \frac{\mu_0 i}{2\pi a}.$$

At $r = b, B \propto r^2 - b^2 = 0$. Finally, for $b = 0$

$$B = \frac{\mu_0 i}{2\pi a^2}\frac{r^2}{r} = \frac{\mu_0 i r}{2\pi a^2}$$

which agrees with Eq. 29-20.

(c) The field is zero for $r < b$ and is equal to Eq. 29-17 for $r > a$, so this along with the result of part (a) provides a determination of B over the full range of values. The graph (with SI units understood) is shown below.

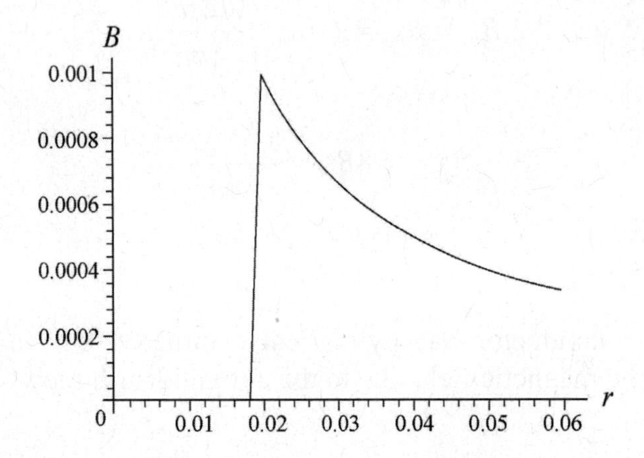

LEARN For $r < b$, the field is zero, and for $r > a$, the field decreases as $1/r$. In the region $b < r < a$, the field increases with r as $r - b^2/r$.

Chapter 30

30-3

THINK Changing the current in the solenoid changes the flux, and therefore, induces a current in the coil.

EXPRESS Using Faraday's law, the total induced emf is given by

$$\varepsilon = -N\frac{d\Phi_B}{dt} = -NA\left(\frac{dB}{dt}\right) = -NA\frac{d}{dt}(\mu_0 ni) = -N\mu_0 nA\frac{di}{dt} = -N\mu_0 n(\pi r^2)\frac{di}{dt}$$

By Ohm's law, the induced current in the coil is $i_{\text{ind}} = |\varepsilon|/R$, where R is the resistance of the coil.

ANALYZE Substituting the values given, we obtain

$$\varepsilon = -N\mu_0 n(\pi r^2)\frac{di}{dt} = -(120)(4\pi\times10^{-7}\,\text{T}\cdot\text{m/A})(22000/\text{m})\,\pi(0.016\text{ m})^2\left(\frac{1.5\text{ A}}{0.025\text{ s}}\right) = 0.16\text{V}.$$

Ohm's law then yields $i_{\text{ind}} = \dfrac{|\varepsilon|}{R} = \dfrac{0.016\text{ V}}{5.3\,\Omega} = 0.030\text{ A}.$

LEARN The direction of the induced current can be deduced from Lenz's law, which states that the direction of the induced current is such that the magnetic field which it produces opposes the change in flux that induces the current.

30-23

THINK Increasing the separation between the two loops changes the flux through the smaller loop and, therefore, induces a current in the smaller loop.

EXPRESS The magnetic flux through a surface is given by $\Phi_B = \int \vec{B}\cdot d\vec{A}$, where $\vec{B}$ is the magnetic field and $d\vec{A}$ is a vector of magnitude dA that is normal to a differential area dA. In the case where $\vec{B}$ is uniform and perpendicular to the plane of the loop, $\Phi_B = BA$.

In the region of the smaller loop the magnetic field produced by the larger loop may be taken to be uniform and equal to its value at the center of the smaller loop, on the axis. Equation 29-27, with $z = x$ (taken to be much greater than R), gives

$$\vec{B} = \frac{\mu_0 i R^2}{2x^3}\hat{\text{i}}$$

where the $+x$ direction is upward in Fig. 30-45. The area of the smaller loop is $A = \pi r^2$.

ANALYZE (a) The magnetic flux through the smaller loop is, to a good approximation, the product of this field and the area of the smaller loop:

$$\Phi_B = BA = \frac{\pi \mu_0 i r^2 R^2}{2x^3}.$$

(b) The emf is given by Faraday's law:

$$\varepsilon = -\frac{d\Phi_B}{dt} = -\left(\frac{\pi\mu_0 i r^2 R^2}{2}\right)\frac{d}{dt}\left(\frac{1}{x^3}\right) = -\left(\frac{\pi\mu_0 i r^2 R^2}{2}\right)\left(-\frac{3}{x^4}\frac{dx}{dt}\right) = \frac{3\pi\mu_0 i r^2 R^2 v}{2x^4}.$$

(c) As the smaller loop moves upward, the flux through it decreases. The induced current will be directed so as to produce a magnetic field that is upward through the smaller loop, in the same direction as the field of the larger loop. It will be counterclockwise as viewed from above, in the same direction as the current in the larger loop.

LEARN The situation in this problem is like that shown in Fig. 30-5(d). The induced magnetic field is in the same direction as the initial magnetic field.

30-31

THINK Thermal energy is generated at the rate given by $P = \varepsilon^2/R$ (see Eq. 27-23), where ε is the emf in the wire and R is the resistance of the wire.

EXPRESS Using Eq. 27-16, the resistance is given by $R = \rho L/A$, where the resistivity is $1.69 \times 10^{-8}\ \Omega\cdot\text{m}$ (by Table 27-1) and $A = \pi d^2/4$ is the cross-sectional area of the wire ($d = 0.00100$ m is the wire thickness). The area *enclosed* by the loop is

$$A_{\text{loop}} = \pi r_{\text{loop}}^2 = \pi\left(\frac{L}{2\pi}\right)^2$$

since the length of the wire ($L = 0.500$ m) is the circumference of the loop. This enclosed area is used in Faraday's law to give the induced emf:

$$\varepsilon = -\frac{d\Phi_B}{dt} = -A_{\text{loop}}\frac{dB}{dt} = -\frac{L^2}{4\pi}\frac{dB}{dt}.$$

ANALYZE The rate of change of the field is $dB/dt = 0.0100$ T/s. Thus, we obtain

$$P = \frac{|\varepsilon|^2}{R} = \frac{(L^2/4\pi)^2(dB/dt)^2}{\rho L/(\pi d^2/4)} = \frac{d^2L^3}{64\pi\rho}\left(\frac{dB}{dt}\right)^2 = \frac{(1.00\times10^{-3}\text{ m})^2(0.500\text{ m})^3}{64\pi(1.69\times10^{-8}\Omega\cdot\text{m})}(0.0100\text{ T/s})^2$$
$$= 3.68\times10^{-6}\text{ W}.$$

LEARN The rate of thermal energy generated is proportional to $(dB/dt)^2$.

30-37

THINK Changing magnetic field induces an electric field.

EXPRESS The induced electric field is given by Eq. 30-20:

$$\oint \vec{E}\cdot d\vec{s} = -\frac{d\Phi_B}{dt}.$$

ANALYZE (a) The point at which we are evaluating the field is inside the solenoid, so

$$E(2\pi r) = -(\pi r^2)\frac{dB}{dt} \quad\Rightarrow\quad E = -\frac{1}{2}\frac{dB}{dt}r.$$

The magnitude of the induced electric field is

$$|E| = \frac{1}{2}\frac{dB}{dt}r = \frac{1}{2}\left(6.5\times10^{-3}\text{ T/s}\right)\left(0.0220\text{ m}\right) = 7.15\times10^{-5}\text{ V/m}.$$

(b) Now the point at which we are evaluating the field is outside the solenoid, so

$$E(2\pi r) = -(\pi R^2)\frac{dB}{dt} \quad\Rightarrow\quad E = -\frac{1}{2}\frac{dB}{dt}\frac{R^2}{r}.$$

The magnitude of the induced field is

$$|E| = \frac{1}{2}\frac{dB}{dt}\frac{R^2}{r} = \frac{1}{2}\left(6.5\times10^{-3}\text{ T/s}\right)\frac{\left(0.0600\text{ m}\right)^2}{0.0820\text{ m}} = 1.43\times10^{-4}\text{ V/m}.$$

LEARN The magnitude of the induced electric field as a function of r is shown next:

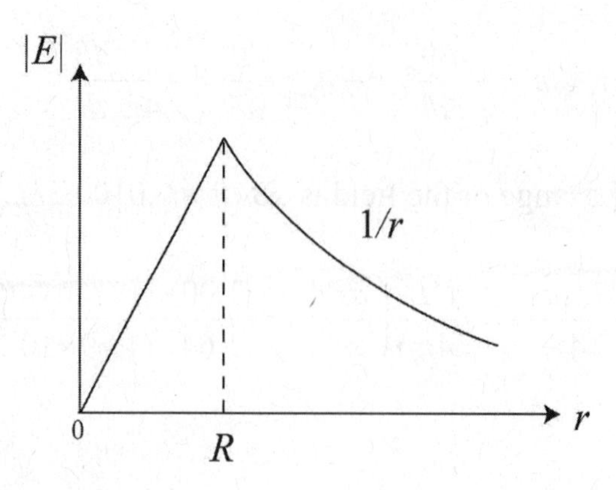

Inside the solenoid, $r < R$, the field $|E|$ is linear in r. However, outside the solenoid, $r > R$, $|E| \sim 1/r$.

30-53

THINK The inductor in the *RL* circuit initially acts to oppose changes in current through it.

EXPRESS If the battery is switched into the circuit at $t = 0$, then the current at a later time t is given by

$$i = \frac{\varepsilon}{R}\left(1 - e^{-t/\tau_L}\right),$$

where $\tau_L = L/R$.

(a) We want to find the time at which $i = 0.800\varepsilon/R$. This means

$$0.800 = 1 - e^{-t/\tau_L} \Rightarrow e^{-t/\tau_L} = 0.200.$$

Taking the natural logarithm of both sides, we obtain $-(t/\tau_L) = \ln(0.200) = -1.609$. Thus,

$$t = 1.609\tau_L = \frac{1.609L}{R} = \frac{1.609(6.30\times10^{-6}\,\text{H})}{1.20\times10^{3}\,\Omega} = 8.45\times10^{-9}\,\text{s}.$$

(b) At $t = 1.0\,\tau_L$ the current in the circuit is

$$i = \frac{\varepsilon}{R}\left(1 - e^{-1.0}\right) = \left(\frac{14.0\,\text{V}}{1.20\times10^{3}\Omega}\right)(1 - e^{-1.0}) = 7.37\times10^{-3}\ \text{A}.$$

LEARN At $t = 0$, the current in the circuit is zero. However, after a very long time, the inductor acts like an ordinary connecting wire, so the current is

$$i_0 = \frac{\varepsilon}{R} = \frac{14.0\,\text{V}}{1.20\times10^{3}\Omega} = 0.0117\ \text{A}.$$

The current as a function of t/τ_L is plotted below.

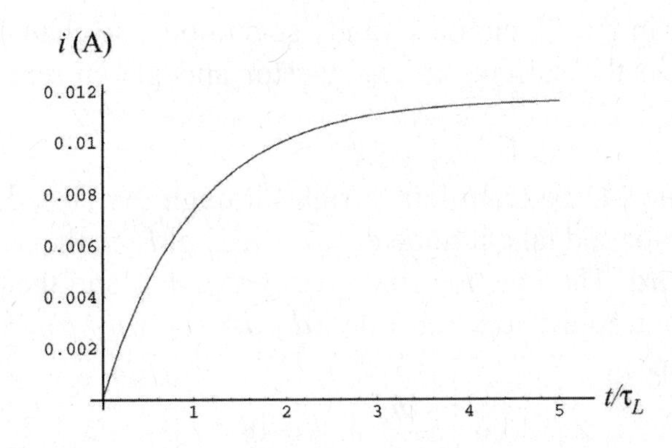

30-55

THINK The inductor in the *RL* circuit initially acts to oppose changes in current through it.

EXPRESS Starting with zero current at $t = 0$ (the moment the switch is closed) the current in the circuit increases according to

$$i = \frac{\varepsilon}{R}\left(1 - e^{-t/\tau_L}\right),$$

where $\tau_L = L/R$ is the inductive time constant and ε is the battery emf.

ANALYZE To calculate the time at which $i = 0.9990\varepsilon/R$, we solve for t:

$$0.990\frac{\varepsilon}{R} = \frac{\varepsilon}{R}\left(1 - e^{-t/\tau_L}\right) \Rightarrow \ln(0.0010) = -\frac{t}{\tau_L} \Rightarrow \frac{t}{\tau_L} = 6.91.$$

LEARN At $t = 0$, the current in the circuit is zero. However, after a very long time, the inductor acts like an ordinary connecting wire, so the current is $i_0 = \varepsilon/R$. The current (in terms of i/i_0) as a function of t/τ_L is plotted below.

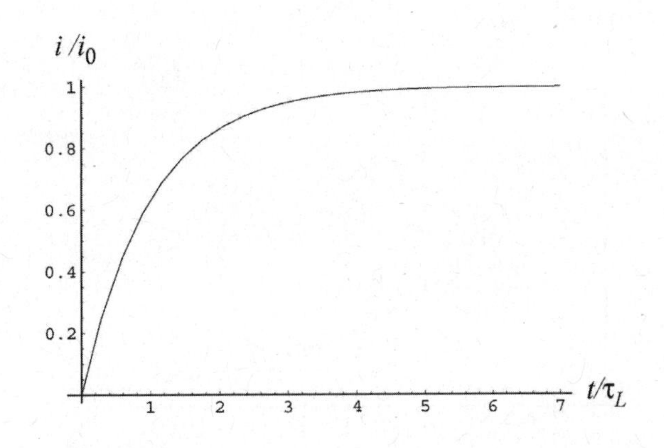

30-59

THINK The inductor in the *RL* circuit initially acts to oppose changes in current through it. We are interested in the currents in the resistor and the current in the inductor as a function of time.

EXPRESS We assume i to be from left to right through the closed switch. We let i_1 be the current in the resistor and take it to be downward. Let i_2 be the current in the inductor, also assumed downward. The junction rule gives $i = i_1 + i_2$ and the loop rule gives $i_1R - L(di_2/dt) = 0$. According to the junction rule, $(di_1/dt) = -(di_2/dt)$. We substitute into the loop equation to obtain

$$L\frac{di_1}{dt} + i_1R = 0.$$

This equation is similar to Eq. 30-46, and its solution is the function given as Eq. 30-47: $i_1 = i_0e^{-Rt/L}$, where i_0 is the current through the resistor at $t = 0$, just after the switch is closed. Now just after the switch is closed, the inductor prevents the rapid build-up of current in its branch, so at that moment $i_2 = 0$ and $i_1 = i$. Thus $i_0 = i$.

ANALYZE (a) The currents in the resistor and the inductor as a function of time are:

$$i_1 = ie^{-Rt/L}, \quad i_2 = i - i_1 = i\left(1 - e^{-Rt/L}\right).$$

(b) When $i_2 = i_1$, we have

$$e^{-Rt/L} = 1 - e^{-Rt/L} \Rightarrow e^{-Rt/L} = \frac{1}{2}.$$

Taking the natural logarithm of both sides and using $\ln(1/2) = -\ln 2$, we obtain

$$\left(\frac{Rt}{L}\right) = \ln 2 \Rightarrow t = \frac{L}{R}\ln 2.$$

LEARN A plot of i_1/i (solid line, for resistor) and i_2/i (dashed line, for inductor) as a function of t/τ_L is shown below.

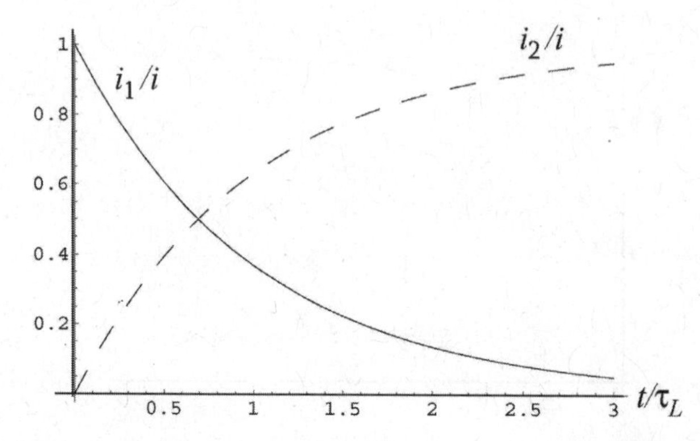

30-61

THINK Inductance L is related to the inductive time constant of an RL circuit by $L=\tau_L R$, where R is the resistance in the circuit. The energy stored by an inductor carrying current i is given by $U_B = Li^2/2$.

EXPRESS If the battery is applied at time $t = 0$ the current is given by

$$i=\frac{\varepsilon}{R}\left(1-e^{-t/\tau_L}\right),$$

where ε is the emf of the battery, R is the resistance, and τ_L is the inductive time constant (L/R). This leads to

$$e^{-t/\tau_L}=1-\frac{iR}{\varepsilon}\Rightarrow -\frac{t}{\tau_L}=\ln\left(1-\frac{iR}{\varepsilon}\right).$$

Since

$$\ln\left(1-\frac{iR}{\varepsilon}\right)=\ln\left[1-\frac{\left(2.00\times10^{-3}\,\text{A}\right)\left(10.0\times10^{3}\,\Omega\right)}{50.0\,\text{V}}\right]=-0.5108,$$

the inductive time constant is $\tau_L = t/0.5108 = (5.00\times10^{-3}\text{ s})/0.5108 = 9.79\times10^{-3}$ s.

ANALYZE (a) The inductance is

$$L=\tau_L R=\left(9.79\times10^{-3}\,\text{s}\right)\left(10.0\times10^{3}\,\Omega\right)=97.9\,\text{H}.$$

(b) The energy stored in the coil is

$$U_B=\frac{1}{2}Li^2=\frac{1}{2}(97.9\,\text{H})\left(2.00\times10^{-3}\,\text{A}\right)^2=1.96\times10^{-4}\,\text{J}.$$

LEARN Note the similarity between $U_B=\frac{1}{2}Li^2$ and $U_C=\frac{q^2}{2C}$, the electric energy stored in a capacitor.

30-67

THINK The magnetic energy density is given by $u_B = B^2/2\mu_0$, where B is the magnitude of the magnetic field at that point.

EXPRESS Inside a solenoid, the magnitude of the magnetic field is $B = \mu_0 ni$, where

$$n = (950 \text{ turns})/(0.850 \text{ m}) = 1.118 \times 10^3 \text{ m}^{-1}.$$

Thus, the energy density is

$$u_B = \frac{B^2}{2\mu_0} = \frac{(\mu_0 ni)^2}{2\mu_0} = \frac{1}{2}\mu_0 n^2 i^2.$$

Since the magnetic field is uniform inside an ideal solenoid, the total energy stored in the field is $U_B = u_B \mathcal{V}$, where $\mathcal{V}$ is the volume of the solenoid.

ANALYZE (a) Substituting the values given, we find the magnetic energy density to be

$$u_B = \frac{1}{2}\mu_0 n^2 i^2 = \frac{1}{2}\left(4\pi \times 10^{-7} \text{ T}\cdot\text{m/A}\right)\left(1.118 \times 10^3 \text{ m}^{-1}\right)^2 (6.60 \text{ A})^2 = 34.2 \text{ J/m}^3 .$$

(b) The volume $\mathcal{V}$ is calculated as the product of the cross-sectional area and the length. Thus,

$$U_B = \left(34.2 \text{ J/m}^3\right)\left(17.0 \times 10^{-4} \text{ m}^2\right)(0.850 \text{ m}) = 4.94 \times 10^{-2} \text{ J} .$$

LEARN Note the similarity between $u_B = \dfrac{B^2}{2\mu_0}$, the energy density at a point in a magnetic field, and $u_E = \dfrac{1}{2}\varepsilon_0 E^2$, the energy density at a point in an electric field. Both quantities are proportional to the square of the fields.

30-73

THINK If two coils are near each other, mutual induction can take place whereby a changing current in one coil can induce an emf in the other.

EXPRESS The mutual inductance is given by

$$\varepsilon_1 = -M\frac{di_2}{dt}$$

where ε_1 is the induced emf in coil 1 due to the changing current in coil 2. The flux linkage in coil 2 is $N_2\Phi_{21} = Mi_1$.

ANALYZE (a) From the equation above, we find the mutual inductance to be

$$M = \frac{|\varepsilon_1|}{di_2/dt} = \frac{25.0 \text{ mV}}{15.0 \text{ A/s}} = 1.67 \text{ mH}.$$

(b) Similarly, the flux linkage in coil 2 is

$$N_2\Phi_{21} = Mi_1 = (1.67\,\text{mH})(3.60\,\text{A}) = 6.00\,\text{mWb}.$$

LEARN The emf induced in one coil is proportional to the rate at which current in the other coil is changing:

$$\varepsilon_1 = -M_{12}\frac{di_2}{dt}, \quad \varepsilon_2 = -M_{21}\frac{di_1}{dt}.$$

The proportionality constants, M_{12} and M_{21}, are the same, $M_{12} = M_{21} = M$, so we simply write

$$\varepsilon_1 = -M\frac{di_2}{dt}, \quad \varepsilon_2 = -M\frac{di_1}{dt}.$$

30-77

THINK To find the equivalent inductance, we calculate the total emf across both coils.

EXPRESS We assume the current to be changing at (nonzero) a rate *di*/*dt*. The induced emf's can take on the following form:

$$\varepsilon_1 = -(L_1 \pm M)\frac{di}{dt}, \quad \varepsilon_2 = -(L_2 \pm M)\frac{di}{dt}$$

The relative sign between *L* and *M* depends on how the coils are connected, as we shall see below.

ANALYZE (a) The connection is shown in Fig. 30-68. First consider coil 1. The magnetic field due to the current in that coil points to the right. The magnetic field due to the current in coil 2 also points to the right. When the current increases, both fields increase and both changes in flux contribute emfs in the same direction. Thus, the induced emfs are

$$\varepsilon_1 = -(L_1 + M)\frac{di}{dt}, \quad \varepsilon_2 = -(L_2 + M)\frac{di}{dt}.$$

Therefore, the total emf across both coils is

$$\varepsilon = \varepsilon_1 + \varepsilon_2 = -(L_1 + L_2 + 2M)\frac{di}{dt}$$

which is exactly the emf that would be produced if the coils were replaced by a single coil with inductance $L_{eq} = L_1 + L_2 + 2M$.

(b) We imagine reversing the leads of coil 2 so the current enters at the back of the coil rather than the front (as pictured in Fig. 30-68). Then the field produced by coil 2 at the site of coil 1 is opposite to the field produced by coil 1 itself. The fluxes have opposite signs. An increasing current in coil 1 tends to increase the flux in that coil, but an increasing current in coil 2 tends to decrease it. The emf across coil 1 is

$$\varepsilon_1 = -(L_1 - M)\frac{di}{dt}.$$

Similarly, the emf across coil 2 is

$$\varepsilon_2 = -(L_2 - M)\frac{di}{dt}.$$

The total emf across both coils is

$$\varepsilon = -(L_1 + L_2 - 2M)\frac{di}{dt}.$$

This is the same as the emf that would be produced by a single coil with inductance

$$L_{eq} = L_1 + L_2 - 2M.$$

LEARN The sign of the mutual inductance term is determined by the senses of the coil winding. The induced emfs can either reinforce one another ($L + M$), or oppose one another ($L - M$).

30-79

THINK The inductor in the *RL* circuit initially acts to oppose changes in current through it.

EXPRESS When the switch *S* is just closed, $V_1 = \varepsilon$ and no current flows through the inductor. A long time later, the currents have reached their equilibrium values and the inductor acts as an ordinary connecting wire; we can solve the multi-loop circuit problem by applying Kirchhoff's junction and loop rules.

ANALYZE (a) Applying the loop rule to the left loop gives $\varepsilon - i_1 R_1 = 0$, so

$$i_1 = \varepsilon/R_1 = 10\text{ V}/5.0\ \Omega = 2.0\text{ A}.$$

(b) Since now $\varepsilon_L = \varepsilon$, we have $i_2 = 0$.

(c) The junction rule gives $i_s = i_1 + i_2 = 2.0\text{ A} + 0 = 2.0\text{ A}$.

(d) Since $V_L = \varepsilon$, the potential difference across resistor 2 is $V_2 = \varepsilon - \varepsilon_L = 0$.

(e) The potential difference across the inductor is $V_L = \varepsilon = 10$ V.

(f) The rate of change of current is $\dfrac{di_2}{dt} = \dfrac{V_L}{L} = \dfrac{\varepsilon}{L} = \dfrac{10\text{ V}}{5.0\text{ H}} = 2.0\text{ A/s}$.

(g) After a long time, we still have $V_1 = \varepsilon$, so $i_1 = 2.0$ A.

(h) Since now $V_L = 0$, $i_2 = \varepsilon/R_2 = 10\ \text{V}/10\ \Omega = 1.0$ A.

(i) The current through the switch is now $i_s = i_1 + i_2 = 2.0\ \text{A} + 1.0\ \text{A} = 3.0$ A.

(j) Since $V_L = 0$, $V_2 = \varepsilon - V_L = \varepsilon = 10$ V.

(k) With the inductor acting as an ordinary connecting wire, we have $V_L = 0$.

(l) The rate of change of current in resistor 2 is $\dfrac{di_2}{dt} = \dfrac{V_L}{L} = 0$.

LEARN In analyzing an *RL* circuit immediately after closing the switch and a very long time after that, there is no need to solve any differential equation.

30-81

THINK The magnetic flux through the area enclosed by the loop varies as the loop moves from region 1 to region 2, that have different magnetic fields.

EXPRESS Using Ohm's law, we relate the induced current to the emf and (the absolute value of) Faraday's law:

$$i = \frac{|\varepsilon|}{R} = \frac{1}{R}\left|\frac{d\Phi}{dt}\right|.$$

As the loop is crossing the boundary between regions 1 and 2 (so that "x" amount of its length is in region 2 while "$D - x$" amount of its length remains in region 1) the flux is

$$\Phi_B = xHB_2 + (D - x)HB_1 = DHB_1 + xH(B_2 - B_1)$$

which means

$$\frac{d\Phi_B}{dt} = \frac{dx}{dt}H(B_2 - B_1) = vH(B_2 - B_1) \Rightarrow \quad i = vH(B_2 - B_1)/R.$$

Similar considerations hold (replacing "B_1" with 0 and "B_2" with B_1) for the loop crossing initially from the zero-field region (to the left of Fig. 30-72(a)) into region 1.

ANALYZE (a) In this latter case, appeal to Fig. 30-72(b) leads to

$$3.0 \times 10^{-6}\ \text{A} = (0.40\ \text{m/s})(0.015\ \text{m})\, B_1/(0.020\ \Omega)$$

which yields $B_1 = 10\ \mu\text{T}$.

(b) Lenz's law considerations lead us to conclude that the direction of the region 1 field is *out of the page*.

(c) Similarly, $i = vH(B_2 - B_1)/R$ leads to $B_2 = 3.3\mu\text{T}$.

(d) The direction of $\vec{B}_2$ is out of the page.

LEARN The induced current is non-zero only when the loop is crossing a boundary. If the loop is entirely within one region (region 1, region 2, or outside), the change in magnetic flux is zero, and the corresponding induced current is also zero.

30-85

THINK Changing magnetic field induces an electric field.

EXPRESS The induced electric field is given by Eq. 30-20:

$$\oint \vec{E}\cdot d\vec{s} = -\frac{d\Phi_B}{dt}.$$

The electric field lines are circles that are concentric with the cylindrical region. Thus,

$$E(2\pi r) = -(\pi r^2)\frac{dB}{dt} \quad \Rightarrow \quad E = -\frac{1}{2}\frac{dB}{dt}r.$$

The force on the electron is $\vec{F} = -e\vec{E}$, so by Newton's second law, the acceleration is $\vec{a} = -e\vec{E}/m$.

ANALYZE (a) At point a,

$$E = -\frac{r}{2}\left(\frac{dB}{dt}\right) = -\frac{1}{2}(5.0\times10^{-2}\,\text{m})(-10\times10^{-3}\,\text{T/s}) = 2.5\times10^{-4}\ \text{V/m}.$$

With the normal taken to be into the page, in the direction of the magnetic field, the positive direction for $\vec{E}$ is clockwise. Thus, the direction of the electric field at point a is to the left, that is $\vec{E} = -(2.5\times10^{-4}\ \text{V/m})\hat{\text{i}}$. The resulting acceleration is

$$\vec{a}_a = \frac{-e\vec{E}}{m} = \frac{(-1.60\times10^{-19}\,\text{C})(-2.5\times10^{-4}\ \text{V/m})}{9.11\times10^{-31}\,\text{kg}}\hat{\text{i}} = (4.4\times10^{7}\ \text{m/s}^2)\hat{\text{i}}.$$

The acceleration is to the right.

(b) At point b we have $r_b = 0$, so the acceleration is zero.

(c) The electric field at point c has the same magnitude as the field in a, but with its direction reversed. Thus, the acceleration of the electron released at point c is

$$\vec{a}_c = -\vec{a}_a = -(4.4\times10^7\ \text{m/s}^2)\hat{\text{i}}.$$

LEARN Inside the cylindrical region, the induced electric field increases with r. Therefore, the greater the value of r, the greater the magnitude of acceleration.

30-87

THINK Changing the area of the loop changes the flux through it. An induced emf is produced to oppose this change.

EXPRESS The magnetic flux through the loop is $\Phi_B = BA$, where B is the magnitude of the magnetic field and A is the area of the loop. According to Faraday's law, the magnitude of the average induced emf is

$$\varepsilon_{\text{avg}} = \left|\frac{-d\Phi_B}{dt}\right| = \left|\frac{\Delta\Phi_B}{\Delta t}\right| = \frac{B|\Delta A|}{\Delta t}.$$

ANALYZE (a) substituting the values given, we obtain

$$\varepsilon_{\text{avg}} = \frac{B|\Delta A|}{\Delta t} = \frac{(2.0\ \text{T})(0.20\ \text{m})^2}{0.20\ \text{s}} = 0.40\text{V}.$$

(b) The average induced current is $i_{\text{avg}} = \dfrac{\varepsilon_{\text{avg}}}{R} = \dfrac{0.40\,\text{V}}{20\times10^{-3}\,\Omega} = 20\,\text{A}$.

LEARN By Lenz's law, the more rapidly the area is changing, the greater the induced current in the loop.

30-91

THINK We have an RL circuit in which the inductor is in series with the battery.

EXPRESS As the switch closes at $t = 0$, the current being zero in the inductor serves as an initial condition for the building-up of current in the circuit.

ANALYZE (a) At $t = 0$, the current through the battery is also zero.

(b) With no current anywhere in the circuit at $t = 0$, the loop rule requires the emf of the inductor ε_L to cancel that of the battery ($\varepsilon = 40$ V). Thus, the absolute value of Eq. 30-35 yields

$$\frac{di_{\text{bat}}}{dt} = \frac{|\varepsilon_L|}{L} = \frac{40\text{ V}}{0.050\text{ H}} = 8.0\times10^2\text{ A/s}.$$

(c) With $\tau_L = L/R = 5 \times 10^{-6}$ s, we have $t/\tau_L = 3/5$, and we apply Eq. 30-41:

$$i_{\text{bat}} = \frac{\varepsilon}{R}\left(1-e^{-3/5}\right) \approx 1.8\times10^{-3}\text{ A}.$$

(d) The rate of change of the current is figured from the loop rule (and Eq. 30-35):

$$\varepsilon - i_{\text{bat}}R - |\varepsilon_L| = 0.$$

Using the values from part (c), we obtain $|\varepsilon_L| \approx 22$ V. Then,

$$\frac{di_{\text{bat}}}{dt} = \frac{|\varepsilon_L|}{L} = \frac{22\text{ V}}{0.050\text{ H}} \approx 4.4\times10^2\text{ A/s}.$$

(e) As $t \to \infty$, the circuit reaches a steady-state condition, so that $di_{\text{bat}}/dt = 0$ and $\varepsilon_L = 0$. The loop rule then leads to

$$\varepsilon - i_{\text{bat}}R - |\varepsilon_L| = 0 \;\Rightarrow\; i_{\text{bat}} = \frac{40\text{ V}}{10000\,\Omega} = 4.0\times10^{-3}\text{ A}.$$

(f) As $t \to \infty$, the circuit reaches a steady-state condition, $di_{\text{bat}}/dt = 0$.

LEARN In summary, at $t = 0$ immediately after the switch is closed, the inductor opposes any change in current, and with the inductor and the battery being connected in series, the induced emf in the inductor is equal to the emf of the battery, $\varepsilon_L = \varepsilon$. A long time later after all the currents have reached their steady-state values, $\varepsilon_L = 0$, and the inductor can be treated as an ordinary connecting wire. In this limit, the circuit can be analyzed as if L were not present.

Chapter 31

31-7
THINK This problem explores the analogy between an oscillating LC system and an oscillating mass–spring system.

EXPRESS Table 31-1 provides a comparison of energies in the two systems. From the table, we see the following correspondences:

$$x \leftrightarrow q, \;\; k \leftrightarrow \frac{1}{C}, \;\; m \leftrightarrow L, \;\; v = \frac{dx}{dt} \leftrightarrow \frac{dq}{dt} = i,$$
$$\frac{1}{2}kx^2 \leftrightarrow \frac{q^2}{2C}, \;\; \frac{1}{2}mv^2 \leftrightarrow \frac{1}{2}Li^2.$$

ANALYZE (a) The mass m corresponds to the inductance, so $m = 1.25$ kg.

(b) The spring constant k corresponds to the reciprocal of the capacitance, $1/C$. Since the total energy is given by $U = Q^2/2C$, where Q is the maximum charge on the capacitor and C is the capacitance, we have

$$C = \frac{Q^2}{2U} = \frac{\left(175\times 10^{-6}\text{ C}\right)^2}{2\left(5.70\times 10^{-6}\text{ J}\right)} = 2.69\times 10^{-3}\text{ F}$$

and

$$k = \frac{1}{2.69\times 10^{-3}\text{ m/N}} = 372\text{ N/m}.$$

(c) The maximum displacement corresponds to the maximum charge, so $x_{\max} = 1.75\times 10^{-4}$ m.

(d) The maximum speed $v_{\max}$ corresponds to the maximum current. The maximum current is

$$I = Q\omega = \frac{Q}{\sqrt{LC}} = \frac{175\times 10^{-6}\text{ C}}{\sqrt{(1.25\text{ H})(2.69\times 10^{-3}\text{ F})}} = 3.02\times 10^{-3}\text{ A}.$$

Consequently, $v_{\max} = 3.02 \times 10^{-3}$ m/s.

LEARN The correspondences suggest that an oscillating LC system is mathematically equivalent to an oscillating mass–spring system. The electrical mechanical analogy can also be seen by comparing their angular frequencies of oscillation:

$$\omega = \sqrt{\frac{k}{m}} \text{ (mass-spring system)}, \quad \omega = \frac{1}{\sqrt{LC}} \text{ } (LC \text{ circuit})$$

31-11

THINK The frequency of oscillation f in an LC circuit is related to the inductance L and capacitance C by $f = 1/2\pi\sqrt{LC}$.

EXPRESS Since $f \sim 1/\sqrt{C}$, the smaller value of C gives the larger value of f, while the larger value of C gives the smaller value of f. Consequently, $f_{max} = 1/2\pi\sqrt{LC_{min}}$, and $f_{min} = 1/2\pi\sqrt{LC_{max}}$.

ANALYZE (a) The ratio of the maximum frequency to the minimum frequency is

$$\frac{f_{max}}{f_{min}} = \frac{\sqrt{C_{max}}}{\sqrt{C_{min}}} = \frac{\sqrt{365\,\text{pF}}}{\sqrt{10\,\text{pF}}} = 6.0.$$

(b) An additional capacitance C is chosen so the desired ratio of the frequencies is

$$r = \frac{1.60\,\text{MHz}}{0.54\,\text{MHz}} = 2.96.$$

Since the additional capacitor is in parallel with the tuning capacitor, its capacitance adds to that of the tuning capacitor. If C is in picofarads (pF), then

$$\frac{\sqrt{C + 365\,\text{pF}}}{\sqrt{C + 10\,\text{pF}}} = 2.96.$$

The solution for C is

$$C = \frac{(365\,\text{pF}) - (2.96)^2(10\,\text{pF})}{(2.96)^2 - 1} = 36\,\text{pF}.$$

(c) We solve $f = 1/2\pi\sqrt{LC}$ for L. For the minimum frequency, C = 365 pF + 36 pF = 401 pF and f = 0.54 MHz. Thus, the inductance is

$$L = \frac{1}{(2\pi)^2 Cf^2} = \frac{1}{(2\pi)^2(401 \times 10^{-12}\,\text{F})(0.54 \times 10^6\,\text{Hz})^2} = 2.2 \times 10^{-4}\,\text{H}.$$

LEARN One could also use the maximum frequency condition to solve for the inductance of the coil in (d). The capacitance is C = 10 pF + 36 pF = 46 pF and f = 1.60 MHz, so

$$L=\frac{1}{(2\pi)^2 Cf^2}=\frac{1}{(2\pi)^2(46\times10^{-12}\text{ F})(1.60\times10^6\text{ Hz})^2}=2.2\times10^{-4}\text{ H}.$$

31-27

THINK With the presence of a resistor in the *RLC* circuit, oscillation is damped, and the total electromagnetic energy of the system is no longer conserved, as some energy is transferred to thermal energy in the resistor.

EXPRESS Let t be a time at which the capacitor is fully charged in some cycle and let $q_{\max 1}$ be the charge on the capacitor then. The energy in the capacitor at that time is

$$U(t)=\frac{q_{\max 1}^2}{2C}=\frac{Q^2}{2C}e^{-Rt/L}$$

where

$$q_{\max 1}=Qe^{-Rt/2L}.$$

One period later the charge on the fully charged capacitor is

$$q_{\max 2}=Qe^{-R(t+T)2/L}$$

where $T=\dfrac{2\pi}{\omega'}$, and the energy is

$$U(t+T)=\frac{q_{\max 2}^2}{2C}=\frac{Q^2}{2C}e^{-R(t+T)/L}.$$

ANALYZE The fractional loss in energy is

$$\frac{|\Delta U|}{U}=\frac{U(t)-U(t+T)}{U(t)}=\frac{e^{-Rt/L}-e^{-R(t+T)/L}}{e^{-Rt/L}}=1-e^{-RT/L}.$$

Assuming that RT/L is very small compared to 1 (which would be the case if the resistance is small), we expand the exponential (see Appendix E). The first few terms are:

$$e^{-RT/L}\approx 1-\frac{RT}{L}+\frac{R^2T^2}{2L^2}+\cdots.$$

If we approximate $\omega\approx\omega'$, then we can write T as $2\pi/\omega$. As a result, we obtain

$$\frac{|\Delta U|}{U} \approx 1-\left(1-\frac{RT}{L}+\cdots\right) \approx \frac{RT}{L}=\frac{2\pi R}{\omega L}.$$

LEARN The ratio $|\Delta U|/U$ can be rewritten as

$$\frac{|\Delta U|}{U}=\frac{2\pi}{Q}$$

where $Q=\omega L/R$ (not to confuse Q with charge) is called the "quality factor" of the oscillating circuit. A high-Q circuit has low resistance and hence, low fractional energy loss.

31-33

THINK Our circuit consists of an ac generator that produces an alternating current, as well as a load that could be purely resistive, capacitive, or inductive. The nature of the load can be determined by the phase angle between the current and the emf.

EXPRESS The generator emf and the current are given by

$$\varepsilon=\varepsilon_m \sin(\omega_d-\pi/4), \quad i(t)=I\sin(\omega_d-3\pi/4).$$

The expressions show that the emf is maximum when $\sin(\omega_d t - \pi/4) = 1$ or

$$\omega_d t - \pi/4 = (\pi/2) \pm 2n\pi \quad [n = \text{integer}].$$

Similarly, the current is maximum when $\sin(\omega_d t - 3\pi/4) = 1$, or

$$\omega_d t - 3\pi/4 = (\pi/2) \pm 2n\pi \quad [n = \text{integer}].$$

ANALYZE (a) The first time the emf reaches its maximum after $t = 0$ is when $\omega_d t - \pi/4 = \pi/2$ (that is, $n = 0$). Therefore,

$$t=\frac{3\pi}{4\omega_d}=\frac{3\pi}{4(350\text{ rad/s})}=6.73\times10^{-3}\text{ s}.$$

(b) The first time the current reaches its maximum after $t = 0$ is when $\omega_d t - 3\pi/4 = \pi/2$, as in part (a) with $n = 0$. Therefore,

$$t=\frac{5\pi}{4\omega_d}=\frac{5\pi}{4(350\text{ rad/s})}=1.12\times10^{-2}\text{ s}.$$

(c) The current lags the emf by $+\pi/2$ rad, so the circuit element must be an inductor.

(d) The current amplitude I is related to the voltage amplitude V_L by $V_L = IX_L$, where X_L is the inductive reactance, given by $X_L = \omega_d L$. Furthermore, since there is only one element in the circuit, the amplitude of the potential difference across the element must be the same as the amplitude of the generator emf: $V_L = \varepsilon_m$. Thus, $\varepsilon_m = I\omega_d L$ and

$$L = \frac{\varepsilon_m}{I\omega_d} = \frac{30.0 \text{ V}}{(620\times10^{-3}\text{A})(350 \text{ rad/s})} = 0.138 \text{ H}.$$

LEARN The current in the circuit can be rewritten as

$$i(t) = I\sin\left(\omega_d - \frac{3\pi}{4}\right) = I\sin\left(\omega_d - \frac{\pi}{4} - \phi\right)$$

where $\phi = +\pi/2$. In a purely inductive circuit, the current lags the voltage by 90°.

31-41

THINK We have a series *RLC* circuit. Since *R*, *L*, and *C* are in series, the same current is driven in all three of them.

EXPRESS The capacitive and the inductive reactances can be written as

$$X_C = \frac{1}{\omega_d C} = \frac{1}{2\pi f_d C}, \quad X_L = \omega_d L = 2\pi f_d L.$$

The impedance of the circuit is $Z = \sqrt{R^2 + (X_L - X_C)^2}$, and the current amplitude is given by $I = \varepsilon_m / Z$.

ANALYZE (a) Substituting the values given, we find the capacitive reactance to be

$$X_C = \frac{1}{2\pi f_d C} = \frac{1}{2\pi(60.0 \text{ Hz})(70.0\times10^{-6}\text{F})} = 37.9\ \Omega.$$

Similarly, the inductive reactance is

$$X_L = 2\pi f_d L = 2\pi(60.0 \text{ Hz})(230\times10^{-3}\text{H}) = 86.7\ \Omega.$$

Thus, the impedance is

$$Z = \sqrt{R^2 + (X_L - X_C)^2} = \sqrt{(200\ \Omega)^2 + (37.9\ \Omega - 86.7\ \Omega)^2} = 206\ \Omega.$$

(b) The phase angle is

$$\phi = \tan^{-1}\left(\frac{X_L - X_C}{R}\right) = \tan^{-1}\left(\frac{86.7\ \Omega - 37.9\ \Omega}{200\ \Omega}\right) = 13.7°.$$

(c) The current amplitude is

$$I = \frac{\varepsilon_m}{Z} = \frac{36.0\ \text{V}}{206\Omega} = 0.175\,\text{A}.$$

(d) We first find the voltage amplitudes across the circuit elements:

$$\begin{aligned} V_R &= IR = (0.175\ \text{A})(200\ \Omega) = 35.0\ \text{V} \\ V_L &= IX_L = (0.175\ \text{A})(86.7\ \Omega) = 15.2\ \text{V} \\ V_C &= IX_C = (0.175\ \text{A})(37.9\ \Omega) = 6.62\,\text{V} \end{aligned}$$

Note that $X_L > X_C$, so that ε_m leads I. The phasor diagram is drawn to scale below.

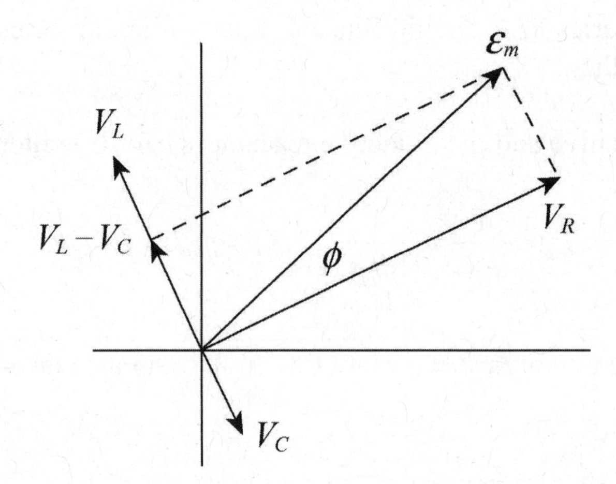

LEARN The circuit in this problem is more inductive since $X_L > X_C$. The phase angle is positive, so the current lags behind the applied emf.

31-47

THINK In a driven *RLC* circuit, the current amplitude is maximum at resonance, where the driven angular frequency is equal to the natural angular frequency.

EXPRESS For a given amplitude ε_m of the generator emf, the current amplitude is given by

$$I = \frac{\varepsilon_m}{Z} = \frac{\varepsilon_m}{\sqrt{R^2 + (\omega_d L - 1/\omega_d C)^2}}.$$

To explicitly show that I is maximum when $\omega_d = \omega = 1/\sqrt{LC}$, we differentiate I with respect to ω_d and set the derivative to zero:

$$\frac{dI}{d\omega_d} = -(E)_m[R^2 + (\omega_d L - 1/\omega_d C)^2]^{-3/2}\left[\omega_d L - \frac{1}{\omega_d C}\right]\left[L + \frac{1}{\omega_d^2 C}\right].$$

The only factor that can equal zero is when $\omega_d L - (1/\omega_d C)$, or $\omega_d = 1/\sqrt{LC} = \omega$.

ANALYZE (a) For this circuit, the driving angular frequency is

$$\omega_d = \frac{1}{\sqrt{LC}} = \frac{1}{\sqrt{(1.00\text{ H})(20.0\times10^{-6}\text{ F})}} = 224\text{ rad/s}.$$

(b) When $\omega_d = \omega$, the impedance is $Z = R$, and the current amplitude is

$$I = \frac{\varepsilon_m}{R} = \frac{30.0\text{ V}}{5.00\ \Omega} = 6.00\text{ A}.$$

(c) We want to find the (positive) values of ω_d for which $I = \varepsilon_m/2R$:

$$\frac{\varepsilon_m}{\sqrt{R^2 + (\omega_d L - 1/\omega_d C)^2}} = \frac{\varepsilon_m}{2R}.$$

This may be rearranged to yield

$$\left(\omega_d L - \frac{1}{\omega_d C}\right)^2 = 3R^2.$$

Taking the square root of both sides (acknowledging the two ± roots) and multiplying by $\omega_d C$, we obtain

$$\omega_d^2(LC) \pm \omega_d\left(\sqrt{3}CR\right) - 1 = 0.$$

Using the quadratic formula, we find the smallest positive solution

$$\omega_2 = \frac{-\sqrt{3}CR + \sqrt{3C^2R^2 + 4LC}}{2LC} = \frac{-\sqrt{3}(20.0\times10^{-6}\text{ F})(5.00\ \Omega)}{2(1.00\text{ H})(20.0\times10^{-6}\text{ F})}$$
$$+\frac{\sqrt{3(20.0\times10^{-6}\text{ F})^2(5.00\ \Omega)^2 + 4(1.00\text{ H})(20.0\times10^{-6}\text{ F})}}{2(1.00\text{ H})(20.0\times10^{-6}\text{ F})}$$
$$= 219\text{ rad/s}.$$

(d) The largest positive solution

$$\omega_1 = \frac{+\sqrt{3}CR + \sqrt{3C^2R^2 + 4LC}}{2LC} = \frac{+\sqrt{3}(20.0\times10^{-6}\text{ F})(5.00\,\Omega)}{2(1.00\text{ H})(20.0\times10^{-6}\text{ F})}$$
$$+\frac{\sqrt{3(20.0\times10^{-6}\text{ F})^2(5.00\ \Omega)^2 + 4(1.00\text{ H})(20.0\times10^{-6}\text{F})}}{2(1.00\text{ H})(20.0\times10^{-6}\text{ F})}$$
$$= 228\text{ rad/s}.$$

(e) The fractional width is

$$\frac{\omega_1 - \omega_2}{\omega} = \frac{228\text{ rad/s} - 219\text{ rad/s}}{224\text{ rad/s}} = 0.040.$$

LEARN The current amplitude as a function of ω_d is plotted below.

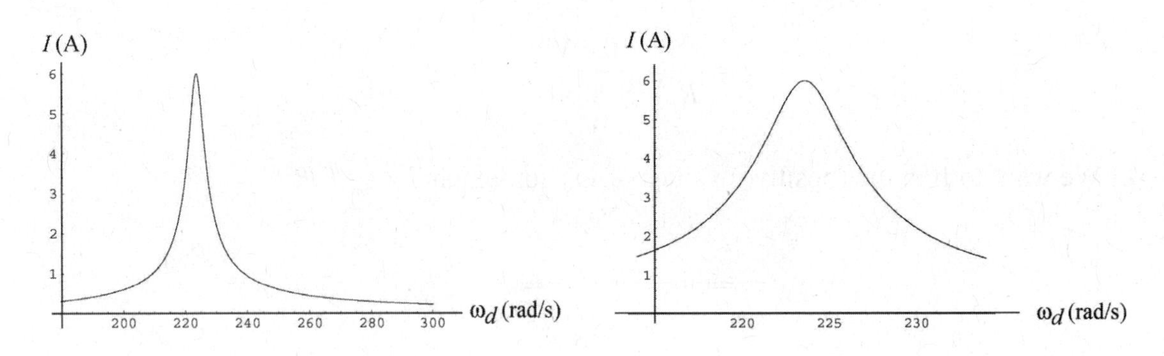

We see that I is a maximum at $\omega_d = \omega = 224$ rad/s, and is at half maximum (3 A) at 219 rad/s and 228 rad/s.

31-51

THINK In a driven *RLC* circuit, the current amplitude is maximum at resonance, where the driven angular frequency is equal to the natural angular frequency. It then falls off rapidly away from resonance.

EXPRESS We use the expressions found in Problem 31-47:

$$\omega_1 = \frac{+\sqrt{3}CR + \sqrt{3C^2R^2 + 4LC}}{2LC}, \quad \omega_2 = \frac{-\sqrt{3}CR + \sqrt{3C^2R^2 + 4LC}}{2LC}.$$

The resonance angular frequency is $\omega = 1/\sqrt{LC}$.

ANALYZE Thus, the fractional half width is

$$\frac{\Delta\omega_d}{\omega}=\frac{\omega_1-\omega_2}{\omega}=\frac{2\sqrt{3}CR\sqrt{LC}}{2LC}=R\sqrt{\frac{3C}{L}}.$$

LEARN Note that the value of $\Delta\omega_d/\omega$ increases linearly with R; that is, the larger the resistance, the broader the peak. As an example, the data of Problem 31-47 gives

$$\frac{\Delta\omega_d}{\omega}=(5.00\ \Omega)\sqrt{\frac{3(20.0\times10^{-6}\text{ F})}{1.00\text{ H}}}=3.87\times10^{-2}.$$

This is in agreement with the result of Problem 31-47. The method used there, however, gives only one significant figure since two numbers close in value are subtracted ($\omega_1-\omega_2$). Here the subtraction is done algebraically, and three significant figures are obtained.

31-53

THINK Energy is supplied by the 120 V rms ac line to keep the air conditioner running.

EXPRESS The impedance of the circuit is $Z=\sqrt{R^2+(X_L-X_C)^2}$, and the average rate of energy delivery is

$$P_{\text{avg}}=I_{\text{rms}}^2R=\left(\frac{\varepsilon_{\text{rms}}}{Z}\right)^2R=\frac{\varepsilon_{\text{rms}}^2R}{Z^2}.$$

ANALYZE (a) Substituting the values given, the impedance is

$$Z=\sqrt{(12.0\,\Omega)^2+(1.30\,\Omega-0)^2}=12.1\,\Omega.$$

(b) The average rate at which energy has been supplied is

$$P_{\text{avg}}=\frac{\varepsilon_{\text{rms}}^2R}{Z^2}=\frac{(120\text{ V})^2(12.0\,\Omega)}{(12.07\,\Omega)^2}=1.186\times10^3\text{ W}\approx1.19\times10^3\text{ W}.$$

LEARN In a steady-state operation, the total energy stored in the capacitor and the inductor stays constant. Thus, the net energy transfer is from the generator to the resistor, where electromagnetic energy is dissipated in the form of thermal energy.

31-61

THINK We have an ac generator connected to a "black box," whose load is of the form of an *RLC* circuit. Given the functional forms of the emf and the current in the circuit, we can deduce the nature of the load.

EXPRESS In general, the driving emf and the current can be written as

$$\varepsilon(t)=\varepsilon_m \sin\omega_d t,\quad i(t)=I\sin(\omega_d t-\phi).$$

Thus, we have $\varepsilon_m = 75$ V, I = 1.20 A, and $\phi = -42°$ for this circuit. The power factor of the circuit is simply given by $\cos\phi$.

ANALYZE (a) With $\phi = -42.0°$, we obtain $\cos\phi = \cos(-42.0°) = 0.743$.

(b) Since the phase constant is negative, $\phi < 0$, $\omega t - \phi > \omega t$. The current leads the emf.

(c) The phase constant is related to the reactance difference by $\tan\phi = (X_L - X_C)/R$. We have

$$\tan\phi = \tan(-42.0°) = -0.900,$$

a negative number. Therefore, $X_L - X_C$ is negative, which implies that $X_C > X_L$. The circuit in the box is predominantly capacitive.

(d) If the circuit were in resonance, X_L would be the same as X_C, then $\tan\phi$ would be zero, and ϕ would be zero as well. Since ϕ is not zero, we conclude the circuit is not in resonance.

(e) Since $\tan\phi$ is negative and finite, neither the capacitive reactance nor the resistance is zero. This means the box must contain a capacitor and a resistor.

(f) The inductive reactance may be zero, so there need not be an inductor.

(g) Yes, there is a resistor.

(h) The average power is

$$P_{\text{avg}} = \frac{1}{2}\varepsilon_m I\cos\phi = \frac{1}{2}(75.0\,\text{V})(1.20\,\text{A})(0.743) = 33.4\,\text{W}.$$

(i) The answers above depend on the frequency only through the phase constant ϕ, which is given. If values were given for *R*, *L*, and *C*, then the value of the frequency would also be needed to compute the power factor.

LEARN The phase constant ϕ allows us to calculate the power factor and deduce the nature of the load in the circuit. In (f) we stated that the inductance may be set to zero. If there is an inductor, then its reactance must be smaller than the capacitive reactance, $X_L < X_C$.

31-63

THINK The transformer in this problem is a step-down transformer.

EXPRESS If N_p is the number of primary turns, and N_s is the number of secondary turns, then the step-down voltage in the secondary circuit is

$$V_s = V_p\left(\frac{N_s}{N_p}\right).$$

By Ohm's law, the current in the secondary circuit is given by $I_s = V_s / R_s$.

ANALYZE (a) The stepped-down voltage is

$$V_s = V_p\left(\frac{N_s}{N_p}\right) = (120\,\text{V})\left(\frac{10}{500}\right) = 2.4\,\text{V}.$$

(b) The current in the secondary is $I_s = \dfrac{V_s}{R_s} = \dfrac{2.4\,\text{V}}{15\,\Omega} = 0.16\,\text{A}.$

We find the primary current from Eq. 31-80:

$$I_p = I_s\left(\frac{N_s}{N_p}\right) = (0.16\,\text{A})\left(\frac{10}{500}\right) = 3.2\times10^{-3}\,\text{A}.$$

(c) As shown above, the current in the secondary is $I_s = 0.16\,\text{A}$.

LEARN In a transformer, the voltages and currents in the secondary circuit are related to that in the primary circuit by

$$V_s = V_p\left(\frac{N_s}{N_p}\right), \qquad I_s = I_p\left(\frac{N_p}{N_s}\right).$$

31-77

THINK The three-phase generator has three ac voltages that are 120° out of phase with each other.

EXPRESS To calculate the potential difference between any two wires, we use the following trigonometric identity:

$$\sin\alpha - \sin\beta = 2\sin\left[(\alpha-\beta)/2\right]\cos\left[(\alpha+\beta)/2\right],$$

where α and β are any two angles.

ANALYZE (a) We consider the following combinations: $\Delta V_{12} = V_1 - V_2$, $\Delta V_{13} = V_1 - V_3$, and $\Delta V_{23} = V_2 - V_3$. For ΔV_{12},

$$\Delta V_{12} = A\sin(\omega_d t) - A\sin(\omega_d t - 120°) = 2A\sin\left(\frac{120°}{2}\right)\cos\left(\frac{2\omega_d t - 120°}{2}\right) = \sqrt{3}A\cos(\omega_d t - 60°)$$

where $\sin 60° = \sqrt{3}/2$. Similarly,

$$\Delta V_{13} = A\sin(\omega_d t) - A\sin(\omega_d t - 240°) = 2A\sin\left(\frac{240°}{2}\right)\cos\left(\frac{2\omega_d t - 240°}{2}\right)$$
$$= \sqrt{3}A\cos(\omega_d t - 120°)$$

and

$$\Delta V_{23} = A\sin(\omega_d t - 120°) - A\sin(\omega_d t - 240°) = 2A\sin\left(\frac{120°}{2}\right)\cos\left(\frac{2\omega_d t - 360°}{2}\right)$$
$$= \sqrt{3}A\cos(\omega_d t - 180°).$$

All three expressions are sinusoidal functions of t with angular frequency ω_d.

(b) We note that each of the above expressions has an amplitude of $\sqrt{3}A$.

LEARN A three-phase generator provides a smoother flow of power than a single-phase generator.

31-79

THINK The total energy in the LC circuit is the sum of electrical energy stored in the capacitor, and the magnetic energy stored in the inductor. Energy is conserved.

EXPRESS Let U_E be the electrical energy in the capacitor and U_B be the magnetic energy in the inductor. The total energy is $U = U_E + U_B$. When $U_E = 0.500U_B$ (at time t), then $U_B = 2.00U_E$ and $U = U_E + U_B = 3.00U_E$. Now, U_E is given by $q^2/2C$, where q is the charge on the capacitor at time t. The total energy U is given by $Q^2/2C$, where Q is the maximum charge on the capacitor.

ANALYZE (a) Solving for q, we obtain

$$\frac{Q^2}{2C} = \frac{3.00q^2}{2C} \Rightarrow q = \frac{Q}{\sqrt{3.00}} = 0.577Q.$$

(b) If the capacitor is fully charged at time $t = 0$, then the time-dependent charge on the capacitor is given by $q = Q\cos\omega t$. This implies that the condition $q = 0.577Q$ is satisfied when $\cos\omega t = 0.557$, or $\omega t = 0.955$ rad. Since $\omega = 2\pi / T$ (where T is the period of oscillation), $t = 0.955T / 2\pi = 0.152T$, or $t / T = 0.152$.

LEARN The fraction of total energy that is of electrical nature at a given time t is given by

$$\frac{U_E}{U} = \frac{(Q^2 / 2C)\cos^2 \omega t}{Q^2 / 2C} = \cos^2 \omega t = \cos^2\left(\frac{2\pi t}{T}\right).$$

A plot of U_E / U as a function of t / T is given below.

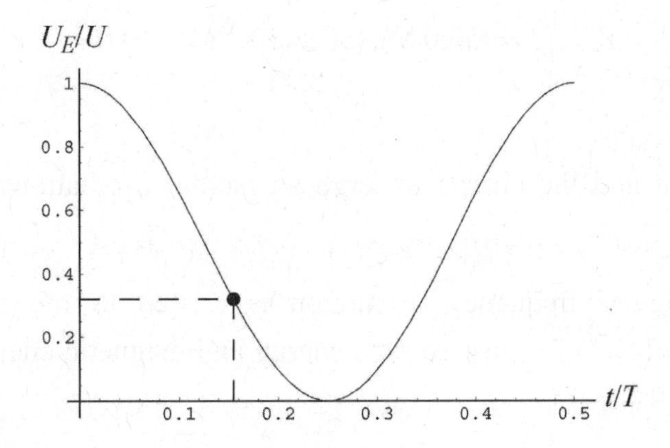

From the plot, we see that $U_E / U = 1/3$ at $t / T = 0.152$.

31-81

THINK Since the current lags the generator emf, the phase angle is positive and the circuit is more inductive than capacitive.

EXPRESS Let V_L be the maximum potential difference across the inductor, V_C be the maximum potential difference across the capacitor, and V_R be the maximum potential difference across the resistor. The phase constant is given by

$$\phi = \tan^{-1}\left(\frac{V_L - V_C}{V_R}\right).$$

The maximum emf is related to the current amplitude by $\varepsilon_m = IZ$, where Z is the impedance.

ANALYZE (a) With $V_C = V_L / 2.00$ and $V_R = V_L / 2.00$, we find the phase constant to be

$$\phi = \tan^{-1}\left(\frac{V_L - V_L/2.00}{V_L/2.00}\right) = \tan^{-1}(1.00) = 45.0°.$$

(b) The resistance is related to the impedance by $R = Z\cos\phi$. Thus,

$$R = \frac{\varepsilon_m \cos\phi}{I} = \frac{(30.0\,\text{V})(\cos 45°)}{300\times10^{-3}\,\text{A}} = 70.7\,\Omega.$$

LEARN With R and I known, the inductive and capacitive reactances are, respectively, $X_L = 2.00R = 141\,\Omega$, and $X_C = R = 70.7\,\Omega$. Similarly, the impedance of the circuit is

$$Z = \frac{\varepsilon_m}{I} = (30.0\text{ V})/(300\times10^{-3}\text{ A}) = 100\,\Omega.$$

31-85

THINK The current and the charge undergo sinusoidal oscillations in the LC circuit. Energy is conserved.

EXPRESS The angular frequency oscillation is related to the capacitance C and inductance L by $\omega = 1/\sqrt{LC}$. The electrical energy and magnetic energy in the circuit as a function of time are given by

$$U_E = \frac{q^2}{2C} = \frac{Q^2}{2C}\cos^2(\omega t + \phi)$$

$$U_B = \frac{1}{2}Li^2 = \frac{1}{2}L\omega^2 Q^2 \sin^2(\omega t + \phi) = \frac{Q^2}{2C}\sin^2(\omega t + \phi).$$

The maximum value of U_E is $Q^2/2C$, which is the total energy in the circuit, U. Similarly, the maximum value of U_B is also $Q^2/2C$, which can also be written as $LI^2/2$ using $I = \omega Q$.

ANALYZE (a) Using the fact that $\omega = 2\pi f$, the inductance is

$$L = \frac{1}{\omega^2 C} = \frac{1}{4\pi^2 f^2 C} = \frac{1}{4\pi^2 (10.4\times10^3\text{ Hz})^2 (340\times10^{-6}\text{ F})} = 6.89\times10^{-7}\text{H}.$$

(b) The total energy may be calculated from the inductor (when the current is at maximum):

$$U = \frac{1}{2}LI^2 = \frac{1}{2}(6.89\times10^{-7}\text{ H})(7.20\times10^{-3}\text{ A})^2 = 1.79\times10^{-11}\text{ J}.$$

(c) We solve for Q from $U = \frac{1}{2}Q^2/C$:

$$Q = \sqrt{2CU} = \sqrt{2(340\times10^{-6}\text{ F})(1.79\times10^{-11}\text{ J})} = 1.10\times10^{-7}\text{ C}.$$

LEARN Figure 31-4 of the textbook illustrates the oscillations of electrical and magnetic energies. The total energy $U = U_E + U_B = Q^2/2C$ remains constant. When U_E is maximum, U_B is zero, and vice versa.

31-89

THINK In this problem, we demonstrate that in a driven *RLC* circuit, the energies stored in the capacitor and the inductor stay constant; however, energy is transferred from the driving emf device to the resistor.

EXPRESS The energy stored in the capacitor is given by $U_E = q^2/2C$. Similarly, the energy stored in the inductor is $U_B = \frac{1}{2}Li^2$. The rate of energy supply by the driving emf device is $P_\varepsilon = i\varepsilon$, where $i = I\sin(\omega_d - \phi)$ and $\varepsilon = \varepsilon_m \sin\omega_d t$. The rate with which energy dissipates in the resistor is $P_R = i^2R$.

ANALYZE (a) Since the charge q is a periodic function of t with period T, so must be U_E. Consequently, U_E will not be changed over one complete cycle. Actually, U_E has period $T/2$, which does not alter our conclusion.

(b) Since the current i is a periodic function of t with period T, so must be U_B.

(c) The energy supplied by the emf device over one cycle is

$$\begin{aligned}U_\varepsilon &= \int_0^T P_\varepsilon dt = I\varepsilon_m \int_0^T \sin(\omega_d t - \phi)\sin(\omega_d t)dt \\ &= I\varepsilon_m \int_0^T [\sin\omega_d t\cos\phi - \cos\omega_d t\sin\phi]\sin(\omega_d t)dt \\ &= \frac{T}{2} I\varepsilon_m \cos\phi,\end{aligned}$$

where we have used

$$\int_0^T \sin^2(\omega_d t)dt = \frac{T}{2}, \qquad \int_0^T \sin(\omega_d t)\cos(\omega_d t)dt = 0.$$

(d) Over one cycle, the energy dissipated in the resistor is

$$U_R = \int_0^T P_R dt = I^2R\int_0^T \sin^2(\omega_d t - \phi)dt = \frac{T}{2}I^2R.$$

(e) Since $\varepsilon_m I \cos\phi = \varepsilon_m I(V_R / \varepsilon_m) = \varepsilon_m I(IR / \varepsilon_m) = I^2 R$, the two quantities are indeed the same.

LEARN In solving for (c) and (d), we could have used Eqs. 31-74 and 31-71: By doing so, we find the energy supplied by the generator to be

$$P_{\text{avg}} T = (I_{\text{rms}} \varepsilon_{\text{rms}} \cos\phi) T = \left(\frac{1}{2} T\right) \varepsilon_m I \cos\phi$$

where we substitute $I_{\text{rms}} = I / \sqrt{2}$ and $\varepsilon_{\text{rms}} = \varepsilon_m / \sqrt{2}$. Similarly, the energy dissipated by the resistor is

$$P_{\text{avg,resistor}} T = (I_{\text{rms}} V_R) T = I_{\text{rms}} (I_{\text{rms}} R) T = \left(\frac{1}{2} T\right) I^2 R.$$

The same results are obtained without any integration.

Chapter 32

32-3

THINK Gauss' law for magnetism states that the net magnetic flux through any closed surface is zero.

EXPRESS Mathematically, Gauss' law for magnetism is expressed as

$$\oint \vec{B}\cdot d\vec{A}=0.$$

Now, our Gaussian surface has the shape of a right circular cylinder with two end caps and a curved surface. Thus,

$$\oint \vec{B}\cdot d\vec{A}=\Phi_1+\Phi_2+\Phi_C,$$

where Φ_1 is the magnetic flux through the first end cap, Φ_2 is the magnetic flux through the second end cap, and Φ_C is the magnetic flux through the curved surface. Over the first end the magnetic field is inward, so the flux is $\Phi_1 = -25.0\ \mu\text{Wb}$. Over the second end the magnetic field is uniform, normal to the surface, and outward, so the flux is $\Phi_2 = AB = \pi r^2 B$, where A is the area of the end and r is the radius of the cylinder.

ANALYZE (a) Substituting the values given, the flux through the second end is

$$\Phi_2=\pi(0.120\,\text{m})^2\left(1.60\times10^{-3}\,\text{T}\right)=+7.24\times10^{-5}\,\text{Wb}=+72.4\,\mu\text{Wb}.$$

Since the three fluxes must sum to zero,

$$\Phi_C=-\Phi_1-\Phi_2=25.0\,\mu\text{Wb}-72.4\,\mu\text{Wb}=-47.4\,\mu\text{Wb}.$$

Thus, the magnitude is $|\Phi_C|=47.4\,\mu\text{Wb}$.

(b) The minus sign in Φ_C indicates that the flux is inward through the curved surface.

LEARN Gauss' law for magnetism implies that magnetic monopoles do not exist; the simplest magnetic structure is a magnetic dipole (having a north pole and a south pole).

32-5
THINK Changing electric flux induces a magnetic field.

EXPRESS Consider a circle of radius *r* between the plates, with its center on the axis of the capacitor. Since there is no current between the capacitor plates, the Ampere-Maxwell's law reduces to

$$\oint \vec{B}\cdot d\vec{A} = \mu_0\varepsilon_0 \frac{d\Phi_E}{dt},$$

where $\vec{B}$ is the magnetic field at points on the circle, and Φ_E is the electric flux through the circle. Since the $\vec{B}$ field on the circle is in the tangential direction, and $\Phi_E = AE = \pi R^2 E,$ where *R* is the radius of the capacitor, we have

$$2\pi r B = \mu_0\varepsilon_0 \pi R^2 \frac{dE}{dt}$$

or

$$B = \frac{\mu_0\varepsilon_0 R^2}{2r}\frac{dE}{dt} \qquad (r \ge R).$$

ANALYZE Solving for *dE*/*dt*, we obtain

$$\frac{dE}{dt} = \frac{2Br}{\mu_0\varepsilon_0 R^2} = \frac{2(2.0\times10^{-7}\ \text{T})(6.0\times10^{-3}\ \text{m})}{(4\pi\times10^{-7}\ \text{T}\cdot\text{m/A})(8.85\times10^{-12}\ \text{C}^2/\text{N}\cdot\text{m}^2)(3.0\times10^{-3}\text{m})^2}$$
$$= 2.4\times10^{13}\ \frac{\text{V}}{\text{m}\cdot\text{s}}.$$

LEARN Outside the capacitor, the induced magnetic field decreases with increased radial distance *r*, from a maximum value at the plate edge $r = R$.

31-15
THINK The displacement current is related to the changing electric flux by $i_d = \varepsilon_0 (d\Phi_E / dt)$.

EXPRESS Let *A* be the area of a plate and *E* be the magnitude of the electric field between the plates. The field between the plates is uniform, so $E = V/d$, where *V* is the potential difference across the plates and *d* is the plate separation.

ANALYZE Thus, the displacement current is

$$i_d = \varepsilon_0 \frac{d\Phi_E}{dt} = \varepsilon_0 \frac{d(EA)}{dt} = \varepsilon_0 A \frac{dE}{dt} = \frac{\varepsilon_0 A}{d}\frac{dV}{dt}.$$

Now, $\varepsilon_0 A/d$ is the capacitance C of a parallel-plate capacitor (not filled with a dielectric), so

$$i_d = C\frac{dV}{dt}.$$

LEARN The real current charging the capacitor is

$$i = \frac{dq}{dt} = \frac{d}{dt}(CV) = C\frac{dV}{dt}.$$

Thus, we see that $i = i_d$.

32-23

THINK The electric field between the plates in a parallel-plate capacitor is changing, so there is a nonzero displacement current $i_d = \varepsilon_0(d\Phi_E/dt)$ between the plates.

EXPRESS Let A be the area of a plate and E be the magnitude of the electric field between the plates. The field between the plates is uniform, so $E = V/d$, where V is the potential difference across the plates and d is the plate separation. The current into the positive plate of the capacitor is

$$i = \frac{dq}{dt} = \frac{d}{dt}(CV) = C\frac{dV}{dt} = \frac{\varepsilon_0 A}{d}\frac{d(Ed)}{dt} = \varepsilon_0 A\frac{dE}{dt} = \varepsilon_0\frac{d\Phi_E}{dt},$$

which is the same as the displacement current.

ANALYZE (a) Thus, at any instant the displacement current i_d in the gap between the plates equals the conduction current i in the wires: $i_d = i = 2.0$ A.

(b) The rate of change of the electric field is

$$\frac{dE}{dt} = \frac{1}{\varepsilon_0 A}\left(\varepsilon_0\frac{d\Phi_E}{dt}\right) = \frac{i_d}{\varepsilon_0 A} = \frac{2.0\,\text{A}}{(8.85\times10^{-12}\,\text{F/m})(1.0\,\text{m})^2} = 2.3\times10^{11}\,\frac{\text{V}}{\text{m}\cdot\text{s}}.$$

(c) The displacement current through the indicated path is

$$i_d' = i_d\left(\frac{d^2}{L^2}\right) = (2.0\text{ A})\left(\frac{0.50\text{m}}{1.0\text{m}}\right)^2 = 0.50\text{ A}.$$

(d) The integral of the field around the indicated path is

$$\oint \vec{B}\cdot d\vec{s} = \mu_0 i_d' = \left(1.26\times10^{-16}\ \text{H/m}\right)\left(0.50\,\text{A}\right) = 6.3\times10^{-7}\ \text{T}\cdot\text{m}.$$

LEARN the displacement through the dashed path is proportional to the area encircled by the path since the displacement current is uniformly distributed over the full plate area.

32-33

THINK An electron in an atom has both orbital angular momentum and spin angular momentum; the *z* components of the angular momenta are quantized.

EXPRESS The *z* component of the orbital angular momentum is give by

$$L_{\text{orb},z} = \frac{m_\ell h}{2\pi}$$

where h is the Planck constant and m_ℓ is the orbital magnetic quantum number. The corresponding z component of the orbital magnetic dipole moment is

$$\mu_{\text{orb},z} = -m_\ell \mu_{\text{B}}$$

where $\mu_{\text{B}} = eh/4\pi m$ is the Bohr magneton. When placed in an external field $\vec{B}_{\text{ext}}$, the energy associated with the orientation of $\vec{\mu}_{\text{orb}}$ is given by

$$U = -\vec{\mu}_{\text{orb}}\cdot\vec{B}_{\text{ext}}.$$

ANALYZE (a) Since $m_\ell = 0$, $L_{\text{orb},z} = m_\ell h/2\pi = 0$.

(b) Since $m_\ell = 0$, $\mu_{\text{orb},z} = -m_\ell \mu_B = 0$.

(c) With $m_\ell = 0$, from Eq. 32-32, the potential energy associated with $\vec{\mu}_{\text{orb}}$ is

$$U_{\text{orb}} = -\mu_{\text{orb},z}B_{\text{ext}} = -m_\ell \mu_B B_{\text{ext}} = 0.$$

(d) Regardless of the value of m_ℓ, we find for the spin part

$$U_{\text{spin}} = -\mu_{s,z}B = \pm\mu_B B = \pm\left(9.27\times10^{-24}\ \text{J/T}\right)\left(35\,\text{mT}\right) = \pm3.2\times10^{-25}\ \text{J}.$$

(e) Now $m_\ell = -3$, so

$$L_{\text{orb},z} = \frac{m_\ell h}{2\pi} = \frac{(-3)\left(6.63\times10^{-27}\ \text{J}\cdot\text{s}\right)}{2\pi} = -3.16\times10^{-34}\ \text{J}\cdot\text{s} \approx -3.2\times10^{-34}\ \text{J}\cdot\text{s}.$$

(f) Similarly, $\mu_{\text{orb},z} = -m_\ell \mu_B = -(-3)\left(9.27\times10^{-24}\ \text{J/T}\right) = 2.78\times10^{-23}\ \text{J/T} \approx 2.8\times10^{-23}\ \text{J/T}$.

(g) The potential energy associated with the electron's orbital magnetic moment is now

$$U_{\text{orb}} = -\mu_{\text{orb},z} B_{\text{ext}} = -\left(2.78\times10^{-23}\ \text{J/T}\right)\left(35\times10^{-3}\ \text{T}\right) = -9.7\times10^{-25}\ \text{J}.$$

(h) On the other hand, the potential energy associated with the electron spin, being independent of m_ℓ, remains the same: $\pm3.2\times10^{-25}$ J.

LEARN Spin is an intrinsic angular momentum that is not associated with the motion of the electron. Its z component is quantized, and can be written as

$$S_z = \frac{m_s h}{2\pi}$$

where $m_s = \pm1/2$ is the spin magnetic quantum number.

32-41

THINK As defined in Eq. 32-38, magnetization is the dipole moment per unit volume.

EXPRESS Let M be the magnetization and $\mathcal{V}$ be the volume of the cylinder ($\mathcal{V} = \pi r^2 L$, where r is the radius of the cylinder and L is its length). The dipole moment is given by $\mu = M\mathcal{V}$.

ANALYZE Substituting the values given, we obtain

$$\mu = M\pi r^2 L = \left(5.30\times10^3\ \text{A/m}\right)\pi\left(0.500\times10^{-2}\ \text{m}\right)^2\left(5.00\times10^{-2}\ \text{m}\right) = 2.08\times10^{-2}\ \text{J/T}.$$

LEARN In a sample with N atoms, the magnetization reaches maximum, or saturation, when all the dipoles are completely aligned, leading to $M_{\text{max}} = N\mu/\mathcal{V}$.

32-45

THINK According to statistical mechanics, the probability of a magnetic dipole moment placed in an external magnetic field having energy U is $P = e^{-U/kT}$, where k is the Boltzmann's constant.

EXPRESS The orientation energy of a dipole in a magnetic field is given by $U = -\vec{\mu}\cdot\vec{B}$. So if a dipole is parallel with $\vec{B}$, then $U = -\mu B$; however, $U = +\mu B$ if the alignment is anti-parallel. We use the notation $P(\mu) = e^{\mu B/kT}$ for the probability of a dipole that is parallel to $\vec{B}$, and $P(-\mu) = e^{-\mu B/kT}$ for the probability of a dipole that is anti-parallel to the field. The magnetization may be thought of as a "weighted average" in terms of these probabilities.

ANALYZE (a) With N atoms per unit volume, we find the magnetization to be

$$M = \frac{N\mu P(\mu) - N\mu P(-\mu)}{P(\mu) + P(-\mu)} = \frac{N\mu\left(e^{\mu B/kT} - e^{-\mu B/kT}\right)}{e^{\mu B/kT} + e^{-\mu B/kT}} = N\mu \tanh\left(\frac{\mu B}{kT}\right).$$

(b) For $\mu B \ll kT$ (that is, $\mu B / kT \ll 1$) we have $e^{\pm\mu B/kT} \approx 1 \pm \mu B/kT$, so

$$M = N\mu \tanh\left(\frac{\mu B}{kT}\right) \approx \frac{N\mu\left[(1+\mu B/kT) - (1-\mu B/kT)\right]}{(1+\mu B/kT) + (1-\mu B/kT)} = \frac{N\mu^2 B}{kT}.$$

(c) For $\mu B \gg kT$ we have $\tanh(\mu B/kT) \approx 1$, so $M = N\mu \tanh\left(\frac{\mu B}{kT}\right) \approx N\mu.$

(d) One can easily plot the tanh function using, for instance, a graphical calculator. One can then note the resemblance between such a plot and Fig. 32-14. By adjusting the parameters used in one's plot, the curve in Fig. 32-14 can reliably be fit with a tanh function.

LEARN As can be seen from Fig. 32-14, the magnetization M is linear in B/kT in the regime $B/T \ll 1$. On the other hand, when $B \gg T$, M approaches a constant.

32-47

THINK In this problem, we model the Earth's magnetic dipole moment with a magnetized iron sphere.

EXPRESS If the magnetization of the sphere is saturated, the total dipole moment is $\mu_{\text{total}} = N\mu$, where N is the number of iron atoms in the sphere and μ is the dipole moment of an iron atom. We wish to find the radius of an iron sphere with N iron atoms. The mass of such a sphere is Nm, where m is the mass of an iron atom. It is also given by $4\pi\rho R^3/3$, where ρ is the density of iron and R is the radius of the sphere. Thus $Nm = 4\pi\rho R^3/3$ and

$$N = \frac{4\pi\rho R^3}{3m}.$$

We substitute this into $\mu_{\text{total}} = N\mu$ to obtain

$$\mu_{\text{total}} = \frac{4\pi\rho R^3 \mu}{3m} \quad \Rightarrow \quad R = \left(\frac{3m\mu_{\text{total}}}{4\pi\rho\mu}\right)^{1/3}.$$

ANALYZE (a) The mass of an iron atom is

$$m = 56\,\text{u} = (56\,\text{u})(1.66\times10^{-27}\ \text{kg/u}) = 9.30\times10^{-26}\ \text{kg}.$$

Therefore, the radius of the iron sphere is

$$R = \left[\frac{3(9.30\times10^{-26}\,\text{kg})(8.0\times10^{22}\ \text{J/T})}{4\pi(14\times10^{3}\ \text{kg/m}^3)(2.1\times10^{-23}\ \text{J/T})}\right]^{1/3} = 1.8\times10^{5}\,\text{m}.$$

(b) The volume of the sphere is $V_s = \frac{4\pi}{3}R^3 = \frac{4\pi}{3}(1.82\times10^5\,\text{m})^3 = 2.53\times10^{16}\,\text{m}^3$ and the volume of the Earth is

$$V_E = \frac{4\pi}{3}R_E^3 = \frac{4\pi}{3}(6.37\times10^6\,\text{m})^3 = 1.08\times10^{21}\,\text{m}^3,$$

so the fraction of the Earth's volume that is occupied by the sphere is

$$\frac{V_s}{V_E} = \frac{2.53\times10^{16}\,\text{m}^3}{1.08\times10^{21}\,\text{m}^3} = 2.3\times10^{-5}.$$

LEARN The finding that $V_s \ll V_E$ makes it unlikely that our simple model of a magnetized iron sphere could explain the origin of Earth's magnetization.

32-49

THINK Exchange coupling is a quantum phenomenon in which electron spins of one atom interact with those of neighboring atoms.

EXPRESS The field of a dipole along its axis is given by Eq. 30-29: $B = \frac{\mu_0}{2\pi}\frac{\mu}{z^3}$, where μ is the dipole moment and z is the distance from the dipole. The energy of a magnetic dipole $\vec{\mu}$ in a magnetic field $\vec{B}$ is given by

$$U = -\vec{\mu}\cdot\vec{B} = -\mu B\cos\phi,$$

where ϕ is the angle between the dipole moment and the field.

ANALYZE (a) Thus, the magnitude of the magnitude field at a distance 10 nm away from the atom is

$$B = \frac{(4\pi\times10^{-7}\ \mathrm{T\cdot m/A})(1.5\times10^{-23}\ \mathrm{J/T})}{2\pi(10\times10^{-9}\ \mathrm{m})} = 3.0\times10^{-6}\ \mathrm{T}.$$

(b) The energy required to turn it end-for-end (from $\phi = 0°$ to $\phi = 180°$) is

$$\Delta U = 2\mu B = 2(1.5\times10^{-23}\ \mathrm{J/T})(3.0\times10^{-6}\mathrm{T}) = 9.0\times10^{-29}\ \mathrm{J} = 5.6\times10^{-10}\ \mathrm{eV}.$$

(c) The mean kinetic energy of translation at room temperature is about 0.04 eV. Thus, if dipole-dipole interactions were responsible for aligning dipoles, collisions would easily randomize the directions of the moments and they would not remain aligned.

LEARN The persistent alignment of magnetic dipole moments despite the randomizing tendency due to thermal agitation is what gives the ferromagnetic materials their permanent magnetism.

32-61

THINK The Earth's magnetic field at a given latitude has both horizontal and vertical components.

EXPRESS Let B_h and B_v be the horizontal and vertical components of the Earth's magnetic field, respectively. Since B_h and B_v are perpendicular to each other, the Pythagorean theorem leads to

$$B = \sqrt{B_h^2 + B_v^2}\ .$$

The tangent of the inclination angle is given by $\tan\phi_i = B_v / B_h$.

ANALYZE (a) Substituting the expression given in the problem statement, we have

$$B = \sqrt{B_h^2 + B_v^2} = \sqrt{\left(\frac{\mu_0\mu}{4\pi r^3}\cos\lambda_m\right)^2 + \left(\frac{\mu_0\mu}{2\pi r^3}\sin\lambda_m\right)^2} = \frac{\mu_0\mu}{4\pi r^3}\sqrt{\cos^2\lambda_m + 4\sin^2\lambda_m}$$
$$= \frac{\mu_0\mu}{4\pi r^3}\sqrt{1+3\sin^2\lambda_m},$$

where $\cos^2\lambda_m + \sin^2\lambda_m = 1$ was used.

(b) The inclination ϕ_i is related to λ_m by $\tan\phi_i = \dfrac{B_v}{B_h} = \dfrac{(\mu_0\mu/2\pi r^3)\sin\lambda_m}{(\mu_0\mu/4\pi r^3)\cos\lambda_m} = 2\tan\lambda_m.$

LEARN At the magnetic equator ($\lambda_m = 0$), $\phi_i = 0°$, and the field is

$$B=\frac{\mu_0\mu}{4\pi r^3}=\frac{\left(4\pi\times10^{-7}\ \mathrm{T\cdot m/A}\right)\left(8.00\times10^{22}\ \mathrm{A\cdot m^2}\right)}{4\pi\left(6.37\times10^{6}\ \mathrm{m}\right)^3}=3.10\times10^{-5}\ \mathrm{T}.$$

32-73

THINK The z component of the orbital angular momentum is give by $L_{\text{orb},z}=m_\ell h/2\pi$, where h is the Planck constant and m_ℓ is the orbital magnetic quantum number.

EXPRESS The "limit" for m_ℓ is 3. This means that the allowed values of m_ℓ are: $0, \pm1, \pm2$, and ±3.

ANALYZE (a) The number of different m_ℓ's is 2(3) + 1 = 7. Since $L_{\text{orb},z} \propto m_\ell$, there are a total of seven different values of $L_{\text{orb},z}$.

(b) Similarly, since $\mu_{\text{orb},z} \propto m_\ell$, there are also a total of seven different values of $\mu_{\text{orb},z}$.

(c) The greatest allowed value of $L_{\text{orb},z}$ is given by $|m_\ell|_{\text{max}}h/2\pi = 3h/2\pi$.

(d) Similar to part (c), since $\mu_{\text{orb},z} = -m_\ell\,\mu_B$, the greatest allowed value of $\mu_{\text{orb},z}$ is given by $|m_\ell|_{\text{max}}\mu_B = 3eh/4\pi m_e$.

(e) From Eqs. 32-23 and 32-29 the z component of the net angular momentum of the electron is given by

$$L_{\text{net},z}=L_{\text{orb},z}+L_{s,z}=\frac{m_\ell h}{2\pi}+\frac{m_s h}{2\pi}.$$

For the maximum value of $L_{\text{net},z}$ let $m_\ell=[m_\ell]_{\text{max}}=3$ and $m_s=\frac{1}{2}$. Thus

$$\left[L_{\text{net},z}\right]_{\text{max}}=\left(3+\frac{1}{2}\right)\frac{h}{2\pi}=\frac{3.5h}{2\pi}.$$

(f) Since the maximum value of $L_{\text{net},z}$ is given by $[m_J]_{\text{max}}h/2\pi$ with $[m_J]_{\text{max}} = 3.5$ (see the last part above), the number of allowed values for the z component of $L_{\text{net},z}$ is given by $2[m_J]_{\text{max}} + 1 = 2(3.5) + 1 = 8$.

LEARN As we shall see in Chapter 40, the allowed values of m_ℓ range from $-\ell$ to $+\ell$, where ℓ is called the orbital quantum number.

Chapter 33

33-5

THINK The frequency of oscillation of the current in the *LC* circuit of the generator is $f = 1/2\pi\sqrt{LC}$, where *C* is the capacitance and *L* is the inductance. This frequency is set to be the frequency of the electromagnetic wave.

EXPRESS If *f* is the frequency and λ is the wavelength of an electromagnetic wave, then $f\lambda = c$. Thus,

$$\frac{\lambda}{2\pi\sqrt{LC}} = c.$$

ANALYZE The solution for *L* is

$$L = \frac{\lambda^2}{4\pi^2 Cc^2} = \frac{\left(550\times10^{-9}\text{ m}\right)^2}{4\pi^2\left(17\times10^{-12}\text{F}\right)\left(2.998\times10^8\text{ m/s}\right)^2} = 5.00\times10^{-21}\text{ H}.$$

This is exceedingly small.

LEARN The frequency is

$$f = \frac{c}{\lambda} = \frac{3.0\times10^8\text{ m/s}}{550\times10^{-9}\text{ m}} = 5.45\times10^{14}\text{ Hz}.$$

The EM wave is in the visible spectrum.

33-19

THINK The plasma completely reflects all the energy incident on it, so the radiation pressure is given by $p_r = 2I/c$, where *I* is the intensity.

EXPRESS The intensity is $I = P/A$, where *P* is the power and *A* is the area intercepted by the radiation.

ANALYZE Thus, the radiation pressure is

$$p_r = \frac{2I}{c} = \frac{2P}{Ac} = \frac{2\left(1.5\times10^{9}\,\text{W}\right)}{\left(1.00\times10^{-6}\text{m}^2\right)\left(2.998\times10^{8}\,\text{m/s}\right)} = 1.0\times10^{7}\,\text{Pa}.$$

LEARN In the case of total absorption, the radiation pressure would be $p_r = I/c$, a factor of 2 smaller than the case of total reflection.

33-25

THINK In this problem we relate radiation pressure to energy density in the incident beam.

EXPRESS Let f be the fraction of the incident beam intensity that is reflected. The fraction absorbed is $1 - f$. The reflected portion exerts a radiation pressure of

$$p_r = \frac{2fI_0}{c}$$

and the absorbed portion exerts a radiation pressure of

$$p_a = \frac{(1-f)I_0}{c},$$

where I_0 is the incident intensity. The factor 2 enters the first expression because the momentum of the reflected portion is reversed. The total radiation pressure is the sum of the two contributions:

$$p_{\text{total}} = p_r + p_a = \frac{2fI_0 + (1-f)I_0}{c} = \frac{(1+f)I_0}{c}.$$

ANALYZE To relate intensity to energy density, we consider a tube with length ℓ and cross-sectional area A, lying with its axis along the propagation direction of an electromagnetic wave. The electromagnetic energy inside is $U = uA\ell$, where u is the energy density. All this energy passes through the end in time $t = \ell / c$, so the intensity is

$$I = \frac{U}{At} = \frac{uA\ell c}{A\ell} = uc.$$

Thus $u = I/c$. The intensity and energy density are positive, regardless of the propagation direction. For the partially reflected and partially absorbed wave, the intensity just outside the surface is

$$I = I_0 + fI_0 = (1 + f)I_0,$$

where the first term is associated with the incident beam and the second is associated with the reflected beam. Consequently, the energy density is

$$u = \frac{I}{c} = \frac{(1+f)I_0}{c},$$

which is the same as the total radiation pressure.

LEARN In the case of total reflection, $f = 1$, and $p_{\text{total}} = p_r = 2I_0 / c$. On the other hand, the energy density is $u = I / c = 2I_0 / c$, which is the same as p_{total}. Similarly, for total absorption, $f = 0$, $p_{\text{total}} = p_a = I_0 / c$, and since $I = I_0$, we have $u = I / c = I_0 / c$, which again is the same as p_{total}.

33-27

THINK Electromagnetic waves travel at the speed of light, and carry both linear momentum and energy.

EXPRESS The speed of the electromagnetic wave is $c = \lambda f$, where λ is the wavelength and f is the frequency of the wave. The angular frequency is $\omega = 2\pi f$, and the angular wave number is $k = 2\pi / \lambda$. The magnetic field amplitude is related to the electric field amplitude by $B_m = E_m / c$. The intensity of the wave is given by Eq. 33-26:

$$I = \frac{1}{c\mu_0} E_{\text{rms}}^2 = \frac{1}{2c\mu_0} E_m^2,$$

where $E_{\text{rms}} = E_m / \sqrt{2}$.

ANALYZE (a) With $\lambda = 3.0$ m, the frequency of the wave is

$$f = \frac{c}{\lambda} = \frac{2.998 \times 10^8 \text{ m/s}}{3.0 \text{ m}} = 1.0 \times 10^8 \text{ Hz}.$$

(b) From the value of f obtained in (a), we find the angular frequency to be

$$\omega = 2\pi f = 2\pi(1.0 \times 10^8 \text{ Hz}) = 6.3 \times 10^8 \text{ rad/s}.$$

(c) The corresponding angular wave number is $k = \dfrac{2\pi}{\lambda} = \dfrac{2\pi}{3.0 \text{ m}} = 2.1 \text{ rad/m}.$

(d) With $E_m = 300$ V/m, the magnetic field amplitude is

$$B_m = \frac{E_m}{c} = \frac{300 \text{ V/m}}{2.998 \times 10^8 \text{ m/s}} = 1.0 \times 10^{-6} \text{ T}.$$

(e) Since $\vec{E}$ is in the positive y direction, $\vec{B}$ must be in the positive z direction so that their cross product $\vec{E} \times \vec{B}$ points in the positive x direction (the direction of propagation).

(f) The intensity of the wave is

$$I = \frac{E_m^2}{2\mu_0 c} = \frac{(300\,\text{V/m})^2}{2(4\pi \times 10^{-7}\,\text{H/m})(2.998 \times 10^8\,\text{m/s})} = 119\,\text{W/m}^2 \approx 1.2 \times 10^2\,\text{W/m}^2.$$

(g) Since the sheet is perfectly absorbing, the rate per unit area with which momentum is delivered to it is I/c, so

$$\frac{dp}{dt} = \frac{IA}{c} = \frac{(119\ \text{W/m}^2)(2.0\ \text{m}^2)}{2.998 \times 10^8\ \text{m/s}} = 8.0 \times 10^{-7}\ \text{N}.$$

(h) The radiation pressure is $p_r = \dfrac{dp/dt}{A} = \dfrac{8.0 \times 10^{-7}\ \text{N}}{2.0\ \text{m}^2} = 4.0 \times 10^{-7}\ \text{Pa}.$

LEARN The energy density in this case is given by

$$u = \frac{I}{c} = \frac{119\ \text{W/m}^2}{2.998 \times 10^8\ \text{m/s}} = 4.0 \times 10^{-7}\ \text{J/m}^3$$

which is the same as the radiation pressure p_r.

33-29

THINK The laser beam carries both energy and momentum. The total momentum of the spaceship and the light is conserved.

EXPRESS If the beam carries energy U away from the spaceship, then it also carries momentum $p = U/c$ away. By momentum conservation, this is the magnitude of the momentum acquired by the spaceship. If P is the power of the laser, then the energy carried away in time t is $U = Pt$.

ANALYZE We note that there are 86400 seconds in a day. Thus, $p = Pt/c$ and, if m is mass of the spaceship, its speed is

$$v = \frac{p}{m} = \frac{Pt}{mc} = \frac{(10 \times 10^3\ \text{W})(86400\ \text{s})}{(1.5 \times 10^3\ \text{kg})(2.998 \times 10^8\ \text{m/s})} = 1.9 \times 10^{-3}\ \text{m/s}.$$

LEARN As expected, the speed of the spaceship is proportional to the power of the laser beam.

33-33

THINK Unpolarized light becomes polarized when it is sent through a polarizing sheet. In this problem, three polarizing sheets are involved, and we work through the system sheet by sheet, applying either the one-half rule or the cosine-squared rule.

EXPRESS Let I_0 be the intensity of the unpolarized light that is incident on the first polarizing sheet. The transmitted intensity is, by one-half rule, $I_1 = \frac{1}{2}I_0$, and the direction of polarization of the transmitted light is $\theta_1 = 40°$ *counterclockwise* from the *y* axis in the diagram. For the second sheet (and the third one as well), we apply the cosine-squared rule:

$$I_2 = I_1 \cos^2\theta_2'$$

where θ_2' is the angle between the direction of polarization that is incident on that sheet and the polarizing direction of the sheet.

ANALYZE The polarizing direction of the second sheet is $\theta_2 = 20°$ *clockwise* from the *y* axis, so $\theta_2' = 40° + 20° = 60°$. The transmitted intensity is

$$I_2 = I_1 \cos^2 60° = \frac{1}{2} I_0 \cos^2 60°,$$

and the direction of polarization of the transmitted light is 20° clockwise from the *y* axis. The polarizing direction of the third sheet is $\theta_3 = 40°$ *counterclockwise* from the *y* axis. Consequently, the angle between the direction of polarization of the light incident on that sheet and the polarizing direction of the sheet is 20° + 40° = 60°. The transmitted intensity is

$$I_3 = I_2 \cos^2 60° = \frac{1}{2} I_0 \cos^4 60° = 3.1\times10^{-2} I_0.$$

Thus, 3.1% of the light's initial intensity is transmitted.

LEARN When two polarizing sheets are crossed ($\theta = 90°$), no light passes through and the transmitted intensity is zero.

33-37

THINK A polarizing sheet can change the direction of polarization of the incident beam since it allows only the component that is parallel to its polarization direction to pass.

EXPRESS The 90° rotation of the polarization direction cannot be done with a single sheet. If a sheet is placed with its polarizing direction at an angle of 90° to the direction of polarization of the incident radiation, no radiation is transmitted.

ANALYZE (a) The 90° rotation of the polarization direction can be done with two sheets. We place the first sheet with its polarizing direction at some angle θ, between 0° and 90°, to the direction of polarization of the incident radiation. Place the second sheet with its polarizing direction at 90° to the polarization direction of the incident radiation. The transmitted radiation is then polarized at 90° to the incident polarization direction. The intensity is

$$I = I_0 \cos^2\theta \cos^2(90° - \theta) = I_0 \cos^2\theta \sin^2\theta,$$

where I_0 is the incident radiation. If θ is not 0° or 90°, the transmitted intensity is not zero.

(b) Consider n sheets, with the polarizing direction of the first sheet making an angle of $\theta = 90°/n$ relative to the direction of polarization of the incident radiation. The polarizing direction of each successive sheet is rotated $90°/n$ in the same sense from the polarizing direction of the previous sheet. The transmitted radiation is polarized, with its direction of polarization making an angle of 90° with the direction of polarization of the incident radiation. The intensity is

$$I = I_0 \cos^{2n}(90°/n).$$

We want the smallest integer value of n for which this is greater than $0.60I_0$. We start with $n = 2$ and calculate $\cos^{2n}(90°/n)$. If the result is greater than 0.60, we have obtained the solution. If it is less, increase n by 1 and try again. We repeat this process, increasing n by 1 each time, until we have a value for which $\cos^{2n}(90°/n)$ is greater than 0.60. The first one will be $n = 5$.

LEARN The intensities associated with $n = 1$ to 5 are:

$$\begin{aligned}
I_{n=1} &= I_0 \cos^2(90°) = 0 \\
I_{n=2} &= I_0 \cos^4(45°) = I_0/4 = 0.25I_0 \\
I_{n=3} &= I_0 \cos^6(30°) = 0.422I_0 \\
I_{n=4} &= I_0 \cos^8(22.5°) = 0.531I_0 \\
I_{n=5} &= I_0 \cos^{10}(18°) = 0.605I_0.
\end{aligned}$$

Thus, $I > 0.60I_0$ with 5 sheets.

33-53

THINK The angle with which the light beam emerges from the triangular prism depends on the index of refraction of the prism.

EXPRESS Consider diagram (a) shown next. The incident angle is θ and the angle of refraction is θ_2. Since $\theta_2 + \alpha = 90°$ and $\phi + 2\alpha = 180°$, we have

$$\theta_2 = 90° - \alpha = 90° - \frac{1}{2}(180° - \phi) = \frac{\phi}{2}.$$

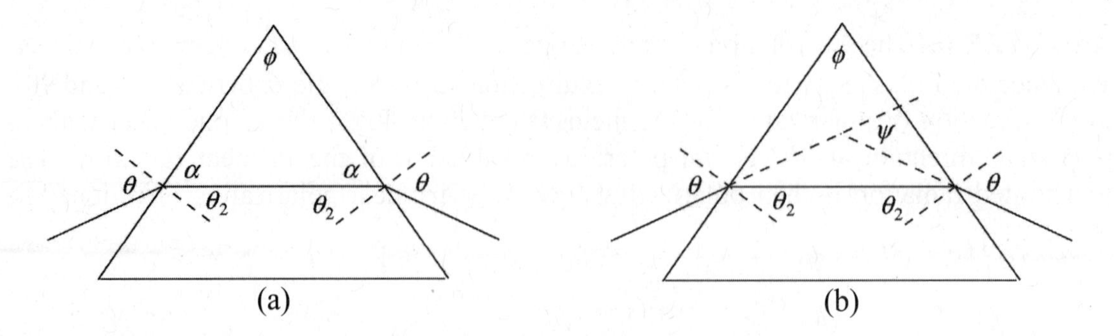

ANALYZE Next, examine diagram (b) and consider the triangle formed by the two normals and the ray in the interior. One can show that ψ is given by

$$\psi = 2(\theta - \theta_2)\,.$$

Upon substituting $\phi/2$ for θ_2, we obtain $\psi = 2(\theta - \phi/2)$ which yields $\theta = (\phi + \psi)/2$. Thus, using the law of refraction, we find the index of refraction of the prism to be

$$n = \frac{\sin\theta}{\sin\theta_2} = \frac{\sin\frac{1}{2}(\phi + \psi)}{\sin\frac{1}{2}\phi}.$$

LEARN The angle ψ is called the deviation angle. Physically, it represents the total angle through which the beam has turned while passing through the prism. This angle is minimum when the beam passes through the prism "symmetrically," as it does in this case. Knowing the value of ϕ and ψ allows us to determine the value of n for the prism material.

33-55

THINK Light is refracted at the air–water interface. To calculate the length of the shadow of the pole, we first calculate the angle of refraction using the Snell's law.

EXPRESS Consider a ray that grazes the top of the pole, as shown in the diagram below.

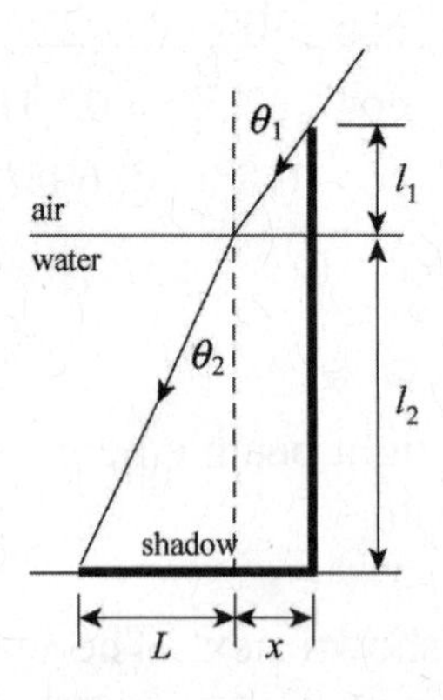

Here θ_1 = 90° – θ = 90° –55° = 35°, $\ell_1 = 0.50$ m, and $\ell_2 = 1.50$ m. The length of the shadow is $d = x + L$.

ANALYZE The distance x is given by

$$x = \ell_1 \tan\theta_1 = (0.50 \text{ m}) \tan 35° = 0.35 \text{ m}.$$

According to the law of refraction, $n_2 \sin \theta_2 = n_1 \sin \theta_1$. We take $n_1 = 1$ and $n_2 = 1.33$ (from Table 33-1). Then,

$$\theta_2 = \sin^{-1}\left(\frac{\sin\theta_1}{n_2}\right) = \sin^{-1}\left(\frac{\sin 35.0°}{1.33}\right) = 25.55°.$$

The distance L is given by

$$L = \ell_2 \tan\theta_2 = (1.50 \text{ m}) \tan 25.55° = 0.72 \text{ m}.$$

Thus, the length of the shadow is $d = x + L = 0.35 \text{ m} + 0.72 \text{ m} = 1.07 \text{ m}$.

LEARN If the pool were empty with no water, then $\theta_1 = \theta_2$ and the length of the shadow would be

$$d' = \ell_1 \tan\theta_1 + \ell_2 \tan\theta_1 = (\ell_1 + \ell_2)\tan\theta_1 = (0.50 \text{ m} + 1.50 \text{ m}) \tan 35° = 1.40 \text{ m}$$

by simple geometric consideration.

33-59

THINK Total internal reflection happens when the angle of incidence exceeds a critical angle such that Snell's law gives $\sin\theta_2 > 1$.

EXPRESS When light reaches the interfaces between two materials with indices of refraction n_1 and n_2, if $n_1 > n_2$, and the incident angle exceeds a critical value given by

$$\theta_c = \sin^{-1}\left(\frac{n_2}{n_1}\right),$$

then total internal reflection will occur. In our case, the incident light ray is perpendicular to the face *ab*. Thus, no refraction occurs at the surface *ab*, so the angle of incidence at surface *ac* is $\theta = 90° - \phi$, as shown in the figure below.

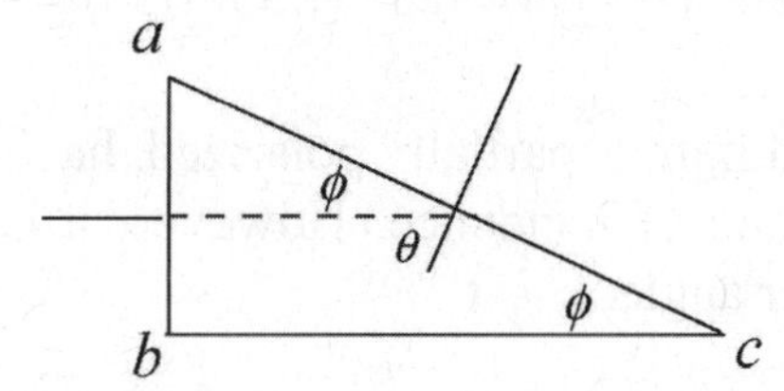

ANALYZE (a) For total internal reflection at the second surface, $n_g \sin(90° - \phi)$ must be greater than n_a. Here n_g is the index of refraction for the glass and n_a is the index of refraction for air. Since sin (90° – ϕ) = cos ϕ, we want the largest value of ϕ for which n_g cos $\phi \geq n_a$. Recall that cos ϕ decreases as ϕ increases from zero. When ϕ has the largest value for which total internal reflection occurs, then $n_g \cos\phi = n_a$, or

$$\phi = \cos^{-1}\left(\frac{n_a}{n_g}\right) = \cos^{-1}\left(\frac{1}{1.52}\right) = 48.9°.$$

The index of refraction for air is taken to be unity.

(b) We now replace the air with water. If n_w = 1.33 is the index of refraction for water, then the largest value of ϕ for which total internal reflection occurs is

$$\phi = \cos^{-1}\left(\frac{n_w}{n_g}\right) = \cos^{-1}\left(\frac{1.33}{1.52}\right) = 29.0°.$$

LEARN Total internal reflection cannot occur if the incident light is in the medium with lower index of refraction. With $\theta_c = \sin^{-1}(n_2/n_1)$, we see that the larger the ratio n_2/n_1, the larger the value of θ_c.

33-69

THINK A reflected wave will be fully polarized if it strikes the boundary at the Brewster angle.

EXPRESS The angle of incidence for which reflected light is fully polarized is given by Eq. 33-48:

$$\theta_B = \tan^{-1}\left(\frac{n_2}{n_1}\right)$$

where n_1 is the index of refraction for the medium of incidence and n_2 is the index of refraction for the second medium. The angle θ_B is called the Brewster angle.

ANALYZE With n_1 = 1.33 and n_2 = 1.53, we obtain

$$\theta_B = \tan^{-1}(n_2/n_1) = \tan^{-1}(1.53/1.33) = 49.0°.$$

LEARN In general, reflected light is partially polarized, having components both parallel and perpendicular to the plane of incidence. However, it can be completely polarized when incident at the Brewster angle.

33-71

THINK All electromagnetic waves, including visible light, travel at the same speed c in vacuum.

EXPRESS The time for light to travel a distance d in free space is $t = d/c$, where c is the speed of light (3.00×10^8 m/s).

ANALYZE (a) We take d to be 150 km = 150×10^3 m. Then,

$$t = \frac{d}{c} = \frac{150 \times 10^3 \text{ m}}{3.00 \times 10^8 \text{ m/s}} = 5.00 \times 10^{-4} \text{ s}.$$

(b) At full moon, the Moon and Sun are on opposite sides of Earth, so the distance traveled by the light is

$$d = (1.5 \times 10^8 \text{ km}) + 2\,(3.8 \times 10^5 \text{ km}) = 1.51 \times 10^8 \text{ km} = 1.51 \times 10^{11} \text{ m}.$$

The time taken by light to travel this distance is

$$t = \frac{d}{c} = \frac{1.51 \times 10^{11} \text{ m}}{3.00 \times 10^8 \text{ m/s}} = 500 \text{ s} = 8.4 \text{ min}.$$

(c) We take d to be $2(1.3 \times 10^9 \text{ km}) = 2.6 \times 10^{12}$ m. Then,

$$t = \frac{d}{c} = \frac{2.6 \times 10^{12} \text{ m}}{3.00 \times 10^8 \text{ m/s}} = 8.7 \times 10^3 \text{ s} = 2.4 \text{ h}.$$

(d) We take d to be 6500 ly and the speed of light to be 1.00 ly/y. Then,

$$t = \frac{d}{c} = \frac{6500 \text{ ly}}{1.00 \text{ ly/y}} = 6500 \text{ y}.$$

The explosion took place in the year 1054 – 6500 = –5446 or 5446 B.C.

LEARN Since the speed c is constant, the travel time is proportional to the distance. The radio signals at 150 km away reach you almost instantly.

33-73

THINK The electric and magnetic components of the electromagnetic waves are always in phase, perpendicular to each other, and perpendicular to the direction of propagation of the wave.

EXPRESS The electric and magnetic fields can be written as sinusoidal functions of position and time as:

$$E = E_m \sin(kx + \omega t), \quad B = B_m \sin(kx + \omega t)$$

where E_m and B_m are the amplitudes of the fields, and ω and k are the angular frequency and angular wave number of the wave, respectively. The two amplitudes are related by Eq. 33-4: $E_m / B_m = c$, where c is the speed of the wave.

ANALYZE (a) From $kc = \omega$ where $k = 1.00 \times 10^6$ m^{-1}, we obtain $\omega = 3.00 \times 10^{14}$ rad/s. The magnetic field amplitude is, from Eq. 33-5,

$$B_m = E_m/c = (5.00 \text{ V/m})/c = 1.67 \times 10^{-8} \text{ T}.$$

From the argument of the sinusoidal function for E, we see that the direction of propagation is in the $-z$ direction. Since $\vec{E} = E_y \hat{\text{j}}$, and that $\vec{B}$ is perpendicular to $\vec{E}$ and $\vec{E} \times \vec{B}$,, we conclude that the only nonzero component of $\vec{B}$ is B_x, so that we have

$$B_x = (1.67 \times 10^{-8} \text{ T}) \sin[(1.00 \times 10^6 \text{ /m})z + (3.00 \times 10^{14} \text{ /s})t].$$

(b) The wavelength is $\lambda = 2\pi/k = 6.28 \times 10^{-6}$ m.

(c) The period is $T = 2\pi/\omega = 2.09 \times 10^{-14}$ s.

(d) The intensity is

$$I = \frac{1}{c\mu_0} \left(\frac{5.00 \text{ V/m}}{\sqrt{2}} \right)^2 = 0.0332 \text{ W/m}^2.$$

(e) As noted in part (a), the only nonzero component of $\vec{B}$ is B_x. The magnetic field oscillates along the x axis.

(f) The wavelength found in part (b) places this in the infrared portion of the spectrum.

LEARN Electromagnetic wave is a transverse wave. Knowing the functional form of the electric field and the direction of propagation allows us to determine the corresponding magnetic field, and vice versa.

33-75

THINK Total internal reflection happens when the angle of incidence exceeds a critical angle such that Snell's law gives $\sin\theta_2 > 1$.

EXPRESS When light reaches the interfaces between two materials with indices of refraction n_1 and n_2, if $n_1 > n_2$, and the incident angle exceeds a critical value given by

$$\theta_c = \sin^{-1}\left(\frac{n_2}{n_1}\right),$$

then total internal reflection will occur.

Referring to Fig. 33-65, let $\theta_1 = 45°$ be the angle of incidence at the first surface and θ_2 be the angle of refraction there. Let θ_3 be the angle of incidence at the second surface. The condition for total internal reflection at the second surface is

$$n \sin \theta_3 \geq 1.$$

We want to find the smallest value of the index of refraction n for which this inequality holds. The law of refraction, applied to the first surface, yields

$$n \sin \theta_2 = \sin \theta_1.$$

Consideration of the triangle formed by the surface of the slab and the ray in the slab tells us that $\theta_3 = 90° - \theta_2$. Thus, the condition for total internal reflection becomes

$$1 \leq n \sin(90° - \theta_2) = n \cos \theta_2.$$

Squaring this equation and using $\sin^2 \theta_2 + \cos^2 \theta_2 = 1$, we obtain $1 \leq n^2 (1 - \sin^2 \theta_2)$. Substituting $\sin \theta_2 = (1/n) \sin \theta_1$ now leads to

$$1 \leq n^2\left(1 - \frac{\sin^2 \theta_1}{n^2}\right) = n^2 - \sin^2 \theta_1.$$

The smallest value of n for which this equation is true is given by $1 = n^2 - \sin^2 \theta_1$. We solve for n:

$$n = \sqrt{1 + \sin^2 \theta_1} = \sqrt{1 + \sin^2 45°} = 1.22.$$

LEARN With $n = 1.22$, we have $\theta_2 = \sin^{-1}[(1/1.22)\sin 45°] = 35°$, which gives $\theta_3 = 90° - 35° = 55°$ as the angle of incidence at the second surface. We can readily verify that $n \sin \theta_3 = (1.22) \sin 55° = 1$, meeting the threshold condition for total internal reflection.

33-79

THINK We apply law of refraction to both interfaces to calculate the sideway displacement.

EXPRESS Let θ be the angle of incidence and θ_2 be the angle of refraction at the left face of the plate. Let n be the index of refraction of the glass. Then, the law of refraction yields

$$\sin\theta = n\sin\theta_2.$$

The angle of incidence at the right face is also θ_2. If θ_3 is the angle of emergence there, then

$$n\sin\theta_2 = \sin\theta_3.$$

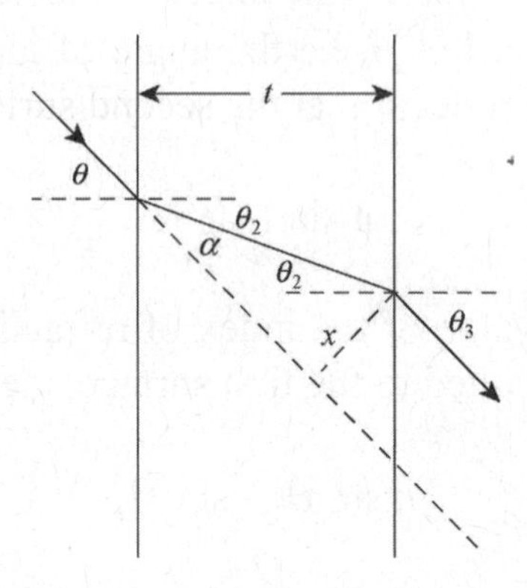

ANALYZE (a) Combining the two expressions gives $\sin\theta_3 = \sin\theta$, which implies that $\theta_3 = \theta$. Thus, the emerging ray is parallel to the incident ray.

(b) We wish to derive an expression for x in terms of θ. If D is the length of the ray in the glass, then $D\cos\theta_2 = t$ and $D = t/\cos\theta_2$. The angle α in the diagram equals $\theta - \theta_2$ and

$$x = D\sin\alpha = D\sin(\theta - \theta_2).$$

Thus,

$$x = \frac{t\sin(\theta-\theta_2)}{\cos\theta_2}.$$

If all the angles θ, θ_2, θ_3, and $\theta - \theta_2$ are small and measured in radians, then $\sin\theta \approx \theta$, $\sin\theta_2 \approx \theta_2$, $\sin(\theta - \theta_2) \approx \theta - \theta_2$, and $\cos\theta_2 \approx 1$. Thus $x \approx t(\theta - \theta_2)$. The law of refraction applied to the point of incidence at the left face of the plate is now $\theta \approx n\theta_2$, so $\theta_2 \approx \theta/n$ and

$$x \approx t\left(\theta - \frac{\theta}{n}\right) = \frac{(n-1)t\theta}{n}.$$

LEARN The thicker the glass, the greater the displacement x. Note in the limit $n = 1$ (no glass), $x = 0$, as expected.

33-83

THINK The index of refraction encountered by light generally depends on the wavelength of the light.

EXPRESS The critical angle for total internal reflection is given by sin $\theta_c = 1/n$. With an index of refraction n = 1.456 at the red end, the critical angle is θ_c = 43.38° for red. Similarly, with n = 1.470 at the blue end, the critical angle is θ_c = 42.86° for blue.

ANALYZE (a) An angle of incidence of θ_1 = 42.00° is less than the critical angles for both red and blue light, so the refracted light is white.

(b) An angle of incidence of θ_1 = 43.10° is slightly less than the critical angle for red light but greater than the critical angle for blue light, so the refracted light is dominated by the red end.

(c) An angle of incidence of θ_1 = 44.00° is greater than the critical angles for both red and blue light, so there is no refracted light.

LEARN The dependence of the index of refraction of fused quartz on wavelength is shown in Fig. 33-18. From the figure, we see that the index of refraction is greater for a shorter wavelength. Such dependence results in the spreading of light as it enters or leaves quartz, a phenomenon called "chromatic dispersion."

33-87

THINK Since the radar beam is emitted uniformly over the hemisphere, the source power is also the same everywhere within the hemisphere.

EXPRESS The intensity of the beam is given by

$$I = \frac{P}{A} = \frac{P}{2\pi r^2}$$

where $A = 2\pi r^2$ is the area of a hemisphere. The power of the aircraft's reflection is equal to the product of the intensity at the aircraft's location and its cross-sectional area: $P_r = IA_r$. The intensity is related to the amplitude of the electric field by Eq. 33-26: $I = E_{\rm rms}^2 / c\mu_0 = E_m^2 / 2c\mu_0$.

ANALYZE (a) Substituting the values given we get

$$I = \frac{P}{2\pi r^2} = \frac{180\times 10^3 \text{ W}}{2\pi(90\times 10^3 \text{ m})^2} = 3.5\times 10^{-6} \text{ W/m}^2.$$

(b) The power of the aircraft's reflection is

$$P_r = IA_r = (3.5\times 10^{-6} \text{ W/m}^2)(0.22 \text{ m}^2) = 7.8\times 10^{-7} \text{ W}.$$

(c) Back at the radar site, the intensity is

$$I_r = \frac{P_r}{2\pi r^2} = \frac{7.8\times10^{-7}\ \text{W}}{2\pi(90\times10^3\ \text{m})^2} = 1.5\times10^{-17}\ \text{W/m}^2.$$

(d) From $I_r = E_m^2/2c\mu_0$, we find the amplitude of the electric field to be

$$E_m = \sqrt{2c\mu_0 I_r} = \sqrt{2(3.0\times10^8\ \text{m/s})(4\pi\times10^{-7}\ \text{T}\cdot\text{m/A})(1.5\times10^{-17}\ \text{W/m}^2)}$$
$$= 1.1\times10^{-7}\ \text{V/m}.$$

(e) The rms value of the magnetic field is

$$B_{\text{rms}} = \frac{E_{\text{rms}}}{c} = \frac{E_m}{\sqrt{2}c} = \frac{1.1\times10^{-7}\ \text{V/m}}{\sqrt{2}(3.0\times10^8\ \text{m/s})} = 2.5\times10^{-16}\,\text{T}.$$

LEARN The intensity due to a power source decreases with the square of the distance. Also, as emphasized in Sample Problem — "Light wave: rms values of the electric and magnetic fields," one cannot compare the values of the two fields because they are measured in different units. Both components are on the same basis from the perspective of wave propagation, and they have the same average energy.

33-109

THINK The general wave equation is of the form

$$\frac{\partial^2 y}{\partial x^2} = \frac{1}{v^2}\frac{\partial^2 y}{\partial t^2}$$

where y is the wave function, and v is the speed of the wave. From Maxwell's equations, one can show that both electric field and magnetic field satisfy the wave equation. The verification requires partial differentiation of E and B with respect to t and x.

EXPRESS From Eq. 33-1, differentiating $E(x,t) = E_m \sin(kx-\omega t)$ twice with respect to x and t gives

$$\frac{\partial^2 E}{\partial t^2} = \frac{\partial^2}{\partial t^2}\left[E_m \sin(kx-\omega t)\right] = -\omega^2 E_m \sin(kx-\omega t),$$
$$\frac{\partial^2 E}{\partial x^2} = \frac{\partial^2}{\partial x^2}\left[E_m \sin(kx-\omega t)\right] = -k^2 E_m \sin(kx-\omega t).$$

ANALYZE (a) Using $\omega = kc$, we find

$$c^2\frac{\partial^2 E}{\partial x^2} = -k^2c^2 \sin(kx-\omega t) = -\omega^2 E_m \sin(kx-\omega t),$$

or

$$\frac{\partial^2 E}{\partial t^2} = c^2 \frac{\partial^2 E}{\partial x^2}.$$

Analogously, one can show that Eq. 33-2 satisfies

$$\frac{\partial^2 B}{\partial t^2} = c^2 \frac{\partial^2 B}{\partial x^2}.$$

(b) From $E = E_m f(kx \pm \omega t)$,

$$\frac{\partial^2 E}{\partial t^2} = E_m \frac{\partial^2 f(kx \pm \omega t)}{\partial t^2} = \omega^2 E_m \left.\frac{d^2 f}{du^2}\right|_{u=kx \pm \omega t}$$

and

$$c^2 \frac{\partial^2 E}{\partial x^2} = c^2 E_m \frac{\partial^2 f(kx \pm \omega t)}{\partial t^2} = c^2 E_m k^2 \left.\frac{d^2 f}{du^2}\right|_{u=kx \pm \omega t}$$

Since $\omega = ck$ the right-hand sides of these two equations are equal. Therefore,

$$\frac{\partial^2 E}{\partial t^2} = c^2 \frac{\partial^2 E}{\partial x^2}.$$

Changing E to B and repeating the derivation above shows that $B = B_m f(kx \pm \omega t)$ satisfies

$$\frac{\partial^2 B}{\partial t^2} = c^2 \frac{\partial^2 B}{\partial x^2}.$$

LEARN Electric and magnetic fields are components of the electromagnetic wave; the fields are perpendicular to each other and also perpendicular to the direction of propagation. The two fields are in phase at all times, with $E_m = cB_m$.

Chapter 34

34-5.
THINK This problem involves refraction at air–water interface and reflection from a plane mirror at the bottom of the pool.

EXPRESS We apply the law of refraction, assuming all angles are in radians:

$$\frac{\sin\theta}{\sin\theta'}=\frac{n_w}{n_{\text{air}}},$$

which in our case reduces to $\theta' \approx \theta/n_w$ (since both θ and θ' are small, and $n_{\text{air}} \approx 1$). We refer to our figure on the right.

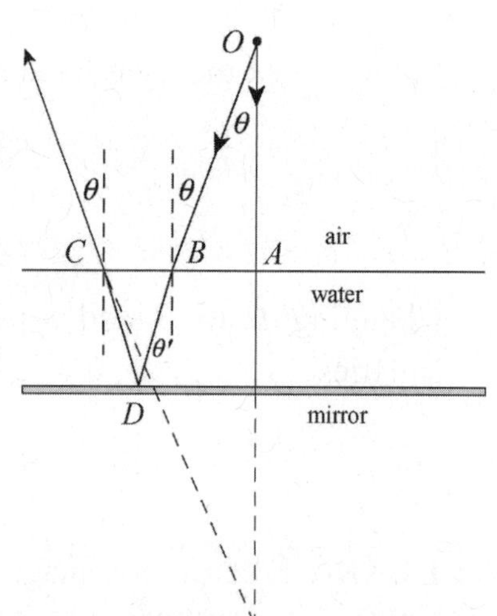

The object O is a vertical distance d_1 above the water, and the water surface is a vertical distance d_2 above the mirror. We are looking for a distance d (treated as a positive number) below the mirror where the image I of the object is formed. In the triangle OAB

$$|AB|=d_1\tan\theta\approx d_1\theta,$$

and in the triangle CBD

$$|BC|=2d_2\tan\theta'\approx 2d_2\theta'\approx\frac{2d_2\theta}{n_w}.$$

Finally, in the triangle ACI, we have $|AI| = d + d_2$.

ANALYZE Therefore,

$$d=|AI|-d_2=\frac{|AC|}{\tan\theta}-d_2\approx\frac{|AB|+|BC|}{\theta}-d_2=\left(d_1\theta+\frac{2d_2\theta}{n_w}\right)\frac{1}{\theta}-d_2=d_1+\frac{2d_2}{n_w}-d_2$$

$$=250\text{ cm}+\frac{2(200\text{ cm})}{1.33}-200\text{ cm}=351\text{ cm}.$$

LEARN If the pool were empty without water, then $\theta=\theta'$, and the distance would be

$$d = d_1 + 2d_2 - d_2 = d_1 + d_2 .$$

This is precisely what we expect from a plane mirror.

34-9

THINK A concave mirror has a positive value of focal length.

EXPRESS For spherical mirrors, the focal length f is related to the radius of curvature r by $f = r/2$. The object distance p, the image distance i, and the focal length f are related by Eq. 34-4:

$$\frac{1}{p} + \frac{1}{i} = \frac{1}{f}.$$

The value of i is positive for real images and negative for virtual images.

The corresponding lateral magnification is

$$m = -\frac{i}{p}.$$

The value of m is positive for upright (not inverted) images, and negative for inverted images. Real images are formed on the same side as the object, while virtual images are formed on the opposite side of the mirror.

ANALYZE (a) With $f = +12$ cm and $p = +18$ cm, the radius of curvature is $r = 2f = 2(12 \text{ cm}) = +24$ cm.

(b) The image distance is $i = \dfrac{pf}{p-f} = \dfrac{(18 \text{ cm})(12 \text{ cm})}{18 \text{ cm} - 12 \text{ cm}} = 36$ cm.

(c) The lateral magnification is $m = -i/p = -(36 \text{ cm})/(18 \text{ cm}) = -2.0$.

(d) Since the image distance i is positive, the image is real (R).

(e) Since the magnification m is negative, the image is inverted (I).

(f) A real image is formed on the <u>same</u> side as the object.

LEARN The situation in this problem is similar to that illustrated in Fig. 34-10(c). The object is outside the focal point, and its image is real and inverted.

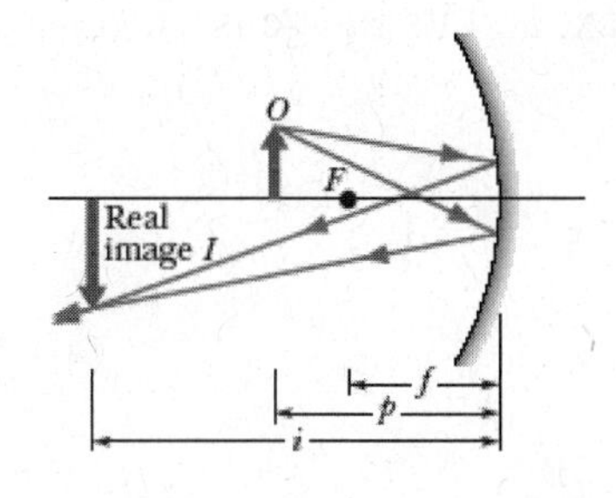

34-11

THINK A convex mirror has a negative value of focal length.

EXPRESS For spherical mirrors, the focal length f is related to the radius of curvature r by $f = r/2$. The object distance p, the image distance i, and the focal length f are related by Eq. 34-4:

$$\frac{1}{p}+\frac{1}{i}=\frac{1}{f}.$$

The value of i is positive for real images and negative for virtual images.

The corresponding lateral magnification is

$$m=-\frac{i}{p}.$$

The value of m is positive for upright (not inverted) images, and negative for inverted images. Real images are formed on the same side as the object, while virtual images are formed on the opposite side of the mirror.

ANALYZE (a) With $f = -10$ cm and $p = +8$ cm, the radius of curvature is $r = 2f = -20$ cm.

(b) The image distance is $i=\dfrac{pf}{p-f}=\dfrac{(8\text{ cm})(-10\text{ cm})}{8\text{ cm}-(-10)\text{ cm}}=-4.44$ cm.

(c) The lateral magnification is $m = -i/p = -(-4.44\text{ cm})/(8.0\text{ cm}) = +0.56$.

(d) Since the image distance is negative, the image is virtual (V).

(e) The magnification m is positive, so the image is upright [not inverted] (NI).

(f) A virtual image is formed on the opposite side of the mirror from the object.

LEARN The situation in this problem is similar to that illustrated in Fig. 34-11(c). The mirror is convex, and its image is virtual and upright.

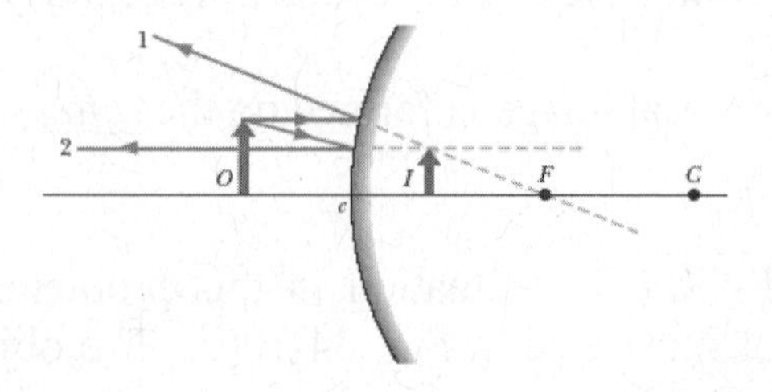

34-13

THINK A concave mirror has a positive value of focal length.

EXPRESS For spherical mirrors, the focal length f is related to the radius of curvature r by $f = r/2$.

The object distance p, the image distance i, and the focal length f are related by Eq. 34-4:

$$\frac{1}{p}+\frac{1}{i}=\frac{1}{f}.$$

The value of i is positive for real images and negative for virtual images.

The corresponding lateral magnification is $m=-i/p$. The value of m is positive for upright (not inverted) images, and is negative for inverted images. Real images are formed on the same side as the object, while virtual images are formed on the opposite side of the mirror.

ANALYZE With $f = +18$ cm and $p = +12$ cm, the radius of curvature is

$$r = 2f = +36 \text{ cm}.$$

(b) Equation 34-9 yields $i = pf/(p-f) = -36$ cm.

(c) Then, by Eq. 34-7, $m = -i/p = +3.0$.

(d) Since the image distance is negative, the image is virtual (V).

(e) The magnification computation produced a positive value, so it is upright [not inverted] (NI).

(f) A virtual image is formed on the opposite side of the mirror from the object.

LEARN The situation in this problem is similar to that illustrated in Fig. 34-11(a). The mirror is concave, and its image is virtual, enlarged, and upright.

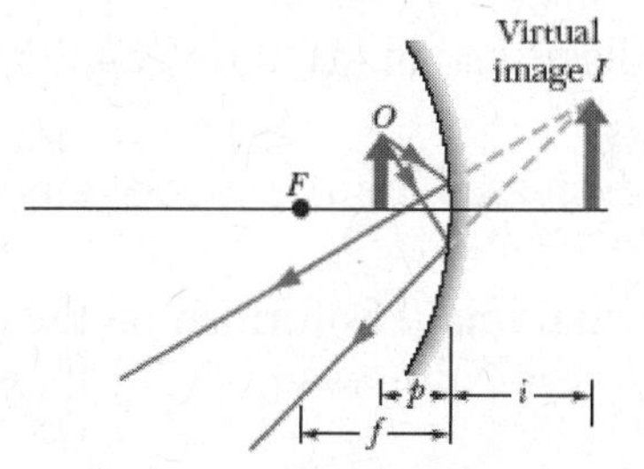

34-23

THINK A positive value for the magnification means that the image is upright (not inverted).

EXPRESS For spherical mirrors, the focal length f is related to the radius of curvature r by $f = r/2$. The object distance p, the image distance i, and the focal length f are related by Eq. 34-4:

$$\frac{1}{p}+\frac{1}{i}=\frac{1}{f}.$$

The value of i is positive for real images, and negative for virtual images. The corresponding lateral magnification is $m = -i/p$. The value of m is positive for upright (not inverted) images, and is negative for inverted images. Real images are formed on the same side as the object, while virtual images are formed on the opposite side of the mirror.

ANALYZE (a) The magnification is given by $m = -i/p$. Since $p > 0$, a positive value for m means that the image distance (i) is negative, implying a virtual image. Looking at the discussion of mirrors in Sections 34-3 and 34-4, we see that a positive magnification of magnitude less than unity is only possible for convex mirrors.

(b) With $i = -mp$, we may write

$$p = f(1-1/m).$$

For $0 < m < 1$, a positive value for p can be obtained only if $f < 0$. Thus, with a minus sign, we have $f = -30$ cm.

(c) The radius of curvature is $r = 2f = -60$ cm.

(d) The object distance is

$$p = f(1-1/m) = (-30 \text{ cm})(1 - 1/0.20) = +120 \text{ cm} = 1.2 \text{ m}.$$

(e) The image distance is $i = -mp = -(0.20)(120 \text{ cm}) = -24$ cm.

(f) The magnification is $m = +0.20$, as given in the Table.

(g) As discussed in (a), the image is virtual (V).

(h) As discussed in (a), the image is upright, or not inverted (NI).

(i) A virtual image is formed on the opposite side of the mirror from the object.

LEARN The situation in this problem is similar to that illustrated in Fig. 34-11(c). The mirror is convex, and its image is virtual and upright.

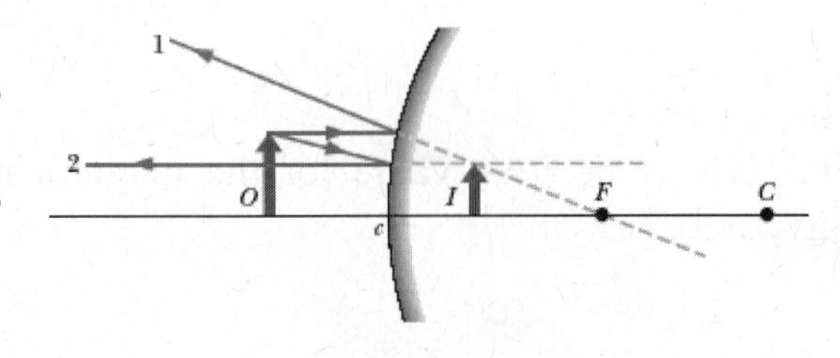

34-29

THINK A convex mirror has a negative value of focal length.

EXPRESS For spherical mirrors, the focal length f is related to the radius of curvature r by $f = r/2$. The object distance p, the image distance i, and the focal length f are related by Eq. 34-4:

$$\frac{1}{p}+\frac{1}{i}=\frac{1}{f}.$$

The value of i is positive for real images, and negative for virtual images. The corresponding lateral magnification is $m = -i/p$. The value of m is positive for upright (not inverted) images, and is negative for inverted images. Real images are formed on the same side as the object, while virtual images are formed on the opposite side of the mirror.

ANALYZE (a) The mirror is convex, as given.

(b) Since the mirror is convex, the radius of curvature is negative, so $r = -40$ cm. Then, the focal length is $f = r/2 = (-40\text{ cm})/2 = -20$ cm.

(c) The radius of curvature is $r = -40$ cm.

(d) The fact that the mirror is convex also means that we need to insert a minus sign in front of the "4.0" value given for i, since the image in this case must be virtual. Equation 34-4 leads to

$$p=\frac{if}{i-f}=\frac{(-4.0\text{ cm})(-20\text{ cm})}{-4.0\text{ cm}-(-20\text{ cm})}=5.0\text{ cm}.$$

(e) As noted above, $i = -4.0$ cm.

(f) The magnification is $m = -i/p = -(-4.0\text{ cm})/(5.0\text{ cm}) = +0.80$.

(g) The image is virtual (V) since $i < 0$.

(h) The image is upright or not inverted (NI).

(i) A virtual image is formed on the opposite side of the mirror from the object.

LEARN The situation in this problem is similar to that illustrated in Fig. 34-11(c). The mirror is convex, and its image is virtual and upright.

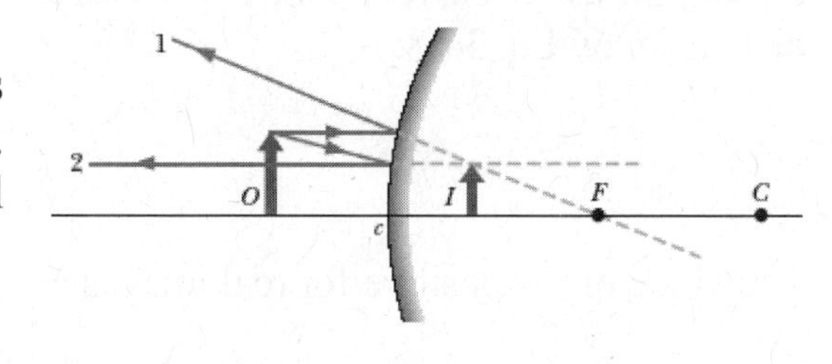

34-33

THINK An image is formed by refraction through a spherical surface. A negative value for the image distance implies that the image is virtual.

EXPRESS Let n_1 be the index of refraction of the material where the object is located, n_2 be the index of refraction of the material on the other side of the refracting surface, and r be the radius of curvature of the surface. The image distance i is related to the object distance p by Eq. 34-8:

$$\frac{n_1}{p}+\frac{n_2}{i}=\frac{n_2-n_1}{r}.$$

The value of i is positive for real images and negative for virtual images.

ANALYZE In addition to n_1 =1.0, we are given (a) n_2= 1.5, (b) p = +10 cm, and (d) $i=-13$ cm.

(c) Equation 34-8 yields

$$r=(n_2-n_1)\left(\frac{n_1}{p}+\frac{n_2}{i}\right)^{-1}=(1.5-1.0)\left(\frac{1.0}{10\text{ cm}}+\frac{1.5}{-13\text{ cm}}\right)^{-1}=-32.5\text{ cm}\approx-33\text{ cm}.$$

(e) The image is virtual (V) and upright.

(f) The object and its image are on the same side.

LEARN The ray diagram for this problem is similar to the one shown in Fig. 34-12(e). Here refraction always directs the ray away from the central axis; the images are always virtual, regardless of the object distance.

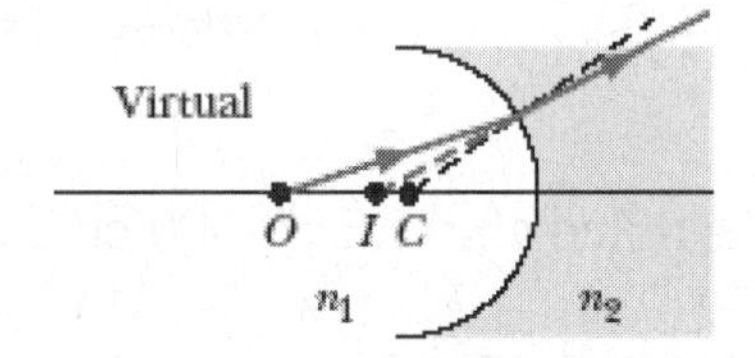

34-35

THINK An image is formed by refraction through a spherical surface. Whether the image is real or virtual depends on the relative values of n_1 and n_2, and on the geometry.

EXPRESS Let n_1 be the index of refraction of the material where the object is located, n_2 be the index of refraction of the material on the other side of the refracting surface, and r be the radius of curvature of the surface. The image distance i is related to the object distance p by Eq. 34-8:

$$\frac{n_1}{p}+\frac{n_2}{i}=\frac{n_2-n_1}{r}.$$

The value of i is positive for real images and negative for virtual images.

ANALYZE In addition to $n_1 = 1.5$, we are also given (a) $n_2 = 1.0$, (b) $p = +70$ cm, and (c) $r = +30$ cm. Notice that $n_2 < n_1$.

(d) We manipulate Eq. 34-8 to find the image distance:

$$i = n_2\left(\frac{n_2 - n_1}{r} - \frac{n_1}{p}\right)^{-1} = 1.0\left(\frac{1.0 - 1.5}{30\text{ cm}} - \frac{1.5}{70\text{ cm}}\right)^{-1} = -26\text{ cm}.$$

(e) The image is virtual (V) and upright.

(f) The object and its image are on the same side.

LEARN The ray diagram for this problem is similar to the one shown in Fig. 34-12(f). Here refraction always directs the ray away from the central axis; the images are always virtual, regardless of the object distance.

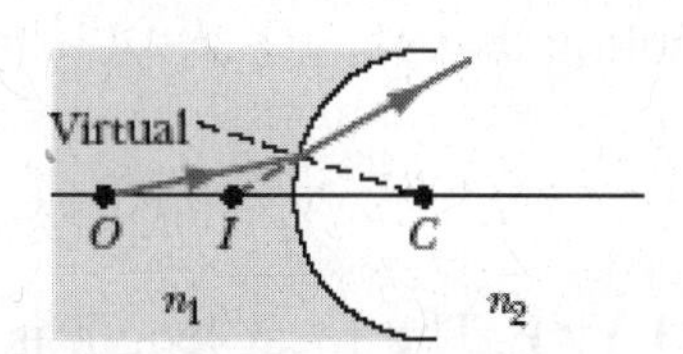

34-47

THINK Our lens is of double-convex type. We apply lens maker's equation to analyze the problem.

EXPRESS The lens maker's equation is given by Eq. 34-10:

$$\frac{1}{f} = (n-1)\left(\frac{1}{r_1} - \frac{1}{r_2}\right)$$

where f is the focal length, n is the index of refraction, r_1 is the radius of curvature of the first surface encountered by the light, and r_2 is the radius of curvature of the second surface. Since one surface has twice the radius of the other and since one surface is convex to the incoming light while the other is concave, set $r_2 = -2r_1$ to obtain

$$\frac{1}{f} = (n-1)\left(\frac{1}{r_1} + \frac{1}{2r_1}\right) = \frac{3(n-1)}{2r_1}.$$

ANALYZE (a) We solve for the smaller radius r_1:

$$r_1 = \frac{3(n-1)f}{2} = \frac{3(1.5-1)(60\text{ mm})}{2} = 45\text{ mm}.$$

(b) The magnitude of the larger radius is $|r_2| = 2r_1 = 90$ mm.

LEARN An image of an object can be formed with a lens because it can bend the light rays, but the bending is possible only if the index of refraction of the lens is different from that of its surrounding medium.

34-49

THINK The image is formed on the screen, so the sum of the object distance and the image distance is equal to the distance between the slide and the screen.

EXPRESS Using Eq. 34-9:

$$\frac{1}{f}=\frac{1}{p}+\frac{1}{i}$$

and noting that $p + i = d = 44$ cm, we obtain

$$p^2 - dp + df = 0.$$

ANALYZE The focal length is $f = 11$ cm. Solving the quadratic equation, we find the solution to p to be

$$p=\frac{1}{2}(d\pm\sqrt{d^2-4df}) = 22\text{ cm}\pm\frac{1}{2}\sqrt{(44\text{ cm})^2-4(44\text{ cm})(11\text{ cm})} = 22\text{ cm}.$$

LEARN Since $p > f$, the object is outside the focal length. The image distance is $i = d - p = 44 - 22 = 22$ cm.

34-53

THINK For a diverging (D) lens, the focal length value is negative.

EXPRESS The object distance p, the image distance i, and the focal length f are related by Eq. 34-9:

$$\frac{1}{f}=\frac{1}{p}+\frac{1}{i}.$$

The value of i is positive for real images and negative for virtual images. The corresponding lateral magnification is $m = -i/p$. The value of m is positive for upright (not inverted) images, and is negative for inverted images.

ANALYZE For this lens, we have $f = -12$ cm and $p = +8.0$ cm.

(a) The image distance is $i=\dfrac{pf}{p-f}=\dfrac{(8.0\text{ cm})(-12\text{ cm})}{8.0\text{ cm}-(-12)\text{ cm}}=-4.8\text{ cm}.$

(b) The magnification is $m = -i/p = -(-4.8\text{ cm})/(8.0\text{ cm}) = +0.60.$

(c) The fact that the image distance is a negative value means the image is virtual (V).

(d) A positive value of magnification means the image is not inverted (NI).

(e) The image is on the same side as the object.

LEARN The ray diagram for this problem is similar to the one shown in Fig. 34-16(c). The lens is diverging, forming a virtual image with the same orientation as the object, and on the same side as the object.

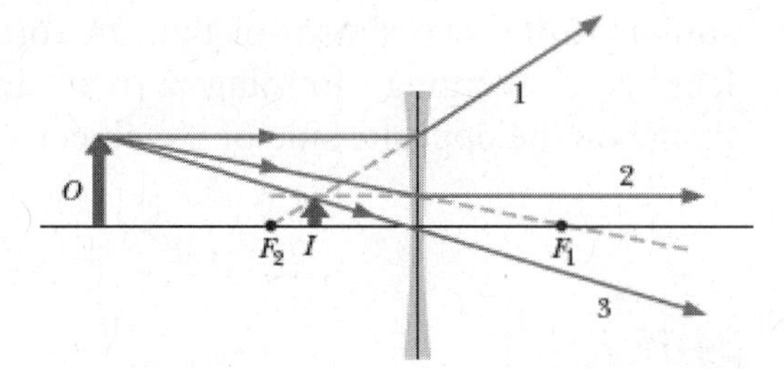

34-59

THINK Since r_1 is positive and r_2 is negative, our lens is of double-convex type. We apply lens maker's equation to analyze the problem.

EXPRESS The lens maker's equation is given by Eq. 34-10:

$$\frac{1}{f}=(n-1)\left(\frac{1}{r_1}-\frac{1}{r_2}\right)$$

where f is the focal length, n is the index of refraction, r_1 is the radius of curvature of the first surface encountered by the light and r_2 is the radius of curvature of the second surface. The object distance p, the image distance i, and the focal length f are related by Eq. 34-9:

$$\frac{1}{f}=\frac{1}{p}+\frac{1}{i}.$$

ANALYZE For this lens, we have $r_1 = +30$ cm, $r_2 = -42$ cm, $n = 1.55$, and $p = +75$ cm.

(a) The focal length is

$$f=\frac{r_1 r_2}{(n-1)(r_2-r_1)}=\frac{(+30\text{ cm})(-42\text{ cm})}{(1.55-1)(-42\text{ cm}-30\text{ cm})}=+31.8\text{ cm}.$$

Thus, the image distance is

$$i=\frac{pf}{p-f}=\frac{(75\text{ cm})(31.8\text{ cm})}{75\text{ cm}-31.8\text{ cm}}=+55\text{ cm}.$$

(b) Equation 34-7 gives $m=-i/p=-(55\text{ cm})/(75\text{ cm})=-0.74$.

(c) The fact that the image distance is a positive value means the image is real (R).

(d) The fact that the magnification is a negative value means the image is inverted (I).

(e) The image is on the side opposite from the object.

LEARN The ray diagram for this problem is similar to the one shown in Fig. 34-16(a). The lens is converging, forming a real, inverted image on the opposite side of the object.

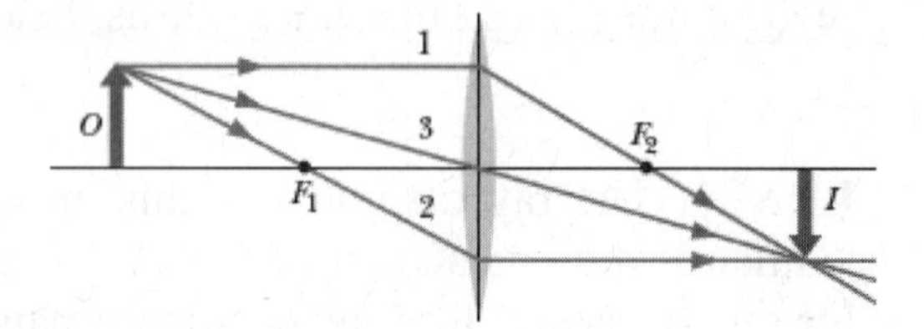

34-75

THINK Since the image is on the same side as the object, it must be a virtual image.

EXPRESS The object distance p, the image distance i, and the focal length f are related by Eq. 34-9:

$$\frac{1}{f}=\frac{1}{p}+\frac{1}{i}.$$

The value of i is positive for real images and negative for virtual images. The corresponding lateral magnification is $m=-i/p$. The value of m is positive for upright (not inverted) images, and is negative for inverted images.

ANALYZE (a) Since the image is virtual (on the same side as the object), the image distance i is negative. By substituting $i=fp/(p-f)$ into $m=-i/p$, we obtain

$$m=-\frac{i}{p}=-\frac{f}{p-f}.$$

The fact that the magnification is less than 1.0 implies that f must be negative. This means that the lens is of the diverging (D) type.

(b) Thus, the focal length is $f=-10$ cm.

(d) The image distance is

$$i=\frac{pf}{p-f}=\frac{(5.0\text{ cm})(-10\text{ cm})}{5.0\text{ cm}-(-10\text{ cm})}=-3.3\text{ cm}.$$

(e) The magnification is $m=-i/p=-(-3.3\text{ cm})/(5.0\text{ cm})=+0.67$.

(f) The fact that the image distance i is a negative value means the image is virtual (V).

(g) A positive value of magnification means the image is not inverted (NI).

LEARN The ray diagram for this problem is similar to the one shown in Fig. 34-16(c). The lens is diverging, forming a virtual image with the same orientation as the object, and on the same side as the object.

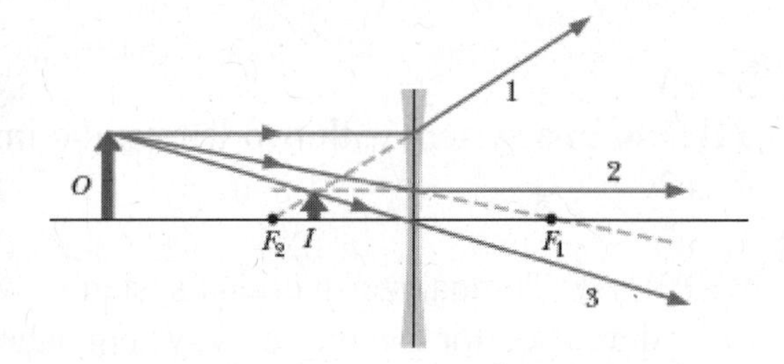

34-77

THINK A positive value for the magnification m means that the image is upright (not inverted). In addition, $m > 1$ indicates that the image is enlarged.

EXPRESS The object distance p, the image distance i, and the focal length f are related by Eq. 34-9: $\frac{1}{f}=\frac{1}{p}+\frac{1}{i}$. The value of i is positive for real images and negative for virtual images. The corresponding lateral magnification is $m=-i/p$. The value of m is positive for upright (not inverted) images, and is negative for inverted images.

ANALYZE (a) Combining Eqs. 34-7 and 34-9, we find the focal length to be

$$f=\frac{p}{1-1/m}=\frac{16\text{ cm}}{1-1/1.25}=80\text{ cm}.$$

Since the value of f is positive, the lens is of the converging type (C).

(b) From (a), we have $f = +80$ cm.

(d) The image distance is $i=-mp=-(1.25)(16\text{ cm})=-20\text{ cm}$.

(e) The magnification is $m = +\,1.25$, as given.

(f) The fact that the image distance i is a negative value means the image is virtual (V).

(g) A positive value of magnification means the image is not inverted (NI).

(h) The image is on the same side as the object.

LEARN The ray diagram for this problem is similar to the one shown in Fig. 34-16(b). The lens is converging. With the object placed inside the focal point ($p < f$), we have a virtual image with the same orientation as the object, and on the same side as the object.

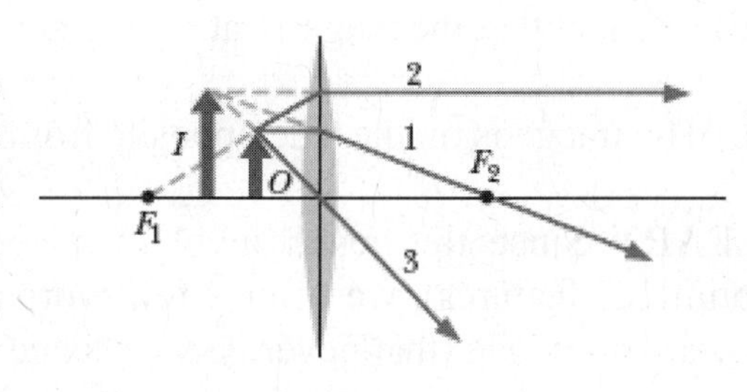

34-83

THINK In a system with two lenses, the image formed by lens 1 serves the "object" for lens 2.

EXPRESS To analyze two-lens systems, we first ignore lens 2, and apply the standard procedure used for a single-lens system. The object distance p_1, the image distance i_1, and the focal length f_1 are related by:

$$\frac{1}{f_1} = \frac{1}{p_1} + \frac{1}{i_1}.$$

Next, we ignore the lens 1 but treat the image formed by lens 1 as the object for lens 2. The object distance p_2 is the distance between lens 2 and the location of the first image. The location of the final image, i_2, is obtained by solving

$$\frac{1}{f_2} = \frac{1}{p_2} + \frac{1}{i_2}$$

where f_2 is the focal length of lens 2.

ANALYZE (a) Since lens 1 is converging, $f_1 = +9$ cm, and we find the image distance to be

$$i_1 = \frac{p_1 f_1}{p_1 - f_1} = \frac{(20 \text{ cm})(9 \text{ cm})}{20 \text{ cm} - 9 \text{ cm}} = 16.4 \text{ cm}.$$

This serves as an "object" for lens 2 (which has $f_2 = +5$ cm) with an object distance given by $p_2 = d - i_1 = -8.4$ cm. The negative sign means that the "object" is behind lens 2. Solving the lens equation, we obtain

$$i_2 = \frac{p_2 f_2}{p_2 - f_2} = \frac{(-8.4 \text{ cm})(5.0 \text{ cm})}{-8.4 \text{ cm} - 5.0 \text{ cm}} = 3.13 \text{ cm}.$$

(b) Te overall magnification is $M = m_1 m_2 = (-i_1 / p_1)(-i_2 / p_2) = i_1 i_2 / p_1 p_2 = -0.31$.

(c) The fact that the (final) image distance is a positive value means the image is real (R).

(d) The fact that the magnification is a negative value means the image is inverted (I).

(e) The image is on the side opposite from the object (relative to lens 2).

LEARN Since this result involves a negative value for p_2 (and perhaps other "non-intuitive" features), we offer a few words of explanation: lens 1 is converging the rays toward an image (that never gets a chance to form due to the intervening presence of lens 2) that would be real and inverted (and 8.4 cm beyond lens 2's location). Lens 2, in a sense, just causes these rays to converge a little more rapidly, and causes the image to form a little closer (to the lens system) than if lens 2 were not present.

34-89

THINK The compound microscope shown in Fig. 34-20 consists of an objective and an eyepiece. It's used for viewing small objects that are very close to the objective.

EXPRESS Let f_{ob} be the focal length of the objective, and f_{ey} be the focal length of the eyepiece. The distance between the two lenses is

$$L = s + f_{ob} + f_{ey},$$

where s is the tube length. The magnification of the objective is

$$m = -\frac{i}{p} = -\frac{s}{f_{ob}}$$

and the angular magnification produced by the eyepiece is $m_\theta = (25\text{ cm})/f_{ey}$.

ANALYZE (a) The tube length is

$$s = L - f_{ob} - f_{ey} = 25.0\text{ cm} - 4.00\text{ cm} - 8.00\text{ cm} = 13.0\text{ cm}.$$

(b) We solve $(1/p) + (1/i) = (1/f_{ob})$ for p. The image distance is

$$i = f_{ob} + s = 4.00\text{ cm} + 13.0\text{ cm} = 17.0\text{ cm},$$

so

$$p = \frac{if_{ob}}{i - f_{ob}} = \frac{(17.0\text{ cm})(4.00\text{ cm})}{17.0\text{ cm} - 4.00\text{ cm}} = 5.23\text{ cm}.$$

(c) The magnification of the objective is $m = -\frac{i}{p} = -\frac{17.0\text{ cm}}{5.23\text{ cm}} = -3.25.$

(d) The angular magnification of the eyepiece is $m_\theta = \frac{25\text{ cm}}{f_{ey}} = \frac{25\text{ cm}}{8.00\text{ cm}} = 3.13.$

(e) The overall magnification of the microscope is $M = mm_\theta = (-3.25)(3.13) = -10.2.$

LEARN The objective produces a real image I of the object inside the focal point of the eyepiece ($i > f_{ey}$). Image I then serves as the object for the eyepiece, which produces a virtual image I' seen by the observer.

34-91

THINK This problem is about human eyes. We model the cornea and eye lens as a single effective thin lens, with image formed at the retina.

EXPRESS When the eye is relaxed, its lens focuses far-away objects on the retina, a distance i behind the lens. We set $p = \infty$ in the thin lens equation to obtain $1/i = 1/f$, where f is the focal length of the relaxed effective lens. Thus, $i = f = 2.50$ cm. When the eye focuses on closer objects, the image distance i remains the same but the object distance and focal length change.

ANALYZE (a) If p is the new object distance and f' is the new focal length, then

$$\frac{1}{p}+\frac{1}{i}=\frac{1}{f'}.$$

We substitute $i = f$ and solve for f': $f' = \dfrac{pf}{f+p} = \dfrac{(40.0\text{ cm})(2.50\text{ cm})}{40.0\text{ cm}+2.50\text{ cm}} = 2.35$ cm.

(b) Consider the lens maker's equation

$$\frac{1}{f}=(n-1)\left(\frac{1}{r_1}-\frac{1}{r_2}\right)$$

where r_1 and r_2 are the radii of curvature of the two surfaces of the lens and n is the index of refraction of the lens material. For the lens pictured in Fig. 34-46, r_1 and r_2 have about the same magnitude, r_1 is positive, and r_2 is negative. Since the focal length decreases, the combination $(1/r_1) - (1/r_2)$ must increase. This can be accomplished by decreasing the magnitudes of both radii.

LEARN When focusing on an object near the eye, the lens bulges a bit (smaller radius of curvature), and its focal length decreases.

34-101

THINK In this problem we convert the Gaussian form of the thin-lens formula to the Newtonian form.

EXPRESS For a thin lens, the Gaussian form of the thin-lens formula gives $(1/p) + (1/i) = (1/f)$, where p is the object distance, i is the image distance, and f is the focal length. To convert the formula to the Newtonian form, let $p = f + x$, where x is positive if the object is outside the focal point and negative if it is inside. In addition, let $i = f + x'$, where x' is positive if the image is outside the focal point and negative if it is inside.

ANALYZE From the Gaussian form, we solve for I and obtain: .

$$i = \frac{fp}{p-f}.$$

Substituting $p = f + x$ gives

$$i = \frac{f(f+x)}{x}.$$

With $i = f + x'$, we have

$$x' = i - f = \frac{f(f+x)}{x} - f = \frac{f^2}{x}$$

which leads to $xx' = f^2$.

LEARN The Newtonain form is equivalent to the Gaussian form, and it provides another convenient way to analyze problems involving thin lenses.

34-103

THINK Two lenses in contact can be treated as one single lens with an effective focal length.

EXPRESS We place an object far away from the composite lens and find the image distance *i*. Since the image is at a focal point, $i = f$, where f equals the effective focal length of the composite. The final image is produced by two lenses, with the image of the first lens being the object for the second. For the first lens, $(1/p_1) + (1/i_1) = (1/f_1)$, where f_1 is the focal length of this lens and i_1 is the image distance for the image it forms. Since $p_1 = \infty$, $i_1 = f_1$. The thin-lens equation, applied to the second lens, is $(1/p_2) + (1/i_2) = (1/f_2)$, where p_2 is the object distance, i_2 is the image distance, and f_2 is the focal length. If the thickness of the lenses can be ignored, the object distance for the second lens is $p_2 = -i_1$. The negative sign must be used since the image formed by the first lens is beyond the second lens if i_1 is positive. This means the object for the second lens is virtual and the object distance is negative. If i_1 is negative, the image formed by the first lens is in front of the second lens and p_2 is positive.

ANALYZE In the thin-lens equation, we replace p_2 with $-f_1$ and i_2 with f to obtain

$$-\frac{1}{f_1} + \frac{1}{f} = \frac{1}{f_2}$$

or

$$\frac{1}{f} = \frac{1}{f_1} + \frac{1}{f_2} = \frac{f_1 + f_2}{f_1 f_2}.$$

Thus, the effective focal length of the system is $f = \dfrac{f_1 f_2}{f_1 + f_2}$.

LEARN The reciprocal of the focal length, $1/f$, is known as the power of the lens, a quantity used by the optometrists to specify the strength of eyeglasses. From the derivation above, we see that when two lenses are in contact, the power of the effective lens is the sum of the two powers.

34-107

THINK The nature of the lenses, whether converging or diverging, can be determined from the magnification and orientation of the images they produce.

EXPRESS By examining the ray diagrams shown in Fig. 34-16(a) – (c), we see that only a converging lens can produce an enlarged, upright image, while the image produced by a diverging lens is always virtual, reduced in size, and not inverted.

ANALYZE (a) In this case $m > +1$ and we know that lens 1 is converging (producing a virtual image), so that our result for focal length should be positive. Since $|P + i_1| = 20$ cm and $i_1 = -2p_1$, we find $p_1 = 20$ cm and $i_1 = -40$ cm. Substituting these into Eq. 34-9,

$$\frac{1}{p_1}+\frac{1}{i_1}=\frac{1}{f_1}$$

leads to

$$f_1=\frac{p_1 i_1}{p_1+i_1}=\frac{(20\text{ cm})(-40\text{ cm})}{20\text{ cm}+(-40\text{ cm})}=+40\text{ cm},$$

which is positive as we expected.

(b) The object distance is $p_1 = 20$ cm, as shown in part (a).

(c) In this case $0 < m < 1$ and we know that lens 2 is diverging (producing a virtual image), so that our result for focal length should be negative. Since $|p + i_2| = 20$ cm and $i_2 = -p_2/2$, we find $p_2 = 40$ cm and $i_2 = -20$ cm. Substituting these into Eq. 34-9 leads to

$$f_2=\frac{p_2 i_2}{p_2+i_2}=\frac{(40\text{ cm})(-20\text{ cm})}{40\text{ cm}+(-20\text{ cm})}=-40\text{ cm},$$

which is negative as we expected.

(d) The object distance is $p_2 = 40$ cm, as shown in part (c).

LEARN The ray diagram for lens 1 is similar to the one shown in Fig. 34-16(b). The lens is converging. With the fly inside the focal point ($p_1 < f_1$), we have a virtual image with the same orientation, and on the same side as the object. On the other hand, the ray diagram for lens 2 is similar to the one shown in Fig. 34-16(c). The lens is diverging, forming a virtual image with the same orientation but smaller in size as the object, and on the same side as the object.

Chapter 35

35-3

THINK The wavelength of the light in a medium depends on the index of refraction of the medium. The nature of the interference, whether constructive or destructive, depends on the phase difference of the two waves.

EXPRESS We take the phases of both waves to be zero at the front surfaces of the layers. The phase of the first wave at the back surface of the glass is given by $\phi_1 = k_1L - \omega t$, where k_1 ($= 2\pi/\lambda_1$) is the angular wave number and λ_1 is the wavelength in glass. Similarly, the phase of the second wave at the back surface of the plastic is given by $\phi_2 = k_2L - \omega t$, where k_2 ($= 2\pi/\lambda_2$) is the angular wave number and λ_2 is the wavelength in plastic. The angular frequencies are the same since the waves have the same wavelength in air and the frequency of a wave does not change when the wave enters another medium. The phase difference is

$$\phi_1 - \phi_2 = (k_1 - k_2)L = 2\pi\left(\frac{1}{\lambda_1} - \frac{1}{\lambda_2}\right)L.$$

Now, $\lambda_1 = \lambda_{air}/n_1$, where λ_{air} is the wavelength in air and n_1 is the index of refraction of the glass. Similarly, $\lambda_2 = \lambda_{air}/n_2$, where n_2 is the index of refraction of the plastic. This means that the phase difference is

$$\phi_1 - \phi_2 = \frac{2\pi}{\lambda_{air}}(n_1 - n_2)L.$$

ANALYZE (a) The value of L that makes this 5.65 rad is

$$L = \frac{(\phi_1 - \phi_2)\lambda_{air}}{2\pi(n_1 - n_2)} = \frac{5.65(400 \times 10^{-9}\,\text{m})}{2\pi(1.60 - 1.50)} = 3.60 \times 10^{-6}\,\text{m}.$$

(b) A phase difference of 5.65 rad is less than 2π rad = 6.28 rad, the phase difference for completely constructive interference, but greater than π rad (= 3.14 rad), the phase difference for completely destructive interference. The interference is, therefore, intermediate, neither completely constructive nor completely destructive. It is, however, closer to completely constructive than to completely destructive.

LEARN The phase difference of two light waves can change when they travel through different materials having different indices of refraction.

35-15

THINK The interference at a point depends on the path-length difference of the light rays reaching that point from the two slits.

EXPRESS The angular positions of the maxima of a two-slit interference pattern are given by

$$\Delta L = d\sin\theta = m\lambda,$$

where ΔL is the path-length difference, d is the slit separation, λ is the wavelength, and m is an integer. If θ is small, sin θ may be approximated by θ in radians. Then, $\theta = m\lambda/d$ to good approximation. The angular separation of two adjacent maxima is $\Delta\theta = \lambda/d$.

ANALYZE Let λ' be the wavelength for which the angular separation is greater by 10.0%. Then, $1.10\lambda/d = \lambda'/d$. or

$$\lambda' = 1.10\lambda = 1.10(589 \text{ nm}) = 648 \text{ nm}.$$

LEARN The angular separation $\Delta\theta$ is proportional to the wavelength of the light. For small θ, we have

$$\Delta\theta' = \left(\frac{\lambda'}{\lambda}\right)\Delta\theta.$$

35-17

THINK Interference maxima occur at angles θ such that d sin $\theta = m\lambda$, where m is an integer.

EXPRESS Since $d = 2.0$ m and $\lambda = 0.50$ m, this means that $\sin\theta = 0.25m$. We want all values of m (positive and negative) for which $|0.25m| \le 1$. These are –4, –3, –2, –1, 0, +1, +2, +3, and +4.

ANALYZE For each of these except –4 and +4, there are two different values for θ. A single value of θ (–90°) is associated with $m = -4$ and a single value (+90°) is associated with $m = +4$. There are 16 different angles in all and, therefore, 16 maxima.

LEARN The angles at which the maxima occur are given by

$$\theta = \sin^{-1}\left(\frac{m\lambda}{d}\right) = \sin^{-1}(0.25m).$$

Similarly, the condition for interference minima (destructive interference) is

$$d\sin\theta = \left(m + \frac{1}{2}\right)\lambda, \quad m = 0,1,2,...$$

35-19

THINK The condition for a maximum in the two-slit interference pattern is $d \sin \theta = m\lambda$, where d is the slit separation, λ is the wavelength, m is an integer, and θ is the angle made by the interfering rays with the forward direction.

EXPRESS If θ is small, $\sin \theta$ may be approximated by θ in radians. Then, $\theta = m\lambda/d$, and the angular separation of adjacent maxima, one associated with the integer m and the other associated with the integer $m + 1$, is given by $\Delta\theta = \lambda/d$. The separation on a screen a distance D away is given by

$$\Delta y = D\,\Delta\theta = \lambda D/d.$$

ANALYZE Thus,

$$\Delta y = \frac{(500 \times 10^{-9}\,\text{m})(5.40\,\text{m})}{1.20 \times 10^{-3}\,\text{m}} = 2.25 \times 10^{-3}\,\text{m} = 2.25\ \text{mm}.$$

LEARN For small θ, the spacing is nearly uniform. However, away from the center of the pattern, θ increases and the spacing gets larger.

35-29

THINK The intensity is proportional to the square of the resultant field amplitude.

EXPRESS Let the electric field components of the two waves be written as

$$E_1 = E_{10} \sin \omega t$$
$$E_2 = E_{20} \sin(\omega t + \phi),$$

where $E_{10} = 1.00$, $E_{20} = 2.00$, and $\phi = 60°$. The resultant field is $E = E_1 + E_2$. We use phasor diagram to calculate the amplitude of E.

ANALYZE The phasor diagram is shown below.
The resultant amplitude E_m is given by the trigonometric law of cosines:

$$E_m^2 = E_{10}^2 + E_{20}^2 - 2E_{10}E_{20}\cos(180° - \phi).$$

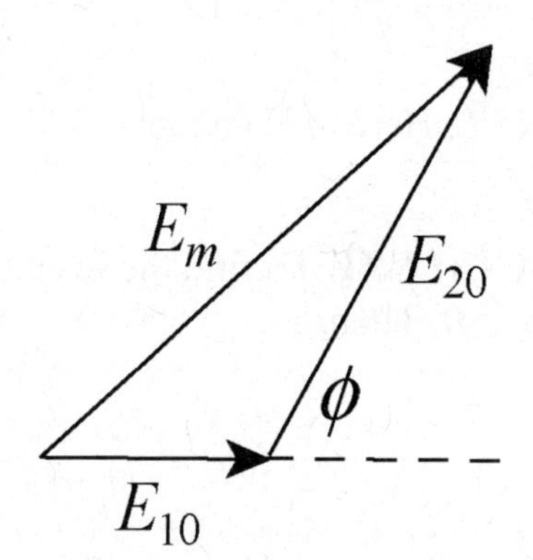

Thus,

$$E_m = \sqrt{(1.00)^2 + (2.00)^2 - 2(1.00)(2.00)\cos 120°} = 2.65.$$

LEARN Summing over the horizontal components of the two fields gives

$$\sum E_h = E_{10}\cos 0 + E_{20}\cos 60° = 1.00 + (2.00)\cos 60° = 2.00 .$$

Similarly, the sum over the vertical components is

$$\sum E_v = E_{10}\sin 0 + E_{20}\sin 60° = 1.00\sin 0° + (2.00)\sin 60° = 1.732 .$$

The resultant amplitude is

$$E_m = \sqrt{(2.00)^2 + (1.732)^2} = 2.65 ,$$

which agrees with what we found above. The phase angle relative to the phasor representing E_1 is

$$\beta = \tan^{-1}\left(\frac{1.732}{2.00}\right) = 40.9° .$$

Thus, the resultant field can be written as $E = (2.65)\sin(\omega t + 40.9°)$.

35-35

THINK For complete destructive interference, we want the waves reflected from the front and back of the coating to differ in phase by an odd multiple of π rad.

EXPRESS Each wave is incident on a medium of higher index of refraction from a medium of lower index, so both suffer phase changes of π rad on reflection. If L is the thickness of the coating, the wave reflected from the back surface travels a distance $2L$ farther than the wave reflected from the front. The phase difference is $2L(2\pi/\lambda_c)$, where λ_c is the wavelength in the coating. If n is the index of refraction of the coating, $\lambda_c = \lambda/n$, where λ is the wavelength in vacuum, and the phase difference is $2nL(2\pi/\lambda)$. We solve

$$2nL\left(\frac{2\pi}{\lambda}\right) = (2m+1)\pi$$

for L. Here m is an integer. The result is $L = \dfrac{(2m+1)\lambda}{4n}$.

ANALYZE To find the least thickness for which destructive interference occurs, we take $m = 0$. Then,

$$L = \frac{\lambda}{4n} = \frac{600\times10^{-9}\,\text{m}}{4(1.25)} = 1.20\times10^{-7}\,\text{m}.$$

LEARN A light ray reflected by a material changes phase by π rad (or 180°) if the refractive index of the material is greater than that of the medium in which the light is traveling.

35-47

THINK For a complete destructive interference, we want the waves reflected from the front and back of material 2 of refractive index n_2 to differ in phase by an odd multiple of π rad.

EXPRESS In this setup, we have $n_2 < n_1$, so there is no phase change from the first surface. On the other hand $n_2 < n_3$, so there is a phase change of π rad from the second surface. Since the second wave travels an extra distance of $2L$, the phase difference is

$$\phi = \frac{2\pi}{\lambda_2}(2L) + \pi ,$$

where $\lambda_2 = \lambda / n_2$ is the wavelength in medium 2. The condition for destructive interference is

$$\frac{2\pi}{\lambda_2}(2L) + \pi = (2m+1)\pi,$$

or

$$2L = m\frac{\lambda}{n_2} \quad \Rightarrow \lambda = \frac{2Ln_2}{m}, \quad m = 0, 1, 2, \ldots .$$

ANALYZE Thus, we have

$$\lambda = \begin{cases} 2Ln_2 = 2(380 \text{ nm})(1.34) = 1018 \text{ nm} \ \ (m=1) \\ Ln_2 = (380 \text{ nm})(1.34) = 509 \text{ nm} \ \ (m=2). \end{cases}$$

For the wavelength to be in the visible range, we choose $m = 2$ with $\lambda = 509$ nm.

LEARN In this setup, the condition for *constructive* interference is

$$\frac{2\pi}{\lambda_2}(2L) + \pi = 2m\pi,$$

or

$$2L = \left(m + \frac{1}{2}\right)\frac{\lambda}{n_2}, \quad m = 0, 1, 2, \ldots .$$

35-51

THINK For a complete destructive interference, we want the waves reflected from the front and back of material 2 of refractive index n_2 to differ in phase by an odd multiple of π rad.

EXPRESS In this setup, we have $n_1 < n_2$ and $n_2 < n_3$, which means that both waves are incident on a medium of higher refractive index from a medium of lower refractive index. Thus, in both cases, there is a phase change of π rad from both surfaces. Since the second wave travels an additional distance of $2L$, the phase difference is

$$\phi = \frac{2\pi}{\lambda_2}(2L)$$

where $\lambda_2 = \lambda / n_2$ is the wavelength in medium 2. The condition for destructive interference is

$$\frac{2\pi}{\lambda_2}(2L) = (2m+1)\pi,$$

or

$$2L = \left(m + \frac{1}{2}\right)\frac{\lambda}{n_2} \quad \Rightarrow \lambda = \frac{4Ln_2}{2m+1}, \qquad m = 0,1,2,...$$

ANALYZE Thus,

$$\lambda = \begin{cases} 4Ln_2 = 4(210 \text{ nm})(1.46) = 1226 \text{ nm} \ \ (m=0) \\ 4Ln_2/3 = 4(210 \text{ nm})(1.46)/3 = 409 \text{ nm} \ \ (m=1) \end{cases}$$

For the wavelength to be in the visible range, we choose $m = 1$ with $\lambda = 409$ nm.

LEARN In this setup, the condition for *constructive* interference is

$$\frac{2\pi}{\lambda_2}(2L) = 2m\pi,$$

or

$$2L = m\frac{\lambda}{n_2}, \qquad m = 0, 1, 2, \ldots .$$

35-55

THINK The index of refraction of oil is greater than that of the air, but smaller than that of the water.

EXPRESS Let the indices of refraction of the air, oil, and water be n_1, n_2, and n_3, respectively. Since $n_1 < n_2$ and $n_2 < n_3$, there is a phase change of π rad from both surfaces. Since the second wave travels an additional distance of $2L$, the phase difference is

$$\phi = \frac{2\pi}{\lambda_2}(2L)$$

where $\lambda_2 = \lambda / n_2$ is the wavelength in the oil. The condition for constructive interference is

$$\frac{2\pi}{\lambda_2}(2L) = 2m\pi,$$

or

$$2L = m\frac{\lambda}{n_2}, \qquad m = 0, 1, 2, \ldots$$

ANALYZE (a) For $m = 1, 2, \ldots$, maximum reflection occurs for wavelengths

$$\lambda = \frac{2n_2L}{m} = \frac{2(1.20)(460\,\text{nm})}{m} = 1104\,\text{nm}\,,\ 552\,\text{nm},\ 368\,\text{nm}\ldots$$

We note that only the 552 nm wavelength falls within the visible light range.

(b) Maximum transmission into the water occurs for wavelengths for which reflection is a minimum. The condition for such destructive interference is given by

$$2L = \left(m + \frac{1}{2}\right)\frac{\lambda}{n_2} \quad \Rightarrow \quad \lambda = \frac{4n_2L}{2m+1}$$

which yields λ = 2208 nm, 736 nm, 442 nm … for the different values of *m*. We note that only the 442 nm wavelength (blue) is in the visible range, though we might expect some red contribution since the 736 nm is very close to the visible range.

LEARN A light ray reflected by a material changes phase by π rad (or 180°) if the refractive index of the material is greater than that of the medium in which the light is traveling. Otherwise, there is no phase change. Note that refraction at an interface does not cause a phase shift.

35-59
THINK Maximum transmission means constructive interference.

EXPRESS As shown in Fig. 35-43, one wave travels a distance of $2L$ further than the other. This wave is reflected twice, once from the back surface (between materials 2 and 3), and once from the front surface (between materials 1 and 2). Since $n_2 > n_3$, there is no phase change at the back-surface reflection. On the other hand, since $n_2 < n_1$, there is a phase change of π rad due to the front-surface reflection. The phase difference of the two waves as they leave material 2 is

$$\phi = \frac{2\pi}{\lambda_2}(2L) + \pi$$

where $\lambda_2 = \lambda / n_2$ is the wavelength in material 2. The condition for constructive interference is

$$\frac{2\pi}{\lambda_2}(2L) + \pi = 2m\pi,$$

or

$$2L = \left(m + \frac{1}{2}\right)\frac{\lambda}{n_2} \quad \Rightarrow \lambda = \frac{4Ln_2}{2m+1}, \qquad m = 0,1,2,...$$

ANALYZE Thus, we have

$$\lambda = \begin{cases} 4Ln_2 = 4(415 \text{ nm})(1.59) = 2639 \text{ nm} \ \ (m=0) \\ 4Ln_2 / 3 = 4(415 \text{ nm})(1.59)/3 = 880 \text{ nm} \ \ (m=1) \\ 4Ln_2 / 5 = 4(415 \text{ nm})(1.59)/5 = 528 \text{ nm} \ \ (m=2) \end{cases}.$$

For the wavelength to be in the visible range, we choose $m = 2$ with $\lambda = 528$ nm.

LEARN Similarly, the condition for destructive interference is

$$\frac{2\pi}{\lambda_2}(2L) + \pi = (2m+1)\pi,$$

or

$$2L = m\frac{\lambda}{n_2} \quad \Rightarrow \lambda = \frac{2Ln_2}{m}, \quad m = 0, 1, 2,....$$

35-73

THINK A light ray reflected by a material changes phase by π rad (or 180°) if the refractive index of the material is greater than that of the medium in which the light is traveling.

EXPRESS Consider the interference of waves reflected from the top and bottom surfaces of the air film. The wave reflected from the upper surface does not change phase on reflection but the wave reflected from the bottom surface changes phase by π rad. At a place where the thickness of the air film is L, the condition for fully constructive interference is $2L = \left(m + \tfrac{1}{2}\right)\lambda$ where λ (= 683 nm) is the wavelength and m is an integer.

ANALYZE For $L = 48$ μm, we find the value of m to be

$$m = \frac{2L}{\lambda} - \frac{1}{2} = \frac{2(4.80\times10^{-5}\text{m})}{683\times10^{-9}\text{m}} - \frac{1}{2} = 140.$$

At the thin end of the air film, there is a bright fringe. It is associated with $m = 0$. There are, therefore, 140 bright fringes in all.

LEARN The number of bright fringes increases with L, but decreases with λ.

35-75

THINK The formation of Newton's rings is due to the interference between the rays reflected from the flat glass plate and the curved lens surface.

EXPRESS Consider the interference pattern formed by waves reflected from the upper and lower surfaces of the air wedge. The wave reflected from the lower surface undergoes a π rad phase change while the wave reflected from the upper surface does not. At a place where the thickness of the wedge is d, the condition for a maximum in intensity is $2d = (m+\frac{1}{2})\lambda$, where λ is the wavelength in air and m is an integer. Therefore,

$$d = (2m + 1)\lambda/4.$$

ANALYZE As the geometry of Fig. 35-46 shows, $d = R - \sqrt{R^2 - r^2}$, where R is the radius of curvature of the lens and r is the radius of a Newton's ring. Thus, $(2m+1)\lambda/4 = R - \sqrt{R^2 - r^2}$. First, we rearrange the terms so the equation becomes

$$\sqrt{R^2 - r^2} = R - \frac{(2m+1)\lambda}{4}.$$

Next, we square both sides, rearrange to solve for r^2, then take the square root. We get

$$r = \sqrt{\frac{(2m+1)R\lambda}{2} - \frac{(2m+1)^2\lambda^2}{16}}.$$

If R is much larger than a wavelength, the first term dominates the second and

$$r = \sqrt{\frac{(2m+1)R\lambda}{2}}.$$

LEARN Similarly, one may show that the radii of the dark fringes are given by

$$r = \sqrt{mR\lambda}.$$

35-81

THINK There is a small difference between the wavelength in air and the wavelength in vacuum.

EXPRESS Let ϕ_1 be the phase difference of the waves in the two arms when the tube has air in it, and let ϕ_2 be the phase difference when the tube is evacuated. If λ is the wavelength in vacuum, then the wavelength in air is λ/n, where n is the index of refraction of air. This means

$$\phi_1 - \phi_2 = 2L\left[\frac{2\pi n}{\lambda} - \frac{2\pi}{\lambda}\right] = \frac{4\pi(n-1)L}{\lambda}$$

where L is the length of the tube. The factor 2 arises because the light traverses the tube twice, once on the way to a mirror and once after reflection from the mirror. Each shift by one fringe corresponds to a change in phase of 2π rad, so if the interference pattern shifts by N fringes as the tube is evacuated, then

$$\frac{4\pi(n-1)L}{\lambda} = 2N\pi .$$

ANALYZE Solving for n, we obtain $n = 1 + \frac{N\lambda}{2L} = 1 + \frac{60(500\times10^{-9}\text{ m})}{2(5.0\times10^{-2}\text{ m})} = 1.00030$.

LEARN The interferometer provides an accurate way to measure the refractive index of the air (and other gases as well).

35-85

THINK The angle between adjacent fringes depends on the wavelength of the light and the distance between the slits.

EXPRESS The angular positions of the maxima of a two-slit interference pattern are given by $\Delta L = d\sin\theta = m\lambda$, where ΔL is the path-length difference, d is the slit separation, λ is the wavelength, and m is an integer. If θ is small, sin θ may be approximated by θ in radians. Then, $\theta = m\lambda/d$ to good approximation. The angular separation of two adjacent maxima is $\Delta\theta = \lambda/d$. When the arrangement is immersed in water, the wavelength changes to $\lambda' = \lambda/n$, and the equation above becomes

$$\Delta\theta' = \frac{\lambda'}{d}.$$

ANALYZE Dividing the equation by $\Delta\theta = \lambda/d$, we obtain

$$\frac{\Delta\theta'}{\Delta\theta} = \frac{\lambda'}{\lambda} = \frac{1}{n}.$$

Therefore, with $n = 1.33$ and $\Delta\theta = 0.30°$, we find $\Delta\theta' = 0.23°$.

LEARN The angular separation decreases with increasing index of refraction; the greater the value of n, the smaller the value of $\Delta\theta$.

35-87

THINK For a completely destructive interference, the intensity produced by the two waves is zero.

EXPRESS When the interference between two waves is completely destructive, their phase difference is given by

$$\phi = (2m+1)\pi, \quad m = 0, 1, 2, \ldots.$$

The equivalent condition is that their path-length difference is an odd multiple of $\lambda/2$, where λ is the wavelength of the light.

ANALYZE (a) Looking at Fig. 35-52, we see that half of the periodic pattern is of length ΔL = 750 nm (judging from the maximum at x = 0 to the minimum at x = 750 nm); this suggests that $\Delta L = \lambda/2$, and the wavelength (the full length of the periodic pattern) is λ = $2\Delta L$ = 1500 nm. Thus, a maximum should be reached again at x = 1500 nm (and at x = 3000 nm, x = 4500 nm, …).

(b) From our discussion in part (a), we expect a minimum to be reached at odd multiple of $\lambda/2$, or x = 750 nm + n(1500 nm), where n = 1, 2, 3 … . For instance, for n = 1 we would find the minimum at x = 2250 nm.

(c) With λ = 1500 nm found in part (a), we can express x = 1200 nm as x = 1200/1500 = 0.80 wavelength.

LEARN For a completely destructive interference, the phase difference between two light sources is an odd multiple of π; however, for a completely constructive interference, the phase difference is a multiple of 2π.

35-89

THINK Since the index of refraction of water is greater than that of air, the wave that is reflected from the water surface suffers a phase change of π rad on reflection.

EXPRESS Suppose the wave that goes directly to the receiver travels a distance L_1 and the reflected wave travels a distance L_2. The last wave suffers a phase change on reflection of half a wavelength since water has higher refractive index than air. To obtain constructive interference at the receiver, the difference $L_2 - L_1$ must be an odd multiple of a half wavelength.

ANALYZE Consider the diagram below.

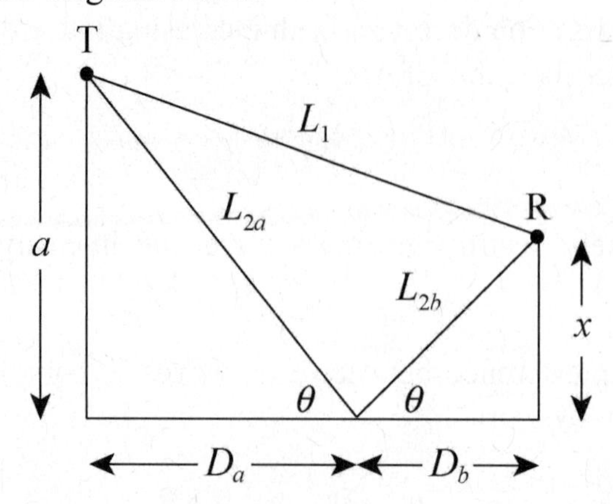

The right triangle on the left, formed by the vertical line from the water to the transmitter T, the ray incident on the water, and the water line, gives $D_a = a/\tan\theta$. The right triangle on the right, formed by the vertical line from the water to the receiver R, the reflected ray, and the water line leads to $D_b = x/\tan\theta$. Since $D_a + D_b = D$,

$$\tan\theta = \frac{a+x}{D}.$$

We use the identity $\sin^2\theta = \tan^2\theta/(1+\tan^2\theta)$ to show that

$$\sin\theta = (a+x)/\sqrt{D^2+(a+x)^2}.$$

This means

$$L_{2a} = \frac{a}{\sin\theta} = \frac{a\sqrt{D^2+(a+x)^2}}{a+x}$$

and

$$L_{2b} = \frac{x}{\sin\theta} = \frac{x\sqrt{D^2+(a+x)^2}}{a+x}.$$

Therefore,

$$L_2 = L_{2a} + L_{2b} = \frac{(a+x)\sqrt{D^2+(a+x)^2}}{a+x} = \sqrt{D^2+(a+x)^2}.$$

Using the binomial theorem, with D^2 large and $a^2 + x^2$ small, we approximate this expression: $L_2 \approx D + (a+x)^2/2D$. The distance traveled by the direct wave is $L_1 = \sqrt{D^2+(a-x)^2}$. Using the binomial theorem, we approximate this expression: $L_1 \approx D + (a-x)^2/2D$. Thus,

$$L_2 - L_1 \approx D + \frac{a^2 + 2ax + x^2}{2D} - D - \frac{a^2 - 2ax + x^2}{2D} = \frac{2ax}{D}.$$

Setting this equal to $\left(m+\frac{1}{2}\right)\lambda$, where m is zero or a positive integer, we find $x = \left(m+\frac{1}{2}\right)\left(\lambda D/2a\right)$.

LEARN Similarly, the condition for destructive interference is

$$L_2 - L_1 \approx \frac{2ax}{D} = m\lambda,$$

or

$$x = m\frac{\lambda D}{2a}, \quad m = 0, 1, 2, \ldots.$$

35-93

THINK Knowing the slit separation and the distance between interference fringes allows us to calculate the wavelength of the light used.

EXPRESS The condition for a minimum in the two-slit interference pattern is $d \sin \theta = (m + ½)\lambda$, where d is the slit separation, λ is the wavelength, m is an integer, and θ is the angle made by the interfering rays with the forward direction. If θ is small, $\sin \theta$ may be approximated by θ in radians. Then, $\theta = (m + ½)\lambda/d$, and the distance from the minimum to the central fringe is

$$y = D\tan\theta \approx D\sin\theta \approx D\theta = \left(m + \frac{1}{2}\right)\frac{D\lambda}{d},$$

where D is the distance from the slits to the screen. For the first minimum $m = 0$ and for the tenth one, $m = 9$. The separation is

$$\Delta y = \left(9 + \frac{1}{2}\right)\frac{D\lambda}{d} - \frac{1}{2}\frac{D\lambda}{d} = \frac{9D\lambda}{d}.$$

ANALYZE We solve for the wavelength:

$$\lambda = \frac{d\Delta y}{9D} = \frac{(0.15 \times 10^{-3}\,\text{m})(18 \times 10^{-3}\,\text{m})}{9(50 \times 10^{-2}\,\text{m})} = 6.0 \times 10^{-7}\,\text{m} = 600\text{ nm}.$$

LEARN The distance between two adjacent dark fringes, one associated with the integer m and the other associated with the integer $m + 1$, is

$$\Delta y = D\theta = D\lambda/d.$$

35-95

THINK The dark band corresponds to a completely destructive interference.

EXPRESS When the interference between two waves is completely destructive, their phase difference is given by

$$\phi = (2m+1)\pi, \quad m = 0, 1, 2, ...$$

The equivalent condition is that their path-length difference is an odd multiple of $\lambda/2$, where λ is the wavelength of the light.

ANALYZE (a) A path-length difference of $\lambda/2$ produces the first dark band, of $3\lambda/2$ produces the second dark band, and so on. Therefore, the fourth dark band corresponds to a path-length difference of $7\lambda/2 = 1750$ nm $= 1.75$ μm.

(b) In the small angle approximation (which we assume holds here), the fringes are equally spaced, so that if Δy denotes the distance from one maximum to the next, then the distance from the middle of the pattern to the fourth dark band must be 16.8 mm = 3.5 Δy. Therefore, we obtain Δy = (16.8 mm)/3.5 = 4.8 mm.

LEARN The distance from the *m*th maximum to the central fringe is

$$y_{\text{bright}} = D\tan\theta \approx D\sin\theta \approx D\theta = m\frac{D\lambda}{d}.$$

Similarly, the distance from the *m*th minimum to the central fringe is

$$y_{\text{dark}} = \left(m+\frac{1}{2}\right)\frac{D\lambda}{d}.$$

35-97

THINK The intensity of the light observed in the interferometer depends on the phase difference between the two waves.

EXPRESS Let the position of the mirror measured from the point at which $d_1 = d_2$ be x. We assume the beam-splitting mechanism is such that the two waves interfere constructively for $x = 0$ (with some beam-splitters, this would not be the case). We can adapt Eq. 35-23 to this situation by incorporating a factor of 2 (since the interferometer utilizes directly reflected light in contrast to the double-slit experiment) and eliminating the $\sin\theta$ factor. Thus, the path difference is $2x$, and the phase difference between the two light paths is $\Delta\phi = 2(2\pi x/\lambda) = 4\pi x/\lambda$.

ANALYZE From Eq. 35-22, we see that the intensity is proportional to $\cos^2(\Delta\phi/2)$. Thus, writing $4I_0$ as I_m, we find

$$I = I_m \cos^2\left(\frac{\Delta\phi}{2}\right) = I_m \cos^2\left(\frac{2\pi x}{\lambda}\right).$$

LEARN The intensity I / I_m as a function of x/λ is plotted below.

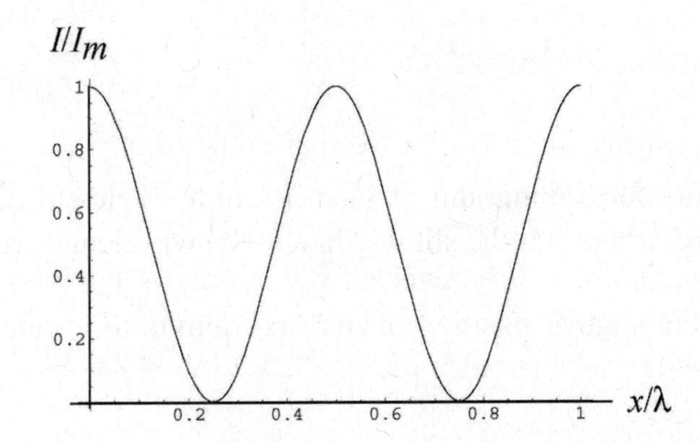

From the figure, we see that the intensity is at a maximum when

$$x = \frac{m}{2}\lambda, \quad m = 0, 1, 2, \ldots.$$

Similarly, the condition for minima is

$$x = \frac{1}{4}(2m+1)\lambda, \quad m = 0, 1, 2, \ldots.$$

Chapter 36

36-9

THINK The condition for a minimum of intensity in a single-slit diffraction pattern is given by $a \sin \theta = m\lambda$, where a is the slit width, λ is the wavelength, and m is an integer.

EXPRESS To find the angular position of the first minimum to one side of the central maximum, we set $m = 1$:

$$\theta_1 = \sin^{-1}\left(\frac{\lambda}{a}\right) = \sin^{-1}\left(\frac{589\times 10^{-9}\text{ m}}{1.00\times 10^{-3}\text{ m}}\right) = 5.89\times 10^{-4}\text{ rad}.$$

If D is the distance from the slit to the screen, the distance on the screen from the center of the pattern to the minimum is

$$y_1 = D\tan\theta_1 = (3.00\text{ m})\tan\left(5.89\times 10^{-4}\text{ rad}\right) = 1.767\times 10^{-3}\text{ m}.$$

To find the second minimum, we set $m = 2$:

$$\theta_2 = \sin^{-1}\left(\frac{2\left(589\times 10^{-9}\text{ m}\right)}{1.00\times 10^{-3}\text{ m}}\right) = 1.178\times 10^{-3}\text{ rad}.$$

ANALYZE The distance from the center of the pattern to this second minimum is

$$y_2 = D\tan\theta_2 = (3.00\text{ m})\tan(1.178\times 10^{-3}\text{ rad}) = 3.534\times 10^{-3}\text{ m}.$$

The separation of the two minima is

$$\Delta y = y_2 - y_1 = 3.534\text{ mm} - 1.767\text{ mm} = 1.77\text{ mm}.$$

LEARN The angles θ_1 and θ_2 found above are quite small. In the small-angle approximation, $\sin\theta \approx \tan\theta \approx \theta$, and the separation between two adjacent diffraction minima can be approximated as

$$\Delta y = D(\tan\theta_{m+1} - \tan\theta_m) \approx D(\theta_{m+1} - \theta_m) = \frac{D\lambda}{a}.$$

36-15

THINK The relative intensity in a single-slit diffraction depends on the ratio a/λ, where a is the slit width and λ is the wavelength.

EXPRESS The intensity for a single-slit diffraction pattern is given by

$$I = I_m \frac{\sin^2 \alpha}{\alpha^2}$$

where I_m is the maximum intensity and $\alpha = (\pi a/\lambda)\sin\theta$. The angle θ is measured from the forward direction.

ANALYZE (a) We require $I = I_m/2$, so

$$\sin^2 \alpha = \frac{1}{2}\alpha^2 .$$

(b) We evaluate $\sin^2 \alpha$ and $\alpha^2/2$ for $\alpha = 1.39$ rad and compare the results. To be sure that 1.39 rad is closer to the correct value for α than any other value with three significant digits, we could also try 1.385 rad and 1.395 rad.

(c) Since $\alpha = (\pi a/\lambda)\sin \theta$,

$$\theta = \sin^{-1}\left(\frac{\alpha\lambda}{\pi a}\right).$$

Now $\alpha/\pi = 1.39/\pi = 0.442$, so

$$\theta = \sin^{-1}\left(\frac{0.442\lambda}{a}\right).$$

The angular separation of the two points of half intensity, one on either side of the center of the diffraction pattern, is

$$\Delta\theta = 2\theta = 2\sin^{-1}\left(\frac{0.442\lambda}{a}\right).$$

(d) For $a/\lambda = 1.0$, $\Delta\theta = 2\sin^{-1}(0.442/1.0) = 0.916\,\text{rad} = 52.5°$.

(e) For $a/\lambda = 5.0$, $\Delta\theta = 2\sin^{-1}(0.442/5.0) = 0.177\,\text{rad} = 10.1°$.

(f) For $a/\lambda = 10$, $\Delta\theta = 2\sin^{-1}(0.442/10) = 0.0884\,\text{rad} = 5.06°$.

LEARN As shown in Fig. 36-8, the wider the slit is (relative to the wavelength), the narrower is the central diffraction maximum.

36-21

THINK We apply the Rayleigh criterion to estimate the linear separation between the two objects.

EXPRESS If L is the distance from the observer to the objects, then the smallest separation D they can have and still be resolvable is $D = L\theta_R$, where θ_R is measured in radians.

ANALYZE (a) With small-angle approximation, $\theta_R = 1.22\lambda/d$, where λ is the wavelength and d is the diameter of the aperture. Thus,

$$D = \frac{1.22\,L\lambda}{d} = \frac{1.22(8.0\times10^{10}\text{ m})(550\times10^{-9}\text{ m})}{5.0\times10^{-3}\text{ m}} = 1.1\times10^{7}\text{ m} = 1.1\times10^{4}\text{ km}\,.$$

This distance is greater than the diameter of Mars; therefore, one part of the planet's surface cannot be resolved from another part.

(b) Now $d = 5.1$ m and $D = \dfrac{1.22(8.0\times10^{10}\text{ m})(550\times10^{-9}\text{ m})}{5.1\text{m}} = 1.1\times10^{4}\text{ m} = 11\text{ km}\,.$

LEARN By the Rayleigh criterion for resolvability, two objects can be resolved only if their angular separation at the observer is greater than $\theta_R = 1.22\lambda/d$.

36-23

THINK We apply the Rayleigh criterion to determine the conditions that allow the headlights to be resolved.

EXPRESS By the Rayleigh criteria, two point sources can be resolved if the central diffraction maximum of one source is centered on the first minimum of the diffraction pattern of the other. Thus, the angular separation (in radians) of the sources must be at least $\theta_R = 1.22\lambda/d$, where λ is the wavelength and d is the diameter of the aperture.

ANALYZE (a) For the headlights of this problem,

$$\theta_R = \frac{1.22(550\times10^{-9}\text{ m})}{5.0\times10^{-3}\text{ m}} = 1.34\times10^{-4}\text{ rad},$$

or 1.3×10^{-4} rad, in two significant figures.

(b) If L is the distance from the headlights to the eye when the headlights are just resolvable and D is the separation of the headlights, then $D = L\theta_R$, where the small angle approximation is made. This is valid for θ_R in radians. Thus,

$$L = \frac{D}{\theta_R} = \frac{1.4\,\text{m}}{1.34\times10^{-4}\ \text{rad}} = 1.0\times10^{4}\ \text{m} = 10\,\text{km}\ .$$

LEARN A distance of 10 km far exceeds what human eyes can resolve. In reality, our visual resolvability depends on other factors such as the relative brightness of the source and their surroundings, turbulence in the air between the lights and the eyes, and the health of one's vision.

36-31

THINK We apply the Rayleigh criterion to calculate the angular width of the central maxima.

EXPRESS The first minimum in the diffraction pattern is at an angular position θ, measured from the center of the pattern, such that $\sin\theta = 1.22\lambda/d$, where λ is the wavelength and d is the diameter of the antenna. If f is the frequency, then the wavelength is

$$\lambda = \frac{c}{f} = \frac{3.00\times10^{8}\ \text{m/s}}{220\times10^{9}\ \text{Hz}} = 1.36\times10^{-3}\ \text{m}\ .$$

ANALYZE (a) Thus, we have

$$\theta = \sin^{-1}\left(\frac{1.22\lambda}{d}\right) = \sin^{-1}\left(\frac{1.22(1.36\times10^{-3}\ \text{m})}{55.0\times10^{-2}\ \text{m}}\right) = 3.02\times10^{-3}\ \text{rad}\ .$$

The angular width of the central maximum is twice this, or 6.04×10^{-3} rad (0.346°).

(b) Now $\lambda = 1.6$ cm and $d = 2.3$ m, so

$$\theta = \sin^{-1}\left(\frac{1.22(1.6\times10^{-2}\ \text{m})}{2.3\,\text{m}}\right) = 8.5\times10^{-3}\ \text{rad}\ .$$

The angular width of the central maximum is 1.7×10^{-2} rad (or 0.97°).

LEARN Using small-angle approximation, we can write the angular width as

$$2\theta \approx 2\left(\frac{1.22\lambda}{d}\right) = \frac{2.44\lambda}{d}\ .$$

36-43

THINK For relatively wide slits, the interference of light from two slits produces bright fringes that do not all have the same intensity; instead, the intensities are modified by diffraction of light passing through each slit.

EXPRESS The angular positions θ of the bright interference fringes are given by $d \sin \theta = m\lambda$, where d is the slit separation, λ is the wavelength, and m is an integer. The first diffraction minimum occurs at the angle θ_1 given by $a \sin \theta_1 = \lambda$, where a is the slit width. The diffraction peak extends from $-\theta_1$ to $+\theta_1$, so we should count the number of values of m for which $-\theta_1 < \theta < +\theta_1$, or, equivalently, the number of values of m for which

$$-\sin \theta_1 < \sin \theta < +\sin \theta_1.$$

The intensity at the screen is given by

$$I = I_m\left(\cos^2 \beta\right)\left(\frac{\sin \alpha}{\alpha}\right)^2$$

where $\alpha = (\pi a/\lambda)\sin \theta$, $\beta = (\pi d/\lambda)\sin \theta$, and I_m is the intensity at the center of the pattern.

ANALYZE (a) The condition above means $-1/a < m/d < 1/a$, or $-d/a < m < +d/a$. Now

$$d/a = (0.150 \times 10^{-3}\ \text{m})/(30.0 \times 10^{-6}\ \text{m}) = 5.00,$$

so the values of m are $m = -4, -3, -2, -1, 0, +1, +2, +3$, and $+4$. There are 9 fringes.

(b) For the third bright interference fringe, $d \sin \theta = 3\lambda$, so $\beta = 3\pi$ rad and $\cos^2 \beta = 1$. Similarly, $\alpha = 3\pi a/d = 3\pi/5.00 = 0.600\pi$ rad and

$$\left(\frac{\sin \alpha}{\alpha}\right)^2 = \left(\frac{\sin 0.600\pi}{0.600\pi}\right)^2 = 0.255\ .$$

The intensity ratio is $I/I_m = 0.255$.

LEARN The expression for intensity contains two factors: (1) the interference factor $\cos^2 \beta$ due to the interference between two slits with separation d, and (2) the diffraction factor $[(\sin\alpha)/\alpha]^2$ which arises due to diffraction by a single slit of width a. In the limit $a \to 0$, $(\sin\alpha)/\alpha \to 1$, and we recover Eq. 35-22 for the interference between two slits of vanishingly narrow slits separated by d. Similarly, setting $d = 0$ or equivalently, $\beta = 0$, we recover Eq. 36-5 for the diffraction of a single slit of width a. A plot of the relative intensity is shown next.

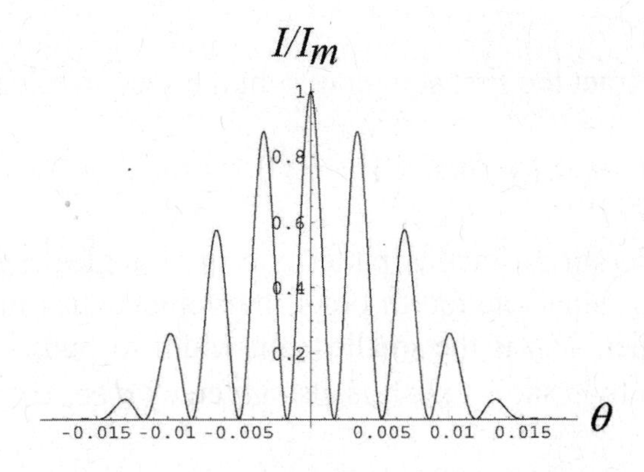

36-47

THINK Diffraction lines occur at angles θ such that $d \sin \theta = m\lambda$, where d is the grating spacing, λ is the wavelength, and m is an integer.

EXPRESS The ruling separation is

$$d = 1/(400\ \text{mm}^{-1}) = 2.5 \times 10^{-3}\ \text{mm}.$$

Notice that for a given order, the line associated with a long wavelength is produced at a greater angle than the line associated with a shorter wavelength. We take λ to be the longest wavelength in the visible spectrum (700 nm) and find the greatest integer value of m such that θ is less than 90°. That is, find the greatest integer value of m for which $m\lambda < d$.

ANALYZE Since

$$\frac{d}{\lambda} = \frac{2.5\times10^{-6}\ \text{m}}{700\times10^{-9}\ \text{m}} \approx 3.57,$$

that value is $m = 3$. There are three complete orders on each side of the $m = 0$ order. The second and third orders overlap.

LEARN From $\theta = \sin^{-1}(m\lambda / d)$, the condition for maxima or lines, we see that for a given diffraction grating, the angle from the central axis to any line depends on the wavelength of the light being used.

36-49

THINK Maxima of a diffraction grating pattern occur at angles θ given by $d \sin \theta = m\lambda$, where d is the slit separation, λ is the wavelength, and m is an integer.

EXPRESS If two lines are adjacent, then their order numbers differ by unity. Let m be the order number for the line with $\sin \theta = 0.2$ and $m + 1$ be the order number for the line with $\sin \theta = 0.3$. Then,

$$0.2d = m\lambda, \quad 0.3d = (m + 1)\lambda.$$

ANALYZE (a) We subtract the first equation from the second to obtain $0.1d = \lambda$, or

$$d = \lambda/0.1 = (600 \times 10^{-9}\text{m})/0.1 = 6.0 \times 10^{-6}\ \text{m}.$$

(b) Minima of the single-slit diffraction pattern occur at angles θ given by $a \sin\theta = m\lambda$, where a is the slit width. Since the fourth-order interference maximum is missing, it must fall at one of these angles. If a is the smallest slit width for which this order is missing, the angle must be given by $a \sin\theta = \lambda$. It is also given by $d \sin\theta = 4\lambda$, so

$$a = d/4 = (6.0 \times 10^{-6}\ \text{m})/4 = 1.5 \times 10^{-6}\ \text{m}.$$

(c) First, we set $\theta = 90°$ and find the largest value of m for which $m\lambda < d \sin\theta$. This is the highest order that is diffracted toward the screen. The condition is the same as $m < d/\lambda$ and since

$$d/\lambda = (6.0 \times 10^{-6}\ \text{m})/(600 \times 10^{-9}\ \text{m}) = 10.0,$$

the highest order seen is the $m = 9$ order. The fourth and eighth orders are missing, so the observable orders are $m = 0, 1, 2, 3, 5, 6, 7$, and 9. Thus, the largest value of the order number is $m = 9$.

(d) Using the result obtained in (c), the second largest value of the order number is $m = 7$.

(e) Similarly, the third largest value of the order number is $m = 6$.

LEARN Interference maxima occur when $d \sin\theta = m\lambda$, while the condition for diffraction minima is $a \sin\theta = m'\lambda$. Thus, a particular interference maximum with order m may coincide with the diffraction minimum of order m'. The value of m is given by

$$\frac{d\sin\theta}{a\sin\theta} = \frac{m\lambda}{m'\lambda} \Rightarrow m = \left(\frac{d}{a}\right)m'.$$

Since $m = 4$ when $m' = 1$, we conclude that $d/a = 4$. Thus, $m = 8$ would correspond to the second diffraction minimum ($m' = 2$).

36-55

THINK If a grating just resolves two wavelengths whose average is λ_{avg} and whose separation is $\Delta\lambda$, then its resolving power is defined by $R = \lambda_{avg}/\Delta\lambda$.

EXPRESS As shown in Eq. 36-32, the resolving power can also be written as Nm, where N is the number of rulings in the grating and m is the order of the lines.

ANALYZE Thus $\lambda_{avg}/\Delta\lambda = Nm$ and

$$N = \frac{\lambda_{\text{avg}}}{m\Delta\lambda} = \frac{656.3\,\text{nm}}{(1)(0.18\,\text{nm})} = 3.65\times10^3 \text{ rulings.}$$

LEARN A large N (more rulings) means greater resolving power.

36-75

THINK Maxima of a diffraction grating pattern occur at angles θ given by $d \sin \theta = m\lambda$, where d is the slit separation, λ is the wavelength, and m is an integer.

EXPRESS The ruling separation is given by

$$d = \frac{1}{200 \text{ mm}^{-1}} = 5.00\times10^{-3} \text{ mm} = 5.00\times10^{-6} \text{ m} = 5000 \text{ nm}.$$

Letting $d \sin \theta = m\lambda$, we solve for λ:

$$\lambda = \frac{d\sin\theta}{m} = \frac{(5000 \text{ nm})(\sin 30°)}{m} = \frac{2500\,\text{nm}}{m}$$

where $m = 1, 2, 3 \ldots$. In the visible light range m can assume the following values: $m_1 = 4$, $m_2 = 5$ and $m_3 = 6$.

(a) The longest wavelength corresponds to $m_1 = 4$ with $\lambda_1 = 2500$ nm/4 = 625 nm.

(b) The second longest wavelength corresponds to $m_2 = 5$ with $\lambda_2 = 2500$ nm/5 = 500 nm.

(c) The third longest wavelength corresponds to $m_3 = 6$ with $\lambda_3 = 2500$ nm/6 = 416 nm.

LEARN As shown above, only three values of m give wavelengths that are in the visible spectrum. Note that if the light incident on the diffraction grating is not monochromatic, a *spectrum* would be observed since the grating spreads out light into its component wavelengths.

36-77

THINK The condition for a minimum of intensity in a single-slit diffraction pattern is given by $a \sin \theta = m\lambda$, where a is the slit width, λ is the wavelength, and m is an integer.

EXPRESS As a slit is narrowed, the pattern spreads outward, so the question about "minimum width" suggests that we are looking at the lowest possible values of m (the label for the minimum produced by light $\lambda = 600$ nm) and m' (the label for the minimum produced by light $\lambda' = 500$ nm). Since the angles are the same, then Eq. 36-3 leads to

$$m\lambda = m'\lambda'$$

which leads to the choices $m = 5$ and $m' = 6$.

ANALYZE We find the slit width from Eq. 36-3:

$$a = \frac{m\lambda}{\sin\theta} = \frac{5(600\times10^{-9}\,\text{m})}{\sin(1.00\times10^{-9}\,\text{rad})} = 3.00\times10^{-3}\ \text{m}\,.$$

LEARN The intensities of the diffraction are shown next (solid line for orange light, and dashed line for blue-green light). The angle $\theta = 0.001$ rad corresponds to $m = 5$ for the orange light, but $m' = 6$ for the blue-green light.

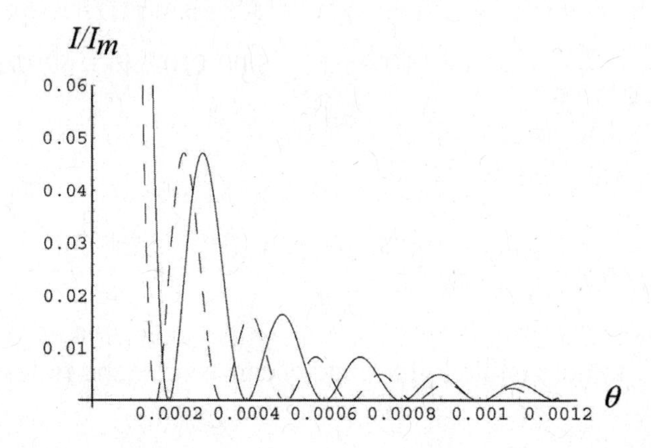

36-79

THINK We relate the resolving power of a diffraction grating to the frequency range.

EXPRESS Since the resolving power of a grating is given by $R = \lambda/\Delta\lambda$ and by Nm, the range of wavelengths that can just be resolved in order m is $\Delta\lambda = \lambda/Nm$. Here N is the number of rulings in the grating and λ is the average wavelength. The frequency f is related to the wavelength by $f\lambda = c$, where c is the speed of light. This means $f\Delta\lambda + \lambda\Delta f = 0$, so

$$\Delta\lambda = -\frac{\lambda}{f}\Delta f = -\frac{\lambda^2}{c}\Delta f$$

where $f = c/\lambda$ is used. The negative sign means that an increase in frequency corresponds to a decrease in wavelength.

ANALYZE (a) Equating the two expressions for $\Delta\lambda$, we have

$$\frac{\lambda^2}{c}\Delta f = \frac{\lambda}{Nm}$$

and

$$\Delta f = \frac{c}{Nm\lambda}.$$

(b) The difference in travel time for waves traveling along the two extreme rays is $\Delta t = \Delta L/c$, where ΔL is the difference in path length. The waves originate at slits that are separated by $(N - 1)d$, where d is the slit separation and N is the number of slits, so the path difference is $\Delta L = (N - 1)d \sin \theta$ and the time difference is

$$\Delta t = \frac{(N-1)d \sin\theta}{c}.$$

If N is large, this may be approximated by $\Delta t = (Nd/c)\sin \theta$. The lens does not affect the travel time.

(c) Substituting the expressions we derived for Δt and Δf, we obtain

$$\Delta f \, \Delta t = \left(\frac{c}{Nm\lambda}\right)\left(\frac{Nd\sin\theta}{c}\right) = \frac{d\sin\theta}{m\lambda} = 1.$$

The condition $d \sin \theta = m\lambda$ for a diffraction line is used to obtain the last result.

LEARN We take Δf to be positive and interpret it as the range of frequencies that can be resolved.

36-83

THINK For relatively wide slits, we consider both the interference of light from two slits, as well as the diffraction of light passing through each slit.

EXPRESS The central diffraction envelope spans the range $-\theta_1 < \theta < +\theta_1$ where $\theta_1 = \sin^{-1}(\lambda / a)$ is the angle that corresponds to the first diffraction minimum. The maxima in the double-slit pattern are at

$$\theta_m = \sin^{-1}\frac{m\lambda}{d},$$

so that our range specification becomes

$$-\sin^{-1}\left(\frac{\lambda}{a}\right) < \sin^{-1}\left(\frac{m\lambda}{d}\right) < +\sin^{-1}\left(\frac{\lambda}{a}\right),$$

which we change (since sine is a monotonically increasing function in the fourth and first quadrants, where all these angles lie) to

$$-\frac{\lambda}{a} < \frac{m\lambda}{d} < +\frac{\lambda}{a}.$$

The equation above sets the range of allowable values of m.

ANALYZE (a) Rewriting the equation as $-d/a < m < +d/a$, noting that $d/a = (14\ \mu\text{m})/(2.0\ \mu\text{m}) = 7$, we arrive at the result $-7 < m < +7$, or (since m must be an integer) $-6 \le m \le +6$, which amounts to 13 distinct values for m. Thus, 13 maxima are within the central envelope.

(b) The range (within *one* of the first-order envelopes) is now

$$-\sin^{-1}\left(\frac{\lambda}{a}\right) < \sin^{-1}\left(\frac{m\lambda}{d}\right) < +\sin^{-1}\left(\frac{2\lambda}{a}\right),$$

which leads to $d/a < m < 2d/a$ or $7 < m < 14$. Since m is an integer, this means $8 \le m \le 13$ which includes 6 distinct values for m in that one envelope. If we were to include the total from both first-order envelopes, the result would be 12, but the wording of the problem implies 6 should be the answer (just one envelope).

LEARN The intensity of the double-slit interference experiment is plotted below. The central diffraction envelope contains 13 maxima, and the first-order envelope has 6 on each side (excluding the very small peak corresponding to $m = 7$).

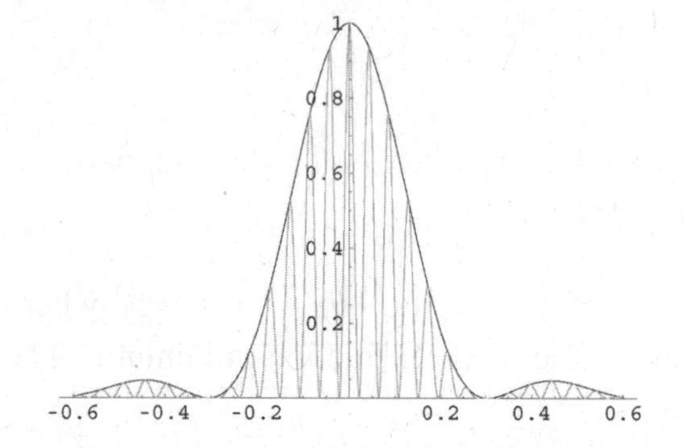

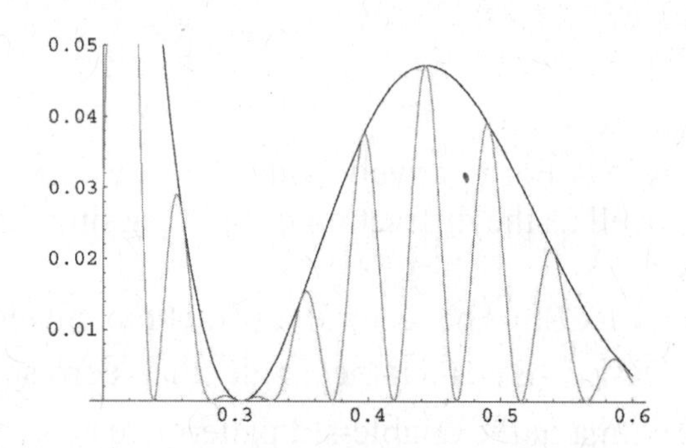

36-95

THINK We use phasors to explore how doubling slit width changes the intensity of the central maximum of diffraction and the energy passing through the slit.

EXPRESS We imagine dividing the original slit into N strips and represent the light from each strip, when it reaches the screen, by a phasor. Then, at the central maximum in the diffraction pattern, we would add the N phasors, all in the same direction and each with the same amplitude. We would find that the intensity there is proportional to N^2.

ANALYZE If we double the slit width, we need $2N$ phasors if they are each to have the amplitude of the phasors we used for the narrow slit. The intensity at the central maximum is proportional to $(2N)^2$ and is, therefore, four times the intensity for the narrow slit. The energy reaching the screen per unit time, however, is only twice the energy reaching it per unit time when the narrow slit is in place. The energy is simply redistributed. For example, the central peak is now half as wide and the integral of the intensity over the peak is only twice the analogous integral for the narrow slit.

LEARN From the discussion above, we see that the intensity of the central maximum increases as N^2. The dependence arises from the following two considerations: (1) The total power reaching the screen is proportional to N, and (2) the width of each maximum (distance between two adjacent minima) is proportional to $1/N$.

36-101

THINK Dispersion is a measure of the separation between diffracted light of different wavelengths. The angular dispersion of a grating is given by $D = d\theta/d\lambda$, where θ is the angular position of a line associated with wavelength λ.

EXPRESS The angular position and wavelength are related by $\mathbf{d}\sin\theta = m\lambda$, where $\mathbf{d}$ is the slit separation (which we made boldfaced in order not to confuse it with the d used in the derivative, below) and m is an integer. We differentiate this expression with respect to θ to obtain

$$\mathbf{d}\cos\theta d\theta = m d\lambda,$$

or

$$D = \frac{d\theta}{d\lambda} = \frac{m}{\mathbf{d}\cos\theta}.$$

ANALYZE Using $m = (\mathbf{d}/\lambda)\sin\theta$, we obtain $D = \dfrac{\mathbf{d}\sin\theta}{\mathbf{d}\lambda\cos\theta} = \dfrac{\tan\theta}{\lambda}$.

LEARN From the expression $D = \dfrac{m}{\mathbf{d}\cos\theta}$, we see that the angular dispersion increases with the order m.

Chapter 37

37-9

THINK Due to Lorentz contraction, the length of the moving spaceship is measured to be shorter by a stationary observer.

EXPRESS Let the rest length of the spaceship be L_0. The length measured by the timing station is

$$L = L_0\sqrt{1-(v/c)^2}.$$

ANALYZE (a) The rest length is $L_0 = 130$ m. With $v = 0.740c$, we obtain

$$L = L_0\sqrt{1-(v/c)^2} = (130\,\text{m})\sqrt{1-(0.740)^2} = 87.4\,\text{m}.$$

(b) The time interval for the passage of the spaceship is

$$\Delta t = \frac{L}{v} = \frac{87.4\,\text{m}}{(0.740)(3.00\times10^8\,\text{m/s})} = 3.94\times10^{-7}\,\text{s}.$$

LEARN The length of the spaceship appears to be contracted by a factor of

$$\gamma = \frac{1}{\sqrt{1-(v/c)^2}} = \frac{1}{\sqrt{1-(0.740)^2}} = 1.487.$$

37-17

THINK We apply Lorentz transformation to calculate x' and t' according to an observer in S'.

EXPRESS The proper time is not measured by clocks in either frame S or frame S' since a single clock at rest in either frame cannot be present at the origin and at the event. The full Lorentz transformation must be used:

$$x' = \gamma(x - vt), \quad t' = \gamma(t - \beta x/c)$$

where $\beta = v/c = 0.950$ and

$$\gamma = \frac{1}{\sqrt{1-\beta^2}} = \frac{1}{\sqrt{1-(0.950)^2}} = 3.20256 .$$

ANALYZE (a) Thus, the spatial coordinate in S' is

$$x' = \gamma(x - vt) = (3.20256)\left[100\times10^3\,\text{m} - (0.950)(2.998\times10^8\,\text{m/s})(200\times10^{-6}\text{s})\right]$$
$$= 1.38\times10^5\,\text{m} = 138\ \text{km}.$$

(b) The temporal coordinate in S' is

$$t' = \gamma(t - \beta x/c) = (3.20256)\left[200\times10^{-6}\text{s} - \frac{(0.950)(100\times10^3\,\text{m})}{2.998\times10^8\,\text{m/s}}\right]$$
$$= -3.74\times10^{-4}\,\text{s} = -374\,\mu\text{s} .$$

LEARN The time and the location of the collision recorded by an observer S' are different than that by another observer in S.

37-27

THINK We apply relativistic velocity transformation to calculate the velocity of the particle with respect to frame S.

EXPRESS We assume S' is moving in the $+x$ direction. Let u' be the velocity of the particle as measured in S' and v be the velocity of S' relative to S, the velocity of the particle as measured in S is given by Eq. 37-29:

$$u = \frac{u'+v}{1+u'v/c^2}.$$

ANALYZE With $u' = +0.40c$ and $v = +0.60c$, we obtain

$$u = \frac{u'+v}{1+u'v/c^2} = \frac{0.40c+0.60c}{1+(0.40c)(+0.60c)/c^2} = 0.81c .$$

LEARN The classical Galilean transformation would have given

$$u = u' + v = 0.40c + 0.60c = 1.0c.$$

37-31

THINK Both the spaceship and the micrometeorite are moving relativistically, and we apply relativistic speed transformation to calculate the velocity of the micrometeorite relative to the spaceship.

EXPRESS Let S be the reference frame of the micrometeorite, and S' be the reference frame of the spaceship. We assume S to be moving in the $+x$ direction. Let u be the velocity of the micrometeorite as measured in S and v be the velocity of S' relative to S, the velocity of the micrometeorite as measured in S' can be solved by using Eq. 37-29:

$$u = \frac{u' + v}{1 + u'v/c^2} \Rightarrow u' = \frac{u - v}{1 - uv/c^2}.$$

ANALYZE The problem indicates that $v = -0.82c$ (spaceship velocity) and $u = +0.82c$ (micrometeorite velocity). We solve for the velocity of the micrometeorite relative to the spaceship:

$$u' = \frac{u - v}{1 - uv/c^2} = \frac{0.82c - (-0.82c)}{1 - (0.82)(-0.82)} = 0.98c$$

or 2.94×10^8 m/s. Using Eq. 37-10, we conclude that observers on the ship measure a transit time for the micrometeorite (as it passes along the length of the ship) equal to

$$\Delta t = \frac{d}{u'} = \frac{350\,\text{m}}{2.94 \times 10^8\,\text{m/s}} = 1.2 \times 10^{-6}\,\text{s}.$$

LEARN The classical Galilean transformation would have given

$$u' = u - v = 0.82c - (-0.82c) = 1.64c,$$

which exceeds c and therefore, is physically impossible.

37-35

THINK This problem deals with the Doppler effect of light. The source is the spaceship that is moving away from the Earth, where the detector is located.

EXPRESS With the source and the detector separating, the frequency received is given directly by Eq. 37-31:

$$f = f_0 \sqrt{\frac{1-\beta}{1+\beta}}$$

where f_0 is the frequency in the frames of the spaceship, $\beta = v/c$, and v is the speed of the spaceship relative to the Earth.

ANALYZE With $\beta = 0.90$ and $f_0 = 100$ MHz, we obtain

$$f = f_0 \sqrt{\frac{1-\beta}{1+\beta}} = (100\,\text{MHz}) \sqrt{\frac{1-0.9000}{1+0.9000}} = 22.9\,\text{MHz}.$$

LEARN Since the source is moving away from the detector, $f < f_0$. Note that in the low speed limit, $\beta \ll 1$, Eq. 37-31 can be approximated as

$$f \approx f_0\left(1-\beta+\frac{1}{2}\beta^2\right).$$

37-39

THINK This problem deals with the Doppler effect of light. The source is the spaceship that is moving away from the Earth, where the detector is located.

EXPRESS With the source and the detector separating, the frequency received is given directly by Eq. 37-31:

$$f = f_0\sqrt{\frac{1-\beta}{1+\beta}}$$

where f_0 is the frequency in the frames of the spaceship, $\beta = v/c$, and v is the speed of the spaceship relative to the Earth. The frequency and the wavelength are related by $f\lambda = c$. Thus, if λ_0 is the wavelength of the light as seen on the spaceship, using $c = f_0\lambda_0 = f\lambda$, then the wavelength detected on Earth would be

$$\lambda = \lambda_0\left(\frac{f_0}{f}\right) = \lambda_0\sqrt{\frac{1+\beta}{1-\beta}}.$$

ANALYZE (a) With $\lambda_0 = 450$ nm and $\beta = 0.20$, we obtain

$$\lambda = (450\,\text{nm})\sqrt{\frac{1+0.20}{1-0.20}} = 550\text{ nm}.$$

(b) This is in the green-yellow portion of the visible spectrum.

LEARN Since $\lambda_0 = 450$ nm, the color of the light as seen on the spaceship is violet-blue. With $\lambda > \lambda_0$, this Doppler shift is a red shift.

37-41

THINK The electron is moving at a relativistic speed since its kinetic energy greatly exceeds its rest energy.

EXPRESS The kinetic energy of the electron is given by Eq. 37-52:

$$K = E - mc^2 = \gamma mc^2 - mc^2 = mc^2(\gamma - 1).$$

Thus, $\gamma = (K/mc^2) + 1$. Similarly, by inverting the Lorentz factor $\gamma = 1/\sqrt{1-\beta^2}$, we obtain $\beta = \sqrt{1-(1/\gamma)^2}$.

ANALYZE (a) Table 37-3 gives mc^2 = 511 keV = 0.511 MeV for the electron rest energy, so the Lorentz factor is

$$\gamma = \frac{K}{mc^2} + 1 = \frac{100\,\text{MeV}}{0.511\,\text{MeV}} + 1 = 196.695.$$

(b) The speed parameter is

$$\beta = \sqrt{1 - \frac{1}{(196.695)^2}} = 0.999987.$$

Thus, the speed of the electron is 0.999987*c*, or 99.9987% of the speed of light.

LEARN The classical expression $K = mv^2/2$, for kinetic energy, is adequate only when the speed of the object is well below the speed of light.

37-47

THINK As a consequence of the theory of relativity, mass can be considered as another form of energy.

EXPRESS The mass of an object and its equivalent energy is given by

$$E_0 = mc^2.$$

ANALYZE The energy equivalent of one tablet is

$$E_0 = mc^2 = (320 \times 10^{-6}\text{ kg})(3.00 \times 10^8\text{ m/s})^2 = 2.88 \times 10^{13}\text{ J}.$$

This provides the same energy as

$$(2.88 \times 10^{13}\text{ J})/(3.65 \times 10^7\text{ J/L}) = 7.89 \times 10^5\text{ L}$$

of gasoline. The distance the car can go is

$$d = (7.89 \times 10^5\text{ L})(12.75\text{ km/L}) = 1.01 \times 10^7\text{ km}.$$

LEARN The distance is roughly 250 times larger than the circumference of Earth (see Appendix C). However, this is possible only if the mass–energy conversion were perfect.

37-71

THINK We calculate the relative speed of the satellites using both the Galilean transformation and the relativistic speed transformation.

EXPRESS Let v be the speed of the satellites relative to Earth. As they pass each other in opposite directions, their relative speed is given by $v_{\text{rel}, c} = 2v$ according to the classical Galilean transformation. On the other hand, applying relativistic velocity transformation gives

$$v_{\text{rel}} = \frac{2v}{1 + v^2/c^2}.$$

ANALYZE (a) With $v = 27000$ km/h, we obtain

$$v_{\text{rel}, c} = 2v = 2(27000 \text{ km/h}) = 5.4 \times 10^4 \text{ km/h}.$$

(b) We can express c in these units by multiplying by 3.6: $c = 1.08 \times 10^9$ km/h. The fractional error is

$$\frac{v_{\text{rel},c} - v_{\text{rel}}}{v_{\text{rel},c}} = 1 - \frac{1}{1 + v^2/c^2} = 1 - \frac{1}{1 + [(27000 \text{ km/h})/(1.08 \times 10^9 \text{ km/h})]^2} = 6.3 \times 10^{-10}.$$

LEARN Since the speeds of the satellites are well below the speed of light, calculating their relative speed using the classical Galilean transformation is adequate.

37-73

THINK The work done to the proton is equal to its change in kinetic energy.

EXPRESS The kinetic energy of the proton is given by Eq. 37-52:

$$K = E - mc^2 = \gamma mc^2 - mc^2 = mc^2(\gamma - 1)$$

where $\gamma = 1/\sqrt{1-\beta^2}$ is the Lorentz factor.

Let v_1 be the initial speed and v_2 be the final speed of the proton. The work required is

$$W = \Delta K = mc^2(\gamma_2 - 1) - mc^2(\gamma_1 - 1) = mc^2(\gamma_2 - \gamma_1) = mc^2\Delta\gamma.$$

ANALYZE When $\beta_2 = 0.9860$, we have $\gamma_2 = 5.9972$, and when $\beta_1 = 0.9850$, we have $\gamma_1 = 5.7953$. Thus, $\Delta\gamma = 0.202$ and the change in kinetic energy (equal to the work) becomes (using Eq. 37-52)

$$W = \Delta K = (mc^2)\Delta\gamma = (938 \text{ MeV})(5.9972 - 5.7953) = 189 \text{ MeV}$$

where $mc^2 = 938$ MeV has been used (see Table 37-3).

LEARN Using the classical expression $K_c = mv^2/2$ for kinetic energy, one would have obtain

$$W_c = \Delta K_c = \frac{1}{2}m(v_2^2 - v_1^2) = \frac{1}{2}mc^2(\beta_2^2 - \beta_1^2) = \frac{1}{2}(938 \text{ MeV})\left[(0.9860)^2 - (0.9850)^2\right]$$
$$= 0.924 \text{ MeV}$$

which is substantially lowered than that calculated using relativistic formulation.

37-75

THINK The electron is moving toward the Earth at a relativistic speed since $E \gg mc^2$, where mc^2 is the rest energy of the electron.

EXPRESS The energy of the electron is given by

$$E = \gamma mc^2 = \frac{mc^2}{\sqrt{1-(v/c)^2}}.$$

With $E = 1533$ MeV and $mc^2 = 0.511$ MeV (see Table 37-3), we obtain

$$v = c\sqrt{1-\left(\frac{mc^2}{E}\right)^2} = c\sqrt{1-\left(\frac{0.511 \text{ MeV}}{1533 \text{ MeV}}\right)^2} = 0.99999994c \approx c.$$

Thus, in the rest frame of Earth, it took the electron 26 y to reach us. In order to transform to its own "clock" it's useful to compute γ directly from Eq. 37-48:

$$\gamma = \frac{E}{mc^2} = \frac{1533 \text{ MeV}}{0.511 \text{ MeV}} = 3000$$

though if one is careful one can also get this result from $\gamma = 1/\sqrt{1-(v/c)^2}$.

ANALYZE Then, Eq. 37-7 leads to

$$\Delta t_0 = \frac{\Delta t}{\gamma} = \frac{26\,\text{y}}{3000} = 0.0087\,\text{y}$$

so that the electron "concludes" the distance he traveled is only 0.0087 light-years.

LEARN In the rest frame of the electron, the Earth appears to be rushing toward the electron with a speed of $0.99999994c$. Thus, the electron starts its journey from a distance of 0.0087 light-years away.

37-79

THINK The electron is moving at a relativistic speed since its total energy E is much greater than mc^2, the rest energy of the electron.

EXPRESS To calculate the momentum of the electron, we use Eq. 37-54:

$$(pc)^2 = K^2 + 2Kmc^2 .$$

ANALYZE With $K = 2.00$ MeV and $mc^2 = 0.511$ MeV (see Table 37-3), we have

$$pc = \sqrt{K^2 + 2Kmc^2} = \sqrt{(2.00\text{ MeV})^2 + 2(2.00\text{ MeV})(0.511\text{ MeV})}$$

This readily yields $p = 2.46$ MeV/c.

LEARN Classically, the electron momentum is

$$p_c = \sqrt{2Km} = \frac{\sqrt{2Kmc^2}}{c} \frac{\sqrt{2(2.00\text{ MeV})(0.511\text{ MeV})}}{c} = 1.43\text{ MeV}/c$$

which is smaller than that obtained using relativistic formulation.

Chapter 38

38-11

THINK The rate of photon emission is the number of photons emitted per unit time.

EXPRESS Let R be the photon emission rate and E be the energy of a single photon. The power output of a lamp is given by $P = RE$, where we assume that all the power goes into photon production. Now, $E = hf = hc/\lambda$, where h is the Planck constant, f is the frequency of the light emitted, and λ is the wavelength. Thus

$$P = \frac{Rhc}{\lambda} \Rightarrow R = \frac{\lambda P}{hc}.$$

ANALYZE (a) The fact that $R \sim \lambda$ means that the lamp that emits light with the longer wavelength (the 700 nm infrared lamp) emits more photons per unit time. The energy of each photon is less, so it must emit photons at a greater rate.

(b) Let R be the rate of photon production for the 700 nm lamp. Then,

$$R = \frac{\lambda P}{hc} = \frac{(700\,\text{nm})(400\,\text{J/s})}{(1.60\times10^{-19}\,\text{J/eV})(1240\ \text{eV}\cdot\text{nm})} = 1.41\times10^{21}\ \text{photon/s}.$$

LEARN With $P = Rhc/\lambda$, we readily see that when the rate of photon emission is held constant, the shorter the wavelength, the greater the power, or rate of energy emission.

38-15

THINK The energy of an incident photon is $E = hf$, where h is the Planck constant, and f is the frequency of the electromagnetic radiation.

EXPRESS The kinetic energy of the most energetic electron emitted is

$$K_m = E - \Phi = (hc/\lambda) - \Phi,$$

where Φ is the work function for sodium, and $f = c/\lambda$, where λ is the wavelength of the photon.

The stopping potential V_{stop} is related to the maximum kinetic energy by $eV_{stop} = K_m$, so

$$eV_{stop} = (hc/\lambda) - \Phi$$

and

$$\lambda = \frac{hc}{eV_{stop} + \Phi} = \frac{1240\,\text{eV}\cdot\text{nm}}{5.0\,\text{eV} + 2.2\,\text{eV}} = 170\,\text{nm}.$$

Here $eV_{stop} = 5.0$ eV and $hc = 1240$ eV·nm are used.

LEARN The cutoff frequency for this problem is

$$f_0 = \frac{\Phi}{h} = \frac{(2.2\text{ eV})(1.6\times10^{-19}\text{ J/eV})}{6.626\times10^{-34}\text{ J}\cdot\text{s}} = 5.3\times10^{14}\text{ Hz}.$$

38-23

THINK The kinetic energy K_m of the fastest electron emitted is given by

$$K_m = hf - \Phi,$$

where Φ is the work function of aluminum, and f is the frequency of the incident radiation.

EXPRESS Since $f = c/\lambda$, where λ is the wavelength of the photon, the above expression can be rewritten as

$$K_m = (hc/\lambda) - \Phi.$$

ANALYZE (a) Thus, the kinetic energy of the fastest electron is

$$K_m = \frac{1240\,\text{eV}\cdot\text{nm}}{200\,\text{nm}} - 4.20\,\text{eV} = 2.00\text{ eV},$$

where we have used $hc = 1240$ eV·nm.

(b) The slowest electron just breaks free of the surface and so has zero kinetic energy.

(c) The stopping potential V_{stop} is given by $K_m = eV_{stop}$, so

$$V_{stop} = K_m/e = (2.00\text{ eV})/e = 2.00\text{ V}.$$

(d) The value of the cutoff wavelength is such that $K_m = 0$. Thus, $hc/\lambda_0 = \Phi$, or

$$\lambda_0 = hc/\Phi = (1240\text{ eV}\cdot\text{nm})/(4.2\text{ eV}) = 295\text{ nm}.$$

LEARN If the wavelength is longer than λ_0, the photon energy is less than Φ and a photon does not have sufficient energy to knock even the most energetic electron out of the aluminum sample.

38-27

THINK The scattering between a photon and an electron initially at rest results in a change or photon's wavelength, or Compton shift.

EXPRESS When a photon scatters off from an electron initially at rest, the change in wavelength is given by

$$\Delta\lambda = (h/mc)(1 - \cos\phi),$$

where m is the mass of an electron and ϕ is the scattering angle.

ANALYZE (a) The Compton wavelength of the electron is $h/mc = 2.43 \times 10^{-12}$ m = 2.43 pm. Therefore, we find the shift to be

$$\Delta\lambda = (h/mc)(1 - \cos\phi) = (2.43 \text{ pm})(1 - \cos 30°) = 0.326 \text{ pm}.$$

The final wavelength is

$$\lambda' = \lambda + \Delta\lambda = 2.4 \text{ pm} + 0.326 \text{ pm} = 2.73 \text{ pm}.$$

(b) With $\phi = 120°$, $\Delta\lambda = (2.43 \text{ pm})(1 - \cos 120°) = 3.645$ pm and

$$\lambda' = 2.4 \text{ pm} + 3.645 \text{ pm} = 6.05 \text{ pm}.$$

LEARN The wavelength shift is greatest when $\phi = 180°$, where $\cos 180° = -1$. At this angle, the photon is scattered back along its initial direction of travel, and $\Delta\lambda = 2h/mc$.

38-47

THINK The de Broglie wavelength of the electron is given by $\lambda = h/p$, where p is the momentum of the electron.

EXPRESS The momentum of the electron can be written as

$$p = m_e v = \sqrt{2m_e K} = \sqrt{2m_e eV},$$

where V is the accelerating potential and e is the fundamental charge. Thus,

$$\lambda = \frac{h}{p} = \frac{h}{\sqrt{2m_e eV}}.$$

ANALYZE With $V = 25.0$ kV, we obtain

$$\lambda = \frac{h}{\sqrt{2m_e eV}} = \frac{6.626\times10^{-34}\,\text{J}\cdot\text{s}}{\sqrt{2(9.109\times10^{-31}\,\text{kg})(1.602\times10^{-19}\,\text{C})(25.0\times10^{3}\,\text{V})}}$$
$$= 7.75\times10^{-12}\,\text{m} = 7.75\,\text{pm}.$$

LEARN The wavelength is of the same order as the Compton wavelength of the electron. Increasing the potential difference V would make the wavelength even smaller.

38-49

THINK The de Broglie wavelength of the sodium ion is given by $\lambda = h/p$, where p is the momentum of the ion.

EXPRESS The kinetic energy acquired is $K = qV$, where q is the charge on an ion and V is the accelerating potential. Thus, the momentum of an ion is $p = \sqrt{2mK}$, and the corresponding de Broglie wavelength is $\lambda = \dfrac{h}{p} = \dfrac{h}{\sqrt{2mK}}$.

ANALYZE (a) The kinetic energy of the ion is

$$K = qV = (1.60 \times 10^{-19}\ \text{C})(300\ \text{V}) = 4.80 \times 10^{-17}\ \text{J}.$$

The mass of a single sodium atom is, from Appendix F,

$$m = (22.9898\ \text{g/mol})/(6.02 \times 10^{23}\ \text{atom/mol}) = 3.819 \times 10^{-23}\ \text{g} = 3.819 \times 10^{-26}\ \text{kg}.$$

Thus, the momentum of a sodium ion is

$$p = \sqrt{2mK} = \sqrt{2\left(3.819\times10^{-26}\ \text{kg}\right)\left(4.80\times10^{-17}\ \text{J}\right)} = 1.91\times10^{-21}\ \text{kg}\cdot\text{m/s}.$$

(b) The de Broglie wavelength is

$$\lambda = \frac{h}{p} = \frac{6.63\times10^{-34}\,\text{J}\cdot\text{s}}{1.91\times10^{-21}\text{kg}\cdot\text{m/s}} = 3.46\times10^{-13}\,\text{m}.$$

LEARN The greater the potential difference, the greater the kinetic energy and momentum, and hence, the smaller the de Broglie wavelength.

38-51

THINK The de Broglie wavelength of a particle is given by $\lambda = h/p$, where p is the momentum of the particle.

EXPRESS Let K be the kinetic energy of the electron, in units of electron volts (eV). Since $K = p^2/2m$, the electron momentum is $p=\sqrt{2mK}$. Thus, the de Broglie wavelength is

$$\lambda=\frac{h}{p}=\frac{h}{\sqrt{2mK}}=\frac{6.626\times10^{-34}\ \text{J}\cdot\text{s}}{\sqrt{2(9.109\times10^{-31}\ \text{kg})(1.602\times10^{-19}\ \text{J/eV})K}}=\frac{1.226\times10^{-9}\ \text{m}\cdot\text{eV}^{1/2}}{\sqrt{K}}$$
$$=\frac{1.226\ \text{nm}\cdot\text{eV}^{1/2}}{\sqrt{K}}.$$

ANALYZE With $\lambda = 590$ nm, the above equation can be inverted to give

$$K=\left(\frac{1.226\,\text{nm}\cdot\text{eV}^{1/2}}{\lambda}\right)^2=\left(\frac{1.226\,\text{nm}\cdot\text{eV}^{1/2}}{590\,\text{nm}}\right)^2=4.32\times10^{-6}\ \text{eV}.$$

LEARN The analytical expression shows that the kinetic energy is proportional to $1/\lambda^2$. This is so because $K\sim p^2$, while $p\sim 1/\lambda$.

38-61

THINK In this problem we solve a special case of the Schrödinger's equation where the potential energy is $U(x)=U_0=\text{constant}$.

EXPRESS For $U = U_0$, Schrödinger's equation becomes

$$\frac{d^2\psi}{dx^2}+\frac{8\pi^2 m}{h^2}[E-U_0]\psi=0.$$

We substitute $\psi=\psi_0 e^{ikx}$.

ANALYZE The second derivative is

$$\frac{d^2\psi}{dx^2}=-k^2\psi_0 e^{ikx}=-k^2\psi.$$

The result is

$$-k^2\psi+\frac{8\pi^2 m}{h^2}[E-U_0]\psi=0.$$

Solving for k, we obtain

$$k=\sqrt{\frac{8\pi^2 m}{h^2}[E-U_0]}=\frac{2\pi}{h}\sqrt{2m[E-U_0]}.$$

LEARN Another way to realize this is to note that with a constant potential energy $U(x)=U_0$, we can simply redefine the total energy as $E'=E-U_0$, and the Schrödinger's equation looks just like the free-particle case:

$$\frac{d^2\psi}{dx^2}+\frac{8\pi^2 mE'}{h^2}\psi=0.$$

The solution is $\psi=\psi_0\exp(ik'x)$, where

$$k'^2=\frac{8\pi^2 mE'}{h^2} \quad\Rightarrow\quad k=\frac{2\pi}{h}\sqrt{2mE'}=\frac{2\pi}{h}\sqrt{2m(E-U_0)}\,.$$

38-64

THINK The angular wave number k is related to the wavelength λ by $k = 2\pi/\lambda$.

EXPRESS The wavelength is related to the particle momentum p by $\lambda = h/p$, so $k = 2\pi p/h$. Now, the kinetic energy K and the momentum are related by $K = p^2/2m$, where m is the mass of the particle.

ANALYZE Thus, we have $p=\sqrt{2mK}$ and

$$k=\frac{2\pi}{\lambda}=\frac{2\pi p}{h}=\frac{2\pi\sqrt{2mK}}{h}.$$

LEARN The expression obtained above applies to the case of a free particle only. In the presence of interaction, the potential energy is nonzero, and the functional form of k will change. For example, as shown in Problem 38-57, when $U(x)=U_0$, the angular wave number becomes

$$k=\frac{2\pi}{h}\sqrt{2m(E-U_0)}\,.$$

38-77

THINK Even though $E<U_b$, barrier tunneling can still take place quantum mechanically with finite probability.

EXPRESS If m is the mass of the particle and E is its energy, then the transmission coefficient for a barrier of height U_b and width L is given by

$$T=e^{-2bL},$$

where

$$b=\sqrt{\frac{8\pi^2 m(U_b-E)}{h^2}}.$$

If the change ΔU_b in U_b is small (as it is), the change in the transmission coefficient is given by

$$\Delta T=\frac{dT}{dU_b}\Delta U_b=-2LT\frac{db}{dU_b}\Delta U_b.$$

Now,

$$\frac{db}{dU_b}=\frac{1}{2\sqrt{U_b-E}}\sqrt{\frac{8\pi^2 m}{h^2}}=\frac{1}{2(U_b-E)}\sqrt{\frac{8\pi^2 m(U_b-E)}{h^2}}=\frac{b}{2(U_b-E)}.$$

Thus,

$$\Delta T=-LTb\frac{\Delta U_b}{U_b-E}.$$

ANALYZE (a) With

$$b=\sqrt{\frac{8\pi^2\left(9.11\times10^{-31}\text{ kg}\right)(6.8\text{ eV}-5.1\text{ eV})\left(1.6022\times10^{-19}\text{ J/eV}\right)}{\left(6.6261\times10^{-34}\text{ J}\cdot\text{s}\right)^2}}=6.67\times10^{9}\text{ m}^{-1},$$

we have $bL=(6.67\times10^{9}\text{ m}^{-1})(750\times10^{-12}\text{ m}^{-1})=5.0$, and

$$\frac{\Delta T}{T}=-bL\frac{\Delta U_b}{U_b-E}=-(5.0)\frac{(0.010)(6.8\text{eV})}{6.8\text{eV}-5.1\text{eV}}=-0.20\,.$$

There is a 20% decrease in the transmission coefficient.

(b) The change in the transmission coefficient is given by

$$\Delta T=\frac{dT}{dL}\Delta L=-2be^{-2bL}\Delta L=-2bT\Delta L$$

and

$$\frac{\Delta T}{T}=-2b\Delta L=-2\left(6.67\times10^{9}\text{ m}^{-1}\right)(0.010)\left(750\times10^{-12}\text{ m}\right)=-0.10\,.$$

There is a 10% decrease in the transmission coefficient.

(c) The change in the transmission coefficient is given by

$$\Delta T=\frac{dT}{dE}\Delta E=-2Le^{-2bL}\frac{db}{dE}\Delta E=-2LT\frac{db}{dE}\Delta E.$$

Now, $db/dE = -db/dU_b = -b/2(U_b - E)$, so

$$\frac{\Delta T}{T} = bL\frac{\Delta E}{U_b - E} = (5.0)\frac{(0.010)(5.1\text{eV})}{6.8\text{eV} - 5.1\text{eV}} = 0.15\,.$$

There is a 15% increase in the transmission coefficient.

LEARN Increasing the barrier height or the barrier thickness reduces the probability of transmission, while increasing the kinetic energy of the electron increases the probability.

38-90

THINK We apply Heisenberg's uncertainty principle to calculate the uncertainty in position.

EXPRESS The uncertainty principle states that $\Delta x \Delta p \geq \hbar$, where Δx and Δp represent the intrinsic uncertainties in measuring the position and momentum, respectively. The uncertainty in the momentum is

$$\Delta p = m\,\Delta v = (0.50\text{ kg})(1.0\text{ m/s}) = 0.50\ \text{kg}\cdot\text{m/s},$$

where Δv is the uncertainty in the velocity.

ANALYZE Solving the uncertainty relationship $\Delta x \Delta p \geq \hbar$ for the minimum uncertainty in the coordinate x, we obtain

$$\Delta x = \frac{\hbar}{\Delta p} = \frac{0.60\,\text{J}\cdot\text{s}}{2\pi(0.50\,\text{kg}\cdot\text{m/s})} = 0.19\,\text{m}.$$

LEARN Heisenberg's uncertainty principle implies that it is impossible to simultaneously measure a particle's position and momentum with infinite accuracy.

Chapter 39

39-15

THINK The probability of finding an electron in an interval is given by $P = \int |\psi|^2\, dx$, where the integral is over the interval.

EXPRESS If the interval width Δx is small, the probability can be approximated by $P = |\psi|^2\, \Delta x$, where the wave function is evaluated for the center of the interval. For an electron trapped in an infinite well of width L, the ground state probability density is

$$|\psi|^2 = \frac{2}{L}\sin^2\left(\frac{\pi x}{L}\right),$$

so

$$P = \left(\frac{2\Delta x}{L}\right)\sin^2\left(\frac{\pi x}{L}\right).$$

ANALYZE (a) We take $L = 100$ pm, $x = 25$ pm, and $\Delta x = 5.0$ pm. Then,

$$P = \left[\frac{2(5.0\,\text{pm})}{100\,\text{pm}}\right]\sin^2\left[\frac{\pi(25\,\text{pm})}{100\,\text{pm}}\right] = 0.050.$$

(b) We take $L = 100$ pm, $x = 50$ pm, and $\Delta x = 5.0$ pm. Then,

$$P = \left[\frac{2(5.0\,\text{pm})}{100\,\text{pm}}\right]\sin^2\left[\frac{\pi(50\,\text{pm})}{100\,\text{pm}}\right] = 0.10.$$

(c) We take $L = 100$ pm, $x = 90$ pm, and $\Delta x = 5.0$ pm. Then,

$$P = \left[\frac{2(5.0\,\text{pm})}{100\,\text{pm}}\right]\sin^2\left[\frac{\pi(90\,\text{pm})}{100\,\text{pm}}\right] = 0.0095.$$

LEARN The probability as a function of x is plotted next. As expected, the probability of detecting the electron is highest near the center of the well at $x = L/2 = 50$ pm.

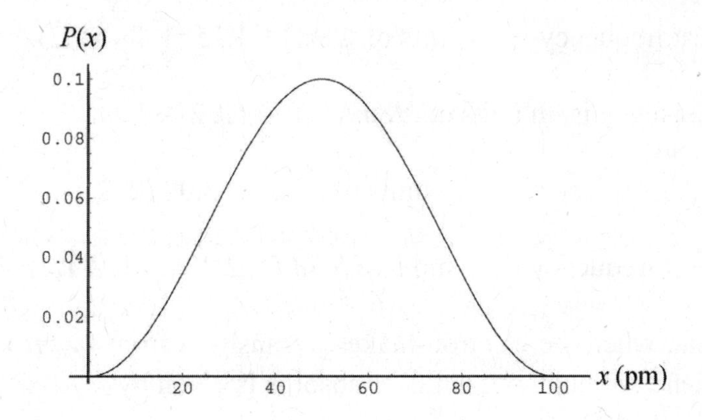

39-27

THINK The energy levels of an electron trapped in a regular corral with widths L_x and L_y are given by Eq. 39-20:

$$E_{n_x,n_y} = \frac{h^2}{8m}\left[\frac{n_x^2}{L_x^2} + \frac{n_y^2}{L_y^2}\right].$$

EXPRESS With $L_x = L$ and $L_y = 2L$, we have

$$E_{n_x,n_y} = \frac{h^2}{8m}\left[\frac{n_x^2}{L_x^2} + \frac{n_y^2}{L_y^2}\right] = \frac{h^2}{8mL^2}\left[n_x^2 + \frac{n_y^2}{4}\right].$$

Thus, in units of $h^2/8mL^2$, the energy levels are given by $n_x^2 + n_y^2/4$. The lowest five levels are $E_{1,1} = 1.25$, $E_{1,2} = 2.00$, $E_{1,3} = 3.25$, $E_{2,1} = 4.25$, and $E_{2,2} = E_{1,4} = 5.00$. It is clear that there are no other possible values for the energy less than 5.

The frequency of the light emitted or absorbed when the electron goes from an initial state i to a final state f is $f = (E_f - E_i)/h$, and in units of $h/8mL^2$ is simply the difference in the values of $n_x^2 + n_y^2/4$ for the two states. The possible frequencies are as follows:

$$\begin{aligned}&0.75(1,2\to1,1), 2.00(1,3\to1,1), 3.00(2,1\to1,1),\\ &3.75(2,2\to1,1), 1.25(1,3\to1,2), 2.25(2,1\to1,2), 3.00(2,2\to1,2), 1.00(2,1\to1,3),\\ &1.75(2,2\to1,3), 0.75(2,2\to2,1),\end{aligned}$$

all in units of $h/8mL^2$.

ANALYZE (a) From the above, we see that there are 8 different frequencies.

(b) The lowest frequency is, in units of $h/8mL^2$, 0.75 (2, 2→2,1).

(c) The second lowest frequency is, in units of $h/8mL^2$, 1.00 (2, 1→1,3).

(d) The third lowest frequency is, in units of $h/8mL^2$, 1.25 (1, 3$\rightarrow$1,2).

(e) The highest frequency is, in units of $h/8mL^2$, 3.75 (2, 2$\rightarrow$1,1).

(f) The second highest frequency is, in units of $h/8mL^2$, 3.00 (2, 2$\rightarrow$1,2) or (2, 1$\rightarrow$1,1).

(g) The third highest frequency is, in units of $h/8mL^2$, 2.25 (2, 1$\rightarrow$1,2).

LEARN In general, when the electron makes a transition from (n_x, n_y) to a higher level (n'_x, n'_y), the frequency of photon it emits or absorbs is given by

$$f = \frac{\Delta E}{h} = \frac{E_{n'_x,n'_y} - E_{n_x,n_y}}{h} = \frac{h}{8mL^2}\left(n_x'^2 + \frac{n_y'^2}{4}\right) - \frac{h}{8mL^2}\left(n_x^2 + \frac{n_y^2}{4}\right)$$

$$= \frac{h}{8mL^2}\left[\left(n_x'^2 - n_x^2\right) + \frac{1}{4}\left(n_y'^2 - n_y^2\right)\right].$$

39-31

THINK The Lyman series is associated with transitions to or from the $n = 1$ level of the hydrogen atom, while the Balmer series is for transitions to or from the $n = 2$ level.

EXPRESS The energy E of the photon emitted when a hydrogen atom jumps from a state with principal quantum number n' to a state with principal quantum number $n < n'$ is given by

$$E = A\left(\frac{1}{n^2} - \frac{1}{n'^2}\right).$$

where A = 13.6 eV. The frequency f of the electromagnetic wave is given by $f = E/h$ and the wavelength is given by $\lambda = c/f$. Thus,

$$\frac{1}{\lambda} = \frac{f}{c} = \frac{E}{hc} = \frac{A}{hc}\left(\frac{1}{n^2} - \frac{1}{n'^2}\right).$$

ANALYZE The shortest wavelength occurs at the series limit, for which $n' = \infty$. For the Balmer series, $n = 2$ and the shortest wavelength is $\lambda_B = 4hc/A$. For the Lyman series, $n = 1$ and the shortest wavelength is $\lambda_L = hc/A$. The ratio is $\lambda_B/\lambda_L = 4.0$.

LEARN The energy of the photon emitted associated with the transition of an electron from $n' = \infty \rightarrow n = 2$ (to become bound) is

$$E_{\infty\rightarrow 2} = \frac{13.6\text{ eV}}{2^2} = 3.4\text{ eV}.$$

Similarly, the energy associated with the transition of an electron from $n' = \infty \to n = 1$ (to become bound) is

$$E_{1\to\infty} = \frac{13.6\text{ eV}}{1^2} = 13.6\text{ eV}.$$

39-37

THINK The energy of the hydrogen atom is quantized.

EXPRESS If kinetic energy is not conserved, some of the neutron's initial kinetic energy could be used to excite the hydrogen atom. The least energy that the hydrogen atom can accept is the difference between the first excited state ($n = 2$) and the ground state ($n = 1$). Since the energy of a state with principal quantum number n is $-(13.6\text{ eV})/n^2$, the smallest excitation energy is

$$\Delta E = E_2 - E_1 = \frac{-13.6\text{ eV}}{(2)^2} - \frac{-13.6\text{ eV}}{(1)^2} = 10.2\text{ eV}.$$

ANALYZE The neutron, with a kinetic energy of 6.0 eV, does not have sufficient kinetic energy to excite the hydrogen atom, so the hydrogen atom is left in its ground state and all the initial kinetic energy of the neutron ends up as the final kinetic energies of the neutron and atom. The collision must be elastic.

LEARN The minimum kinetic energy the neutron must have in order to excite the hydrogen atom is 10.2 eV.

39-39

THINK The radial probability function for the ground state of hydrogen is

$$P(r) = (4r^2/a^3)e^{-2r/a},$$

where a is the Bohr radius.

EXPRESS We want to evaluate the integral $\int_0^\infty P(r)\,dr$. Equation 15 in the integral table of Appendix E is an integral of this form: $\int_0^\infty x^n e^{-ax}\,dx = \frac{n!}{a^{n+1}}$.

ANALYZE We set $n = 2$ and replace a in the given formula with $2/a$ and x with r. Then

$$\int_0^\infty P(r)dr = \frac{4}{a^3}\int_0^\infty r^2 e^{-2r/a}dr = \frac{4}{a^3}\frac{2}{(2/a)^3} = 1.$$

LEARN The integral over the radial probability function $P(r)$ must be equal to 1. This means that in a hydrogen atom, the electron must be somewhere in the space surrounding the nucleus.

39-45

THINK The probability density is given by $|\psi_{n\ell m_\ell}(r,\theta)|^2$, where $\psi_{n\ell m_\ell}(r,\theta)$ is the wave function.

EXPRESS To calculate $|\psi_{n\ell m_\ell}|^2 = \psi^*_{n\ell m_\ell}\psi_{n\ell m_\ell}$, we multiply the wave function by its complex conjugate. If the function is real, then $\psi^*_{n\ell m_\ell} = \psi_{n\ell m_\ell}$. Note that $e^{+i\phi}$ and $e^{-i\phi}$ are complex conjugate of each other, and $e^{i\phi}e^{-i\phi} = e^0 = 1$.

ANALYZE (a) ψ_{210} is real. Squaring it gives the probability density:

$$|\psi_{210}|^2 = \frac{r^2}{32\pi a^5} e^{-r/a} \cos^2\theta.$$

(b) Similarly,

$$|\psi_{21+1}|^2 = \frac{r^2}{64\pi a^5} e^{-r/a} \sin^2\theta$$

and

$$|\psi_{21-1}|^2 = \frac{r^2}{64\pi a^5} e^{-r/a} \sin^2\theta.$$

The last two functions lead to the same probability density.

(c) For $m_\ell = 0$, the probability density $|\psi_{210}|^2$ decreases with radial distance from the nucleus. With the $\cos^2\theta$ factor, $|\psi_{210}|^2$ is greatest along the z axis where $\theta = 0$. This is consistent with the dot plot of Fig. 39-23(a).

Similarly, for $m_\ell = \pm 1$, the probability density $|\psi_{21\pm 1}|^2$ decreases with radial distance from the nucleus. With the $\sin^2\theta$ factor, $|\psi_{21\pm 1}|^2$ is greatest in the xy-plane where $\theta =$ 90°. This is consistent with the dot plot of Fig. 39-23(b).

(d) The total probability density for the three states is the sum:

$$|\psi_{210}|^2 + |\psi_{21+1}|^2 + |\psi_{21-1}|^2 = \frac{r^2}{32\pi a^5} e^{-r/a}\left[\cos^2\theta + \frac{1}{2}\sin^2\theta + \frac{1}{2}\sin^2\theta\right] = \frac{r^2}{32\pi a^5} e^{-r/a}.$$

The trigonometric identity $\cos^2\theta + \sin^2\theta = 1$ is used. We note that the total probability density does not depend on θ or ϕ; it is spherically symmetric.

LEARN The wave functions discussed above are for the hydrogen states with $n = 2$ and $\ell = 1$. Since the angular momentum is nonzero, the probability densities are not spherically symmetric, but depend on both r and θ.

39-53

THINK The ground state of the hydrogen atom corresponds to $n = 1,\ \ell = 0,$ and $m_\ell = 0.$

EXPRESS The proposed wave function is

$$\psi = \frac{1}{\sqrt{\pi}a^{3/2}}e^{-r/a}$$

where a is the Bohr radius. Substituting this into the right side of Schrödinger's equation, our goal is to show that the result is zero.

ANALYZE The derivative is

$$\frac{d\psi}{dr} = -\frac{1}{\sqrt{\pi}a^{5/2}}e^{-r/a}$$

so

$$r^2\frac{d\psi}{dr} = -\frac{r^2}{\sqrt{\pi}a^{5/2}}e^{-r/a}$$

and

$$\frac{1}{r^2}\frac{d}{dr}\left(r^2\frac{d\psi}{dr}\right) = \frac{1}{\sqrt{\pi}a^{5/2}}\left[-\frac{2}{r}+\frac{1}{a}\right]e^{-r/a} = \frac{1}{a}\left[-\frac{2}{r}+\frac{1}{a}\right]\psi.$$

The energy of the ground state is given by $E = -me^4/8\varepsilon_0^2h^2$ and the Bohr radius is given by $a = h^2\varepsilon_0/\pi me^2$, so $E = -e^2/8\pi\varepsilon_0 a.$ The potential energy is given by

$$U = -e^2/4\pi\varepsilon_0 r,$$

so

$$\frac{8\pi^2 m}{h^2}[E-U]\psi = \frac{8\pi^2 m}{h^2}\left[-\frac{e^2}{8\pi\varepsilon_0 a}+\frac{e^2}{4\pi\varepsilon_0 r}\right]\psi = \frac{8\pi^2 m}{h^2}\frac{e^2}{8\pi\varepsilon_0}\left[-\frac{1}{a}+\frac{2}{r}\right]\psi$$

$$= \frac{\pi me^2}{h^2\varepsilon_0}\left[-\frac{1}{a}+\frac{2}{r}\right]\psi = \frac{1}{a}\left[-\frac{1}{a}+\frac{2}{r}\right]\psi.$$

The two terms in Schrödinger's equation cancel, and the proposed function ψ satisfies that equation.

LEARN The radial probability density of the ground state of hydrogen atom is given by Eq. 39-44:

$$P(r) = |\psi|^2(4\pi r^2) = \frac{1}{\pi a^3}e^{-2r/a}(4\pi r^2) = \frac{4}{a^3}r^2 e^{-2r/a}.$$

A plot of $P(r)$ is shown in Fig. 39-21.

39-59

THINK For a finite well, the electron matter wave can penetrate the walls of the well. Thus, the wave function outside the well is not zero, but decreases exponentially with distance.

EXPRESS Schrödinger's equation for the region $x > L$ is

$$\frac{d^2\psi}{dx^2}+\frac{8\pi^2 m}{h^2}[E-U_0]\psi=0,$$

where $E - U_0 < 0$. If $\psi^2(x) = Ce^{-2kx}$, then $\psi(x) = \sqrt{C}\,e^{-kx}$.

ANALYZE (a) and (b) Thus,

$$\frac{d^2\psi}{dx^2}=4k^2\sqrt{C}e^{-kx}=4k^2\psi$$

and

$$\frac{d^2\psi}{dx^2}+\frac{8\pi^2 m}{h^2}[E-U_0]\psi=k^2\psi+\frac{8\pi^2 m}{h^2}[E-U_0]\psi.$$

This is zero provided that $k^2=\frac{8\pi^2 m}{h^2}[U_0-E]$. Choosing the positive root, we have

$$k=\frac{2\pi}{h}\sqrt{2m(U_0-E)}.$$

LEARN Note that the quantity $U_0 - E$ is positive, so k is real and the proposed function satisfies Schrödinger's equation. If k is negative, however, the proposed function would be physically unrealistic. It would increase exponentially with x. Since the integral of the probability density over the entire x axis must be finite, ψ diverging as $x\to\infty$ would be unacceptable.

Chapter 40

40-9

THINK Knowing the value of ℓ, the orbital quantum number, allows us to determine the magnitudes of the angular momentum and the magnetic dipole moment.

EXPRESS The magnitude of the orbital angular momentum is

$$L = \sqrt{\ell(\ell+1)}\hbar.$$

Similarly, with $\vec{\mu}_{\text{orb}} = -\dfrac{e}{2m}\vec{L}$, the magnitude of $\vec{\mu}_{\text{orb}}$ is

$$\mu_{\text{orb}} = \frac{e\hbar}{2m}\sqrt{\ell(\ell+1)} = \mu_{\text{B}},$$

where $\mu_{\text{B}} = e\hbar/2m$ is the Bohr magneton.

ANALYZE (a) For $\ell = 3$, we have

$$L = \sqrt{\ell(\ell+1)}\hbar = \sqrt{3(3+1)}\hbar = \sqrt{12}\hbar.$$

So the multiple is $\sqrt{12} \approx 3.46$.

(b) The magnitude of the orbital dipole moment is

$$\mu_{\text{orb}} = \sqrt{\ell(\ell+1)}\mu_B = \sqrt{12}\mu_B.$$

So the multiple is $\sqrt{12} \approx 3.46$.

(c) The largest possible value of m_ℓ is $m_\ell = \ell = 3$.

(d) We use $L_z = m_\ell\hbar$ to calculate the z component of the orbital angular momentum. The multiple is $m_\ell = 3$.

(e) We use $\mu_z = -m_\ell\mu_B$ to calculate the z component of the orbital magnetic dipole moment. The multiple is $-m_\ell = -3$.

(f) We use $\cos\theta = m_\ell / \sqrt{\ell(\ell+1)}$ to calculate the angle between the orbital angular momentum vector and the *z* axis. For $\ell = 3$ and $m_\ell = 3$, we have $\cos\theta = 3/\sqrt{12} = \sqrt{3}/2$, or $\theta = 30.0°$.

(g) For $\ell = 3$ and $m_\ell = 2$, we have $\cos\theta = 2/\sqrt{12} = 1/\sqrt{3}$, or $\theta = 54.7°$.

(h) For $\ell = 3$ and $m_\ell = -3$, $\cos\theta = -3/\sqrt{12} = -\sqrt{3}/2$, or $\theta = 150°$.

LEARN Neither $\vec{L}$ nor $\vec{\mu}_{\text{orb}}$ can be measured in any way. We can, however, measure their *z* components.

40-11

THINK We can only measure one component of $\vec{L}$, say L_z, but not all three components.

EXPRESS Since $L^2 = L_x^2 + L_y^2 + L_z^2$, $\sqrt{L_x^2 + L_y^2} = \sqrt{L^2 - L_z^2}$. Replacing L^2 with $\ell(\ell+1)\hbar^2$ and L_z with $m_\ell\hbar$, we obtain

$$\sqrt{L_x^2 + L_y^2} = \hbar\sqrt{\ell(\ell+1) - m_\ell^2}\ .$$

ANALYZE For a given value of ℓ, the greatest that m_ℓ can be is ℓ, so the smallest that $\sqrt{L_x^2 + L_y^2}$ can be is $\hbar\sqrt{\ell(\ell+1) - \ell^2} = \hbar\sqrt{\ell}$. The smallest possible magnitude of m_ℓ is zero, so the largest $\sqrt{L_x^2 + L_y^2}$ can be is $\hbar\sqrt{\ell(\ell+1)}$. Thus,

$$\hbar\sqrt{\ell} \le \sqrt{L_x^2 + L_y^2} \le \hbar\sqrt{\ell(\ell+1)}\ .$$

LEARN Once we have chosen to measure $\vec{L}$ along the *z* axis, the *x*- and *y*-components cannot be measured with infinite certainty.

40-13

THINK A gradient magnetic field gives rise to a magnetic force on the silver atom.

EXPRESS The force on the silver atom is given by

$$F_z = -\frac{dU}{dz} = -\frac{d}{dz}(-\mu_z B) = \mu_z \frac{dB}{dz}$$

where μ_z is the z component of the magnetic dipole moment of the silver atom, and B is the magnetic field. The acceleration is

$$a = \frac{F_z}{M} = \frac{\mu_z (dB/dz)}{M},$$

where M is the mass of a silver atom.

ANALYZE Using the data given in Sample Problem —"Beam separation in a Stern-Gerlach experiment," we obtain

$$a = \frac{\left(9.27\times10^{-24}\text{ J/T}\right)\left(1.4\times10^{3}\text{ T/m}\right)}{1.8\times10^{-25}\text{ kg}} = 7.2\times10^{4}\text{ m/s}^{2}.$$

LEARN The deflection of the silver atom is due to the interaction between the magnetic dipole moment of the atom and the magnetic field. However, if the field is uniform, then $dB/dz = 0$, and the silver atom will pass the poles undeflected.

40-23

THINK With eight electrons, the ground-state energy of the system is the sum of the energies of the individual electrons in the system's ground-state configuration.

EXPRESS In terms of the quantum numbers n_x, n_y, and n_z, the single-particle energy levels are given by

$$E_{n_x,n_y,n_z} = \frac{h^2}{8mL^2}\left(n_x^2 + n_y^2 + n_z^2\right).$$

The lowest single-particle level corresponds to $n_x = 1$, $n_y = 1$, and $n_z = 1$ and is $E_{1,1,1} = 3(h^2/8mL^2)$. There are two electrons with this energy, one with spin up and one with spin down. The next lowest single-particle level is three-fold degenerate in the three integer quantum numbers. The energy is

$$E_{1,1,2} = E_{1,2,1} = E_{2,1,1} = 6(h^2/8mL^2).$$

Each of these states can be occupied by a spin up and a spin down electron, so six electrons in all can occupy the states. This completes the assignment of the eight electrons to single-particle states.

ANALYZE The ground-state energy of the system is

$$E_{gr} = (2)(3)(h^2/8mL^2) + (6)(6)(h^2/8mL^2) = 42(h^2/8mL^2).$$

Thus, the multiple of $h^2/8mL^2$ is 42.

LEARN We summarize the ground-state configuration and the energies (in multiples of $h^2/8mL^2$) in the chart below:

n_x	n_y	n_z	m_s	energy
1	1	1	−1/2, + 1/2	3 + 3
1	1	2	−1/2, + 1/2	6 + 6
1	2	1	−1/2, + 1/2	6 + 6
2	1	1	−1/2, + 1/2	6 + 6
			total	42

40-27

THINK The four quantum numbers (n, ℓ, m_ℓ, m_s) identify the quantum states of individual electrons in a multi-electron atom.

EXPRESS A lithium atom has three electrons. The first two electrons have quantum numbers (1, 0, 0, $\pm 1/2$). All states with principal quantum number $n = 1$ are filled. The next lowest states have $n = 2$.

The orbital quantum number can have the values $\ell = 0$ or 1 and of these, the $\ell = 0$ states have the lowest energy. The magnetic quantum number must be $m_\ell = 0$ since this is the only possibility if $\ell = 0$. The spin quantum number can have either of the values $m_s = -\frac{1}{2}$ or $+\frac{1}{2}$. Since there is no external magnetic field, the energies of these two states are the same.

ANALYZE (a) Therefore, in the ground state, the quantum numbers of the third electron are either $n = 2, \ell = 0, m_\ell = 0, m_s = -\frac{1}{2}$ or $n = 2, \ell = 0, m_\ell = 0, m_s = +\frac{1}{2}$. That is, $(n, \ell, m_\ell, m_s) =$ (2,0,0, +1/2) and (2,0,0, −1/2).

(b) The next lowest state in energy is an $n = 2$, $\ell = 1$ state. All $n = 3$ states are higher in energy. The magnetic quantum number can be $m_\ell = -1$, 0, or +1; the spin quantum number can be $m_s = -\frac{1}{2}$ or $+\frac{1}{2}$. Thus, (n, ℓ, m_ℓ, m_s) = (2,1,1, +1/2), (2,1,1, −1/2), (2,1,0,+1/2), (2,1,0,−1/2), (2,1,−1,+1/2) and (2,1,−1,−1/2).

LEARN No two electrons can have the same set of quantum numbers, as required by the Pauli exclusion principle.

40-35

THINK X-rays are produced when a solid target (silver in this case) is bombarded with electrons whose kinetic energies are in the keV range.

EXPRESS The wavelength is $\lambda_{\min} = hc/K_0$, where K_0 is the initial kinetic energy of the incident electron.

ANALYZE (a) With $hc = 1240 \text{ eV}\cdot\text{nm}$, we obtain

$$\lambda_{\min} = \frac{hc}{K_0} = \frac{1240\,\text{eV}\cdot\text{nm}}{35\times10^3\,\text{eV}} = 3.54\times10^{-2}\text{ nm} = 35.4\,\text{pm}.$$

(b) A K_α photon results when an electron in a target atom jumps from the *L*-shell to the *K*-shell. The energy of this photon is

$$E = 25.51 \text{ keV} - 3.56 \text{ keV} = 21.95 \text{ keV}$$

and its wavelength is

$$\lambda_{K\alpha} = hc/E = (1240 \text{ eV}\cdot\text{nm})/(21.95\times10^3 \text{ eV}) = 5.65\times10^{-2} \text{ nm} = 56.5 \text{ pm}.$$

(c) A K_β photon results when an electron in a target atom jumps from the *M*-shell to the *K*-shell. The energy of this photon is 25.51 keV – 0.53 keV = 24.98 keV and its wavelength is

$$\lambda_{K\beta} = (1240\,\text{eV}\cdot\text{nm})/(24.98\times10^3 \text{ eV}) = 4.96\times10^{-2} \text{ nm} = 49.6 \text{ pm}.$$

LEARN Note that the cut-off wavelength $\lambda_{\min}$ is characteristic of the incident electrons, not of the target material.

40-39

THINK The frequency of an x-ray emission is proportional to $(Z - 1)^2$, where *Z* is the atomic number of the target atom.

EXPRESS The ratio of the wavelength λ_{Nb} for the K_α line of niobium to the wavelength λ_{Ga} for the K_α line of gallium is given by

$$\lambda_{Nb}/\lambda_{Ga} = (Z_{Ga}-1)^2/(Z_{Nb}-1)^2,$$

where Z_{Nb} is the atomic number of niobium (41) and Z_{Ga} is the atomic number of gallium (31). Thus,

$$\lambda_{Nb}/\lambda_{Ga} = (30)^2/(40)^2 = 9/16 \approx 0.563.$$

LEARN The frequency of the K_α line is given by Eq. 40-26:

$$f = (2.46\times10^{15} \text{ Hz})(Z-1)^2.$$

40-51

THINK The number of atoms in a state with energy E is proportional to $e^{-E/kT}$, where T is the temperature on the Kelvin scale and k is the Boltzmann constant.

EXPRESS Thus, the ratio of the number of atoms in the thirteenth excited state to the number in the eleventh excited state is

$$\frac{n_{13}}{n_{11}} = \frac{e^{-E_{13}/kT}}{e^{-E_{11}/kT}} = e^{-(E_{13}-E_{11})/kT} = e^{-\Delta E/kT},$$

where $\Delta E = E_{13} - E_{11}$ is the difference in the energies:

$$\Delta E = E_{13} - E_{11} = 2(1.2\text{ eV}) = 2.4\text{ eV}.$$

ANALYZE For the given temperature, $kT = (8.62 \times 10^{-2}\text{ eV/K})(2000\text{ K}) = 0.1724\text{ eV}$. Hence,

$$\frac{n_{13}}{n_{11}} = e^{-2.4/0.1724} = 9.0 \times 10^{-7}.$$

LEARN The 13th excited state has higher energy than the 11th excited state. Therefore, we expect fewer atoms to be in the 13th excited state.

40-69

THINK The intensity at the target is given by $I = P/A$, where P is the power output of the source and A is the area of the beam at the target. We want to compute I and compare the result with 10^8 W/m^2.

EXPRESS The laser beam spreads because diffraction occurs at the aperture of the laser. Consider the part of the beam that is within the central diffraction maximum. The angular position of the edge is given by $\sin\theta = 1.22\lambda/d$, where λ is the wavelength and d is the diameter of the aperture. At the target, a distance D away, the radius of the beam is $r = D\tan\theta$. Since θ is small, we may approximate both $\sin\theta$ and $\tan\theta$ by θ, in radians. Then,

$$r = D\theta = 1.22D\lambda/d.$$

ANALYZE (a) Thus, we find the intensity to be

$$I = \frac{P}{\pi r^2} = \frac{Pd^2}{\pi(1.22D\lambda)^2} = \frac{(5.0\times10^6\text{ W})(4.0\text{m})^2}{\pi\left[1.22(3000\times10^3\text{ m})(3.0\times10^{-6}\text{ m})\right]^2} = 2.1\times10^5\text{ W/m}^2,$$

not great enough to destroy the missile.

(b) We solve for the wavelength in terms of the intensity and substitute $I = 1.0\times10^8$ W/m^2:

$$\lambda = \frac{d}{1.22D}\sqrt{\frac{P}{\pi I}} = \frac{4.0\,\text{m}}{1.22(3000\times10^3\,\text{m})}\sqrt{\frac{5.0\times10^6\,\text{W}}{\pi(1.0\times10^8\,\text{W/m}^2)}} = 1.40\times10^{-7}\,\text{m} = 140\,\text{nm}.$$

LEARN The wavelength corresponds to the x-rays on the electromagnetic spectrum.

40-73

THINK One femtosecond (fs) is equal to 10^{-15} s.

EXPRESS The length of the pulse's wave train is given by $L = c\Delta t$, where Δt is the duration of the laser. Thus, the number of wavelengths contained in the pulse is

$$N = \frac{L}{\lambda} = \frac{c\Delta t}{\lambda}.$$

ANALYZE (a) With $\lambda = 500$ nm and $\Delta t = 10\times10^{-15}$ s, we have

$$N = \frac{L}{\lambda} = \frac{(3.0\times10^8\text{ m/s})(10\times10^{-15}\text{ s})}{500\times10^{-9}\text{ m}} = 6.0.$$

(b) We solve for X from 10 fm/1 m = 1 s/X:

$$X = \frac{(1\text{ s})(1\text{ m})}{10\times10^{-15}\text{ m}} = \frac{1\text{ s}}{(10\times10^{-15})(3.15\times10^7\text{ s/y})} = 3.2\times10^6\text{ y}.$$

LEARN Femtosecond lasers have important applications in areas such as micro-machining and optical data storage.

Chapter 41

41-7

THINK This problem deals with occupancy probability *P*(*E*), the probability that an energy level will be occupied by an electron.

EXPRESS A plot of *P*(*E*) as a function of *E* is shown in Fig. 41-7. From the figure, we see that at $T = 0$ K, *P*(*E*) is unity for $E \le E_F$, where E_F is the Fermi energy, and zero for $E > E_F$. On the other hand, the probability that a state with energy *E* is occupied at temperature *T* is given by

$$P(E) = \frac{1}{e^{(E-E_F)/kT}+1}$$

where *k* is the Boltzmann constant and E_F is the Fermi energy.

ANALYZE (a) At absolute temperature $T = 0$, the probability is zero that any state with energy above the Fermi energy is occupied.

(b) Now, $E - E_F = 0.0620$ eV, and

$$(E-E_F)/kT = (0.0620\text{eV})/(8.62\times10^{-5}\text{ eV/K})(320\text{ K}) = 2.248\,.$$

We find *P*(*E*) to be

$$P(E) = \frac{1}{e^{2.248}+1} = 0.0955\,.$$

See Appendix B for the value of *k*.

LEARN When $E = E_F$, the occupancy probability is $P(E_F) = 0.5$. Thus, one may think of the Fermi energy as the energy of a quantum state that has a probability 0.5 of being occupied by an electron.

41-9

THINK According to Appendix F the molar mass of silver is $M = 107.870$ g/mol and the density is $\rho = 10.49$ g/cm^3. Silver is monovalent.

EXPRESS The mass of a silver atom is, dividing the molar mass by Avogadro's number:

$$M_0 = \frac{M}{N_A} = \frac{107.870\times10^{-3}\ \text{kg/mol}}{6.022\times10^{23}\ \text{mol}^{-1}} = 1.791\times10^{-25}\ \text{kg}\ .$$

Since silver is monovalent, there is one valence electron per atom (see Eq. 41-2).

ANALYZE (a) The number density is

$$n = \frac{\rho}{M_0} = \frac{10.49\times10^{-3}\ \text{kg/m}^3}{1.791\times10^{-25}\ \text{kg}} = 5.86\times10^{28}\ \text{m}^{-3}\ .$$

This is the same as the number density of conduction electrons.

(b) The Fermi energy is

$$E_F = \frac{0.121h^2}{m}n^{2/3} = \frac{(0.121)(6.626\times10^{-34}\ \text{J}\cdot\text{s})^2}{9.109\times10^{-31}\ \text{kg}} = (5.86\times10^{28}\ \text{m}^{-3})^{2/3}$$
$$= 8.80\times10^{-19}\ \text{J} = 5.49\text{eV}.$$

(c) Since $E_F = \frac{1}{2}mv_F^2$,

$$v_F = \sqrt{\frac{2E_F}{m}} = \sqrt{\frac{2(8.80\times10^{-19}\ \text{J})}{9.109\times10^{-31}\ \text{kg}}} = 1.39\times10^6\ \text{m/s}\ .$$

(d) The de Broglie wavelength is

$$\lambda = \frac{h}{mv_F} = \frac{6.626\times10^{-34}\ \text{J}\cdot\text{s}}{(9.109\times10^{-31}\ \text{kg})(1.39\times10^6\text{m/s})} = 5.22\times10^{-10}\ \text{m}.$$

LEARN Once the number density of conduction electrons is known, the Fermi energy for a particular metal can be calculated using Eq. 41-9.

41-15

THINK The Fermi-Dirac occupation probability is given by $P_{\text{FD}} = 1/\left(e^{\Delta E/kT}+1\right)$, and the Boltzmann occupation probability is given by $P_B = e^{-\Delta E/kT}$.

EXPRESS Let *f* be the fractional difference between the two probabilities. Then

$$f = \frac{P_B - P_{FD}}{P_B} = \frac{e^{-\Delta E/kT} - \frac{1}{e^{\Delta E/kT}+1}}{e^{-\Delta E/kT}}.$$

Using a common denominator and a little algebra yields

$$f = \frac{e^{-\Delta E/kT}}{e^{-\Delta E/kT}+1}.$$

The solution for $e^{-\Delta E/kT}$ is

$$e^{-\Delta E/kT} = \frac{f}{1-f}.$$

We take the natural logarithm of both sides and solve for T. The result is

$$T = \frac{\Delta E}{k \ln\left(\frac{f}{1-f}\right)}.$$

ANALYZE (a) Letting f equal 0.01, we evaluate the expression for T:

$$T = \frac{(1.00\,\text{eV})(1.60\times10^{-19}\,\text{J/eV})}{(1.38\times10^{-23}\,\text{J/K})\ln\left(\frac{0.010}{1-0.010}\right)} = 2.50\times10^{3}\,\text{K}.$$

(b) We set f equal to 0.10 and evaluate the expression for T:

$$T = \frac{(1.00\,\text{eV})(1.60\times10^{-19}\,\text{J/eV})}{(1.38\times10^{-23}\,\text{J/K})\ln\left(\frac{0.10}{1-0.10}\right)} = 5.30\times10^{3}\,\text{K}.$$

LEARN The fractional difference as a function of T is plotted below:

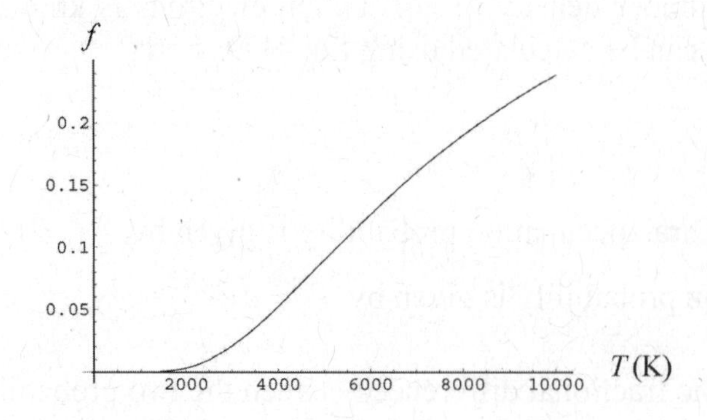

With a given ΔE, the difference increases with T.

41-31

THINK The valence band and the conduction band are separated by an energy gap.

EXPRESS Since the electron jumps from the conduction band to the valence band, the energy of the photon equals the energy gap between those two bands. The photon energy is given by $hf = hc/\lambda$, where f is the frequency of the electromagnetic wave and λ is its wavelength.

ANALYZE (a) Thus, $E_g = hc/\lambda$ and

$$\lambda = \frac{hc}{E_g} = \frac{(6.63\times10^{-34}\,\text{J}\cdot\text{s})(2.998\times10^{8}\,\text{m/s})}{(5.5\,\text{eV})(1.60\times10^{-19}\,\text{J/eV})} = 2.26\times10^{-7}\,\text{m} = 226\,\text{nm}\,.$$

(b) These photons are in the ultraviolet portion of the electromagnetic spectrum.

LEARN Note that photons from other transitions have a greater energy, so their waves have shorter wavelengths.

41-35

THINK Doping silicon with phosphorus increases the number of electrons in the conduction band.

EXPRESS Sample Problem — "Doping silicon with phosphorus" gives the fraction of silicon atoms that must be replaced by phosphorus atoms. We find the number the silicon atoms in 1.0 g, then the number that must be replaced, and finally the mass of the replacement phosphorus atoms. The molar mass of silicon is $M_{\text{Si}} = 28.086$ g/mol, so the mass of one silicon atom is

$$m_{0,\text{Si}} = M_{\text{Si}}/N_A = (28.086\text{ g/mol})/(6.022\times10^{23}\text{ mol}^{-1}) = 4.66\times10^{-23}\text{ g}$$

and the number of atoms in 1.0 g is

$$N_{\text{Si}} = m_{\text{Si}}/m_{0,\text{Si}} = (1.0\text{ g})/(4.66\times10^{-23}\text{ g}) = 2.14\times10^{22}.$$

According to the Sample Problem, one of every 5×10^{6} silicon atoms is replaced with a phosphorus atom. This means there will be

$$N_{\text{P}} = (2.14\times10^{22})/(5\times10^{6}) = 4.29\times10^{15}$$

phosphorus atoms in 1.0 g of silicon.

ANALYZE The molar mass of phosphorus is $M_{\text{P}} = 30.9758$ g/mol so the mass of a phosphorus atom is

$$m_{0,\mathrm{P}} = M_{\mathrm{P}} / N_A = (30.9758 \text{ g/mol})/(6.022 \times 10^{-23} \text{ mol}^{-1}) = 5.14 \times 10^{-23} \text{ g}.$$

The mass of phosphorus that must be added to 1.0 g of silicon is

$$m_{\mathrm{P}} = N_{\mathrm{P}} m_{0,\mathrm{P}} = (4.29 \times 10^{15})(5.14 \times 10^{-23} \text{ g}) = 2.2 \times 10^{-7} \text{ g}.$$

LEARN The phosphorus atom is a *donor* atom since it donates an electron to the conduction band. Semiconductors doped with donor atoms are called *n*-type semiconductors.

41-39

THINK The valence band and the conduction band are separated by an energy gap E_g. An electron must acquire E_g in order to make the transition to the conduction band.

EXPRESS Since the energy received by each electron is exactly E_g, the difference in energy between the bottom of the conduction band and the top of the valence band, the number of electrons that can be excited across the gap by a single photon of energy E is

$$N = E / E_g.$$

ANALYZE With $E_g = 1.1$ eV and $E = 662$ keV, we obtain

$$N = (662 \times 10^3 \text{ eV})/(1.1 \text{ eV}) = 6.0 \times 10^5.$$

Since each electron that jumps the gap leaves a hole behind, this is also the number of electron-hole pairs that can be created.

LEARN The wavelength of the photon is

$$\lambda = \frac{hc}{E} = \frac{1240 \text{ nm}\cdot\text{eV}}{662\times 10^3 \text{ eV}} = 1.87\times 10^{-3} \text{ nm} = 1.87 \text{ pm}.$$

41-45

THINK We differentiate the occupancy probability $P(E)$ with respect to E to explore the properties of $P(E)$.

EXPRESS The probability that a state with energy E is occupied at temperature T is given by

$$P(E) = \frac{1}{e^{(E-E_F)/kT} + 1}$$

where k is the Boltzmann constant and E_F is the Fermi energy.

ANALYZE (a) The derivative of $P(E)$ is

$$\frac{dP}{dE}=\frac{-1}{[e^{(E-E_F)/kT}+1]^2}\frac{d}{dE}e^{(E-E_F)/kT}=\frac{-1}{[e^{(E-E_F)/kT}+1]^2}\frac{1}{kT}e^{(E-E_F)/kT}.$$

For $E = E_F$, we readily obtain the desired result:

$$\left.\frac{dP}{dE}\right|_{E=E_F}=\frac{-1}{[e^{(E_E-E_F)/kT}+1]^2}\frac{1}{kT}e^{(E_F-E_F)/kT}=-\frac{1}{4kT}.$$

(b) The equation of a line may be written as $y = m(x - x_0)$ where $m = -1/4kT$ is the slope, and x_0 is the x-intercept (which is what we are asked to solve for). It is clear that $P(E_F) = 1/2$, so our equation of the line, evaluated at $x = E_F$, becomes

$$1/2 = (-1/4kT)(E_F - x_0),$$

which leads to $x_0 = E_F + 2kT$.

LEARN The straight line can be rewritten as

$$y=\frac{1}{2}-\frac{1}{4kT}(E-E_F).$$

A plot of $P(E)$ (solid line) and $y(E)$ (dashed line) in units of E_F/kT. The straight line passes the horizontal axis at $E/E_F = 3$.

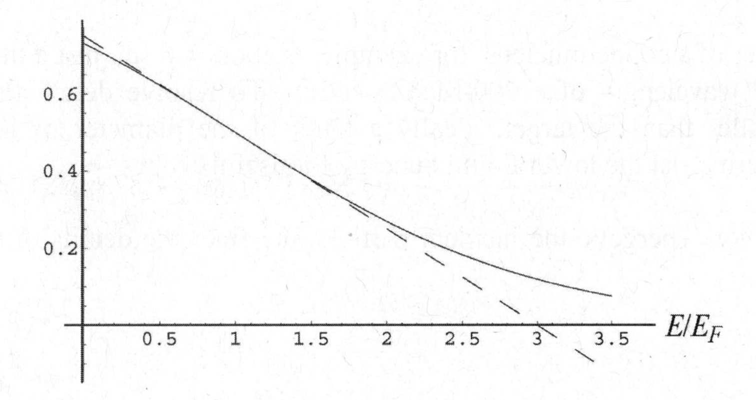

Chapter 42

42-11

THINK To resolve the detail of a nucleus, the de Broglie wavelength of the probe must be smaller than the size of the nucleus.

EXPRESS The de Broglie wavelength is given by $\lambda = h/p$, where p is the magnitude of the momentum. Since the kinetic energy K of the electron is much greater than its rest energy, relativistic formulation must be used. The kinetic energy and the momentum are related by Eq. 37-54:

$$pc = \sqrt{K^2 + 2Kmc^2}.$$

ANALYZE (a) With $K = 200$ MeV and $mc^2 = 0.511$ MeV, we obtain

$$pc = \sqrt{K^2 + 2Kmc^2} = \sqrt{(200\,\text{MeV})^2 + 2(200\,\text{MeV})(0.511\,\text{MeV})} = 200.5\,\text{MeV}.$$

Thus,

$$\lambda = \frac{hc}{pc} = \frac{1240\ \text{eV}\cdot\text{nm}}{200.5\times10^6\ \text{eV}} = 6.18\times10^{-6}\,\text{nm} \approx 6.2\ \text{fm}.$$

(b) The diameter of a copper nucleus, for example, is about 8.6 fm, just a little larger than the de Broglie wavelength of a 200-MeV electron. To resolve detail, the wavelength should be smaller than the target, ideally a tenth of the diameter or less. 200-MeV electrons are perhaps at the lower limit in energy for useful probes.

LEARN The more energetic the incident particle, the finer the details of the target that can be probed.

42-21

THINK Binding energy is the difference in mass energy between a nucleus and its individual nucleons.

EXPRESS If a nucleus contains Z protons and N neutrons, its binding energy is given by Eq. 42-7:

$$\Delta E_{\text{be}} = \sum(mc^2) - Mc^2 = \left(Zm_H + Nm_n - M\right)c^2,$$

where m_H is the mass of a hydrogen atom, m_n is the mass of a neutron, and M is the mass of the atom containing the nucleus of interest.

ANALYZE (a) If the masses are given in atomic mass units, then mass excesses are defined by $\Delta_H = (m_H - 1)c^2, \Delta_n = (m_n - 1)c^2$, and $\Delta = (M - A)c^2$. This means $m_H c^2 = \Delta_H + c^2$, $m_n c^2 = \Delta_n + c^2$, and $Mc^2 = \Delta + Ac^2$. Thus,

$$\Delta E_{\text{be}} = (Z\Delta_H + N\Delta_n - \Delta) + (Z + N - A)c^2 = Z\Delta_H + N\Delta_n - \Delta,$$

where $A = Z + N$ is used.

(b) For ${}^{197}_{79}\text{Au}$, $Z = 79$ and $N = 197 - 79 = 118$. Hence,

$$\Delta E_{\text{be}} = (79)(7.29\,\text{MeV}) + (118)(8.07\,\text{MeV}) - (-31.2\,\text{MeV}) = 1560\,\text{MeV}.$$

This means the binding energy per nucleon is $\Delta E_{\text{ben}} = (1560\,\text{MeV})/197 = 7.92\,\text{MeV}$.

LEARN Using mass excesses (Δ_H, Δ_n, and Δ) instead of actual masses provides another convenient way of calculating the binding energy of a nucleus.

42-23

THINK The binding energy is given by

$$\Delta E_{\text{be}} = \left[Zm_H + (A - Z)m_n - M_{\text{Pu}}\right]c^2,$$

where Z is the atomic number (number of protons), A is the mass number (number of nucleons), m_H is the mass of a hydrogen atom, m_n is the mass of a neutron, and M_{Pu} is the mass of a ${}^{239}_{94}\text{Pu}$ atom.

EXPRESS In principle, nuclear masses should be used, but the mass of the Z electrons included in Zm_H is canceled by the mass of the Z electrons included in M_{Pu}, so the result is the same. First, we calculate the mass difference in atomic mass units:

$$\Delta m = (94)(1.00783\text{ u}) + (239 - 94)(1.00867\text{ u}) - (239.05216\text{ u}) = 1.94101\text{ u}.$$

Since the mass energy of 1 u is equivalent to 931.5 MeV,

$$\Delta E_{\text{be}} = (1.94101\text{ u})(931.5\text{ MeV/u}) = 1808\text{ MeV}.$$

ANALYZE With 239 nucleons, the binding energy per nucleon is

$$\Delta E_{\text{ben}} = E/A = (1808\text{ MeV})/239 = 7.56\text{ MeV}.$$

The result is the same as that given in Table 42-1.

LEARN An alternative way to calculate binding energy is to use mass excesses, as discussed in Problem 21. The formula is

$$\Delta E_{\text{be}} = Z\Delta_H + N\Delta_n - \Delta_{239},$$

where $\Delta_H = (m_H - 1)c^2$, $\Delta_n = (m_n - 1)c^2$, and $\Delta_{239} = (M_{\text{Pu}} - 239\text{ u})c^2$.

42-29

THINK Half-life is the time is takes for the number of radioactive nuclei to decrease to half of its initial value.

EXPRESS The half-life $T_{1/2}$ and the disintegration constant λ are related by

$$T_{1/2} = (\ln 2)/\lambda.$$

ANALYZE (a) With $\lambda = 0.0108\text{ h}^{-1}$, we obtain

$$T_{1/2} = (\ln 2)/(0.0108\text{ h}^{-1}) = 64.2\text{ h}.$$

(b) At time t, the number of undecayed nuclei remaining is given by

$$N = N_0\, e^{-\lambda t} = N_0\, e^{-(\ln 2)t/T_{1/2}}.$$

We substitute $t = 3T_{1/2}$ to obtain

$$\frac{N}{N_0} = e^{-3\ln 2} = 0.125.$$

In each half-life, the number of undecayed nuclei is reduced by half. At the end of one half-life, $N = N_0/2$, at the end of two half-lives, $N = N_0/4$, and at the end of three half-lives, $N = N_0/8 = 0.125N_0$.

(c) We use

$$N = N_0\, e^{-\lambda t}.$$

Since 10.0 d is 240 h, $\lambda t = (0.0108\text{ h}^{-1})(240\text{ h}) = 2.592$ and

$$\frac{N}{N_0} = e^{-2.592} = 0.0749.$$

LEARN The fraction of the Hg sample remaining as a function of time (measured in days) is plotted below.

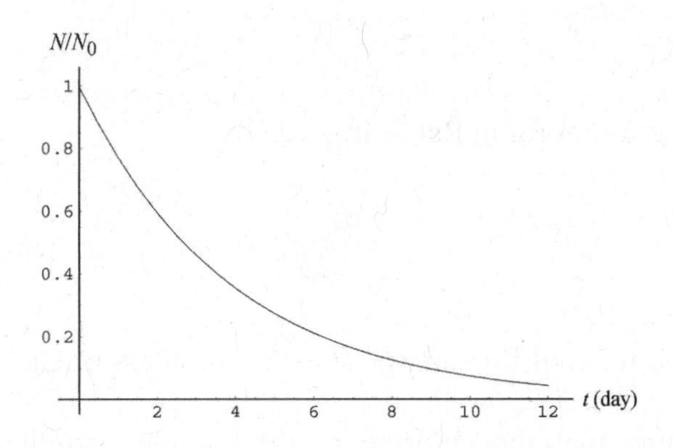

42-35

THINK We modify Eq. 42-11 to take into consideration the rate of production of the radionuclide.

EXPRESS If N is the number of undecayed nuclei present at time t, then

$$\frac{dN}{dt} = R - \lambda N$$

where R is the rate of production by the cyclotron and λ is the disintegration constant. The second term gives the rate of decay. Note the sign difference between R and λN.

ANALYZE (a) Rearrange the equation slightly and integrate:

$$\int_{N_0}^{N} \frac{dN}{R - \lambda N} = \int_0^t dt$$

where N_0 is the number of undecayed nuclei present at time $t = 0$. This yields

$$-\frac{1}{\lambda} \ln \frac{R - \lambda N}{R - \lambda N_0} = t.$$

We solve for N:

$$N = \frac{R}{\lambda} + \left(N_0 - \frac{R}{\lambda} \right) e^{-\lambda t}.$$

After many half-lives, the exponential is small and the second term can be neglected. Then, $N = R/\lambda$.

(b) The result $N = R/\lambda$ holds regardless of the initial value N_0, because the dependence on N_0 shows up only in the second term, which is exponentially suppressed at large t.

LEARN At times that are long compared to the half-life, the rate of production equals the rate of decay and N is a constant. The nuclide is in secular equilibrium with its source.

42-47

THINK The mass fraction of Ra in $RaCl_2$ is given by

$$\frac{M_{Ra}}{M_{Ra}+2M_{Cl}}$$

where M_{Ra} is the molar mass of Ra and M_{Cl} is the molar mass of Cl.

EXPRESS We assume that the chlorine in the sample had the naturally occurring isotopic mixture, so the average molar mass is 35.453 g/mol, as given in Appendix F. Then, the mass of ^{226}Ra was

$$m=\frac{226}{226+2(35.453)}(0.10\,\text{g})=76.1\times10^{-3}\,\text{g}.$$

ANALYZE (a) The mass of a ^{226}Ra nucleus is (226 u)(1.661×10^{-24} g/u) = 3.75×10^{-22} g, so the number of ^{226}Ra nuclei present was

$$N=(76.1\times10^{-3}\,\text{g})/(3.75\times10^{-22}\,\text{g})=2.03\times10^{20}.$$

(b) The decay rate is given by

$$R=N\lambda=(N\ln 2)/T_{1/2},$$

where λ is the disintegration constant, $T_{1/2}$ is the half-life, and N is the number of nuclei. The relationship $\lambda=(\ln 2)/T_{1/2}$ is used. Thus,

$$R=\frac{(2.03\times10^{20})\ln 2}{(1600\,\text{y})(3.156\times10^{7}\,\text{s/y})}=2.79\times10^{9}\,\text{s}^{-1}.$$

LEARN Radium has 33 different known isotopes, four of which naturally occurring. ^{226}Ra, with a half-life of 1600 years, is the most stable isotope of radium.

42-49

THINK The time for half the original ^{238}U nuclei to decay is equal to 4.5×10^{9} y, which is the half-life of ^{238}U.

EXPRESS The fraction of undecayed nuclei remaining after time t is given by

$$\frac{N}{N_0}=e^{-\lambda t}=e^{-(\ln 2)t/T_{1/2}}$$

where λ is the disintegration constant and $T_{1/2} = (\ln 2)/\lambda$ is the half-life.

(a) For ^{244}Pu at $t = 4.5 \times 10^9$ y,

$$\lambda t = \frac{(\ln 2)t}{T_{1/2}} = \frac{(\ln 2)(4.5\times10^9\,\text{y})}{8.0\times10^7\,\text{y}} = 39$$

and the fraction remaining is

$$\frac{N}{N_0} = e^{-39.0} \approx 1.2\times10^{-17}.$$

(b) For ^{248}Cm at $t = 4.5 \times 10^9$ y,

$$\frac{(\ln 2)t}{T_{1/2}} = \frac{(\ln 2)(4.5\times10^9\,\text{y})}{3.4\times10^5\,\text{y}} = 9170$$

and the fraction remaining is

$$\frac{N}{N_0} = e^{-9170} = 3.31\times10^{-3983}.$$

For any reasonably sized sample this is less than one nucleus and may be taken to be zero. A standard calculator probably cannot evaluate e^{-9170} directly. Our recommendation is to treat it as $(e^{-91.70})^{100}$.

LEARN Since $(T_{1/2})_{^{248}\text{Cm}} < (T_{1/2})_{^{244}\text{Pu}} < (T_{1/2})_{^{238}\text{U}}$, with $N/N_0 = e^{-(\ln 2)t/T_{1/2}}$, we have

$$(N/N_0)_{^{248}\text{Cm}} < (N/N_0)_{^{244}\text{Pu}} < (N/N_0)_{^{238}\text{U}}.$$

42-53

THINK The energy released in the decay is the disintegration energy:

$$Q = M_i c^2 - M_f c^2 = (M_i - M_f)c^2 = -\Delta M\, c^2,$$

where $\Delta M = M_f - M_i$ is the change in mass due to the decay.

EXPRESS Let M_{Cs} be the mass of one atom of $^{137}_{55}$Cs and M_{Ba} be the mass of one atom of $^{137}_{56}$Ba. The energy released is $Q = (M_{\text{Cs}} - M_{\text{Ba}})c^2$.

ANALYZE With M_{Cs} = 136.9071 u and M_{Ba} = 136.9058 u, we obtain

$$Q = [136.9071\,\text{u} - 136.9058\,\text{u}]c^2 = (0.0013\,\text{u})c^2 = (0.0013\,\text{u})(931.5\,\text{MeV/u}) = 1.21\,\text{MeV}.$$

LEARN In calculating Q above, we have used the atomic masses instead of nuclear masses. One can readily show that both lead to the same results. To obtain the nuclear masses, we subtract the mass of 55 electrons from M_{Cs} and the mass of 56 electrons from M_{Ba}. The energy released is

$$Q = [(M_{Cs} - 55m) - (M_{Ba} - 56m) - m]\, c^2,$$

where m is the mass of an electron (the last term in the bracket comes from the beta decay). Once cancellations have been made, $Q = (M_{Cs} - M_{Ba})c^2$, which is the same as before.

42-59

THINK The beta decay of ^{32}P is given by $^{32}\text{P} \to {}^{32}\text{S} + e^- + \bar{\nu}$. However, since the electron has the maximum possible kinetic energy, no (anti)neutrino is emitted.

EXPRESS Since momentum is conserved, the momentum of the electron and the momentum of the residual sulfur nucleus are equal in magnitude and opposite in direction. If p_e is the momentum of the electron and p_S is the momentum of the sulfur nucleus, then $p_S = -p_e$. The kinetic energy K_S of the sulfur nucleus is

$$K_S = p_S^2 / 2M_S = p_e^2 / 2M_S,$$

where M_S is the mass of the sulfur nucleus. Now, the electron's kinetic energy K_e is related to its momentum by the relativistic equation $(p_e c)^2 = K_e^2 + 2K_e mc^2$, where m is the mass of an electron.

ANALYZE With $K_e = 1.71$ MeV, the kinetic energy of the recoiling sulfur nucleus is

$$K_S = \frac{(p_e c)^2}{2M_S c^2} = \frac{K_e^2 + 2K_e mc^2}{2M_S c^2} = \frac{(1.71\,\text{MeV})^2 + 2(1.71\,\text{MeV})(0.511\,\text{MeV})}{2(32\,\text{u})(931.5\,\text{MeV/u})}$$
$$= 7.83 \times 10^{-5}\,\text{MeV} = 78.3\,\text{eV}$$

where $mc^2 = 0.511$ MeV is used for the electron (see Table 37-3).

LEARN The maximum kinetic energy of the electron is equal to the disintegration energy Q: $Q = K_{\text{max}}$. To show this, we use the following data: $M_P = 31.97391$ u and $M_S = 31.97207$ u. The result is

$$Q = [31.97391\,\text{u} - 31.97207\,\text{u}]c^2 = (0.00184\,\text{u})c^2 = (0.00184\,\text{u})(931.5\,\text{MeV/u})$$
$$= 1.71\,\text{MeV}.$$

42-65

THINK The activity of a radioactive sample expressed in curie (Ci) can be converted to SI units (Bq) as

$$1 \text{ curie} = 1 \text{ Ci} = 3.7 \times 10^{10} \text{ Bq} = 3.7 \times 10^{10} \text{ disintegrations/s}.$$

EXPRESS The decay rate R is related to the number of nuclei N by $R = \lambda N$, where λ is the disintegration constant. The disintegration constant is related to the half-life $T_{1/2}$ by

$$\lambda = \frac{\ln 2}{T_{1/2}} \Rightarrow N = \frac{R}{\lambda} = \frac{RT_{1/2}}{\ln 2}.$$

Since 1 Ci = 3.7×10^{10} disintegrations/s,

$$N = \frac{(250\,\text{Ci})(3.7\times10^{10}\,\text{s}^{-1}/\text{Ci})(2.7\,\text{d})(8.64\times10^{4}\,\text{s/d})}{\ln 2} = 3.11\times10^{18}.$$

ANALYZE The mass of a ^{198}Au atom is

$$M_0 = (198\text{ u})(1.661 \times 10^{-24}\text{ g/u}) = 3.29 \times 10^{-22}\text{ g},$$

so the mass required is

$$M = N M_0 = (3.11 \times 10^{18})(3.29 \times 10^{-22}\text{ g}) = 1.02 \times 10^{-3}\text{ g} = 1.02\text{ mg}.$$

LEARN The ^{198}Au atom undergoes beta decay and emit an electron:

$$^{198}\text{Au} \rightarrow {}^{198}\text{Hg} + e^{-} + \bar{\nu}.$$

42-73

THINK A generalized formation reaction can be written as $X + x \rightarrow Y$, where X is the target nucleus, x is the incident light particle, and Y is the excited compound nucleus (^{20}Ne).

EXPRESS We assume that X is initially at rest. Then, conservation of energy yields

$$m_X c^2 + m_x c^2 + K_x = m_Y c^2 + K_Y + E_Y$$

where m_X, m_x, and m_Y are masses, K_x and K_Y are kinetic energies, and E_Y is the excitation energy of Y. Conservation of momentum yields

$$p_x = p_Y.$$

Now,

$$K_Y = \frac{p_Y^2}{2m_Y} = \frac{p_x^2}{2m_Y} = \left(\frac{m_x}{m_Y}\right)K_x$$

so

$$m_X c^2 + m_x c^2 + K_x = m_Y c^2 + (m_x / m_Y)K_x + E_Y$$

and

$$K_x = \frac{m_Y}{m_Y - m_x}\left[(m_Y - m_X - m_x)c^2 + E_Y\right].$$

ANALYZE (a) Let x represent the alpha particle and X represent the ^{16}O nucleus. Then,

$$\begin{aligned}(m_Y - m_X - m_x)c^2 &= (19.99244\text{ u} - 15.99491\text{ u} - 4.00260\text{ u})(931.5\text{ MeV/u})\\ &= -4.722\text{ MeV}\end{aligned}$$

and

$$K_\alpha = \frac{19.99244\text{ u}}{19.99244\text{ u} - 4.00260\text{ u}}(-4.722\text{ MeV} + 25.0\text{ MeV}) = 25.35\text{ MeV} \approx 25.4\text{ MeV}.$$

(b) Let x represent the proton and X represent the ^{19}F nucleus. Then,

$$\begin{aligned}(m_Y - m_X - m_x)c^2 &= (19.99244\text{ u} - 18.99841\text{ u} - 1.00783\text{ u})(931.5\text{ MeV/u})\\ &= -12.85\text{ MeV}\end{aligned}$$

and

$$K_\alpha = \frac{19.99244\text{ u}}{19.99244\text{ u} - 1.00783\text{ u}}(-12.85\text{ MeV} + 25.0\text{ MeV}) = 12.80\text{ MeV}.$$

(c) Let x represent the photon and X represent the ^{20}Ne nucleus. Since the mass of the photon is zero, we must rewrite the conservation of energy equation: if E_γ is the energy of the photon, then

$$E_\gamma + m_X c^2 = m_Y c^2 + K_Y + E_Y.$$

Since $m_X = m_Y$, this equation becomes $E_\gamma = K_Y + E_Y$. Since the momentum and energy of a photon are related by $p_\gamma = E_\gamma / c$, the conservation of momentum equation becomes $E_\gamma / c = p_Y$. The kinetic energy of the compound nucleus is

$$K_Y = \frac{p_Y^2}{2m_Y} = \frac{E_\gamma^2}{2m_Y c^2}.$$

We substitute this result into the conservation of energy equation to obtain

$$E_\gamma = \frac{E_\gamma^2}{2m_Y c^2} + E_Y.$$

This quadratic equation has the solutions

$$E_\gamma = m_Y c^2 \pm \sqrt{\left(m_Y c^2\right)^2 - 2m_Y c^2 E_Y}.$$

If the problem is solved using the relativistic relationship between the energy and momentum of the compound nucleus, only one solution would be obtained, the one corresponding to the negative sign above. Since

$$m_Y c^2 = (19.99244 \text{ u})(931.5 \text{ MeV/u}) = 1.862 \times 10^4 \text{ MeV},$$

we have

$$\begin{aligned} E_\gamma &= \left(1.862\times10^4 \text{ MeV}\right) - \sqrt{\left(1.862\times10^4 \text{ MeV}\right)^2 - 2\left(1.862\times10^4 \text{ MeV}\right)(25.0 \text{ MeV})} \\ &= 25.0 \text{ MeV}. \end{aligned}$$

LEARN In part (c), the kinetic energy of the compound nucleus is

$$K_Y = \frac{E_\gamma^2}{2m_Y c^2} = \frac{(25.0 \text{ MeV})^2}{2(1.862\times10^4 \text{ MeV})} = 0.0168 \text{ MeV}$$

which is very small compared to E_Y = 25.0 MeV. Essentially all of the photon energy goes to excite the nucleus.

42-75

THINK We represent the unknown nuclide as ${}^A_Z X$, where A and Z are its mass number and atomic number, respectively.

EXPRESS The reaction equation can be written as

$${}^A_Z X + {}^1_0\text{n} \rightarrow {}^0_{-1}\text{e} + 2\,{}^4_2\text{He}.$$

Conservation of charge yields $Z + 0 = -1 + 4$ or $Z = 3$. Conservation of mass number yields $A + 1 = 0 + 8$ or $A = 7$.

ANALYZE According to the periodic table in Appendix G (also see Appendix F), lithium has atomic number 3, so the nuclide must be ${}^7_3\text{Li}$.

LEARN Charge and mass number are conserved in the neutron-capture process. The intermediate nuclide is ^{8}Li, which is unstable and decays (via α and β^- modes) into two ^{4}He's and an electron.

42-77

THINK The decay rate R is proportional to N, the number of radioactive nuclei.

EXPRESS According to Eq. 42-17, $R = \lambda N$, where λ is the decay constant. Since R is proportional to N, then $N/N_0 = R/R_0 = e^{-\lambda t}$. Since $\lambda = (\ln 2)/T_{1/2}$, the solution for t is

$$t = -\frac{1}{\lambda}\ln\left(\frac{R}{R_0}\right) = -\frac{T_{1/2}}{\ln 2}\ln\left(\frac{R}{R_0}\right).$$

ANALYZE With $T_{1/2} = 5730$ y and $R/R_0 = 0.020$, we obtain

$$t = -\frac{T_{1/2}}{\ln 2}\ln\left(\frac{R}{R_0}\right) = -\frac{5730\,\text{y}}{\ln 2}\ln(0.020) = 3.2\times 10^4\ \text{y}.$$

LEARN Radiocarbon dating based on the decay of ^{14}C is one of the most widely used dating method in estimating the age of organic remains.

42-79

THINK The count rate in the area in question is given by $R = \lambda N$, where λ is the decay constant and N is the number of radioactive nuclei.

EXPRESS Since the spreading is assumed uniform, the count rate $R = 74{,}000$/s is given by

$$R = \lambda N = \lambda(M/m)(a/A),$$

where M is the mass of ^{90}Sr produced, m is the mass of a single ^{90}Sr nucleus, A is the area over which fall out occurs, and a is the area in question. Since $\lambda = (\ln 2)/T_{1/2}$, the solution for a is

$$a = A\left(\frac{m}{M}\right)\left(\frac{R}{\lambda}\right) = \frac{AmRT_{1/2}}{M\ln 2}.$$

ANALYZE The molar mass of ^{90}Sr is 90g/mol. With $M = 400$ g and $A = 2000$ km^2, we find the area to be

$$a = \frac{AmRT_{1/2}}{M\ln 2} = \frac{(2000\times 10^6\ \text{m}^2)(90\,\text{g/mol})(74{,}000/\text{s})(29\,\text{y})(3.15\times 10^7\ \text{s/y})}{(400\,\text{g})(6.02\times 10^{23}/\text{mol})(\ln 2)}$$
$$= 7.3\times 10^{-2}\ \text{m}^{-2} = 730\,\text{cm}^2.$$

LEARN The Chernobyl nuclear accident in 1986 contaminated a very large area with ^{90}Sr.

Chapter 43

43-15

THINK One megaton of TNT releases 2.6×10^{28} MeV of energy. The energy released in each fission event is about 200 MeV.

EXPRESS The energy yield of the bomb is

$$E = (66 \times 10^{-3} \text{ megaton})(2.6 \times 10^{28} \text{ MeV/ megaton}) = 1.72 \times 10^{27} \text{ MeV}.$$

At 200 MeV per fission event, the total number of fission events taking place is

$$(1.72 \times 10^{27} \text{ MeV})/(200 \text{ MeV}) = 8.58 \times 10^{24}.$$

Now, since only 4.0% of the ^{235}U nuclei originally present undergo fission, there must have been $(8.58 \times 10^{24})/(0.040) = 2.14 \times 10^{26}$ nuclei originally present.

ANALYZE (a) The mass of ^{235}U originally present was

$$(2.14 \times 10^{26})(235 \text{ u})(1.661 \times 10^{-27} \text{ kg/u}) = 83.7 \text{ kg} \approx 84 \text{ kg}.$$

(b) Two fragments are produced in each fission event, so the total number of fragments is

$$2(8.58 \times 10^{24}) = 1.72 \times 10^{25} \approx 1.7 \times 10^{25}.$$

(c) One neutron produced in a fission event is used to trigger the next fission event, so the average number of neutrons released to the environment in each event is 1.5. The total number released is $(8.58 \times 10^{24})(1.5) = 1.29 \times 10^{25} \approx 1.3 \times 10^{25}$.

LEARN When one ^{235}U nucleus undergoes fission, the neutrons it produces (an average number of 2.5 neutrons per fission) can trigger other ^{235}U nuclei to fission, thereby setting up a chain reaction that allows an enormous amount of energy to be released.

43-17

THINK We represent the unknown fragment as $^{A}_{Z}X$, where A and Z are its mass number and atomic number, respectively. Charge and mass number are conserved in the neutron-capture process.

EXPRESS The reaction can be written as

$$^{235}_{92}\text{U} + {}^{1}_{0}\text{n} \rightarrow {}^{82}_{32}\text{Ge} + {}^{A}_{Z}X\,.$$

Conservation of charge yields 92 + 0 = 32 + Z, so Z = 60. Conservation of mass number yields 235 + 1 = 83 + A, so A = 153.

ANALYZE (a) Looking in Appendix F or G for nuclides with Z = 60, we find that the unknown fragment is $^{153}_{60}\text{Nd}$.

(b) We neglect the small kinetic energy and momentum carried by the neutron that triggers the fission event. Then,

$$Q = K_{\text{Ge}} + K_{\text{Nd}},$$

where K_{Ge} is the kinetic energy of the germanium nucleus and K_{Nd} is the kinetic energy of the neodymium nucleus. Conservation of momentum yields $\vec{p}_{\text{Ge}} + \vec{p}_{\text{Nd}} = 0$. Now, we can write the classical formula for kinetic energy in terms of the magnitude of the momentum vector:

$$K = \frac{1}{2}mv^2 = \frac{p^2}{2m}$$

which implies that

$$K_{\text{Nd}} = \frac{p_{\text{Nd}}^2}{2M_{\text{Nd}}} = \frac{p_{\text{Ge}}^2}{2M_{\text{Nd}}} = \frac{M_{\text{Ge}}}{M_{\text{Nd}}}\frac{p_{\text{Ge}}^2}{2M_{\text{Ge}}} = \frac{M_{\text{Ge}}}{M_{\text{Nd}}}K_{\text{Ge}}\,.$$

Thus, the energy equation becomes

$$Q = K_{\text{Ge}} + \frac{M_{\text{Ge}}}{M_{\text{Nd}}}K_{\text{Ge}} = \frac{M_{\text{Nd}} + M_{\text{Ge}}}{M_{\text{Nd}}}K_{\text{Ge}}$$

and

$$K_{\text{Ge}} = \frac{M_{\text{Nd}}}{M_{\text{Nd}} + M_{\text{Ge}}}Q = \frac{153\text{ u}}{153\text{ u} + 83\text{ u}}(170\text{ MeV}) = 110\text{ MeV}.$$

(c) Similarly,

$$K_{\text{Nd}} = \frac{M_{\text{Ge}}}{M_{\text{Nd}} + M_{\text{Ge}}}Q = \frac{83\text{ u}}{153\text{ u} + 83\text{ u}}(170\text{ MeV}) = 60\text{ MeV}.$$

(d) The initial speed of the germanium nucleus is

$$v_{\text{Ge}} = \sqrt{\frac{2K_{\text{Ge}}}{M_{\text{Ge}}}} = \sqrt{\frac{2(110\times10^6\text{ eV})(1.60\times10^{-19}\text{ J/eV})}{(83\text{ u})(1.661\times10^{-27}\text{ kg/u})}} = 1.60\times10^7\text{ m/s}.$$

(e) The initial speed of the neodymium nucleus is

$$v_{\rm Nd}=\sqrt{\frac{2K_{\rm Nd}}{M_{\rm ND}}}=\sqrt{\frac{2(60\times10^{6}\ {\rm eV})(1.60\times10^{-19}\ {\rm J/eV})}{(153\ {\rm u})(1.661\times10^{-27}\ {\rm kg/u})}}=8.69\times10^{6}\ {\rm m/s}.$$

LEARN By momentum conservation, the two fragments fly apart in opposite directions.

43-23

THINK The neutron generation time $t_{\rm gen}$ in a reactor is the average time needed for a fast neutron emitted in a fission event to be slowed to thermal energies by the moderator and then initiate another fission event.

EXPRESS Let P_0 be the initial power output, P be the final power output, k be the multiplication factor, t be the time for the power reduction, and $t_{\rm gen}$ be the neutron generation time. Then, according to the result of Problem 43-19,

$$P=P_0k^{t/t_{\rm gen}}.$$

ANALYZE We divide by P_0, take the natural logarithm of both sides of the equation and solve for ln k:

$$\ln k=\frac{t_{\rm gen}}{t}\ln\left(\frac{P}{P_0}\right)=\frac{1.3\times10^{-3}\ {\rm s}}{2.6\ {\rm s}}\ln\left(\frac{350\ {\rm MW}}{1200\ {\rm MW}}\right)=-0.0006161.$$

Hence, $k=e^{-0.0006161}=0.99938$.

LEARN The power output as a function of time is plotted below:

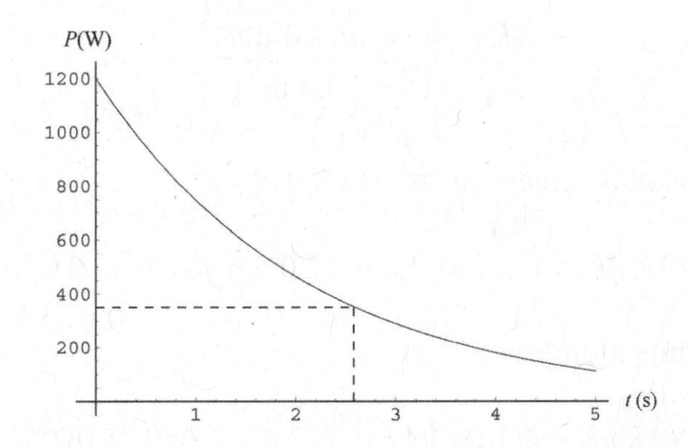

Since the multiplication factor k is smaller than 1, the output decreases with time.

43-25

THINK Momentum is conserved in the collision process. In addition, energy is also conserved since the collision is elastic.

EXPRESS Let v_{ni} be the initial velocity of the neutron, v_{nf} be its final velocity, and v_f be the final velocity of the target nucleus. Then, since the target nucleus is initially at rest, conservation of momentum yields

$$m_n v_{ni} = m_n v_{nf} + m v_f$$

and conservation of energy yields

$$\frac{1}{2} m_n v_{ni}^2 = \frac{1}{2} m_n v_{nf}^2 + \frac{1}{2} m v_f^2 .$$

We solve these two equations simultaneously for v_f. This can be done, for example, by using the conservation of momentum equation to obtain an expression for v_{nf} in terms of v_f and substituting the expression into the conservation of energy equation. We solve the resulting equation for v_f. We obtain $v_f = 2m_n v_{ni}/(m + m_n)$.

ANALYZE (a) The energy lost by the neutron is the same as the energy gained by the target nucleus, so

$$\Delta K = \frac{1}{2} m v_f^2 = \frac{1}{2} \frac{4 m_n^2 m}{(m+m_n)^2} v_{ni}^2 .$$

The initial kinetic energy of the neutron is $K = \frac{1}{2} m_n v_{ni}^2$, so

$$\frac{\Delta K}{K} = \frac{4 m_n m}{(m+m_n)^2} .$$

(b) The mass of a neutron is 1.0 u and the mass of a hydrogen atom is also 1.0 u. (Atomic masses can be found in Appendix G.) Thus,

$$\frac{\Delta K}{K} = \frac{4(1.0\ \text{u})(1.0\ \text{u})}{(1.0\ \text{u} + 1.0\ \text{u})^2} = 1.0.$$

(c) Similarly, the mass of a deuterium atom is 2.0 u, so

$$(\Delta K)/K = 4(1.0\ \text{u})(2.0\ \text{u})/(2.0\ \text{u} + 1.0\ \text{u})^2 = 0.89.$$

(d) The mass of a carbon atom is 12 u, so

$$(\Delta K)/K = 4(1.0\ \text{u})(12\ \text{u})/(12\ \text{u} + 1.0\ \text{u})^2 = 0.28.$$

(e) The mass of a lead atom is 207 u, so

$$(\Delta K)/K = 4(1.0\ \text{u})(207\ \text{u})/(207\ \text{u} + 1.0\ \text{u})^2 = 0.019.$$

(f) During each collision, the energy of the neutron is reduced by the factor 1 – 0.89 = 0.11. If E_i is the initial energy, then the energy after n collisions is given by $E = (0.11)^n E_i$. We take the natural logarithm of both sides and solve for n. The result is

$$n = \frac{\ln(E/E_i)}{\ln 0.11} = \frac{\ln(0.025\text{ eV}/1.00\text{ eV})}{\ln 0.11} = 7.9 \approx 8.$$

The energy first falls below 0.025 eV on the eighth collision.

LEARN The fractional kinetic energy loss as a function of the mass of the stationary atom (in units of m/m_n) is plotted below.

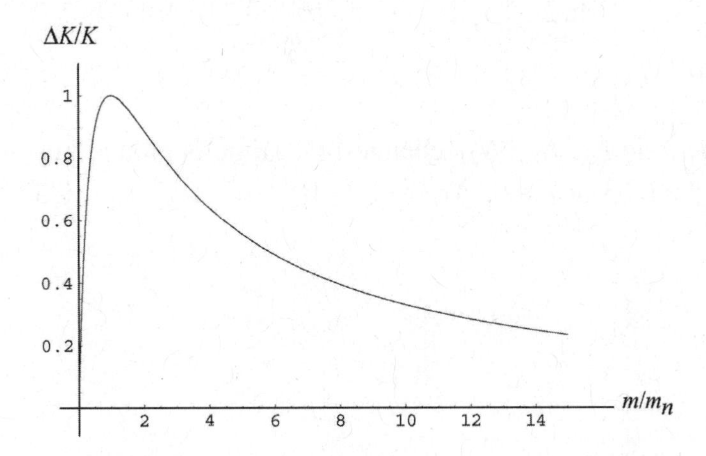

From the plot, it is clear that the energy loss is greatest ($\Delta K/K = 1$) when the atom has the same as the neutron.

43-29

THINK With a shorter half-life, ^{235}U has a greater decay rate than ^{238}U. Thus, if the ore contains only 0.72% of ^{235}U today, then the concentration must be higher in the far distant past.

EXPRESS Let t be the present time and $t = 0$ be the time when the ratio of ^{235}U to ^{238}U was 3.0%. Let N_{235} be the number of ^{235}U nuclei present in a sample now and $N_{235,0}$ be the number present at $t = 0$. Let N_{238} be the number of ^{238}U nuclei present in the sample now and $N_{238,0}$ be the number present at $t = 0$. The law of radioactive decay holds for each species, so

$$N_{235} = N_{235,0}e^{-\lambda_{235}t}$$

and

$$N_{238} = N_{238,0}e^{-\lambda_{238}t}.$$

Dividing the first equation by the second, we obtain

$$r = r_0 e^{-(\lambda_{235}-\lambda_{238})t}$$

where $r = N_{235}/N_{238}$ (= 0.0072) and $r_0 = N_{235,0}/N_{238,0}$ (= 0.030). We solve for t:

$$t = -\frac{1}{\lambda_{235}-\lambda_{238}} \ln\left(\frac{r}{r_0}\right).$$

ANALYZE Now we use $\lambda_{235} = (\ln 2)/T_{1/2_{235}}$ and $\lambda_{238} = (\ln 2)/T_{1/2_{238}}$ to obtain

$$t = \frac{T_{1/2_{235}} T_{1/2_{238}}}{(T_{1/2_{238}} - T_{1/2_{235}})\ln 2} \ln\left(\frac{r}{r_0}\right) = -\frac{(7.0\times10^8\text{ y})(4.5\times10^9\text{ y})}{(4.5\times10^9\text{ y} - 7.0\times10^8\text{ y})\ln 2} \ln\left(\frac{0.0072}{0.030}\right)$$
$$= 1.7\times10^9\text{ y}.$$

LEARN How the ratio $r = N_{235}/N_{238}$ changes with time is plotted below. In the plot, we take the ratio to be 0.03 at $t = 0$. At $t = 1.7\times10^9$ y or $t/T_{1/2,238} = 0.378$, r is reduced to 0.072.

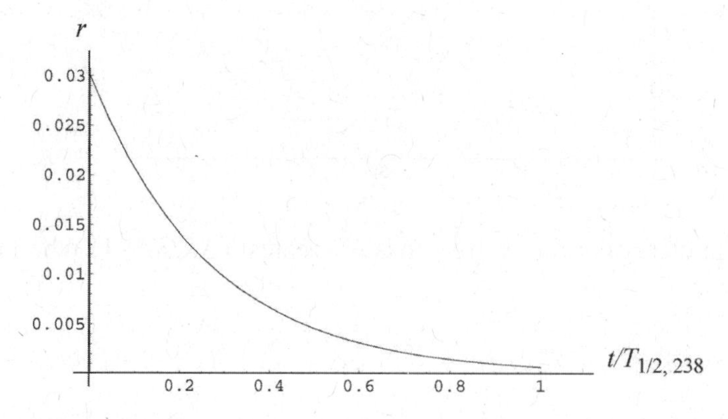

43-31

THINK Coulomb repulsion acts to prevent two charged particles from coming close enough to be within the range of their attractive nuclear force.

EXPRESS We take the height of the Coulomb barrier to be the value of the kinetic energy K each deuteron must initially have if they are to come to rest when their surfaces touch. If r is the radius of a deuteron, conservation of energy yields

$$2K = \frac{1}{4\pi\varepsilon_0}\frac{e^2}{2r}.$$

ANALYZE With $r = 2.1$ fm, we have

$$K = \frac{1}{4\pi\varepsilon_0}\frac{e^2}{4r} = (8.99\times10^9\ \text{V}\cdot\text{m/C})\frac{(1.60\times10^{-19}\ \text{C})^2}{4(2.1\times10^{-15}\ \text{m})} = 2.74\times10^{-14}\ \text{J} = 170\ \text{keV}.$$

LEARN The height of the Coulomb barrier depends on the charges and radii of the two interacting nuclei. Increasing the charge raises the barrier.

43-47

THINK The energy released by burning 1 kg of carbon is 3.3×10^7 J.

EXPRESS The mass of a carbon atom is $(12.0 \text{ u})(1.661 \times 10^{-27} \text{ kg/u}) = 1.99 \times 10^{-26}$ kg, so the number of carbon atoms in 1.00 kg of carbon is

$$(1.00 \text{ kg})/(1.99 \times 10^{-26} \text{ kg}) = 5.02 \times 10^{25}.$$

ANALYZE (a) The heat of combustion per atom is

$$(3.3 \times 10^7 \text{ J/kg})/(5.02 \times 10^{25} \text{ atom/kg}) = 6.58 \times 10^{-19} \text{ J/atom}.$$

This is 4.11 eV/atom.

(b) In each combustion event, two oxygen atoms combine with one carbon atom, so the total mass involved is $2(16.0 \text{ u}) + (12.0 \text{ u}) = 44$ u. This is

$$(44 \text{ u})(1.661 \times 10^{-27} \text{ kg/u}) = 7.31 \times 10^{-26} \text{ kg}.$$

Each combustion event produces 6.58×10^{-19} J so the energy produced per unit mass of reactants is $(6.58 \times 10^{-19} \text{ J})/(7.31 \times 10^{-26} \text{ kg}) = 9.00 \times 10^6$ J/kg.

(c) If the Sun were composed of the appropriate mixture of carbon and oxygen, the number of combustion events that could occur before the Sun burns out would be

$$(2.0 \times 10^{30} \text{ kg})/(7.31 \times 10^{-26} \text{ kg}) = 2.74 \times 10^{55}.$$

The total energy released would be

$$E = (2.74 \times 10^{55})(6.58 \times 10^{-19} \text{ J}) = 1.80 \times 10^{37} \text{ J}.$$

If P is the power output of the Sun, the burn time would be

$$t = \frac{E}{P} = \frac{1.80 \times 10^{37} \text{ J}}{3.9 \times 10^{26} \text{ W}} = 4.62 \times 10^{10} \text{ s} = 1.46 \times 10^3 \text{ y},$$

or 1.5×10^3 y, to two significant figures.

LEARN The Sun burns not coal but hydrogen via the proton-proton cycle in which the fusion of hydrogen nuclei into helium nuclei take place. The mechanism of thermonuclear fusion reactions allows the Sun to radiate energy at a rate of 3.9×10^{26} W for several billion years.

Chapter 44

44-11

THINK The conservation laws we shall examine are associated with energy, momentum, angular momentum, charge, baryon number, and the three lepton numbers.

EXPRESS In all particle interactions, the net lepton number for each family (L_e for electron, L_μ for muon, and L_τ for tau) is separately conserved. Conservation of baryon number implies that a process cannot occur if the net baryon number is changed.

ANALYZE (a) For the process $\mu^- \to e^- + \nu_\mu$, the rest energy of the muon is 105.7 MeV, the rest energy of the electron is 0.511 MeV, and the rest energy of the neutrino is zero. Thus, the total rest energy before the decay is greater than the total rest energy after. The excess energy can be carried away as the kinetic energies of the decay products and energy can be conserved. Momentum is conserved if the electron and neutrino move away from the decay in opposite directions with equal magnitudes of momenta. Since the orbital angular momentum is zero, we consider only spin angular momentum. All the particles have spin $\hbar/2$. The total angular momentum after the decay must be either $\hbar$ (if the spins are aligned) or zero (if the spins are anti-aligned). Since the spin before the decay is $\hbar/2$ angular momentum cannot be conserved. The muon has charge $-e$, the electron has charge $-e$, and the neutrino has charge zero, so the total charge before the decay is $-e$ and the total charge after is $-e$. Charge is conserved. All particles have baryon number zero, so baryon number is conserved. The muon lepton number of the muon is +1, the muon lepton number of the muon neutrino is +1, and the muon lepton number of the electron is 0. Muon lepton number is conserved. The electron lepton numbers of the muon and muon neutrino are 0 and the electron lepton number of the electron is +1. Electron lepton number is not conserved. The laws of conservation of angular momentum and electron lepton number are not obeyed and this decay does not occur.

(b) We analyze the decay $\mu^- \to e^+ + \nu_e + \bar{\nu}_\mu$ in the same way. We find that charge and the muon lepton number L_μ are not conserved.

(c) For the process $\mu^+ \to \pi^+ + \nu_\mu$, we find that energy cannot be conserved because the mass of muon is less than the mass of a pion. Also, muon lepton number L_μ is not conserved.

LEARN In all three processes considered, since the initial particle is stationary, the question associated with energy conservation amounts to asking whether the initial mass energy is sufficient to produce the mass energies and kinetic energies of the decayed products.

44-30

THINK A baryon is made up of three quarks.

EXPRESS The quantum numbers of the up, down, and strange quarks are (see Table 44-5) as follows:

Particle	Charge q	Strangeness S	Baryon number B
Up (u)	+2/3	0	+1/3
Down (d)	−1/3	0	+1/3
Strange (s)	−1/3	−1	+1/3

ANALYZE (a) To obtain a strangeness of –2, two of them must be s quarks. Each of these has a charge of $-e/3$, so the sum of their charges is $-2e/3$. To obtain a total charge of e, the charge on the third quark must be $5e/3$. There is no quark with this charge, so the particle cannot be constructed. In fact, such a particle has never been observed.

(b) Again the particle consists of three quarks (and no antiquarks). To obtain a strangeness of zero, none of them may be s quarks. We must find a combination of three u and d quarks with a total charge of $2e$. The only such combination consists of three u quarks.

LEARN The baryon with three u quarks is Δ^{++}.

44-43

THINK The radius of the orbit is still given by 1.50×10^{11} km, the original Earth-Sun distance.

EXPRESS The gravitational force on Earth is only due to the mass M within Earth's orbit. If r is the radius of the orbit, R is the radius of the new Sun, and M_S is the mass of the Sun, then

$$M=\left(\frac{r}{R}\right)^3 M_s=\left(\frac{1.50\times10^{11}\text{ m}}{5.90\times10^{12}\text{ m}}\right)^3(1.99\times10^{30}\text{ kg})=3.27\times10^{25}\text{ kg}.$$

The gravitational force on Earth is given by GMm/r^2, where m is the mass of Earth and G is the universal gravitational constant. Since the centripetal acceleration is given by v^2/r, where v is the speed of Earth, $GMm/r^2=mv^2/r$ and

$$v = \sqrt{\frac{GM}{r}}.$$

ANALYZE (a) Substituting the values given, we obtain

$$v = \sqrt{\frac{GM}{r}} = \sqrt{\frac{(6.67\times10^{-11}\ \text{m}^3/\text{s}^2\cdot\text{kg})(3.27\times10^{25}\ \text{kg})}{1.50\times10^{11}\ \text{m}}} = 1.21\times10^2\ \text{m/s}\,.$$

(b) The ratio of the speeds is

$$\frac{v}{v_0} = \frac{1.21\times10^2\ \text{m/s}}{2.98\times10^4\ \text{m/s}} = 0.00405.$$

(c) The period of revolution is

$$T = \frac{2\pi r}{v} = \frac{2\pi(1.50\times10^{11}\ \text{m})}{1.21\times10^2\ \text{m/s}} = 7.82\times10^9\ \text{s} = 247\ \text{y}\,.$$

LEARN An alternative ways to calculate the speed ratio and the periods are as follows. Since $v \sim \sqrt{M}$, the ratio of the speeds can be obtained as

$$\frac{v}{v_0} = \sqrt{\frac{M}{M_S}} = \left(\frac{r}{R}\right)^{3/2} = \left(\frac{1.50\times10^{11}\ \text{m}}{5.90\times10^{12}\ \text{m}}\right)^{3/2} = 0.00405.$$

In addition, since $T \sim 1/v \sim 1/\sqrt{M}$, we have

$$T = T_0\sqrt{\frac{M_S}{M}} = T_0\left(\frac{R}{r}\right)^{3/2} = (1\ \text{y})\left(\frac{5.90\times10^{12}\ \text{m}}{1.50\times10^{11}\ \text{m}}\right)^{3/2} = 247\ \text{y}.$$

44-45

THINK A meson is made up of a quark and an antiquark.

EXPRESS Only the strange quark has nonzero strangeness; an s quark has strangeness $S = -1$ and charge $q = -1/3$, while an $\bar{s}$ quark has strangeness $S = +1$ and charge $q = +1/3$.

ANALYZE (a) In order to obtain $S = -1$ we need to combine s with some non-strange antiquark (which would have the negative of the quantum numbers listed in Table 44-5). The difficulty is that the charge of the strange quark is $-1/3$, which means that (to obtain a total charge of $+1$) the antiquark would have to have a charge of $+\frac{4}{3}$. Clearly, there are no such antiquarks in our list. Thus, a meson with $S = -1$ and $q = +1$ cannot be formed with the quarks/antiquarks of Table 44-5.

(b) Similarly, one can show that, since no quark has $q = -\frac{4}{3}$, there cannot be a meson with $S = +1$ and $q = -1$.

LEARN Quarks and antiquarks can be combined to form baryons and mesons, but not all combinations are allowed because of the constraint from the quantum numbers.

44-47

THINK Pair annihilation is a process in which a particle and its antiparticle collide and annihilate each other.

EXPRESS The energy released would be twice the rest energy of Earth, or $E = 2M_E c^2$.

ANALYZE The mass of the Earth is $M_E = 5.98 \times 10^{24}$ kg (found in Appendix C). Thus, the energy released is

$$E = 2M_E c^2 = 2(5.98 \times 10^{24}\ \text{kg})(2.998 \times 10^8\ \text{m/s})^2 = 1.08 \times 10^{42}\ \text{J}.$$

LEARN As in the case of annihilation between an electron and a positron, the total energy of the Earth and the anti-Earth after the annihilation would appear as electromagnetic radiation.

44-51

THINK This problem deals with the phenomenon of cosmological redshift, which is due to the expansion of the universe.

EXPRESS Increasing the separation between Earth and a galaxy results in the observed spectral lines shifting toward longer (redder) wavelengths.

ANALYZE (a) During the time interval Δt, the light emitted from galaxy A has traveled a distance $c\Delta t$. Meanwhile, the distance between Earth and the galaxy has expanded from r to $r' = r + r\alpha\,\Delta t$. Let $c\Delta t = r' = r + r\alpha\Delta t$, which leads to

$$\Delta t = \frac{r}{c - r\alpha}.$$

(b) The detected wavelength λ' is longer than λ by $\lambda\alpha\Delta t$ due to the expansion of the universe: $\lambda' = \lambda + \lambda\alpha\Delta t$. Thus,

$$\frac{\Delta\lambda}{\lambda} = \frac{\lambda' - \lambda}{\lambda} = \alpha\Delta t = \frac{\alpha r}{c - \alpha r}.$$

(c) We use the binomial expansion formula (see Appendix E):

$$(1 \pm x)^n = 1 \pm \frac{nx}{1!} + \frac{n(n-1)x^2}{2!} + \cdots \quad (x^2 < 1)$$

to obtain

$$\frac{\Delta\lambda}{\lambda} = \frac{\alpha r}{c - \alpha r} = \frac{\alpha r}{c}\left(1 - \frac{\alpha r}{c}\right)^{-1} = \frac{\alpha r}{c}\left[1 + \frac{-1}{1!}\left(-\frac{\alpha r}{c}\right) + \frac{(-1)(-2)}{2!}\left(-\frac{\alpha r}{c}\right)^2 + \cdots\right]$$

$$\approx \frac{\alpha r}{c} + \left(\frac{\alpha r}{c}\right)^2 + \left(\frac{\alpha r}{c}\right)^3 .$$

(d) When only the first term in the expansion for $\Delta\lambda/\lambda$ is retained we have

$$\frac{\Delta\lambda}{\lambda} \approx \frac{\alpha r}{c} .$$

(e) We set

$$\frac{\Delta\lambda}{\lambda} = \frac{v}{c} = \frac{Hr}{c}$$

and compare with the result of part (d) to obtain $\alpha = H$.

(f) We use the formula $\Delta\lambda/\lambda = \alpha r/(c - \alpha r)$ to solve for r:

$$r = \frac{c(\Delta\lambda/\lambda)}{\alpha(1 + \Delta\lambda/\lambda)} = \frac{(2.998\times10^8 \text{ m/s})(0.050)}{(0.0218 \text{ m/s}\cdot\text{ly})(1 + 0.050)} = 6.548\times10^8 \text{ ly} \approx 6.5\times10^8 \text{ ly}.$$

(g) From the result of part (a),

$$\Delta t = \frac{r}{c - \alpha r} = \frac{(6.5\times10^8 \text{ ly})(9.46\times10^{15} \text{ m/ly})}{2.998\times10^8 \text{ m/s} - (0.0218 \text{ m/s}\cdot\text{ly})(6.5\times10^8 \text{ ly})} = 2.17\times10^{16} \text{ s},$$

which is equivalent to 6.9×10^8 y.

(h) Letting $r = c\Delta t$, we solve for Δt:

$$\Delta t = \frac{r}{c} = \frac{6.5\times10^8 \text{ ly}}{c} = 6.5\times10^8 \text{ y}.$$

(i) The distance is given by

$$r = c\Delta t = c(6.9\times10^8 \text{ y}) = 6.9\times10^8 \text{ ly}.$$

(j) From the result of part (f),

$$r_B = \frac{c(\Delta\lambda/\lambda)}{\alpha(1+\Delta\lambda/\lambda)} = \frac{(2.998\times10^8\ \text{m/s})(0.080)}{(0.0218\ \text{mm/s}\cdot\text{ly})(1+0.080)} = 1.018\times10^9\ \text{ly} \approx 1.0\times10^9\ \text{ly}.$$

(k) From the formula obtained in part (a),

$$\Delta t_B = \frac{r_B}{c - r_B\alpha} = \frac{(1.0\times10^9\ \text{ly})(9.46\times10^{15}\ \text{m/ly})}{2.998\times10^8\ \text{m/s} - (1.0\times10^9\ \text{ly})(0.0218\,\text{m/s}\cdot\text{ly})} = 3.4\times10^{16}\ \text{s},$$

which is equivalent to 1.1×10^9 y.

(l) At the present time, the separation between the two galaxies A and B is given by $r_{\text{now}} = c\Delta t_B - c\Delta t_A$. Since $r_{\text{now}} = r_{\text{then}} + r_{\text{then}}\alpha\Delta t$, we get

$$r_{\text{then}} = \frac{r_{\text{now}}}{1+\alpha\Delta t} = 3.9\times10^8\ \text{ly}.$$

LEARN Cosmological redshift is a consequence of the expansion of universe and not from the motion of an individual body, as in Doppler shift. Cosmological redshift is determined by how far away the galaxy was when the photons were first emitted: the larger the distance, the longer the emitted photons have traversed through expanding space and hence, the greater cosmological redshift.